W9-AOG-214

MAGNUS, KARRASS, and SOLITAR–Combinatorial Group Theory

MARTIN–Nonlinear Operators and Differential Equations in Banach Spaces

MELZAK–Companion to Concrete Mathematics

MELZAK–Mathematical Ideas, Modeling and Applications (Volume II of Companion to Concrete Mathematics)

MEYER–Introduction to Mathematical Fluid Dynamics

MORSE–Variational Analysis: Critical Extremals and Sturmian Extensions

NAYFEH–Perturbation Methods

NAYFEH and MOOK–Nonlinear Oscillations

ODEN and REDDY–An Introduction to the Mathematical Theory of Finite Elements

PAGE–Topological Uniform Structures

PASSMAN–The Algebraic Structure of Group Rings

PRENTER–Splines and Variational Methods

RIBENBOIM–Algebraic Numbers

RICHTMYER and MORTON–Difference Methods for Initial Value Problems

RIVLIN–The Chebyshev Polynomials

RUDIN–Fourier Analysis on Groups

SAMELSON–An Introduction to Linear

SIEGEL–Topics in Complex Function

      Volume 1–Elliptic Functions

      Volume 2–Automorphic Func

      Volume 3–Abelian Functions

STAKGOLD–Green's Functions and B

STOKER–Differential Geometry

STOKER–Nonlinear Vibrations in Mecl

STOKER–Water Waves

WHITHAM–Linear and Nonlinear Wav

WOUK–A Course of Applied Functiona

DATE DUE

GAYLORD   PRINTED IN U.S.A.

GREEN'S FUNCTIONS AND
BOUNDARY VALUE PROBLEMS

GREEN'S FUNCTIONS AND BOUNDARY VALUE PROBLEMS

IVAR STAKGOLD
University of Delaware

A WILEY-INTERSCIENCE PUBLICATION

JOHN WILEY & SONS, New York • Chichester • Brisbane • Toronto

Library of Congress Cataloging in Publication Data

Stakgold, Ivar.
    Green's functions and boundary value problems.

    (Pure and applied mathematics)
    "A Wiley-Interscience publication."
    Includes index.
    1. Boundary value problems.   2. Green's functions.
3. Mathematical physics.   I. Title.

QA379.S72   515'.35   78-27259
ISBN 0-471-81967-0

Printed in the United States of America

10 9 8 7 6 5 4 3 2 1

For

ALICE and ALISSA

Preface

As a result of graduate-level adoptions of my earlier two-volume book, *Boundary Value Problems of Mathematical Physics*, I received many constructive suggestions from users. One frequent recommendation was to consolidate and reorganize the topics into a single volume that could be covered in a one-year course. Another was to place additional emphasis on modeling and to choose examples from a wider variety of physical applications, particularly some emerging ones. In the meantime my own research interests had turned to nonlinear problems, so that, inescapably, some of these would also have to be included in any revision. The only way to incorporate these changes, as well as others, was to write a new book, whose main thrust, however, remains the systematic analysis of boundary value problems. Of course some topics had to be dropped and others curtailed, but I can only hope that your favorite ones are not among them.

My book is aimed at graduate students in the physical sciences, engineering, and applied mathematics who have taken the typical "methods" course that includes vector analysis, elementary complex variables, and an introduction to Fourier series and boundary value problems. Why go beyond this? A glance at modern publications in science and engineering provides the answer. To the lament of some and the delight of others, much of this literature is deeply mathematical. I am referring not only to areas such as mechanics and electromagnetic theory that are traditionally mathematical but also to relative newcomers to mathematization, such as chemical engineering, materials science, soil mechanics, environmental engineering, biomedical engineering, and nuclear engineering. These fields give rise to challenging mathematical problems whose flavor can be sensed from the following short list of examples; integrodifferential equations of neutron transport theory, combined diffusion and reaction in chemical and environmental engineering, phase transitions in metallurgy, free boundary problems for dams in soil mechanics, propagation of impulses along nerves in biology. It would be irresponsible and foolish to claim that readers of my book will become instantaneous experts in these fields, but they will be prepared to tackle many of the mathematical aspects of the relevant literature.

Next, let me say a few words about the numbering system. The book is divided into ten chapters, and each chapter is divided into sections. Equations do *not* carry a chapter designation. A reference to, say, equation (4.32) is to the thirty-second numbered equation in Section 4 of the chapter you happen to be reading. The same system is used for figures and exercises, the latter being found at the end of sections. The exercises, by the way, are rarely routine and, on occasion, contain substantial extensions of the main text. Examples do not carry any section designation and are numbered consecutively within a section, even though there may be separate clusters of examples within the same section. Some theorems have numbers and others do not; those that do are numbered in a sequence within a section—Theorem 1, Theorem 2, and so on.

A brief description of the book's contents follows. No attempt is made to mention all topics covered; only the general thread of the development is indicated.

Chapter 0 presents background material that consists principally of careful derivations of several of the equations of mathematical physics. Among them are the equations of heat conduction, of neutron transport, and of vibrations of rods. In the last-named derivation an effort is made to show how the usual linear equations for beams and strings can be regarded as first approximations to nonlinear problems. There are also two short sections on modes of convergence and on Lebesgue integration.

Many of the principal ideas related to boundary value problems are introduced on an intuitive level in Chapter 1. A boundary value problem (BVP, for short) consists of a differential equation $Lu = f$ with boundary conditions of the form $Bu = h$. The pair (f, h) is known collectively as the data for the problem, and u is the response to be determined. Green's function is the response when f represents a concentrated unit source and $h = 0$. In terms of Green's function, the BVP with arbitrary data can be solved in a form that shows clearly the dependence of the solution on the data. Various examples are given, including some multidimensional ones, some involving interface conditions, and some initial value problems. The useful notion of a well-posed problem is discussed, and a first look is taken at maximum principles for differential equations.

Chapter 2 deals with the theory of distributions, which provides a rigorous mathematical framework for singular sources such as the point charges, dipoles, line charges, and surface layers of electrostatics. The notion of response to such sources is made precise by defining the distributional solution of a differential equation. The related concepts of weak solution, adjoint, and fundamental solution are also introduced. Fourier series and Fourier transforms are presented in both classical and distributional settings.

Chapter 3 returns to a more detailed study of one-dimensional linear boundary value problems. To an equation of order p there are usually associated p independent boundary conditions involving derivatives of order less than p at the endpoints a and b of a bounded interval. If the corresponding BVP with 0 data has only the trivial solution, then the BVP with arbitrary data has one and only one solution which can be expressed in terms of Green's function. If, however, the BVP with 0 data has a nontrivial solution, certain solvability conditions must be satisfied for the BVP with arbitrary data to have a solution. These statements are formulated precisely in an alternative theorem, which recurs throughout the book in various forms. When the BVP with 0 data has a nontrivial solution, Green's function cannot be constructed in the ordinary way, but some of its properties can be salvaged by using a modified Green's function, defined in Section 5.

Chapter 4 begins the study of Hilbert spaces. A Hilbert space is the proper setting for many of the linear problems of applied analysis. Though its elements may be functions or abstract "vectors," a Hilbert space enjoys all the algebraic and geometric properties of ordinary Euclidean space. A Hilbert space is a linear space equipped with an inner product that induces a natural notion of distance between elements, thereby converting it into a metric space which is required to be complete. Some of the important geometric properties of Hilbert spaces are developed, including the projection theorem and the existence of orthonormal bases for separable spaces. Metric spaces can be useful quite apart from any linear structure. A contraction is a transformation on a metric space that uniformly reduces distances between pairs of points. A contraction on a complete metric space has a unique fixed point that can be calculated by iteration from any initial approximation. Examples demonstrate how to use these ideas to prove uniqueness and constructive existence for certain classes of nonlinear differential equations and integral equations.

Chapter 5 examines the theory of linear operators on a separable Hilbert space, particularly integral and differential operators, the latter being unbounded operators. The principal problem of operator theory is the solution of the equation $Au = f$, where A is a linear operator and f an element of the space. A thorough discussion of this problem leads again to adjoint operators, solvability conditions, and alternative theorems. Additional insight is obtained by considering the inversion of the equation

$Au - \lambda u = f$, which leads to the idea of the spectrum, a generalization of the more familiar concept of eigenvalue. For compact operators (which include most integral operators) the inversion problem is essentially solved by the Riesz-Schauder theory of Section 7. Section 8 relates the spectrum of symmetric operators to extremal principles for the Rayleigh quotient. Throughout, the theory is illustrated by specific examples.

In Chapter 6 the general ideas of operator theory are specialized to integral equations. Integral equations are particularly important as alternative formulations of boundary value problems. Special emphasis is given to Fredholm equations with symmetric Hilbert-Schmidt kernels. For the corresponding class of operators, the nonzero eigenvalues and associated eigenfunctions can be characterized through successive extremal principles, and it is then possible to give a complete treatment of the inhomogeneous equation. The last section discusses the Ritz procedure for estimating eigenvalues, as well as other approximation methods for eigenvalues and eigenfunctions. There is also a brief introduction to integrodifferential operators in Exercises 5.3 to 5.8.

Chapter 7 extends the Sturm-Liouville theory of second-order ordinary differential equations to the case of singular endpoints. It is shown, beginning with the regular case, how the necessarily discrete spectrum can be constructed from Green's function. A formal extension of this relationship to the singular case makes it possible to calculate the spectrum, which may now be partly continuous. The transition from regular to singular is analyzed rigorously for equations of the first order, but the Weyl classification for second-order equations is given without proof. The eigenfunction expansion in the singular case can lead to integral transforms such as Fourier, Hankel, Mellin, and Weber. It is shown how to use these transforms and their inversion formulas to solve partial differential equations in particular geometries by separation of variables.

Although partial differential equations have appeared frequently as examples in earlier chapters, they are treated more systematically in Chapter 8. Examination of the Cauchy problem—the appropriate generalization of the initial value problem to higher dimensions—gives rise to a natural classification of partial differential equations into hyperbolic, parabolic, and elliptic types. The theory of characteristics for hyperbolic equations is introduced and applied to simple linear and nonlinear examples. In the second and third sections various methods (Green's functions, Laplace transforms, images, etc.) are used to solve BVPs for the wave equation, the heat equation, and Laplace's equation. The simple and double layers of potential theory make it possible to reduce the Dirichlet problem to an integral equation on the boundary of the domain, thereby providing a rather weak existence proof. In Section 4 a stronger existence proof is given, using variational principles. Two-sided bounds for some functionals of physical interest, such as capacity and torsional rigidity, are obtained by introducing complementary principles. Another application involving level-line analysis is also given, and there is a very brief treatment of unilateral constraints and variational inequalities.

Finally, in Chapter 9, a number of methods applicable to nonlinear problems are developed. Section 1 points out some of the features that distinguish nonlinear problems from linear ones and illustrates these differences through some simple examples. In Section 2 the principal qualitative results of branching theory (also known as bifurcation theory) are presented. The phenomenon of bifurcation is understood most easily in terms of the buckling of a rod under compressive thrust. As the thrust is increased beyond a certain critical value, the state of simple compression gives way to the buckled state with its appreciable transverse deflection. Section 3 shows how a variety of linear problems can be handled by perturbation theory (inhomogeneous problems, eigenvalue problems, change in boundary conditions, domain perturbations). These techniques, as well as monotone methods, are then adapted to the solution of nonlinear BVPs. The concluding section discusses the possible loss of stability of the basic steady state when an underlying parameter is allowed to vary.

I have already acknowledged my debt to the students and teachers who were kind enough to comment on my earlier book. There are, however, two colleagues to whom I am particularly grateful: Stuart Antman, who generously contributed the ideas underlying the derivation of the equations for rods in Chapter 0, and W. Edward Olmstead, who suggested some of the examples on contractions in Chapter 4 and on branching in Chapter 9.

IVAR STAKGOLD

Newark, Delaware
April 1979

Contents

GREEN'S FUNCTIONS AND
BOUNDARY VALUE PROBLEMS

0

Preliminaries

As its name and number indicate, this chapter contains background material that has no precise place in the systematic development beginning with Chapter 1. Many readers may prefer to look over this preliminary material rather casually and then refer to it again as need arises.

The principal purpose here is to give fairly careful derivations of some of the equations of mathematical physics that will be studied extensively in the rest of the book. The attention paid to modeling in the present chapter could, regrettably, not be sustained in subsequent ones. Readers who are particularly interested in this aspect of applied mathematics are encouraged to consult the books by Lin and Segel, by Aris (1978), and by Segel.

Two other isolated sections of a mathematical nature complete the chapter. One section reviews some fundamental ideas on convergence. The other presents a short treatment of Lebesgue integration. Although only a few essential properties of this integral are needed in the book, it seemed worthwhile to take a few pages to explain its construction. These limited goals made it convenient to use Tonelli's approach as presented in Silverman's fine translation of Shilov's book.

A few words about terminology are in order. R_n stands for n-dimensional Euclidean space. The definitions below are given for R_3 but are easily modified for R_n. A point in R_3 is identified by its position vector $\mathbf{x} = (x_1, x_2, x_3)$, where x_1, x_2, x_3 are *Cartesian* coordinates; $|\mathbf{x}| = (x_1^2 + x_2^2 + x_3^2)^{1/2}$, where the nonnegative square root is understood; $d\mathbf{x}$ stands for a volume element $dx_1 \, dx_2 \, dx_3$. In later chapters the distinguishing notation for vectors is dropped.

An *open ball* of radius a, centered at the origin, is the set of points $\mathbf{x}$ such that $|\mathbf{x}| < a$. The set $|\mathbf{x}| \leqslant a$ is a *closed ball*, and the set $|\mathbf{x}| = a$ is a *sphere*. In R_2 the words *disk* and *circle* are often substituted for ball and sphere, respectively. An *open set* Ω has the property that, whenever $x \in \Omega$, so does some sufficiently small ball with center at $\mathbf{x}$. A point $\mathbf{x}$ belongs to the

1

boundary Γ of an open set Ω if $\mathbf{x}$ is not in Ω but if every open ball centered at $\mathbf{x}$ contains a point of Ω. The *closure* $\overline{\Omega}$ of Ω is the union of Ω and Γ. These ideas are best illustrated by an egg with a very thin shell. The interior of the egg is an open set Ω, the shell is Γ, and the egg with shell is $\overline{\Omega}$. An open set Ω is *connected* if each pair of points in Ω can be connected by a curve lying entirely in Ω. A *domain* is an open connected set. Thus an open ball is a domain, but the union of two disjoint open balls is not.

The symbol $\doteq$ means "set equal to." It is occasionally used to define a new expression. For instance, in writing $D \doteq dS/dx$ we are defining D as dS/dx which, in turn, is presumably known from earlier discussion.

The terms

$$\inf_{\mathbf{x} \in \Omega} f(\mathbf{x}), \; \sup_{\mathbf{x} \in \Omega} f(\mathbf{x})$$

stand for the infimum (greatest lower bound) and supremum (least upper bound) of the real-valued function f on Ω. For instance, if Ω is the open ball in R_n with radius a and center at the origin, and $f(\mathbf{x}) = |\mathbf{x}|$, then

$$\inf_{\mathbf{x} \in \Omega} f(\mathbf{x}) = 0, \; \sup_{\mathbf{x} \in \Omega} f(\mathbf{x}) = a,$$

even though the supremum is not attained for any element $\mathbf{x}$ in Ω.

1. HEAT CONDUCTION

We shall consider the flow of heat in an inhomogeneous medium occupying the three-dimensional domain Ω with boundary Γ. The temperature $u(\mathbf{x}, t)$ is a scalar function defined for $\mathbf{x}$ in Ω and a time interval $t_1 < t < t_2$. In the general situation there will exist within Ω certain sources of heat, known as *body sources*, whose nature will be specified more precisely later. The discussion of boundary and initial conditions may safely be postponed since they do not affect internal heat balances.

Let R be an *arbitrary* portion of Ω with boundary B. It is vital that R be allowed to range through a variety of subdomains of Ω, so that we can obtain sufficient information for our purposes. A heat balance over R for the time interval $(t, t + dt)$ gives

(1.1) heat produced by body sources

$$= \text{rise in heat content} + \text{outflow of heat through } B.$$

We now make two physical assumptions that have successfully weathered the passage of time:

1. If a material element of volume $d\mathbf{x}$ is raised from the temperature u to the temperature $u + du$, its heat content is raised by $Cdu\,d\mathbf{x}$, where C is

the *specific heat* (which may depend on u) in calories per degree per cubic centimeter.

2. Fourier's law: Consider an element of surface with normal $\mathbf{n}$ and area dS. The amount of heat flowing in time dt across this element is

$$dt\, dS\, \mathbf{n} \cdot (-k \operatorname{grad} u) = -dt\, dS\, k \frac{\partial u}{\partial n},$$

where k is the thermal conductivity (which may depend on u) measured in calories per second per centimeter per degree. The vector $-k \operatorname{grad} u$ is known as the *heat flow vector*, the minus sign being consistent with the fact that heat flows in the direction of decreasing temperature.

Since our medium may be inhomogeneous, both C and k may depend on $\mathbf{x}$ as well as on u. If we let $du = du(\mathbf{x}, t)$ stand for the rise in temperature at $\mathbf{x}$ in the time from t to $t + dt$, the change in heat content of R in that time is given by

$$(1.2) \qquad \int_R C\, du\, d\mathbf{x},$$

and the heat flowing outward through B by

$$(1.3) \qquad -dt \int_B k \frac{\partial u}{\partial n}\, dS,$$

where n is the outward normal to B.

In the absence of sources that liberate heat instantaneously in time, we can write the heat produced by body sources as $F_R(t)\, dt$, where $F_R(t)$ is the *rate*, in calories per second, at which body sources generate heat in the whole of R. After division by dt and passage to the limit, (1.1) takes the form

$$(1.4) \qquad F_R(t) = \int_R C(\mathbf{x}, u) \frac{\partial u}{\partial t}\, d\mathbf{x} - \int_B k(\mathbf{x}, u) \frac{\partial u}{\partial n}\, dS,$$

which we regard as the *primary equation* describing heat conduction. From (1.4) we can specialize in different directions, one of which leads to the familiar but more restrictive partial differential equation of heat conduction.

By excluding spatially "singular" sources (those that are concentrated at points, on curves, or on surfaces), we can write

$$(1.5) \qquad F_R(t) = \int_\Omega f(\mathbf{x}, t)\, d\mathbf{x},$$

where $f(\mathbf{x}, t)$, measured in calories per second per cubic centimeter, is the volume rate at which body sources produce heat at $(\mathbf{x}, t)$. The corresponding temperature $u(\mathbf{x}, t)$ will then be smooth enough so that the divergence theorem can be used on the surface integral in (1.4) to give

$$(1.6) \qquad \int_R \left[C \frac{\partial u}{\partial t} - \operatorname{div}(k \operatorname{grad} u) - f(\mathbf{x}, t) \right] dx = 0,$$

which holds for every portion R of Ω. The integrand is a function of $\mathbf{x}$ and t defined for $\mathbf{x} \in \Omega$ and $t_1 < t < t_2$. Assuming that this function is continuous, we claim it must vanish identically. Indeed, if the function differed from 0 at a point $\mathbf{x}$ at time t, there would exist a neighborhood R of that point such that the function would be of one sign throughout R at time t. The integral in (1.6) would then fail to vanish for that particular R and t. We therefore conclude that

$$(1.7) \qquad C \frac{\partial u}{\partial t} - \operatorname{div}(k \operatorname{grad} u) = f, \qquad \mathbf{x} \in \Omega, \quad t_1 < t < t_2,$$

which is the usual equation of heat conduction. If C and k are *constants*, the equation reduces to

$$(1.8) \qquad \frac{\partial u}{\partial t} - a \Delta u = \frac{f}{C},$$

where $a = k/C$ is the *thermal diffusivity* in square centimeters per second, and $\Delta \doteq \operatorname{div} \operatorname{grad}$ is the Laplacian operator, which has the familiar form

$$\frac{\partial^2}{\partial x_1^2} + \frac{\partial^2}{\partial x_2^2} + \frac{\partial^2}{\partial x_3^2}$$

in Cartesian coordinates.

Remark.   In the simplest case the production term f is prescribed as a function of $\mathbf{x}$ and t. In that case (1.8) is a linear inhomogeneous equation. There are many problems, however, in which the production term also depends on the unknown temperature u. If, for instance, a *chemical reaction* takes place which liberates heat, it is reasonable to assume that the rate at which heat is released is given by the Arrhenius law:

$$(1.9) \qquad A e^{-B/u},$$

where A and B are known positive constants. We must therefore substitute

$$f(\mathbf{x}, t) = A e^{-B/u(\mathbf{x}, t)}$$

on the right sides of (1.7) and (1.8), so that these become nonlinear partial differential equations.

In *steady-state heat conduction* the temperature $u(\mathbf{x})$ is independent of time, and (1.4) reduces to

$$(1.10) \qquad -\int_B k \frac{\partial u}{\partial n} \, dS = F_R,$$

where F_R is the steady heat input per unit time over the domain R bounded by B. If these body sources stem from a volume density $f(\mathbf{x})$, we can use reasoning similar to that yielding (1.7) to conclude that

$$(1.11) \qquad -\operatorname{div}(k \operatorname{grad} u) = f(\mathbf{x}), \qquad \mathbf{x} \in \Omega.$$

To show that (1.10) is indeed more general than (1.11), consider the case of a single source concentrated at $\mathbf{x} = \xi$ and generating p calories per second. Then (1.10) tells us that

$$(1.12) \qquad -\int_B k \frac{\partial u}{\partial n} \, dS = \begin{cases} 0 & \text{if } \xi \text{ is not in } R, \\ p & \text{if } \xi \text{ is in } R. \end{cases}$$

By a now familiar argument, the first line shows that $-\operatorname{div}(k \operatorname{grad} u) = 0$ for $\mathbf{x} \neq \xi$. By specializing the second line to a small sphere centered at ξ, it is possible to extract precise information on the nature of the singularity in u at ξ (see Section 4, Chapter 1, for instance).

Whenever there is possible ambiguity in the interpretation of (1.11), it is wise to return to the integral formulation (1.10) for guidance. As another example, suppose that Ω consists of two media separated by an interface σ. The thermal conductivity k is continuous with the possible exception of a jump discontinuity across σ. Assuming that there are no prescribed sources or any heat losses (caused by films or imperfect fitting) on the interface, we can apply (1.10) to a thin pillbox straddling the interface with its bases parallel to σ. It is permissible to neglect the contribution from the lateral surface of the pillbox to obtain

$$(1.13) \qquad k_+ \frac{\partial u}{\partial n_+} = k_- \frac{\partial u}{\partial n_-},$$

where the subscripts $+$ and $-$ denote the two sides of σ. Equation (1.13) and the continuity of u comprise the interface conditions (see also Section 4, Chapter 1).

Despite their usefulness (1.10) and (1.4) are not general enough to treat certain important idealized singularities. For instance, a dipole located at ξ would apparently go undetected in (1.10) since $F_R = 0$ whether ξ is in R or outside R. We shall see in Chapter 2 that such problems are best handled within the theory of distributions.

Boundary and Initial Conditions

Equation (1.7) alone does not determine the temperature $u(\mathbf{x}, t)$. We must in addition give the initial temperature $u(\mathbf{x}, t_1)$ and a boundary condition on Γ for $t_1 < t < t_2$. This boundary condition is usually of one of the following three types:

$$
(1.14) \quad
\begin{cases}
\text{(i)} & \text{temperature } u \text{ prescribed on } \Omega \text{ for } t_1 < t < t_2, \\[2mm]
\text{(ii)} & \text{heat flow } -k\dfrac{\partial u}{\partial n} \text{ prescribed on } \Gamma \text{ for } t_1 < t < t_2, \\[2mm]
\text{(iii)} & -k\dfrac{\partial u}{\partial n} = \alpha u \text{ on } \Gamma \text{ for } t_1 < t < t_2.
\end{cases}
$$

For $\alpha > 0$ the last condition is Newton's law of cooling, which characterizes radiation into a surrounding medium at uniform temperature [which is then taken to be the datum of temperature in (1.7)]. Thus heat is lost from the surface of the body at a rate proportional to the difference between the surface temperature and the surrounding temperature. If $\alpha < 0$, condition (iii) states that the larger the boundary temperature the more heat enters the body, a circumstance which obviously tends to increase the internal temperature. Newton's law of cooling can be regarded as an approximation to Stefan's radiation law, which has a term βu^4 on the right side of (iii). It is of course possible to have an inhomogeneous version of (iii) if the surface is simultaneously heated by prescribed heat flow.

One-Dimensional Problems

Equation (1.7) can sometimes be reduced to an equation involving derivatives in only one space direction. Let (x_1, x_2, x_3) be Cartesian coordinates; we want to describe classes of problems in which u depends only on x_1 and t.

1. Suppose Ω is the slab $0 < x_1 < a$, $-\infty < x_2, x_3 < \infty$. Assume that the source term depends only on x_1 and t, that the boundary conditions on the

faces of the slab depend only on t, that the initial temperature depends only on x_1, and that k and C depend only on x_1 and u. On physical grounds it is clear that heat flows only in the x_1 direction, so that u will be a function of x_1 and t alone. The differential equation (1.7) then becomes

$$(1.15) \qquad C\frac{\partial u}{\partial t} - \frac{\partial}{\partial x_1}\left(k\frac{\partial u}{\partial x_1}\right) = f(x_1, t), \qquad 0 < x_1 < a, \quad t_1 < t < t_2.$$

The solution $u(x_1, t)$ of the boundary-initial value problem associated with (1.15) obviously also satisfies the related boundary-initial value problem associated with (1.7). If we can prove uniqueness in the latter case, we will have shown that $u(x_1, t)$ is in fact the desired solution of (1.7).

2. Let Ω be a cylindrical rod (of arbitrary cross section) whose axis coincides with the segment $0 < x_1 < a$. In addition to the assumptions in part 1, let us suppose that the *lateral surface of the rod is insulated*. We may imagine the rod as having been punched out of the slab of part 1. Since the temperature in the slab is independent of x_2 and x_3, it must satisfy the condition $\partial u/\partial n = 0$ on the lateral surface of the cylinder, which is the criterion of insulation. Thus the flow in the rod is one-dimensional, and (1.15) holds as before. In particular, the steady-state equation is

$$(1.16) \qquad -\frac{d}{dx_1}\left(k\frac{du}{dx_1}\right) = \tilde{f}(x_1), \qquad 0 < x_1 < a,$$

where $\tilde{f}(x_1)$ is the *volume density* of sources. If A is the cross-sectional area of the rod, we can write $\tilde{f}(x_1) = f(x_1)/A$, where $f(x_1)$ is the source density per unit length of the rod. Equation (1.16) becomes

$$(1.17) \qquad -\frac{d}{dx_1}\left(k\frac{du}{dx_1}\right) = \frac{f(x_1)}{A}, \qquad 0 < x_1 < a.$$

2. DIFFUSION

With a different interpretation of the terms, (1.4) and (1.7) also govern the *concentration* $c(\mathbf{x}, t)$ of a substance diffusing through some medium. The energy balance (1.1) is replaced by a mass balance of the substance in question over R for the time interval $(t, t + dt)$:

(2.1) mass created by body sources

$$= \text{increase of mass} + \text{outflow of mass through } B.$$

We shall assume that the only mechanism causing mass flow is the gradient of the concentration. The equivalent form of Fourier's law is that the amount of substance flowing in time dt across a surface element is

$$(2.2) \qquad dt\, dS\, \mathbf{n} \cdot (-D \operatorname{grad} c) = -dt\, dS\, D \frac{\partial c}{\partial n},$$

where D is known as the *diffusivity* and is measured in square centimeters per second. Thus (2.1) becomes

$$(2.3) \qquad P_R(t) = \int_R \frac{\partial c}{\partial t}\, d\mathbf{x} - \int_B D \frac{\partial c}{\partial n}\, dS,$$

where $P_R(t)$ is the rate at which body sources generate the substance in the whole of R (the mechanism for creating or removing the substance may be a chemical reaction). If P_R can be expressed in terms of a density $p(\mathbf{x}, t)$, (2.3) can be reduced to

$$(2.4) \qquad \frac{\partial c}{\partial t} - \operatorname{div}(D \operatorname{grad} c) = p$$

and, if D is constant, to

$$(2.5) \qquad \frac{\partial c}{\partial t} - D \Delta c = p.$$

The production p may be given as a function of $\mathbf{x}$ and t or may depend on the concentration c. In an absorbing medium it is frequently assumed that the amount of the substance absorbed is proportional to the concentration, so that $p = -q(\mathbf{x})c + n(\mathbf{x}, t)$, where $q(\mathbf{x})$ is a measure of the absorption properties of the medium, which may vary from point to point, and $n(\mathbf{x}, t)$ represents the density of prescribed sources. As a result (2.4) takes the form

$$(2.6) \qquad \frac{\partial c}{\partial t} - \operatorname{div}(D \operatorname{grad} c) + qc = n(\mathbf{x}, t)$$

or, in the steady state,

$$(2.7) \qquad -\operatorname{div}(D \operatorname{grad} c) + q(\mathbf{x})c = n(\mathbf{x}).$$

If the medium itself is moving at a velocity $\mathbf{v}(\mathbf{x}, t)$, there is an additional contribution

$$\int_B c\mathbf{v} \cdot \mathbf{n}\, dS$$

to the outward mass flow through B. Inserting this term in (2.1), with due regard to the minus sign, and again applying the divergence theorem, we find that (2.4) becomes

$$(2.8) \qquad \frac{\partial c}{\partial t} - \operatorname{div}(D \operatorname{grad} c) + \operatorname{div}(vc) = p.$$

3. CHEMICAL ENGINEERING PROBLEMS

We shall consider a number of problems of chemical reaction engineering in which a reactant *disappears* at a volume rate dependent on its concentration c and temperature u. In a first-order irreversible reaction this rate is proportional to c and its temperature dependence follows the Arrhenius law (1.9), so that the *production* rate is

$$(3.1) \qquad p(c, u) = -Ace^{-B/u},$$

where A and B are positive constants. The amount of heat released (or absorbed) in the reaction is proportional to p. For an exothermic reaction, heat is therefore liberated at the rate

$$(3.2) \qquad -hp = hAce^{-B/u},$$

where h is a positive constant.

Continuous-Flow Stirred Tank Reactor

Here the reaction takes place in a tank of volume V which is continuously stirred to maintain uniform reactant concentration $\bar{c}$ and temperature $\bar{u}$. Of course $\bar{c}$ and $\bar{u}$ will depend on time $\bar{t}$, and it is this variation which we wish to determine. The tank is fed by a stream that has volume rate q at temperature u_f and in which the concentration of the reactant is c_f. Products of the reaction are removed at the same volume rate q, so that the tank remains full at all times (see Figure 3.1).

A mass balance per unit time on the amount of reactant in the tank gives

$$(3.3) \qquad V\frac{d\bar{c}}{d\bar{t}} = qc_f - q\bar{c} + Vp,$$

where $p(\bar{c}, \bar{u})$ is given by (3.1). We shall assume that the tank is insulated and that no heat is lost in its walls. A heat balance then gives

$$(3.4) \qquad VC\frac{d\bar{u}}{d\bar{t}} = qCu_f - qC\bar{u} - hVp,$$

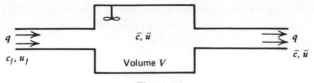

Figure 3.1

where C is the specific heat per unit volume of the reaction mixture, which we take as constant. Introducing the dimensionless variables

$$(3.5) \qquad c = \frac{\bar{c}}{c_f}, \qquad u = \frac{\bar{u}}{u_f}, \qquad t = \frac{\bar{t}q}{V}$$

and the dimensionless constants

$$(3.6) \qquad \alpha = \frac{AV}{q}, \qquad \beta = \frac{hc_f}{u_f C}, \qquad \gamma = \frac{B}{u_f},$$

we find that (3.3) and (3.4) become

$$(3.7) \qquad \frac{dc}{dt} = 1 - c - \alpha c e^{-\gamma/u},$$

$$(3.8) \qquad \frac{du}{dt} = 1 - u + \beta \alpha c e^{-\gamma/u},$$

which form a pair of nonlinear differential equations of the first order for $c(t)$ and $u(t)$. Multiplying (3.7) by β and adding to (3.8) we obtain

$$\frac{d}{dt}(u + \beta c) = (1 + \beta) - (u + \beta c),$$

so that

$$(3.9) \qquad u + \beta c = 1 + \beta + (\text{constant})e^{-t}.$$

If we choose the initial value for $u + \beta c$ to be just its steady-state value $1 + \beta$, the system (3.7)-(3.8) can be reduced to the single nonlinear equation

$$(3.10) \qquad \frac{du}{dt} = 1 - u + \alpha(1 + \beta - u)e^{-\gamma/u}$$

for the dimensionless temperature $u(t)$.

Tubular Reactor

Consider an insulated tubular reactor packed with catalyst pellets of small size, the centerline of the tube lying on the x axis between $x=0$ and $x=l$. A reactant material, from a feed of concentration c_f and temperature u_f, enters the tube at $x=0$ with velocity v in the axial direction. Inside the tube an exothermic reaction is taking place governed by (3.1) and (3.2), where c and u denote the concentration and the temperature, respectively, of the reactant. We shall assume that the space variations of c and u occur only in the axial direction; this is justified on the basis of the turbulent mixing that takes place in the radial direction. The reaction rate p is a pseudohomogeneous rate in which appropriate consideration has been given to the presence of the pellets, to the transfer from stream to catalyst, to the transfer within the catalyst, and so on.

A possible mathematical model for the reactor is so-called plug flow, where the fluid moves down the tube as an identifiable plug with no axial dispersion or backmixing. A more accurate model takes into account the repeated splitting and reforming of the stream that occurs in the interstices of the packed bed, a phenomenon that can be described by including diffusion terms in both the mass and energy balances. Note, however, that the coefficients associated with these terms are not molecular diffusion coefficients but rather dispersion coefficients characterizing the backmixing. A mass balance of the reactant over an infinitesimal section of the tube yields, as in (2.8),

$$(3.11) \qquad \frac{\partial c}{\partial t} = D_c \frac{\partial^2 c}{\partial x^2} - v \frac{\partial c}{\partial x} + p,$$

where the diffusivity D_c has been assumed constant and v is the constant scalar speed of the fluid. In plug flow $D_c = 0$. A heat balance gives

$$(3.12) \qquad C \frac{\partial u}{\partial t} = k \frac{\partial^2 u}{\partial x^2} - vC \frac{\partial u}{\partial x} - hp,$$

where k is an effective thermal conductivity and C is the specific heat per unit volume. We can rewrite (3.12) as

$$(3.13) \qquad \frac{\partial u}{\partial t} = D_u \frac{\partial^2 u}{\partial x^2} - v \frac{\partial u}{\partial x} - \frac{h}{C} p,$$

where $D_u = k/C$ is a thermal diffusivity. Since the same mechanism of turbulent mixing is often responsible for the diffusion of both mass and heat, it may be reasonable to set the *Lewis number* D_c/D_u equal to 1 and to denote the common value of these diffusion coefficients by D.

Just as was done for the stirred tank, it is possible to introduce dimensionless variables and constants that simplify the problem. Probably the most convenient way of doing this is to set

$$u' = \frac{u}{u_f}, \qquad c' = \frac{c}{c_f}, \qquad t' = \frac{tv}{l}, \qquad x' = \frac{x}{l}.$$

In terms of the primed variables (3.11) and (3.13) become (upon *dropping the burdensome primes*)

$$(3.14) \qquad \frac{\partial c}{\partial t} = \frac{1}{\text{Pe}} \frac{\partial^2 c}{\partial x^2} - \frac{\partial c}{\partial x} - \alpha c e^{-\gamma/u},$$

$$(3.15) \qquad \frac{\partial u}{\partial t} = \frac{1}{\text{Pe}} \frac{\partial^2 u}{\partial x^2} - \frac{\partial u}{\partial x} + \beta \alpha c e^{-\gamma/u},$$

where

$$\alpha = \frac{Al}{v}, \qquad \beta = \frac{hc_f}{Cu_f}, \qquad \gamma = \frac{B}{u_f},$$

and the *Peclet number* Pe is defined from

$$(3.16) \qquad \text{Pe} = \frac{lv}{D}.$$

Equations (3.14) and (3.15) are simultaneous nonlinear partial differential equations for the dimensionless temperature and concentration. In some cases it is possible to assume that the reaction is essentially isothermal; then (3.14) becomes a single linear equation.

The initial conditions associated with (3.14) and (3.15) are that $c(x,0)$ and $u(x,0)$ are given in $0 < x < l$. The boundary conditions are easy in the plug-flow case, where it suffices to set $c(0,t)$ and $u(0,t)$ equal to unity. This is inappropriate, however, in the dispersive model, where two boundary conditions have to be specified for c and two conditions for u. The condition on c at the inlet expresses the fact that the flux of reactant vc_f outside the entrance must equal the flux just inside the mouth of the tube. In terms of dimensionless variables this condition is

$$(3.17) \qquad \frac{\partial c}{\partial x}(0,t) - (\text{Pe})c(0,t) = -1.$$

Similarly, the inlet condition for the temperature is

(3.18)
$$\frac{\partial u}{\partial x}(0,t) - (\text{Pe})u(0,t) = -1.$$

At the right end of the tube the conditions are

(3.19)
$$\frac{\partial c}{\partial x} = \frac{\partial u}{\partial x} = 0$$

and are perhaps a little more difficult to justify. Boundary conditions involving $c(0,t)$ and $c(l,t)$ in the same equation occur in the *recycling* problem, in which the inlet mixture is a combination of the feed and of a fraction of the products at the outlet. Then (3.19) becomes

(3.20)
$$\frac{\partial c}{\partial x}(0,t) - (\text{Pe})c(0,t) = -(1-f) - fc(l,t),$$

where f is the recycled fraction.

Catalyst Pellet

Consider a catalyst pellet occupying the domain Ω in R_3 with boundary Γ. Suppose that a first-order, exothermic, irreversible reaction is taking place within the catalyst. Again let D_c and D_u be effective coefficients for the reactant mass flow and heat flow, respectively. The usual balances give

$$\frac{\partial c}{\partial t} = D_c \Delta c + p,$$

$$\frac{\partial u}{\partial t} = D_u \Delta u - \frac{h}{C}p,$$

where p is given by (3.1). One can introduce suitable dimensionless variables to obtain equations similar to (3.14) and (3.15) without the transport term. In terms of these new variables the equations become

(3.21)
$$\frac{\partial c}{\partial t} = \Delta c - \alpha c e^{-\gamma/u},$$

$$\frac{\partial u}{\partial t} = \Delta u + \beta \alpha c e^{-\gamma/u},$$

where the Lewis number D_c/D_u has been taken equal to 1.

For various mathematical aspects of chemical engineering, the reader is referred to the books by Aris (1975), by Denn, and by Perlmutter.

4. THE MOTION OF RODS AND STRINGS

The motion of an elastic body may be described by tracing the history of its particles in time. In some reference configuration such as its unstressed state, the elastic body occupies a certain three-dimensional domain. A typical particle located at $\mathbf{x}_0$ in the reference configuration will be found at $\mathbf{x}$ at time t. Our goal is to formulate a boundary value problem for $\mathbf{x} = \mathbf{x}(\mathbf{x}_0, t)$. We confine ourselves to a special class of bodies known generically as *rods*, a term that subsumes such other terms as "beams," "columns," "struts," and "rings." Even a *string* can be regarded as a special case of a rod that cannot support a moment on any cross section.

What distinguishes a rod from other elastic bodies is its slenderness—length is its only "significant" dimension. More precisely, the elastic behavior of a rod can be completely described by a set of equations having as its only independent variables the time t and the arclength S along a distinguished curve in the reference configuration (usually the centerline of the slender body). The material particles along this distinguished curve form the *axis* of the rod. During the motion the axis deforms, and a particle whose arclength coordinate is S in the reference configuration is then located on the deformed axis at a point $\mathbf{x}(S, t)$.

We shall be principally concerned with the behavior of initially straight rods, symmetric about the x_1-x_2 plane, and subject to loads lying in this plane of symmetry. Under these assumptions it is reasonable to consider only deformations for which the axis of the rod remains in its original plane. The position vector $\mathbf{x}$ of a particle on the deformed axis will therefore have only x_1 and x_2 components, denoted by $u(S, t)$ and $v(S, t)$, respectively, so that we write $\mathbf{x} = u\mathbf{i} + v\mathbf{j}$. The arclength coordinate along the deformed axis is denoted by s, and the angle between the deformed axis and the horizontal by ϕ.

The nature of the forces that cause the motion of the rod also requires some discussion. Some of these forces are distributed; others may be concentrated. Some may be given explicitly either in the reference configuration or in the deformed state, whereas forces of constraint are known only indirectly from their geometrical effects. Some loads depend on the unknown displacement $\mathbf{x}$ (springlike resistance) or on the unknown velocity $\partial \mathbf{x}/\partial t$ (viscous resistance). Without committing ourselves to any special type of loading, we shall denote by Σ_{AB} the resultant of the external forces applied between the points A and B, and by N_{AB} their moment about the x_3 axis.

The cross-sectional force $\mathbf{F}$ and moment M are defined as the resultant force and moment exerted on the material to the left of the cross section by material on the right (see point B of Figure 4.1). A counterclockwise moment is regarded as positive. It may be convenient to express $\mathbf{F}$ either in

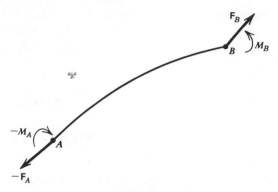

Figure 4.1

terms of its x_1, x_2 components H and V or in terms of its tangential and normal components T and Q:

(4.1) $$\mathbf{F} = H\mathbf{i} + V\mathbf{j} = T\mathbf{e} + Q\mathbf{n},$$

where $\mathbf{e}$ ($= d\mathbf{x}/ds$) is a unit tangential vector and $\mathbf{n}$ a unit normal vector to the deformed axis. The components T and Q are known as the *tension* and *shear*, respectively. We have the obvious relations

(4.2) $$Q = -H\sin\phi + V\cos\phi, \qquad T = H\cos\phi + V\sin\phi.$$

The equations of motion for the portion of the rod between A and B (see Figure 4.1) consist of the two-component linear momentum equation

(4.3) $$\mathbf{F}_B - \mathbf{F}_A + \mathbf{\Sigma}_{AB} = \frac{d}{dt}(\text{linear momentum})$$

and of the one-component angular momentum equation (about the origin)

(4.4) $$\mathbf{k}(M_B - M_A) + \mathbf{x}_B \times \mathbf{F}_B - \mathbf{x}_A \times \mathbf{F}_A + \mathbf{k}N_{AB} = \frac{d}{dt}(\text{angular momentum}).$$

Remark. Although these equations are adequate to describe applied forces that are either distributed or concentrated in space, they must be modified to deal with impulsive forces that cause an instantaneous change in momentum. In that case we would have to use an integrated form over time. For instance, (4.3) would become: net impulse = change in linear momentum.

To derive differential equations from (4.3) and (4.4) we shall have to express the momenta as functions of a space coordinate and time. Choosing the arclength S in the reference configuration as an independent variable, we find that

$$\text{linear momentum} = \int_{S_A}^{S_B} \rho(S) \frac{\partial \mathbf{x}}{\partial t}(S,t)\, dS,$$

$$\text{angular momentum} = \int_{S_A}^{S_B} \rho(S)\mathbf{x} \times \frac{\partial \mathbf{x}}{\partial t}\, dS,$$

where $\rho(S)$ is the density per unit length of the rod in its reference configuration, and S_A and S_B are the arclength coordinates of A and B in the reference configuration.

If the external forces are distributed, we can write

$$(4.5) \qquad \Sigma_{AB} = \int_{S_A}^{S_B} \boldsymbol{\sigma}(S,t)\, dS,$$

where $\boldsymbol{\sigma}$ is the density of loading expressed in the reference configuration. Similarly, we find that

$$\mathbf{k}N_{AB} = \int_{S_A}^{S_B} \mathbf{x} \times \boldsymbol{\sigma}\, dS.$$

Let us set $S_B = S_A + dS$ and $S_A = S$, substitute in (4.3) and (4.4), divide by dS, and let dS tend to 0 to obtain

$$(4.6) \qquad \frac{\partial \mathbf{F}}{\partial S} + \boldsymbol{\sigma} = \rho \frac{\partial^2 \mathbf{x}}{\partial t^2}$$

and

$$\mathbf{k}\frac{\partial M}{\partial S} + \frac{\partial}{\partial S}(\mathbf{x} \times \mathbf{F}) + \mathbf{x} \times \boldsymbol{\sigma} = \rho \mathbf{x} \times \frac{\partial^2 \mathbf{x}}{\partial t^2},$$

which, in view of (4.6) and (4.1), reduces to

$$(4.7) \qquad \frac{\partial M}{\partial S} - H\frac{\partial v}{\partial S} + V\frac{\partial u}{\partial S} = 0.$$

Equation (4.7) can also be written as

$$(4.8) \qquad \frac{\partial M}{\partial S} + Q\delta = 0, \qquad \delta \doteq \frac{\partial s}{\partial S} = \left[\left(\frac{\partial u}{\partial S}\right)^2 + \left(\frac{\partial v}{\partial S}\right)^2\right]^{1/2}.$$

We see that (4.6) and (4.7) are three simultaneous equations for the five unknowns H, V, M, u, v. It is sometimes preferable to use (4.6) and (4.8) as three simultaneous equations for the alternative set of unknowns T, Q, M, u, v. In any event we must supplement our three equations with two constitutive relations between stress and strain. In the case of a string one of these constitutive equations is $M \equiv 0$; since δ never vanishes, we find from (4.8) that the shear Q also vanishes identically in a string.

Before developing a rigorous theory of rods it is perhaps worthwhile to interrupt the discussion to present the usual derivation of the linearized equations for rods and strings. This derivation is unsatisfactory in some respects, particularly because it does not show that the linear equations are a first-order approximation to the full nonlinear problem and therefore does not make it possible to generate a hierarchy of improved approximations.

Classical Derivation of the Linearized Equation

Consider a straight rod which in its stress-free state occupies the interval $(0, L)$ on the x_1 axis. Taking this state as our reference configuration, we have $S = x_1$, and all our equations could be rewritten with x_1 instead of S (but we prefer not to do so). The rod is now given some "small" initial transverse deflection and velocity; furthermore the rod is subject to a transverse loading $\sigma(S, t)\mathbf{j}$ for $t > 0$. Appropriate boundary conditions, to be discussed later, are to be satisfied at the ends of the rod. *We assume that each particle undergoes a purely transverse motion*, so that $u(S, t) = S$. The longitudinal component of (4.6) shows that H does not vary along the rod.

In many beam problems we can take $H \equiv 0$ (for instance, if one end of the beam is free or is supported on frictionless rollers whose motion is confined to the x_1 direction). Equation (4.7) and the transverse component of (4.6) become

$$(4.9) \qquad \frac{\partial M}{\partial S} = -V, \qquad \frac{\partial V}{\partial S} + \sigma = \rho \frac{\partial^2 v}{\partial t^2}.$$

These two equations are supplemented by a single constitutive equation,

$$(4.10) \qquad M = EI \frac{\partial^2 v}{\partial S^2},$$

where E is Young's modulus, I the moment of inertia about the neutral axis, and $\partial^2 v / \partial S^2$ the linearized approximation to the curvature. Combining these equations, we obtain the familiar beam equation

$$(4.11) \qquad -\frac{\partial^2}{\partial S^2} \left(EI \frac{\partial^2 v}{\partial S^2} \right) + \sigma = \rho \frac{\partial^2 v}{\partial t^2}.$$

This equation is supplemented by two initial conditions giving $v(S,0)$ and $(\partial v/\partial t)(S,0)$ and by four boundary conditions (two at each end of the beam). At a *simply supported end* the point of support is hinged so that the ends of the beam can rotate freely during bending; thus the vertical deflection and moment both vanish:

$$(4.12) \qquad v=0, \qquad \frac{\partial^2 v}{\partial S^2}=0 \quad \text{(simply supported)}.$$

At an end built into a wall, the beam cannot rotate, so that the appropriate boundary conditions are

$$(4.13) \qquad v=0, \qquad \frac{\partial v}{\partial S}=0 \quad \text{(built in)}.$$

At a free end the moment and shear vanish:

$$(4.14) \qquad \frac{\partial^2 v}{\partial S^2}=0, \qquad \frac{\partial^3 v}{\partial S^3}=0 \quad \text{(free)}.$$

The problem of transverse vibrations of a string is set in a slightly different framework. The string is tautly stretched between the points $(0,0)$ and $(L,0)$ under tension H_0. This is then the reference configuration, and $S=x_1$ is the x_1 coordinate of a particle in this simple stretched state. We then apply a transverse loading $\sigma(S,t)\mathbf{j}$ and give the string some small initial transverse deflection and velocity. We again assume that each particle undergoes a purely transverse motion and that the longitudinal component H of the tension remains equal to H_0 throughout the motion.

For a string both M and Q vanish, and (4.2) implies that

$$V = H_0 \tan\phi = H_0 \frac{\partial v}{\partial u} = H_0 \frac{\partial v}{\partial S},$$

which when substituted in (4.6) yields the equation

$$(4.15) \qquad H_0 \frac{\partial^2 v}{\partial S^2} + \sigma = \rho \frac{\partial^2 v}{\partial t^2}$$

for the transverse vibrations of a string [see (4.36) for a more careful derivation].

If a beam under tension H_0 is subject to transverse loading, the same line of reasoning shows that the transverse deflection satisfies

$$(4.16) \qquad -\frac{\partial^2}{\partial S^2}\left(EI\frac{\partial^2 v}{\partial S^2}\right) + H_0\frac{\partial^2 v}{\partial S^2} + \sigma = \rho\frac{\partial^2 v}{\partial t^2}.$$

We now return to the mainstream of the discussion.

Constitutive Relations

We shall try to supplement (4.6) and (4.8) by two constitutive relations. We shall assume that the local elongation in the rod depends only on the tension T (axial component of the cross-sectional force) and that the local bending depends only on the moment M on the cross section. Natural strain measures consistent with these assumptions are (see Antman)

$$(4.17) \qquad \delta \doteq \frac{\partial s}{\partial S} = \left(u_S^2 + v_S^2\right)^{1/2} \qquad \text{(for elongation)}$$

and

$$(4.18) \qquad \mu \doteq \frac{\partial \phi}{\partial S} = \frac{u_S v_{SS} - v_S u_{SS}}{u_S^2 + v_S^2} \qquad \text{(for bending)},$$

where we have used a subscript to denote differentiation, and we have appealed to the relation $\tan\phi = v_S / u_S$. Clearly δ is a measure of elongation, since it is the ratio of the length of a deformed element to its length in the reference configuration. The strain ϵ is defined as the change in length of an element divided by its original length: $\epsilon = \delta - 1$. Whereas δ is always positive, ϵ is positive for extension but negative for compression. The bending measure μ is in the nature of a curvature, the actual curvature being $K \doteq \partial\phi/\partial S$ (many authors define the curvature with a minus sign). Thus we have

$$\mu = \frac{\partial\phi}{\partial S} = K\frac{\partial s}{\partial S} = K\delta.$$

One of the reasons for preferring μ to K as a measure of bending is that a circular ring which merely expands under applied forces exhibits a change in K but none in μ, and it seems reasonable to say that no additional bending takes place in the process.

The constitutive laws are taken to be of the form

$$(4.19) \qquad \delta = \hat{\delta}(T, S), \qquad \mu = \hat{\mu}(M, S),$$

where $\hat{\delta}$ and $\hat{\mu}$ are specified functions. Substituting the first of (4.19) in (4.17) gives a relation among T, u and v, whereas substituting the second of (4.19) in (4.18) gives a relation among M, u, and v. These two new relations, together with (4.6) and (4.8), form a set of five equations in the five unknowns T, Q, M, u, v.

Let us now examine the constitutive laws (4.19) in a little more detail. For each fixed S, $\hat{\delta}$ and $\hat{\mu}$ have the general appearance shown in Figure 4.2. If the rod is homogeneous, $\hat{\delta}$ and $\hat{\mu}$ will be independent of S. It is clear from Figure 4.2 that we have tacitly assumed that the reference configuration is the stress-free straight rod. For some purposes, such as the vibrations of a taut string, it is more convenient to use a prestressed state as the reference configuration, in which case a change of variables is necessary to express the constitutive relations in terms of the new variables.

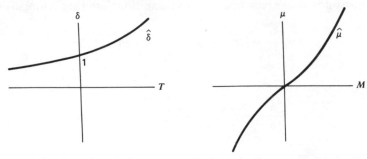

Figure 4.2

The function $\hat{\delta}$ is positive and monotonically increasing in T with $\hat{\delta}(0, S) = 1$ (for the special case of an *inextensible* rod, $\hat{\delta} \equiv 1$). The function $\hat{\mu}$ is odd and increasing in M so that, by continuity, $\hat{\mu}(0, S) = 0$. For a string the second of (4.19) is replaced by $M \equiv 0$, and we then further assume that the first equation can be inverted to give

$$(4.20) \qquad\qquad T = \hat{T}(\delta, S),$$

where $\hat{T}$ is regarded as a *known* function, strictly increasing in δ for each S.

The Vibrating String

We have already noted that both M and Q vanish for a string, so that (4.6) becomes

$$(4.21) \qquad\qquad (T\mathbf{e})_S + \boldsymbol{\sigma} = \rho \mathbf{x}_{tt},$$

where the subscripts denote differentiation. If we take into account the relations

$$(4.22) \quad \begin{aligned} &\mathbf{x} = u\mathbf{i} + v\mathbf{j}, \qquad \delta = \left(u_S^2 + v_S^2\right)^{1/2}, \qquad T = \hat{T}\big[\delta(S,t), S\big], \\ &\mathbf{e} = \mathbf{x}_s = \frac{1}{\delta}\mathbf{x}_S = \frac{1}{\delta}\left(u_S\mathbf{i} + v_S\mathbf{j}\right), \end{aligned}$$

we see that (4.21) becomes a pair of nonlinear partial differential equations for $u(S,t)$ and $v(S,t)$ valid for small and large deflections alike. In any specific problem (4.21) will be supplemented by appropriate initial and boundary conditions.

We shall consider two problems for a string. The first is the extremely simple problem of taking a string whose length in its natural, stress-free state is L and stretching it between the two points $(0,0)$ and $(l,0)$, where $l > L$. With $\sigma = 0$, we are looking for a solution of (4.21)-(4.22) which has the following properties: (a) it is independent of t and has only an x_1 component, so that we write $\mathbf{x} = \mathbf{x}(S) = u(S)\mathbf{i}$, and (b) $\mathbf{x}(0) = 0$, $\mathbf{x}(L) = l\mathbf{i}$. Then $\mathbf{e} = \mathbf{i}$, and (4.21) gives $T_S = 0$. Thus T is a constant along the string, say $T^{(0)}$. When the constitutive law $\delta = \hat{\delta}(T,S)$ is inserted in $u_S = \delta$, we obtain

$$(4.23) \quad u(S) = \int_0^S \hat{\delta}(T^{(0)}, S^*) \, dS^*,$$

so that, since $u(L) = l$, $T^{(0)}$ must satisfy

$$(4.24) \quad l = \int_0^L \hat{\delta}(T^{(0)}, S) \, dS.$$

If $\hat{\delta}$ is a strictly increasing function of T tending to $+\infty$ with T, then (4.24) will have a unique solution $T^{(0)}$ for any given $l > L$. This is the tension required to stretch the string from its natural state to length l. The position of the particles is given by (4.23). We shall refer to this state of the string as the state of *simple stretching*.

Let us now turn from this trivial (though nonlinear) problem to the more difficult one of vibrations. We consider a string which is first stretched between $(0,0)$ and $(l,0)$; then with its endpoints fixed it is subject to an initial displacement and velocity and to external forces for $t > 0$. *It will be convenient to use the state of simple stretching as the reference configuration.* To avoid confusion we label particles in the new reference configuration by the letter X; thus $0 \leqslant X \leqslant l$, and we rewrite (4.21)-(4.22) with X and t as the independent variables:

$$(4.25) \quad [Te]_X + \sigma = \rho \mathbf{x}_{tt}, \qquad t > 0, \quad 0 < X < l,$$

together with

(4.26)   boundary conditions: $\mathbf{x}(0,t)=0, \qquad \mathbf{x}(l,t)= l\mathbf{i}$,

(4.27)   initial conditions: $\mathbf{x}(X,0)=\mathbf{w}(X), \qquad \mathbf{x}_t(X,0)=\mathbf{z}(X)$.

Our measure of elongation is now

(4.28)
$$D \doteq \frac{\partial s}{\partial X} = (u_X^2 + v_X^2)^{1/2},$$

and the constitutive relation (4.20) can be rewritten in the form

(4.29)
$$T = \tilde{T}(D,X),$$

where $\tilde{T}(1,X) = T^{(0)}$, the constant tension corresponding to the state of simple stretching. We also have

(4.30)
$$\mathbf{e} = \mathbf{x}_s = \frac{1}{D}\mathbf{x}_X = \frac{1}{D}(u_X\mathbf{i} + v_X\mathbf{j}).$$

Substitution of (4.28) through (4.30) in (4.25) gives a pair of nonlinear partial differential equations for $u(X,t)$ and $v(X,t)$ which are to be solved under conditions (4.26) and (4.27). Although some static problems such as a cable hanging under its own weight can be solved analytically, this will not be the case for the general dynamic problem. If, however, the forcing terms $\sigma, \mathbf{w}, \mathbf{z}$ are "small," it is possible to use a linearized theory, which we now develop.

Linearized Problem

To treat the problem of small deflections systematically, we introduce an amplitude parameter α (perturbation parameter) in all forcing terms, writing

(4.31)
$$\sigma = \alpha\mathbf{f}, \qquad \mathbf{w} = \alpha\mathbf{W}, \qquad \mathbf{z} = \alpha\mathbf{Z},$$

where $\mathbf{f}, \mathbf{W}, \mathbf{Z}$ are independent of α.

The solution of (4.25) through (4.30) will now depend on the parameter α. Thus $\mathbf{x}, \mathbf{e}, D, T$ will all depend on α. Their behavior for small α is of particular interest. Let us assume that all these quantities can be expanded in power series in α. A typical quantity such as D will have an expansion

(4.32)
$$D(X,t,\alpha) = D^{(0)} + \alpha D^{(1)} + \frac{\alpha^2}{2}D^{(2)} + \cdots,$$

where

$$D^{(k)}(X,t) = \frac{\partial^k D}{\partial \alpha^k}\bigg|_{\alpha=0}.$$

We shall be content with characterizing $x^{(0)}$ and $x^{(1)}$, the first two terms of the alpha expansion of the solution. On setting $\alpha = 0$ in (4.25) through (4.31), we are left with the boundary value problem for the state of simple stretch, which in terms of our reference configuration has the solution

(4.33)
$$x^{(0)} = X\mathbf{i}, \quad u^{(0)}(X) = X, \quad v^{(0)}(X) = 0, \quad \mathbf{e}^{(0)} = \mathbf{i},$$
$$D^{(0)}(X) = 1, \quad T^{(0)}(X) = T^{(0)},$$

where $T^{(0)}$ is the constant tension in simple stretching.

To formulate a boundary value problem for $x^{(1)}(X,t)$ requires more work. We differentiate (4.25) through (4.31) with respect to α and set α equal to 0. The first step gives

$$(T_\alpha \mathbf{e} + \mathbf{e}_\alpha T)_X + \mathbf{f} = \rho x_{tt\alpha}; \quad x_\alpha|_{X=0,l} = 0,$$

$$x_\alpha|_{t=0} = \mathbf{W}, \quad x_{\alpha t}|_{t=0} = \mathbf{Z},$$

and, on setting $\alpha = 0$,

(4.34)
$$(T^{(1)}\mathbf{e}^{(0)} + \mathbf{e}^{(1)} T^{(0)})_X + \mathbf{f} = \rho x_{tt}^{(1)}; \quad x^{(1)}(0,t) = x^{(1)}(l,t) = 0,$$
$$x^{(1)}(X,0) = \mathbf{W}, \quad x_t^{(1)}(X,0) = \mathbf{Z}.$$

To simplify (4.34) we note that from (4.28) we have

$$D^2 = u_X^2 + v_X^2, \quad DD_\alpha = u_X u_{\alpha X} + v_X v_{\alpha X},$$

and hence

$$D^{(0)}D^{(1)} = u_X^{(0)} u_X^{(1)} + v_X^{(0)} v_X^{(1)},$$

which, by (4.33), reduces to

$$D^{(1)} = u_X^{(1)}.$$

A similar calculation starting from (4.30) gives

$$\mathbf{e}^{(1)} = v_X^{(1)}\mathbf{j}.$$

Since $\tilde{T} = T(\tilde{D}, X)$, we have $\tilde{T}_\alpha = T_D D_\alpha$ and therefore

$$T^{(1)} = \tilde{T}_D(1, X) D^{(1)} = \tilde{T}_D(1, X) u_X^{(1)}.$$

Substitution in (4.34) yields separate *linear* boundary value problems for $u^{(1)}$ and $v^{(1)}$:

(4.35)
$$\left[\tilde{T}_D(1, X) u_X^{(1)} \right]_X + \mathbf{f} \cdot \mathbf{i} = \rho u_{tt}^{(1)}; \qquad u^{(1)}(0, t) = u^{(1)}(l, t) = 0,$$
$$u^{(1)}(X, 0) = \mathbf{W} \cdot \mathbf{i}, \quad u_t^{(1)}(X, 0) = \mathbf{Z} \cdot \mathbf{i},$$

and

(4.36)
$$T^{(0)} v_{XX}^{(1)} + \mathbf{f} \cdot \mathbf{j} = \rho v_{tt}^{(1)}; \qquad v^{(1)}(0, t) = v^{(1)}(l, t) = 0,$$
$$v^{(1)}(X, 0) = \mathbf{W} \cdot \mathbf{j}, \quad v_t^{(1)}(X, 0) = \mathbf{Z} \cdot \mathbf{j}.$$

Both (4.35) and (4.36) are *wave equations* which will be useful if $\mathbf{f}$ and ρ can be expressed in terms of the stretched variable X. In fact, $\mathbf{f}$ is usually given directly as a function of X, but ρ is normally given in the natural state in terms of S. This presents no difficulty, however, because $X(S)$ is known explicitly from (4.23) as $\hat{\delta}(T^{(0)}, S)$. Thus, if $\rho(S)$ is the density of the string in its stress-free state, we find that

$$\rho(X) = \tilde{\rho}(S(X)) \frac{dS}{dX}.$$

Equation (4.35) describes the small *longitudinal* vibrations of a string and applies equally well to a rod, since bending effects appear only in the transverse equation. Note that the local stiffness of the string or rod is given by $\tilde{T}_D(1, X)$; this plays the role of Young's modulus, which will vary with position when the body is inhomogeneous.

The *transverse* vibrations of a *string* are governed by (4.36). Note that the coefficient of $v_{XX}^{(1)}$ is constant even though the string may be inhomogeneous.

For the *transverse* vibrations of a beam we return to (4.6), (4.8), and the two constitutive laws (4.19), using a linearization procedure similar to that for the string, but now about the stress-free state; and recalling (4.18), we find that v obeys the *fourth-order linear equation*

(4.37)
$$- \left[\hat{M}_\mu(0, S) v_{SS}^{(1)} \right]_{SS} + \mathbf{f} \cdot \mathbf{j} = \rho v_{tt}^{(1)},$$

which clarifies and generalizes (4.11). In the usual Euler-Bernoulli theory one takes $\hat{M}_\mu = EI$.

In deriving (4.37) we have assumed an initially stress-free rod. If there is a tension $T^{(0)}$ in the beam, a term $T^{(0)}v_{SS}^{(1)}$ must be added to the left side of (4.37) as in (4.16).

In Chapter 9 we consider a fully nonlinear problem: the buckling of an initially straight homogeneous rod under compression.

5. NEUTRON TRANSPORT

As neutrons travel in a medium consisting of fixed nuclei, they frequently collide with some of these nuclei. A neutron is a point particle moving at constant velocity between successive collisions with nuclei. Neutrons do not interact with one another, nor do they affect the positions of the nuclei with which they collide. In a collision a neutron can be absorbed, can be scattered (that is, deflected into a new direction with a possibly different speed), or may enter the nucleus to trigger a fission in which new neutrons are emitted. Fortunately we shall not need a detailed picture of these interactions. The overall process can be described by two "constitutive" functions, one giving the probability of collision and the other the distribution of the products of a collision.

For our purposes we first make the simplifying assumption that all neutrons have the *same speed*; setting $\mathbf{v} = v\boldsymbol{\omega}$, where ω is a unit vector in the direction of motion, we find that $v = |\mathbf{v}|$ is *constant* for all neutrons. Although this assumption appears drastic, it enables us to maintain some of the character of the theory without sinking into a bog of notational and calculational complexity.

The *collision cross section* σ is the *probability per unit path length* that a neutron at $\mathbf{x}$ moving in the direction ω experiences a collision. Since we wish to confine ourselves to isotropic but possibly inhomogeneous media, we shall take σ to be a function of $\mathbf{x}$ only. The quantity $1/\sigma(\mathbf{x})$ is the *mean free path* at $\mathbf{x}$. The frequency of collisions (expected number per unit time per neutron) is then $v\sigma$.

We shall also assume that the scattering and fission products are distributed isotropically. If a neutron experiences a collision at $\mathbf{x}$, the average number of outgoing neutrons is $c(\mathbf{x})$, *the multiplication factor*; the average number of outgoing neutrons in the solid angle $d\omega$ is therefore $[c(\mathbf{x})/4\pi]d\omega$. Even within the theory of isotropic scattering, a more precise approach would be to introduce, instead of the average number of outgoing neutrons, the actual probabilities of $0, 1, 2$, etc., neutrons emerging from a collision. The resulting theory is nonlinear, and, except for an example in Chapter 9, we shall be satisfied with the coarse description using expected values.

Since our interest is not so much in the migration of individual neutrons as in their distribution, it is natural to introduce the *angular density* of neutrons $\psi(\mathbf{x}, \omega, t)$, where $\psi(\mathbf{x}, \omega, t) d\mathbf{x} \, d\omega$ is the expected number of neutrons present at time t in the volume $d\mathbf{x}$ about $\mathbf{x}$ and having directions in the solid angle $d\omega$ about the direction ω. By integrating ψ over the solid angle, we obtain a reduced neutron density N, depending only on $\mathbf{x}$ and t:

$$(5.1) \qquad N(\mathbf{x}, t) = \int_{4\pi} \psi(\mathbf{x}, \omega, t) \, d\omega,$$

where the 4π indicates that the integral is taken over the entire range of the solid angle. Thus $N(\mathbf{x}, t) d\mathbf{x}$ is the expected number of neutrons at time t in the volume $d\mathbf{x}$ about $\mathbf{x}$. Usually $d\omega$ is expressed in spherical coordinates θ and ϕ, so that $d\omega = \sin\theta \, d\theta \, d\phi$, where $0 \leqslant \theta \leqslant \pi, \, -\pi \leqslant \phi \leqslant \pi$. By introducing

$$(5.2) \qquad \mu = \cos\theta,$$

we can rewrite (5.1) as

$$(5.3) \qquad N(\mathbf{x}, t) = \int_{-\pi}^{\pi} d\phi \int_{-1}^{1} d\mu \, \psi(\mathbf{x}, \phi, \mu, t).$$

Although the information contained in the reduced density N is sufficient for many purposes, the fundamental boundary value problem must be formulated for ψ, the angular density.

An important vector quantity related to ψ is the *angular current density*

$$(5.4) \qquad \mathbf{j}(\mathbf{x}, \omega, t) = \mathbf{v}\psi(\mathbf{x}, \omega, t) = v\omega\psi.$$

If dS is a surface element through $\mathbf{x}$ with unit normal $\mathbf{n}$, the expected number of neutrons with directions in $d\omega$ about ω crossing dS per unit time is

$$(5.5) \qquad \mathbf{j} \cdot \mathbf{n} \, dS \, d\omega.$$

Now let R be an *arbitrary* portion of the medium. The boundary of R is denoted by B. For the region R we take a balance *per unit time* of neutrons having directions in $d\omega$ about ω, which we call the *relevant neutrons*. This balance has the general form

$$(5.6) \qquad \text{rate of increase in relevant neutrons} = \text{gains} - \text{losses},$$

whose individual terms will now be analyzed.

1. *Rate of increase in relevant neutrons in R.* By definition this term is

$$(5.7) \qquad d\omega \int_R \frac{\partial \psi}{\partial t}(\mathbf{x}, \omega, t) \, d\mathbf{x}.$$

2. *Loss from leakage through B*. The quantity

(5.8)
$$d\omega \int_B \mathbf{j} \cdot \mathbf{n} \, dS$$

represents the net rate at which relevant neutrons cross B (here $\mathbf{n}$ is the outward normal to R). Of course, if the integral is negative, there is a net flow of relevant neutrons into R. By using the divergence theorem, we transform (5.8) into

(5.9)
$$d\omega \int_R \operatorname{div} \mathbf{j} \, dx = v \, d\omega \int_R \omega \operatorname{grad} \psi \, dx.$$

3. *Collision losses*. Any relevant neutron experiencing a collision in R is regarded as lost (relevant neutrons that emerge as scattering or fission products will be counted later as gains):

(5.10)
$$\text{rate of collision losses} = d\omega \int_R v\sigma(\mathbf{x}) \psi(\mathbf{x}, \omega, t) \, dx.$$

4. *Gains from scattering and fission*. In the element of volume dx there are $v\sigma(\mathbf{x})N(\mathbf{x}, t) \, dx$ total collisions per unit time for all incoming neutrons. These collisions contribute

$$\frac{c(\mathbf{x})}{4\pi} \, d\omega v\sigma(\mathbf{x}) N(\mathbf{x}, t) \, dx$$

emerging relevant neutrons per unit time. The total contribution from all collisions in R is therefore

(5.11)
$$\frac{d\omega}{4\pi} v \int_R c(\mathbf{x})\sigma(\mathbf{x})N(\mathbf{x}, t) \, dx,$$

where we have assumed that fission neutrons are emitted instantaneously. A more careful analysis is needed to treat delayed neutrons.

5. *Gains from independent sources*. We shall assume that there are *prescribed* body sources of density $q(\mathbf{x}, \omega, t)$. Such sources might arise from spontaneous fission or from the action of cosmic rays. The rate of gain of relevant neutrons in R is

(5.12)
$$d\omega \int_R g(\mathbf{x}, \omega, t) \, dx.$$

If we now substitute the five terms (5.7) and (5.9) through (5.12) in the balance equation (5.6), we can drop the common factor $d\omega$, obtaining an equation of the form $\int_R F(\mathbf{x}, \omega, t)\,d\mathbf{x} = 0$. Since R is an *arbitrary* subdomain of the domain Ω occupied by the medium, it follows that the function F, assumed continuous, must vanish identically. This leads to the equation

$$(5.13) \qquad \frac{\partial \psi}{\partial t} = \frac{v}{4\pi} c(\mathbf{x})\sigma(\mathbf{x})N(\mathbf{x}, t) + q(\mathbf{x}, \omega, t) - v\sigma(\mathbf{x})\psi(\mathbf{x}, \omega, t) - \mathrm{div}\,\mathbf{j}.$$

It is worthwhile to write this equation in dimensionless variables. Let $\tilde{\sigma}$ be a typical value of the cross section, for instance, the average value. We define dimensionless space and time variables

$$(5.14) \qquad \mathbf{x}' = \mathbf{x}\tilde{\sigma}, \qquad t' = tv\tilde{\sigma},$$

which has the effect of measuring distance and time in terms of mean free paths and mean free times, respectively. Substituting in (5.14) and *dropping primes*, we obtain

$$(5.15) \qquad \frac{\partial \psi}{\partial t} = \frac{c\sigma^*}{4\pi} N + Q - \sigma^*\psi - \mathrm{div}\,(\omega\psi),$$

where we have set $\sigma^*(\mathbf{x}) = \sigma(\mathbf{x})/\tilde{\sigma}$ and $Q = q/v\tilde{\sigma}$, and where the same symbols ψ, N have been used for the angular densities now expressed in terms of the dimensionless variables. On using (5.1) and (5.4), we find that

$$(5.16) \qquad \frac{\partial \psi}{\partial t} + \omega \cdot \mathrm{grad}\,\psi + \sigma^*(\mathbf{x}) = Q + \frac{c(\mathbf{x})\sigma^*(\mathbf{x})}{4\pi} \int_{4\pi} \psi(\mathbf{x}, \omega, t)\,d\omega,$$

which is a *linear integrodifferential equation* for $\psi(\mathbf{x}, \omega, t)$ to be solved under certain initial and boundary conditions.

In a typical initial-boundary value problem, the angular density is given at time $t = 0$ throughout Ω. At a *free surface* neutrons are permitted to leave the medium but not to enter it. If Γ is the boundary of Ω, this condition is expressed as

$$(5.17) \qquad \psi(\mathbf{x}, \omega, t) = 0, \qquad t > 0, \quad \mathbf{x} \in \Gamma, \quad \omega \cdot \mathbf{n} < 0.$$

The right side of (5.16) is the sum of a prescribed neutron source term Q and of a source term arising from scattering and fission in which the density ψ appears under the integral sign. In a problem of *pure absorption* (5.16) reduces to the partial differential equation

$$(5.18) \qquad \frac{\partial \psi}{\partial t} + \omega \cdot \mathrm{grad}\,\psi + \sigma^*\psi = Q,$$

where ω can be regarded as a parameter, since all differentiations are with respect to space and time. We may therefore solve (5.18) for each value of ω separately, and, as it turns out, the solution can be written as a linear integral operator acting on Q. By treating the right side of (5.16) as an effective source term, the equation can be translated into a pure integral equation.

The terms on the left side of (5.16) and (5.18) represent, respectively, the change of density at a fixed point in x, ω space, a *streaming* contribution corresponding to the unimpeded motion of neutrons traveling in the ω direction, and a loss term due to collisions. The first two terms can be incorporated into a single term, $d\psi/dt$, a *substantial derivative*, that is, a rate of change as it would appear to an observer attached to a packet of neutrons moving in the ω direction.

We shall now simplify (5.18) in the case of slab geometry. Position in the medium is described by Cartesian coordinates x, y, and z. The medium is of infinite extent in x and y, so that we are dealing with a slab $a \leqslant z \leqslant b$ or a half-space $a \leqslant z \leqslant \infty$ or the whole space. To specify the direction of motion ω of neutrons, we use spherical coordinates θ and ϕ. Here θ is the polar angle measured from the z axis, and ϕ is the azimuthal angle. As in (5.2), we introduce $\mu = \cos\theta$, so that the angular density ψ is regarded as a function of x, y, z, ϕ, μ, and t. The material properties σ^* and c, the source term Q, and the initial angular density are assumed to be *independent of x and y*. The source term and initial angular density are also taken to be *independent* of ϕ. Under these conditions it is apparent that ψ will, for $t > 0$, depend only on z, μ, and t. Since $\omega \cdot \text{grad}\,\psi = \mu(\partial\psi/\partial z)$, we can rewrite (5.16) after performing the ϕ integration as an equation for $\psi(z, \mu, t)$:

$$(5.19) \qquad \frac{\partial\psi}{\partial t} + \mu\frac{\partial\psi}{\partial z} + \sigma^*\psi = Q + \frac{c\sigma^*}{2}\int_{-1}^{1}\psi(z, \mu', t)\,d\mu',$$

where v is a given constant, c and σ^* are given functions of z, and Q is a given function of z, μ, t.

A further simplification is possible in the *time-independent* case by introducing the new independent variable

$$Z = \int_0^z \sigma^*(z')\,dz'.$$

Then (5.19) becomes

$$(5.20) \qquad \mu\frac{\partial\psi}{\partial z} + \psi = \frac{c}{2}\int_{-1}^{1}\psi(Z, \mu')\,d\mu' + Q^*(Z, \mu),$$

where $Q^* = Q/\sigma^*$ and ψ is the density expressed in terms of the new variables.

So far we have considered exact forms of the equation of neutron transport appropriate in special geometries with simple constitutive laws. In practice, however, the usual approach is to make the diffusion *theory approximation*, which we now discuss. We start by integrating (5.15) over ω to obtain

$$(5.21) \qquad \frac{\partial N}{\partial t} = -(1-c)\sigma^* N + \tilde{Q} - \operatorname{div}\left(\int_{4\pi} \omega \psi \, d\omega\right),$$

where $\tilde{Q}$ is the integral of Q over ω. Were it not for the divergence term, (5.21) could be regarded as an equation to determine N. The integral in (5.21) is a reduced current which we denote by **J**. In analogy with heat flow and mass flow, it is tempting to assert that **J** and N should be related by *Fick's law*:

$$(5.22) \qquad \mathbf{J} = -D(\mathbf{x})\operatorname{grad} N,$$

where D is a suitable diffusion coefficient. Then (5.21) becomes a diffusion equation for N:

$$(5.23) \qquad \frac{\partial N}{\partial t} - \operatorname{div}(D\operatorname{grad} N) + (1-c)\sigma^* N = Q.$$

Unfortunately both **J** and N are defined in terms of ψ (which is presumably determined as the solution of an initial-boundary value problem), so that we have no control over whether or not (5.22) is satisfied. It turns out that (5.22) and (5.23) are reasonable approximations if (a) σ^*, c, Q^* are slowly varying over distances and times of order 1, (b) c is nearly 1, and (c) we are not near the boundary. (Remember that all distances and times are measured in terms of mean free paths and mean free times.)

6. NOTES ON CONVERGENCE

A sequence of real numbers $\{s_1, \ldots, s_n, \ldots\}$ is said to *converge to s* (or *have the limit s*) if for each $\varepsilon > 0$ there exists an index N such that

$$|s - s_n| < \varepsilon \qquad \text{for all } n > N.$$

This definition is a way of stating in precise mathematical language the intuitive notion that, for n large enough, s_n can be made arbitrarily close to s. To apply the definition, that is, to exhibit N as a function of ε, we must be able to guess the limit s. The following theorem permits us to determine whether a sequence converges without reference to the actual limit.

Theorem (Cauchy criterion). The necessary and sufficient condition for the sequence $\{s_n\}$ to converge is that, for each $\varepsilon > 0$, there exists an index P such that

$$|s_n - s_m| < \varepsilon \qquad \text{for all } m, n > P.$$

Thus, if all its terms beyond a certain index can be made close to each other, a sequence must converge to some limit. This agrees with our mental image of the real line as a continuum without gaps. The Cauchy criterion can be taken as a fundamental postulate of the real number system or can be derived from some other, essentially equivalent axiom such as the Dedekind cut property.

For sequences of *functions* there are many different ways of defining convergence, that is, there are various modes of convergence. We shall deal with a fixed interval E (open or closed) on the x axis and with a sequence $\{s_1(x), \ldots, s_n(x), \ldots\}$ of real-valued functions on E.

Perhaps the most natural (but, as we shall see, not the most useful) way of defining convergence for a sequence of functions is to examine each point x in E individually. With x fixed, the sequence $\{s_n(x)\}$ is a sequence of real numbers, and we certainly know what it means for this sequence to converge to the real number $s(x)$. This gives rise to the definition of *pointwise convergence*: for each x in E and for each $\varepsilon > 0$ there exists an index N (depending on both x and ε) such that

$$|s(x) - s_n(x)| < \varepsilon \qquad \text{for all } n > N.$$

If we look at this definition graphically, then, for each fixed x, the ordinate $s_n(x)$ will be within ε of the ordinate $s(x)$ if n exceeds N. At another point x in E a much larger N may be required to achieve the same degree of approximation. The following example shows some of the deceptive features of pointwise convergence.

Consider the sequence of functions

(6.1) $$s_n(x) = nxe^{-nx} \qquad \text{on } E: 0 \leqslant x \leqslant 1,$$

a few terms of which are graphed in Figure 6.1. Each function $s_n(x)$ has the same maximum value $1/e$ taken on at $x = 1/n$, so that $s_n(x)$ rises sharply to the height $1/e$ and then dies down quickly. Thus each $s_n(x)$ has a bump of the same height, the bump becoming thinner and thinner and moving closer and closer to $x = 0$ as n becomes larger. If we fix $x > 0$, then, by waiting long enough (that is, taking N sufficiently large), the bump will have moved to the left of the fixed value of x and the ordinate $s_n(x)$ at that

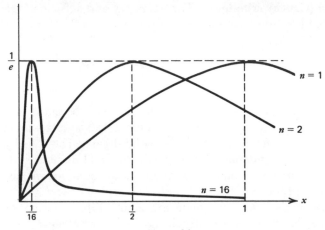

Figure 6.1

fixed point will tend to 0 (of course, the nearer x is to 0, the longer we have to wait to get within a preassigned degree of approximation of the limit). At $x=0$ each $s_n=0$, so clearly the limit is 0. Thus, according to the definition of pointwise convergence, which treats each x individually, the sequence $\{s_n(x)\}$ tends to 0 pointwise on $0 \leqslant x \leqslant 1$. Nevertheless we feel somewhat uncomfortable with the result, since the graph of $s_n(x)$ with its bump of fixed height $1/e$ does not seem like such a good approximation to the limit function, the curve $s \equiv 0$. The situation is even more disturbing for the sequence

$$(6.2) \qquad\qquad s_n(x) = n^\alpha x e^{-nx} \qquad \text{on } E: 0 \leqslant x \leqslant 1,$$

when $\alpha > 1$. Here we still have $s_n(x)$ tending to 0 pointwise on $0 \leqslant x \leqslant 1$, but now the bump in $s_n(x)$ increases in height indefinitely as n tends to infinity. The notion of uniform convergence is perhaps more in keeping with our intuition.

The sequence $\{s_n(x)\}$ is said to *converge uniformly to* $s(x)$ *on E* if for each $\varepsilon > 0$ there exists N (depending only on ε) such that

$$|s_n(x) - s(x)| < \varepsilon \qquad \text{for all } n > N \text{ and all } x \text{ in } E.$$

Remarks

1. If the convergence is uniform and a degree of approximation ε is specified, it is possible to choose a value of N that works for all x in E

simultaneously; this means that all the curves $s_n(x)$ with $n > N$ lie entirely within an ε strip of the limit curve $s(x)$. The maximum deviation between the approximating curve $s_n(x)$ and the limit $s(x)$ therefore approaches 0 as n tends to infinity. Thus we have the equivalent definition: $\{s_n(x)\}$ *converges to $s(x)$ uniformly on E* if the numerical sequence

(6.3)
$$\sup_{x \in E} |s_n(x) - s(x)|$$

tends to 0 as $n \to \infty$. The advantage of this last formulation is that it enables us to use the framework of metric spaces (see Chapter 4) with (6.3) playing the role of the "distance" between the functions $s_n(x)$ and $s(x)$.

2. The Cauchy criterion obviously applies also to uniform convergence: the sequence $\{s_n(x)\}$ converges uniformly on E if, for each $\varepsilon > 0$, there exists an index P such that

$$|s_n(x) - s_m(x)| < \varepsilon \qquad \text{for all } m, n > P \text{ and all } x \text{ in } E.$$

We shall need two fundamental theorems about uniform convergence.

Theorem 1. If the sequence $\{s_n(x)\}$ of continuous functions converges uniformly on E, the limit function $s(x)$ is necessarily continuous.

Theorem 2. If $\{s_n(x)\}$ tends to $s(x)$ uniformly on the *bounded* interval E, and if s_n and s are integrable (say, in the Riemann sense) over E, then

(6.4)
$$\lim_{n \to \infty} \int_E s_n(x)\, dx = \int_E s(x)\, dx.$$

The second of these theorems is on term-by-term integration. A much more powerful theorem of this type is the Lebesgue dominated convergence theorem (see Section 7). In (6.2) the sequence $\{s_n(x)\}$ converges uniformly to 0 on E if $\alpha < 1$. Indeed the maximum of $s_n(x)$ is $n^{\alpha-1}/e$, which tends to 0 as n approaches infinity if $\alpha < 1$. If $\alpha \geqslant 1$, we still have pointwise convergence to 0, but not uniform convergence. Obviously $s_n(x)$ is integrable on E, and so is $s(x)$. A direct calculation shows that

$$\int_0^1 s_n(x)\, dx = n^{\alpha-2} \int_0^n x e^{-x}\, dx,$$

and, since $\int_0^\infty xe^{-x}dx=1$, we have

$$(6.5) \qquad \lim_{n\to\infty}\int_0^1 s_n(x)\,dx=\begin{cases} 0, & \alpha<2, \\ 1, & \alpha=2, \\ \infty, & \alpha>2. \end{cases}$$

This explicit result is predicted by Theorem 2 on uniform convergence only for $\alpha<1$. However, the Lebesgue dominated convergence theorem will give us the result for the whole range $\alpha<2$.

It is also worth emphasizing that the boundedness of the interval is needed in Theorem 2. Consider, for instance, a bounded function $f(x)$ on $0\leqslant x<\infty$ such that $\int_0^\infty f(x)\,dx=1$ (for instance, $f=xe^{-x}$ will do). The sequence $\{s_n(x)\doteq(1/n)f(x/n)\}$ then converges uniformly to 0 on $0\leqslant x<\infty$, yet

$$\int_0^\infty s_n(x)\,dx=\int_0^\infty \frac{1}{n}f\left(\frac{x}{n}\right)dx=\int_0^\infty f(u)\,du=1,$$

which clearly does not tend to 0 as $n\to\infty$.

Uniform convergence on an interval (a,b) is based on the distance function (or metric)

$$(6.6) \qquad d_\infty(u,v)=\sup_{a\leqslant x\leqslant b}|u(x)-v(x)|,$$

which is only one way of measuring how close the two functions $u(x)$ and $v(x)$ are to each other. Other useful distance functions are

$$(6.7) \qquad d_1(u,v)=\int_a^b|u(x)-v(x)|\,dx$$

and

$$(6.8) \qquad d_2(u,v)=\left[\int_a^b|u(x)-v(x)|^2\,dx\right]^{1/2}.$$

In Figure 6.2 we show two functions which are close in all three metrics; in Figure 6.3 u and v are close in the sense of (6.7) and (6.8) but not in the sense of (6.6). Each of these gives rise to its own mode of convergence. We say that $\{s_n(x)\}$ tends to $s(x)$ in the mean (or in L_1) if $\lim_{n\to\infty}d_1(s_n,s)=0$, that is,

$$(6.9) \qquad \lim_{n\to\infty}\int_a^b|s_n(x)-s(x)|\,dx=0.$$

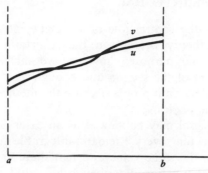

Figure 6.2

The sequence $\{s_n(x)\}$ tends to $s(x)$ in *mean square* (or in L_2) if $\lim_{n\to\infty} d_2(s_n, s) = 0$, that is,

$$(6.10) \qquad \lim_{n\to\infty} \int_a^b |s_n(x) - s(x)|^2\, dx = 0,$$

where we have dropped the square root at no cost.

In (6.2) we see that $\{s_n(x)\}$ tends to 0 in the mean for $\alpha < 2$ and in the mean square for $\alpha < 3/2$.

We note the following theorem. If $\{s_n(x)\}$ is a sequence of continuous functions which tends uniformly to $s(x)$ on the bounded interval $a \leqslant x \leqslant b$, then $\{s_n\}$ also tends to s in L_1 and in L_2.

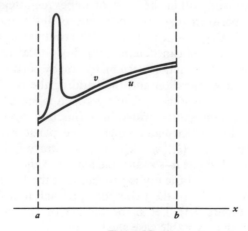

Figure 6.3

7. THE LEBESGUE INTEGRAL

No one will dispute the central role of Lebesgue integration in analysis, and this beautiful theory belongs in the repertoire of every aspiring mathematician. For our purposes, however, we shall need only the few facts listed at the end of the section and enough discussion to give them meaning. We can safely omit proofs because the methods involved are not used elsewhere in the book.

The Lebesgue integral may be viewed as an extension of the Riemann integral in the sense that every Riemann-integrable function is also Lebesgue integrable to the same value and that there exist some functions, such as the Dirichlet function described below, that fail to be Riemann integrable but are Lebesgue integrable. The kind of function that is Lebesgue integrable but not Riemann integrable rarely, if ever, occurs in practice. Thus it is not for computational reasons that the Lebesgue integral is so important. What the Lebesgue integral does is to give structural unity to analysis. From a philosophical point of view, the relationship of Lebesgue-integrable functions to Riemann-integrable functions is similar to that of real numbers to rational numbers. Concrete calculations require only rational numbers, but mathematics needs irrational numbers. The totality of real numbers (rational plus irrational) has an inner consistency absent from the class of rational numbers alone. It is the completeness (see Chapter 4) of the real number system which makes it powerful. Principally this means that when we apply limiting processes in the class of real numbers we remain within the class. Similarly we shall find that for most concrete calculations the notion of Riemann integral is adequate, but theorems involving passage to the limit are more easily formulated and proved within the class of Lebesgue-integrable functions.

The difference between these two concepts of integration is illustrated by the following analogy, which, though not strictly apt, has some anecdotal value. A shopkeeper can determine a day's total receipts either by adding the individual transactions (Riemann) or by sorting bills and coins according to their denomination and then adding the respective contributions (Lebesgue). Obviously the second approach is more efficient!

Consider now a nonnegative real-valued function $f(x)$ defined on the interval $0 \leqslant x \leqslant 1$. In the Riemann scheme one partitions the x interval, then forms the sum $\sum_{k=1}^{n} f(\xi_k)(x_k - x_{k-1})$ for arbitrary ξ_k in $[x_{k-1}, x_k]$, and finally passes to the limit as $n \to \infty$ and the length of the largest subdivision tends to 0. The principal difficulty is proving that the limit exists independently of the choice of ξ_k. In the Lebesgue approach it is the y axis that is partitioned (see Figure 7.1). Let E_i be the set of values of x such that $y_{i-1} \leqslant f(x) < y_i$; in the favorable case shown in the figure, E_i is the union of a finite number of disjoint intervals. We then form the sum $\sum_{i=1}^{n} \eta_i m(E_i)$, where η_i is chosen arbitrarily in $[y_{i-1}, y_i]$ and $m(E_i)$ is the measure of E_i, that is, the sum of the lengths of the disjoint intervals that make up E_i. As the partition is made finer, there is no longer any question as to the existence of the limit of the sum. Indeed, the lower sum $\sum_{i=1}^n y_{i-1} m(E_i)$ is monotonically increasing with n and bounded above, and so must converge. The upper sum $\sum_{i=1}^n y_i m(E_i)$ differs from the lower sum by less than $\max(y_i - y_{i-1}) \sum_{i=1}^n m(E_i)$; since $\sum_{i=1}^n m(E_i) = 1$ and $\max(y_i - y_{i-1})$ tends to 0, the upper sum must also converge to the same value as the lower sum. This common value is the Lebesgue integral

$$\int_0^1 f(x)\, dx.$$

In less favorable cases the set E_i may be much more complicated than shown in the figure. A consistent definition for the measure of sets must be given so that the analysis just presented can be suitably adapted.

Rather than developing the measure-oriented approach to the Lebesgue integral, we prefer to follow the method of Tonelli for constructing the Lebesgue integral of a nonnegative real-valued function $f(x)$ on the interval $[0, 1]$. A set E of points on this interval is said to have *measure less than ε* if E can be contained in a finite or countably infinite set of intervals of total length less than ε. The set has *measure 0* if such a covering can be found for each $\varepsilon > 0$. A function $f(x)$ is *measurable* if, for any $\varepsilon > 0$, we can convert it into a continuous function by changing its values on a set of

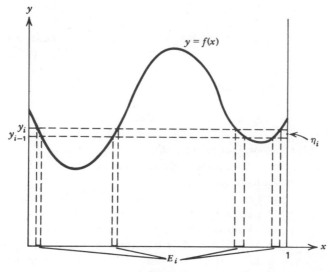

Figure 7.1

measure less than ε. Note that f is surely measurable if it can be converted to a continuous function by altering its values on a set of measure 0.

Example 1. Suppose $f(x)$ is piecewise continuous in $[0, 1]$ with simple jumps at $x_1, \ldots, x_n$. In each of the intervals $[x_i - \varepsilon/2n, x_i + \varepsilon/2n]$ we can replace $f(x)$ by a straight line joining the points $(x_i - \varepsilon/2n, f(x_i - \varepsilon/2n))$ and $(x_i + \varepsilon/2n, f(x_i + \varepsilon/2n))$. The resulting function is continuous, and we have altered the values of the original function over a set of measure $n(\varepsilon/n) = \varepsilon$.

Example 2. The Dirichlet function $f(x)$ has the value 1 when x is rational and 0 when x is irrational. If we change the value of f from 1 to 0 on the set of rationals, we obtain the continuous function that vanishes identically on $[0, 1]$. We claim the set of rationals has measure 0. Since the rational numbers form a countable set, they can be placed into 1-1 correspondence with the positive integers, for instance by ordering them so that they have increasing denominators:

$$0, 1, \frac{1}{2}, \frac{1}{3}, \frac{2}{3}, \frac{1}{4}, \frac{3}{4}, \frac{1}{5}, \frac{2}{5}, \frac{3}{5}, \frac{4}{5}, \frac{1}{6}, \ldots.$$

For each $\varepsilon > 0$, we enclose the kth rational number in this list in an interval of length $\varepsilon/2^k$. The total length of the intervals enclosing all rationals is then

$$\sum_{k=1}^{\infty} \frac{\varepsilon}{2^k} = \varepsilon.$$

Thus for each $\varepsilon > 0$ we can enclose the rational numbers in a countably infinite set of intervals whose total length is ε. Since this can be done for each $\varepsilon > 0$, we conclude that the set of rationals has measure 0.

Therefore the *Dirichlet function is measurable*.

Example 3. Let $f(x) = x^{-\alpha}, 0 < x \leqslant 1$, where $\alpha > 0$; we define $f(0)$ to be 0, but any other value would do as well. Then $f(x)$ is measurable, since it can be converted into a continuous function by replacing it on $0 \leqslant x < \varepsilon$ by the constant function $\varepsilon^{-\alpha}$. Of course there are many other ways in which such a conversion can be accomplished.

The notion of the Lebesgue integral of a nonnegative measurable function $f(x)$ can now be introduced. Let $\{\varepsilon_n\}$ be a sequence of positive numbers such that $\varepsilon_n \to 0$. For each n construct a nonnegative continuous function $f_n(x)$ which differs from $f(x)$ only on a set of measure less than ε_n. Now $f_n(x)$ is certainly Riemann integrable on $0 \leqslant x \leqslant 1$. Let us suppose that the functions $f_n(x)$ can be constructed so that *their integrals in the Riemann sense have a common bound.* Then the functions $f_n(x)$ can always be chosen so that their integrals form a convergent sequence. Since there are many possible choices for the sequence $\{f_n\}$, the limit of the sequence of integrals is not uniquely defined by $f(x)$ but can depend on the choice of the sequence $\{f_n\}$. The greatest lower bound of the possible limits is defined as the *Lebesgue integral* of $f(x)$:

$$\int_0^1 f(x)\,dx.$$

It is easily shown that every Riemann-integrable function is Lebesgue integrable to the same value. If $f(x)$ has an improper Riemann integral, as in Example 3 with $0<\alpha<1$, then $f(x)$ is Lebesgue integrable and the values of the integrals again coincide; if we take $\alpha \geqslant 1$ in Example 3, then neither the improper Riemann integral *nor* the Lebesgue integral exists. Thus the Lebesgue approach does not miraculously reduce infinite areas to finite values. However, the Dirichlet function of Example 2 is Lebesgue integrable to the value 0 but is not Riemann integrable (for any partition each subdivision contains both rational and irrational numbers, so that the Riemann sum can be made either 0 or 1 by choice of ξ_k).

If $f(x)$ can take both positive and negative values, we write $f(x)=f_+(x)-f_-(x)$, where f_+ and f_- are both nonnegative functions defined by

$$f_+ = \begin{cases} f, & x\in E_+, \\ 0, & \text{elsewhere}, \end{cases} \qquad f = \begin{cases} -f, & x\in E_-, \\ 0, & \text{elsewhere}, \end{cases}$$

where E_+ and E_- are the sets on the x axis where f is positive and negative, respectively. It is then possible to define

$$\int_0^1 f\,dx = \int_0^1 f_+\,dx - \int_0^1 f_-\,dx.$$

The Lebesgue integral can also be defined for arbitrary finite intervals or for infinite intervals. The integral has the usual linearity property

$$\int_a^b \left[\alpha f(x) + \beta g(x) \right] dx = \alpha \int_a^b f(x)\,dx + \beta \int_a^b g(x)\,dx.$$

We have already remarked that the importance of the Lebesgue integral lies in the relative impunity with which we can use limiting processes in connection with Lebesgue integration. One of the most important theorems is the following.

Lebesgue Dominated Convergence Theorem. Let $\{s_n(x)\}$ be a sequence of integrable functions over $[a,b]$, which approaches a limit $s(x)$ pointwise except possibly over a set of measure 0. If there exists an integrable function $f(x)$ such that, for all sufficiently large n, $|s_n(x)| \leqslant f(x)$, then $s(x)$ is integrable and

$$\lim_{n\to\infty} \int_a^b s_n(x)\,dx = \int_a^b s(x)\,dx.$$

Note that this theorem is much more powerful than Theorem 2 based on uniform convergence in Section 6. Here we need only pointwise convergence (and then only almost everywhere), the interval does not need to be bounded, and the integrability of the limit is guaranteed by the theorem instead of having to be hypothesized.

Let us apply the theorem to sequence (6.2). If $\alpha < e$, we can show from elementary calculus that $\alpha \log y < y$ for all $y > 0$. Setting $y = nx$, we find $n^\alpha e^{-nx} < x^{-\alpha}$ or

$$(7.1) \qquad\qquad s_n(x) < x^{1-\alpha}, \qquad x > 0.$$

Clearly $x^{1-\alpha}$ is integrable from 0 to 1 if $\alpha < 2$; since $2 \leqslant e$, (7.1) also holds and therefore the Lebesgue theorem yields (6.5) for $\alpha < 2$. In fact, we can refine (7.1) to show that, for $\alpha < 2$, $s_n(x) < f(x)$, $0 < x < \infty$, where $\int_0^\infty f(x)\,dx$ is finite. The Lebesgue theorem then tells us that

$$\lim_{n\to\infty} \int_0^\infty s_n(x)\,dx = 0, \qquad \alpha < 2.$$

The other principal fact we need to know about Lebesgue integration is the completeness of $L_2(a,b)$, the space of real-valued, square integrable functions on $a \leqslant x \leqslant b$. Thus $u(x) \in L_2(a,b)$ if and only if

$$\int_a^b u^2(x)\,dx < \infty.$$

In L_2 we use the distance function (6.8). The significance of the completeness of L_2 will be better appreciated on reading Chapter 4.

REFERENCES AND ADDITIONAL READING

Antman, S., General solutions for plane extensible elasticae having nonlinear stress-strain laws, *Quart. Appl. Math.* **26** (1968), 35.

Aris, R., *The mathematical theory of diffusion and reaction in permeable catalysts*, Vols. 1 and 2, Oxford University Press, Oxford, 1975.

Aris, R., *Mathematical modelling techniques*, Pitman, London, 1978.

Denn, M. M., *Stability of reaction and transport processes*, Prentice-Hall, Englewood Cliffs, N. J., 1974.

Levine, H., *Unidirectional wave motions*, North-Holland, Elsevier, 1978.

Lin, C. C. and Segel, L. A., *Mathematics applied to deterministic problems in the natural sciences*, Macmillan, New York, 1974.

Perlmutter, D. D., *Stability of chemical reactors*, Prentice-Hall, Englewood Cliffs, N. J., 1972.

Segel, L., *Mathematics applied to continuum mechanics*, Macmillan, New York, 1977.

Shilov, G. E., *Linear spaces*, Prentice-Hall, Englewood Cliffs, N. J., 1961.

1
Green's Functions
(Intuitive Ideas)

1. INTRODUCTION AND GENERAL COMMENTS

For the limited purposes of this section we shall look mainly at steady heat flow in a homogeneous medium. Consider first the one-dimensional problem of a thin rod occupying the interval $(0, 1)$ on the x axis. Setting the product of the thermal conductivity and cross-sectional area equal to 1, we find that (1.17), Chapter 0, becomes

$$(1.1) \qquad -\frac{d^2 u}{dx^2} = f(x), \quad 0 < x < 1; \qquad u(0) = \alpha, \quad u(1) = \beta,$$

where $f(x)$ is the prescribed source density (per unit *length* of the rod) of heat and α, β are the prescribed end temperatures. The three quantities $\{f(x); \alpha, \beta\}$ are known collectively as the *data* for the problem. The data consists of the *boundary data* α, β and of the *forcing function* $f(x)$.

We shall be concerned not only with solving (1.1) for specific data but also with finding a suitable form for the solution that will exhibit its dependence on the data. Thus as we change the data our expression for the solution should remain useful. The feature of (1.1) that enables us to achieve this goal is its linearity, as reflected in the *superposition principle*: If $u_1(x)$ is a solution for the data $\{f_1(x); \alpha_1, \beta_1\}$ and $u_2(x)$ for the data $\{f_2(x); \alpha_2, \beta_2\}$, then $Au_1(x) + Bu_2(x)$ is a solution for the data $\{Af_1(x) + Bf_2(x); A\alpha_1 + B\alpha_2, A\beta_1 + B\beta_2\}$. One can extend this principle in an obvious manner to n solutions corresponding to n sets of data. Under mild restrictions it is even possible to extend the superposition principle to infinite sets of data (see Exercise 1.4 for the case of superposition over a continuously varying parameter). In practice, the superposition principle permits us to decompose complicated data into possibly simpler parts, to solve each of the simpler boundary value problems, and then to reassemble these solutions to find the solution of the original problem. One decomposition of the data which is often used is

$$\{f(x);\alpha,\beta\} = \{f(x);0,0\} + \{0;\alpha,\beta\}.$$

The problem with data $\{f(x);0,0\}$ is an *inhomogeneous equation* with *homogeneous* boundary conditions; the problem with data $\{0;\alpha,\beta\}$ is a *homogeneous equation* with *inhomogeneous* boundary conditions. It should be noted that data $\{0;\alpha,\beta\}$ is itself often split up into $\{0;\alpha,0\}$ and $\{0;0,\beta\}$, each of which involves one inhomogeneous and one homogeneous boundary condition.

Later in this section [equation (1.12)] and again in Section 2 [equations (2.9) and (2.10)] we show how the superposition principle or other methods lead to the following form for the solution of (1.1):

(1.2) $$u(x) = \int_0^1 g(x,\xi)f(\xi)\,d\xi + (1-x)\alpha + x\beta,$$

where *Green's function* $g(x,\xi)$ is a function of the real variables x and ξ defined on the square $0 \leqslant x, \xi \leqslant 1$ and is explicitly given by

(1.3) $$g(x,\xi) = x_<(1-x_>) = \begin{cases} x(1-\xi), & 0<x<\xi, \\ \xi(1-x), & \xi<x<1. \end{cases}$$

Here $x_<$ stands for the lesser of the two quantities x and ξ, and $x_>$ for the greater of x and ξ. Since g does not depend on the data, it is clear that (1.2) expresses in a very simple manner the dependence of u on the data $\{f;\alpha,\beta\}$. Symbolically we can write (1.2) as

$$u(x) = F(f,\alpha,\beta),$$

where F is a linear operator transforming the data into the solution.

For specific f the integration in (1.2) can sometimes be performed in closed terms, using elementary integration techniques. One must, however, divide the interval of integration into two parts to take advantage of the simple formulas for g. Since the integration in (1.2) is over ξ, we write $\int_0^1 = \int_0^x + \int_x^1$, where in the interval from 0 to x we have $\xi < x$, so that we must use the second line of (1.3), whereas in the interval from x to 1 the first line of (1.3) applies. The integral term in (1.2) therefore becomes

$$(1-x)\int_0^x \xi f(\xi)\,d\xi + x\int_x^1 (1-\xi)f(\xi)\,d\xi.$$

Turning to the three-dimensional problem of heat conduction in a homogeneous medium of unit thermal conductivity, occupying the domain Ω with boundary Γ, we know, from (1.11), Chapter 0, that the steady temperature $u(x)$ satisfies

$$(1.4) \qquad -\Delta u = f(x), \quad x \in \Omega; \qquad u = h(x), \quad x \in \Gamma.$$

Here $x = (x_1, x_2, x_3)$ is a position *vector* in three-dimensional space. The source density per unit volume $f(x)$ is given for $x \in \Omega$, whereas the boundary temperature $h(x)$ is given for x on the surface Γ.

Note that we are no longer using any distinguishing notation for vectors. The context should make it clear whether a quantity is a vector or a scalar. The differential operator Δ appearing in (1.4) is the *Laplacian*, which, in *Cartesian coordinates*, takes the form $\partial^2/\partial x_1^2 + \partial^2/\partial x_2^2 + \partial^2/\partial x_3^2$, whereas in other coordinate systems it will look quite different. One of the advantages of the notation of (1.4) is that it does not commit us to a particular coordinate system.

In any event the solution of (1.4) can be written in terms of Green's function $g(x, \xi)$ (which is now a function of the six real variables $x_1, x_2, x_3, \xi_1, \xi_2, \xi_3$):

$$(1.5) \qquad u(x) = \int_{\Omega} g(x, \xi) f(\xi)\, d\xi - \int_{\Gamma} \frac{\partial g}{\partial n} h(\xi)\, dS_{\xi},$$

where $d\xi$ is an element of volume integration $(= d\xi_1 d\xi_2 d\xi_3$ if Cartesian coordinates are used), dS_{ξ} is an element of surface integration at the point ξ on Γ, and $\partial/\partial n$ denotes differentiation with respect to ξ in the outward normal direction on Γ.

Thus (1.5) expresses the solution of (1.4) in terms of the data $\{f(x); h(x)\}$ with f the forcing function and h the boundary data; again we see that the superposition principle holds. It remains only to confess that the function $g(x, \xi)$ appearing in (1.5) is usually not known explicitly (unless the domain Ω is of a very simple type such as a ball or parallelepiped); nevertheless one can obtain a great deal of useful information about $g(x, \xi)$. First we point out that $g(x, \xi)$ has a very simple physical interpretation as the temperature at x when the only source is a *concentrated unit source* located at ξ, the boundary being kept at 0 temperature. One can also characterize $g(x, \xi)$ mathematically as the solution of a well-defined boundary value problem; this formulation requires a little delicacy, however, and we shall take up the question in some of the succeeding sections.

The reader may have noticed that in (1.1) the differential equation was formulated on the open interval $0 < x < 1$ rather than on the closed interval

$0 \leqslant x \leqslant 1$; similarly in (1.4) the differential equation held on a *domain* Ω, which, by definition, is an open, connected set (see Section 1, Chapter 0). Why do we insist that Ω be an open set? The reason is to avoid discussing the differential equation on the boundary. Take (1.1), for instance; if we required the differential equation to hold at $x = 1$, we either would have to extend the function u for $x > 1$ (to be able to form the difference quotient at $x = 1$), or would have to use the concept of a one-sided derivative at $x = 1$. For a higher dimensional problem such as (1.4), it is even more awkward to try to use the differential equation on the boundary since this would necessarily require some smoothness for the boundary Γ and the boundary data $h(x)$.

We shall therefore always formulate the differential equation on a domain Ω (open, connected set).

How do we relate the boundary values of $u(x)$ to its interior values? The boundary values of u are given, whereas the interior values are obtained by solving a differential equation with its attendant indeterminacy. To see that some clarification is needed, consider (1.1) when $f(x) \equiv 0$, $0 < x < 1$, and $\alpha = \beta = 0$. We clearly would like the solution $u(x)$ to be identically 0; we want to rule out ridiculous candidates such as

$$v(x) = \begin{cases} 1, & 0 < x < 1, \\ 0, & x = 0, x = 1. \end{cases}$$

This function $v(x)$ satisfies the differential equation $-d^2v/dx^2 = 0$, $0 < x < 1$, and clearly $v(0) = v(1) = 0$; yet $v(x)$ is a spurious solution. We can reject v by requiring that the solution $u(x)$ of (1.1) be continuous in the *closed interval* $0 \leqslant x \leqslant 1$, or, equivalently, by requiring that $\lim_{x \to 0+} u(x) = \alpha$, $\lim_{x \to 1-} u(x) = \beta$.

Similarly in (1.4) *we shall require that the solution $u(x)$ be continuous in the closed region $\bar{\Omega} = \Omega + \Gamma$.* [It is of course understood that the given boundary data $h(x)$ constitutes a continuous function of position on Γ.]

So far we have said nothing about how to decide whether or not a function $u(x)$ satisfies the differential equation $-u'' = f(x)$ in (1.1). At first glance there seems to be little to say: one merely makes sure that $u(x)$ is twice differentiable in $0 < x < 1$ (which implies that u is continuous and has a continuous first derivative) and that the function $-u''(x)$ coincides with the given function $f(x)$ over the whole interval $0 < x < 1$ [in other words, for each x in $0 < x < 1$, the numbers $-u''(x)$ and $f(x)$ should be the same]. This works splendidly if $f(x)$ is continuous, but there are good reasons, both mathematical and physical, for considering forcing functions $f(x)$ that are only piecewise continuous. For instance, one can easily envisage a situation in which the prescribed source density $f(x)$ is a nonzero constant, say 1, in $0 < x < \xi$ and is 0 in $\xi < x < 1$. Note that $f(x)$ is discontinuous at

the point $x = \xi$. One often hears the argument that such functions are inadmissible on physical grounds, that the "real" source density is continuous and merely decreases quickly from the value 1 to 0 in a small neighborhood of the point $x = \xi$. Such philosophical arguments are immaterial; all we care about is that the temperature calculated on the basis of the discontinuous source density should be nearly the same (in some suitable sense) as that calculated on the basis of the continuous density (see Exercises 2.2 and 2.6, for instance).

We shall return to this question in due time, but now let us try to incorporate piecewise continuous forcing functions into our framework at the cost of slightly reinterpreting the meaning of the differential equation $-u'' = f$. We still want u' to be an integral of f, and of course integrals of piecewise continuous functions are well defined and are necessarily *continuous*; the continuity of u' implies that u is continuous. The new feature is that u'' no longer exists at the points where f has jumps. Let x_0 be such a point, and let us try to calculate $u''(x_0)$ by forming the difference quotient for u':

$$\frac{u'(x_0 + \Delta x) - u'(x_0)}{\Delta x} = \frac{-\int_{x_0}^{x_0 + \Delta x} f(x)\, dx}{\Delta x},$$

whose approximate value is $-f(x_0+)$ for $\Delta x > 0$ and $-f(x_0-)$ for $\Delta x < 0$. Thus $u''(x_0)$ cannot exist, no matter how we try to adjust the value of f at x_0, as long as $f(x_0+)$ and $f(x_0-)$ are different. Of course at points where $f(x)$ is continuous we still require that $u''(x)$ exist and satisfy $-u''(x) = f(x)$. We can easily generalize these ideas to an arbitrary linear differential equation of order p:

(1.6) $a_p(x)u^{(p)}(x) + a_{p-1}(x)u^{(p-1)}(x) + \cdots + a_1(x)u'(x) + a_0(x)u(x)$
$$= f(x), \qquad a < x < b.$$

Definition.   Let $f(x)$ be piecewise continuous, and let $a_0(x), \ldots, a_p(x)$ be continuous. A *classical solution* of (1.6) is a function $u(x)$ belonging to $C^{p-1}(a,b)$—the class of functions with continuous derivatives of order $p-1$ on $a < x < b$—such that, at all points of continuity of f, $u^{(p)}(x)$ exists and satisfies the differential equation (1.6).

Remark.   By using the notion of weak solution (see Section 5, Chapter 2), we can give a reasonable interpretation of (1.6) even when f is only integrable. This idea applies also to partial differential equations, where difficulties can arise even if f is continuous.

Let us now solve (1.1) for the very simple piecewise continuous, forcing function

(1.7)
$$f(x,a)=H(x-a)=\begin{cases} 0, & 0<x<a, \\ 1, & a<x<1, \end{cases}$$

where $H(x)$ is the usual Heaviside function, which vanishes for $x<0$ and is equal to 1 for $x>0$ (its value at $x=0$ plays no role in the analysis). In (1.7) x is the primary variable and a is a parameter. We first solve (1.1) when $\alpha=\beta=0$, that is, for data $\{H(x-a);0,0\}$. The solution will be denoted by $u(x,a)$, since it depends not only on x but also on the parameter a. In $0<x<a$ we have $-d^2u/dx^2=0$, whereas in $a<x<1$ we have $-d^2u/dx^2=1$. Integration and use of the boundary conditions gives

$$u=Ax \quad \text{in } (0,a) \quad \text{and} \quad u=-\frac{(x-1)^2}{2}+B(1-x) \quad \text{in } (a,1),$$

where A and B may depend on a but not on x. For u to be a classical solution we must require that u and u' be continuous at $x=a$ (we already have more than enough smoothness in the subintervals $0<x<a$ and $a<x<1$). This gives $A=(1-a)^2/2$ and $B=(1-a^2)/2$, so that

(1.8)
$$u(x,a)=\begin{cases} \dfrac{(a-1)^2}{2}x, & 0<x<a, \\[2mm] -\dfrac{(x-1)^2}{2}+\dfrac{1-a^2}{2}(1-x), & a<x<1, \end{cases}$$

which is plotted in Figure 1.1. Exercises 1.4 and 1.5 show how to use (1.8) to obtain the solution of (1.1) for *arbitrary* f (with $\alpha=\beta=0$).

It is of interest to present another approach to (1.1) with data $\{f(x);0,0\}$, which lends itself to graphical analysis. This method is based on interpreting the problem as the transverse deflection of a taut string with fixed ends. The static version of (4.36), Chapter 0, with $T^{(0)}=1$, $l=1$, $X=x$, $\mathbf{f}\cdot\mathbf{j}=f(x)$, and u instead of v for the transverse deflection, gives us (1.1) with $\alpha=\beta=0$. It then follows that the vertical component of the tension at a point $(x,u(x))$ along the string is just $u'(x)$; thus the reactions at the ends $x=0$ and $x=1$ are $-u'(0)$ and $u(1)$, respectively. By taking moments about the ends of the string, we find that

(1.9) $\quad u'(1)+\displaystyle\int_0^1 xf(x)\,dx=0, \qquad -u'(0)+\displaystyle\int_0^1 (1-x)f(x)\,dx=0,$

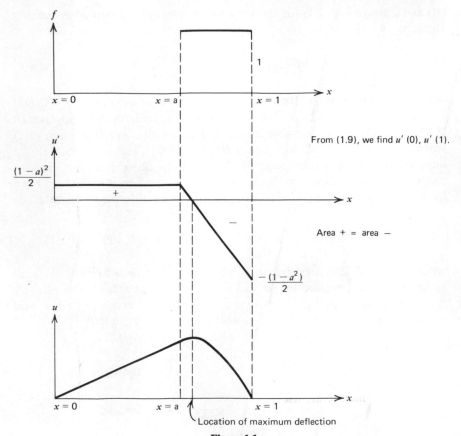

Figure 1.1

which could also be derived without recourse to the physical interpretation by multiplying the differential equation in (1.1) by x and $1-x$, respectively, and then integrating from 0 to 1. In any event we have calculated the reactions at the ends and can now find $u'(x)$ at any point from

$$(1.10) \qquad u'(x) = u'(0) - \int_0^x f(\xi)\,d\xi,$$

which can of course be done graphically. Since $u(0)=0$, we can find $u(x)$ from (1.10) by integrating from 0 to x. This again is easy to do graphically; analytically we find that

$$(1.11) \qquad u(x) = \int_0^x u'(\eta)\,d\eta = u'(0)x - \int_0^x d\eta \int_0^\eta f(\xi)\,d\xi.$$

The iterated integral can be viewed as a double integral over a triangular region in the ξ-η plane; on changing the order of integration, we obtain

$$u(x) = u'(0)x - \int_0^x (x-\xi)f(\xi)\,d\xi$$

(1.12)
$$= x\int_0^1 (1-\xi)f(\xi)\,d\xi - \int_0^x (x-\xi)f(\xi)\,d\xi$$

$$= \int_0^1 g(x,\xi)f(\xi)\,d\xi,$$

where $g(x,\xi)$ is just Green's function as predicted in (1.3). It is then an easy matter to show that (1.2) holds when the data is $\{f;\alpha,\beta\}$ instead of $\{f;0,0\}$. In Figure 1.1 we have illustrated the graphical integration when the data is $\{H(x-a);0,0\}$, the corresponding formula for the deflection being (1.8).

Exercises

1.1. Consider the transverse deflection $u(x)$ of a string satisfying

$$-u'' = f(x), \quad 0<x<1; \qquad u(0)=0, \quad u(1)=0,$$

where
$$f(x) = \begin{cases} x - \frac{1}{4}, & 0<x<\frac{1}{2}, \\ \frac{1}{4}, & \frac{1}{2}<x<1. \end{cases}$$

(a) Find u' at one of the ends, and then carry out graphically two successive integrations to obtain the deflection $u(x)$.

(b) Find $u(x)$ using (1.2) and (1.3). To perform the integration explicitly you must divide the interval into $(0,x)$ and $(x,1)$; in the first subinterval, x is larger than ξ, so that the second line of (1.3) applies. You will then need a further subdivision to handle our specific forcing function. Compare your result with that for part (a).

1.2. The small transverse deflection $u(x)$ of a homogeneous beam of unit length subject to a distributed transverse loading $f(x)$ satisfies

(1.13)
$$\frac{d^4u}{dx^4} = f(x), \qquad 0<x<1,$$

where we have set $EI=1$ in (4.11), Chapter 0. For a beam simply supported at its ends the boundary conditions are

(1.14)
$$u(0) = u''(0) = u(1) = u''(1) = 0.$$

The shear force V and moment M at a cross section satisfy

$$V = -u'''(x), \qquad M = u''(x),$$

where the choice of signs is in accord with the convention used in Section 4, Chapter 0. For (1.13) subject to (1.14), show how to calculate $u'''(0)$. It is therefore straightforward to find $V(x)$ and $M(x)$ by graphical integration. Once $M(x)$ is known, it is easy to calculate $u'(0)$ and hence to proceed in determining $u'(x)$ and $u(x)$ graphically.

1.3. Consider the boundary value problem

$$(1.15) \quad -\frac{d}{dx}\left[k(x)\frac{du}{dx}\right] = f(x), \quad 0<x<1; \qquad u(0)=u(1)=0.$$

where $k(x)>0$ in $0 \leqslant x \leqslant 1$. Let $K(x)$ be a solution of the homogeneous equation satisfying the boundary condition at $x=1$. Show how one can calculate $u'(0)$ by multiplying both sides of the differential equation in (1.15) by $K(x)$ and integrating from $x=0$ to $x=1$.

1.4. For each θ, $\theta_1 < \theta < \theta_2$, denote by $u(x,\theta)$ the solution of the problem

$$-\frac{d^2u}{dx^2} = f(x,\theta), \quad 0<x<1; \qquad u|_{x=0}=\alpha(\theta), \quad u|_{x=1}=\beta(\theta).$$

Show that the function

$$U(x) = \int_{\theta_1}^{\theta_2} u(x,\theta)\,d\theta$$

satisfies

$$-\frac{d^2U}{dx^2} = F(x), \quad 0<x<1; \qquad U|_{x=0}=A, \quad U|_{x=1}=B,$$

where

$$F(x) = \int_{\theta_1}^{\theta_2} f(x,\theta)\,d\theta, \qquad A = \int_{\theta_1}^{\theta_2} \alpha(\theta)\,d\theta, \qquad B = \int_{\theta_1}^{\theta_2} \beta(\theta)\,d\theta.$$

1.5. If f has a continuous derivative on $-\infty<x<\infty$, we can write

$$f(x) = f(x_0) + \int_{x_0}^{x} f'(\xi)\,d\xi, \qquad -\infty<x<\infty.$$

For $x>x_0$ show that we can use the equivalent formula

$$(1.16) \qquad f(x)=f(x_0)H(x-x_0)+\int_{x_0}^{\infty}H(x-\xi)f'(\xi)\,d\xi.$$

Use Exercise 1.4 and (1.8) and (1.16) to find the solution of (1.1) with $\alpha=\beta=0$ in the form (1.2).

2. THE FINITE ROD

Construction of Green's Function

We return to the heat conduction problem (1.1), repeated below for convenience:

$$(2.1) \qquad -\frac{d^2u}{dx^2}=f(x), \quad 0<x<1; \qquad u(0)=\alpha, \quad u(1)=\beta.$$

We want to solve the problem as compactly as possible for arbitrary data $\{f;\alpha,\beta\}$. The differential operator and the boundary operators appearing on the left sides of the equality signs in (2.1) are kept fixed; no one is proposing to solve all differential equations with arbitrary boundary conditions at one stroke!

To solve (2.1) for arbitrary data, we introduce an accessory problem where, instead of a distributed density of sources, there is only a concentrated source of unit strength at $x=\xi$ and where the *boundary data vanishes* (which means in our case that the temperature is 0 at both ends). Physically this accessory problem makes sense, and the resulting steady temperature should be well defined; moreover, it is clear that the temperature cannot vanish identically, since there is a steady nonzero heat input from the source. This temperature (solution of the accessory problem) is known as Green's function and is denoted by $g(x,\xi)$. Here ξ is the position of the source, and x is the observation point. We usually regard ξ as a parameter and x as the running variable; but when we are all through we have a function of two real variables, and we are at liberty to forget the original significance of x and ξ. In any event all differentiations below are with respect to the first variable in g. Let us see whether we can construct g on the basis of the information available so far. Since there are no sources in $0<x<\xi$ and in $\xi<x<1$, we have $-g''=0$ in both intervals. Taking into account the fact that g vanishes at $x=0$ and $x=1$, we find that

$$(2.2) \qquad g=Ax, \quad 0<x<\xi; \qquad g=B(1-x), \quad \xi<x<1.$$

Here A and B are "constants," that is, independent of x; they may, however, depend on the parameter ξ. If at this stage we demanded the continuity of g and g' at $x = \xi$, we would find $A = B = 0$, so that g would vanish identically—which is nonsense! We must *abandon* at $x = \xi$ the requirement that g' be continuous, although we shall still insist on the continuity of g. The jump of g' at $x = \xi$ is easily calculated if we recall the primary integral formulation of the problem of heat conduction in terms of energy balance [see (1.1), and (1.10), Chapter 0]. Consider a thin slice of the rod from $\xi - \varepsilon$ to $\xi + \varepsilon$. The one-dimensional character of the problem means that no heat flows through the lateral surface; since the product of the cross-sectional area and the thermal conductivity is 1 and the amount of heat generated in the slice is 1, we have

$$-g'|_{x=\xi+\varepsilon} + g'|_{x=\xi-\varepsilon} = 1,$$

which, as ε tends to 0, leads to the *jump condition* for g':

(2.3) $$g'|_{x=\xi+} - g'|_{x=\xi-} = -1.$$

Condition (2.3) and the continuity of g at $x = \xi$ enable us to calculate A and B in (2.2) from the simultaneous equations $A = B(1-\xi)$ and $-B-A = -1$. Thus $B = \xi$ and $A = 1 - \xi$, so that

(2.4) $$g(x,\xi) = \begin{cases} (1-\xi)x, & 0 \leqslant x < \xi, \\ (1-x)\xi, & \xi < x \leqslant 1, \end{cases}$$

confirming (1.3). In Figure 2.1a we picture Green's function as a function of x for fixed ξ, and in Figure 2.1b as a function of x and ξ. Thus Figure 2.1a can be viewed as a cross section of the surface in Figure 2.1b.

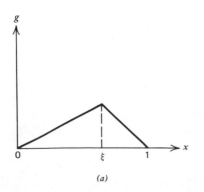

(a)

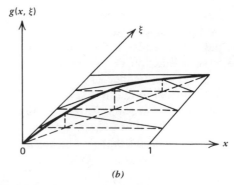

(b)

Figure 2.1

We have therefore characterized Green's function $g(x,\xi)$ both physically and mathematically. Before proceeding to another characterization, based on the delta function, let us recapitulate what has been done so far.

1. *Physical description.* We chose to describe g in terms of heat conduction in a rod: $g(x,\xi)$ is the temperature at x when the only source is a unit concentrated source at ξ, the ends being at 0 temperature. It is also possible to interpret g as the transverse deflection of a string: $g(x,\xi)$ is the deflection at x when the only load is a unit concentrated force at ξ, the ends being kept fixed on the x axis at $x=0$ and $x=1$.

2. *Classical mathematical formulation.* Green's function $g(x,\xi)$ associated with (2.1) satisfies

$$(2.5) \quad \begin{cases} -\dfrac{d^2g}{dx^2}=0, \quad 0<x<\xi, \xi<x<1; \\[2mm] g(0,\xi)=g(1,\xi)=0; \\[2mm] g \text{ continuous at } x=\xi; \qquad \dfrac{dg}{dx}\Big|_{x=\xi+} - \dfrac{dg}{dx}\Big|_{x=\xi-} = -1. \end{cases}$$

In our third formulation we would like to consider (2.5) as a boundary value problem of the form (2.1) with specific data. The boundary data for g clearly vanishes, but what is the forcing function? In (2.1) the forcing function is a source density (per unit length) rather than the concentrated source of the Green's function problem. How can we describe a concentrated source at the point ξ as a density? This is easy to do symbolically but is not so easy within a consistent mathematical framework. Suppose we let $\delta(x)$ be the density corresponding to a concentrated source at $x=0$. We would then need

$$\int_a^b \delta(x)\,dx = \begin{cases} 1 & \text{if } (a,b) \text{ contains the origin,} \\ 0 & \text{otherwise.} \end{cases}$$

Unfortunately, no integrable function satisfies these properties. Nevertheless we shall use $\delta(x)$ symbolically to represent the source density corresponding to a unit source at the origin. This symbolic function is known as the *Dirac delta function*; $\delta(x-\xi)$ is $\delta(x)$ translated ξ units to the right and so must be the source density for a unit source at $x=\xi$.

Perhaps the most natural way to view $\delta(x)$ is as the limit of a sequence of narrow, uniform densities of large magnitude (with total strength unity) such as

$$f_n(x) = \begin{cases} n, & |x| < \frac{1}{2n}, \\ 0, & |x| > \frac{1}{2n}. \end{cases}$$

Thus we may think of $\delta(x)$ as the limit as $n \to \infty$ of $f_n(x)$, and $\delta(x - \xi)$ as the limit of $f_n(x - \xi)$. The *sifting property*

$$(2.6) \qquad \int_a^b \delta(x - \xi)\phi(x)\, dx = \begin{cases} \phi(\xi) & \text{if } a < \xi < b, \\ 0 & \text{if } \xi < a \text{ or } \xi > b, \end{cases}$$

where $\phi(x)$ is an arbitrary function continuous at $x = \xi$, then follows by replacing $\delta(x - \xi)$ by $f_n(x - \xi)$ and proceeding to the limit as $n \to \infty$.

3. *Delta function formulation.* Green's function $g(x, \xi)$ associated with (2.1) satisfies

$$(2.7) \quad -\frac{d^2 g}{dx^2} = \delta(x - \xi), \quad 0 < x < 1, \quad 0 < \xi < 1; \quad g(0, \xi) = g(1, \xi) = 0.$$

At this stage (2.7) is nothing but shorthand for (2.5), but we will develop in Chapter 2 a mathematical framework in which (2.7) will have impeccable standing in its own right.

Solution of the Inhomogeneous Equation

The simple physical interpretation for Green's function guides us in constructing the solution of problem (2.1) with data $\{ f; 0, 0 \}$:

$$(2.8) \qquad -u'' = f(x), \quad 0 < x < 1; \qquad u(0) = u(1) = 0.$$

The idea is to decompose the distributed source $f(x)$ into a number of small concentrated sources located at various points along the rod and then add their individual contributions to the temperature to find u. Divide the interval $(0, 1)$ into n equal parts, calling the center of the kth subinterval ξ_k. The length of each subinterval is $\Delta\xi = 1/n$. It is reasonable to suppose that the temperature corresponding to the distributed density $f(x)$ is closely approximated by the temperature corresponding to small concentrated sources $f(\xi_1)\Delta\xi, \ldots, f(\xi_n)\Delta\xi$, located at $\xi_1, \ldots, \xi_n$, respectively (see Figure 2.2); that is, the temperature for the data $\{ f(x); 0, 0 \}$ is close to the temperature for the data $\{ \sum_{i=1}^n \delta(x - \xi_i) f(\xi_i)\Delta\xi; 0, 0 \}$. According to the principle of superposition extended to concentrated sources, the temperature at x for all the small concentrated sources is

$$\sum_{i=1}^{n} g(x, \xi_i) f(\xi_i) \Delta \xi,$$

which, as $n \to \infty$, tends to

$$(2.9) \qquad u(x) = \int_0^1 g(x, \xi) f(\xi) \, d\xi.$$

Thus our intuitive (or *heuristic*) argument leads us to believe that (2.9) provides a solution to (2.8). Observe that this construction will not work directly for (2.1) with nonzero boundary data, but the solution is easy to determine. Since $\alpha(1 - x) + \beta x$ satisfies (2.1) with data $\{0; \alpha, \beta\}$, the superposition principle shows that

$$(2.10) \qquad u(x) = \int_0^1 g(x, \xi) f(\xi) \, d\xi + \alpha(1 - x) + \beta x$$

satisfies (2.1) with data $\{f; \alpha, \beta\}$.

We must now verify that (2.10) *actually solves* (2.1); *we would also like to show that it is the only solution to the problem and, finally, that* $u(x)$ *depends continuously on the data.*

The rigorous proof below is based, as it must be at this time, on the classical definition (2.5) of g. There will be occasions, however, when we will be satisfied to give merely plausible arguments using the symbolic formulation (2.7), together with the sifting property (2.6) of the delta function.

Verification of Solution

We confine ourselves here to the case where f is continuous, leaving the more general case to Exercise 2.7. Consider first the problem with vanishing boundary data. Clearly (2.9) vanishes at $x = 0$ and 1 because $g(0, \xi) = g(1, \xi) = 0$, so that we only have to show that $-u'' = f$ at each point x,

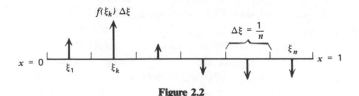

Figure 2.2

$0 < x < 1$. In view of the discontinuity of $g'\ (= dg/dx)$ at $x = \xi$, a certain amount of care is required in differentiating expression (2.9). Let us split the interval of integration into the parts $(0, x)$ and $(x, 1)$, within each of which g and its derivatives are continuous. Then

$$\frac{du}{dx} = \frac{d}{dx}\left[\int_0^x g(x, \xi) f(\xi)\, d\xi + \int_x^1 g(x, \xi) f(\xi)\, d\xi\right],$$

and we now appeal to the classical formula for differentiation under the integral sign

$$(2.11)\quad \frac{d}{dx}\int_{a(x)}^{b(x)} h(x, \xi)\, d\xi = \int_{a(x)}^{b(x)} \frac{\partial h}{\partial x}\, d\xi + h(x, b(x))\frac{db}{dx} - h(x, a(x))\frac{da}{dx}$$

to obtain

$$\frac{du}{dx} = \int_0^x g'(x, \xi) f(\xi)\, d\xi + \int_x^1 g'(x, \xi) f(\xi)\, d\xi$$
$$+ g(x, x-)f(x-) - g(x, x+)f(x+).$$

Here the notation $x-, x+$ serves to distinguish between left- and right-hand values at a possible point of discontinuity of a function. Since $g(x, \xi)$ and $f(\xi)$ are continuous, the distinction is unnecessary in the expression for du/dx, in which the last two terms cancel. A further differentiation leads to

$$\frac{d^2 u}{dx^2} = \int_0^x g''f(\xi)\, d\xi + \int_x^1 g''f(\xi)\, d\xi + g'(x, x-)f(x-) - g'(x, x+)f(x+).$$

The jump property (2.3) of g' can be rewritten as $g'(x, x-) - g'(x, x+) = -1$, and since $g'' = 0$ in the intervals $\xi < x$ and $\xi > x$, it follows that $-u'' = f$ and so (2.9) is a solution of (2.8). Since $\alpha(1 - x) + \beta x$ satisfies (2.1) with data $\{0; \alpha, \beta\}$, we conclude that (2.10) is a solution of (2.1) as required.

Uniqueness

Suppose $u_1(x)$ and $u_2(x)$ satisfy (2.1) for the same data $\{f(x); \alpha, \beta\}$. Then $w(x) = u_1(x) - u_2(x)$ satisfies (2.1) with data $\{0; 0, 0\}$. By definition of the concept of a classical solution, u_1' and u_2' must be continuous and u_1'' and u_2'' exist except at points of discontinuity of f. It follows that w and w' are continuous on $0 \leqslant x \leqslant 1$ and that $w'' = 0$ except possibly at points of discontinuity of f. In each subinterval where f is continuous, w' must be constant; since w' is continuous, the constant is the same in each subinterval and therefore $w = Cx + D$ in $0 < x < 1$. Applying the boundary conditions $w(0) = w(1) = 0$, we find that $w \equiv 0$ in $0 \leqslant x \leqslant 1$.

It is perhaps worth noting that a similar argument shows also that Green's function satisfying (2.5) is unique.

Continuity with Respect to the Data

In most experimental situations the data $\{f(x); \alpha, \beta\}$ is not known precisely. It would be comforting to know that the solution of the boundary value problem is not hypersensitive to small changes in the data. We feel that many physical problems should exhibit this kind of stability. We would like to show that a "small" change in the data leads only to a "small" change in the solution. To make this precise we must introduce a notion of "separation" or "distance" between functions (for real numbers there is no problem: the distance between a and b is $|b - a|$). Here we shall define two different *numerical* measures of the "distance" between functions (this notion was introduced in Section 6, Chapter 0, and will be treated in greater generality in Chapter 4):

$$(2.12) \qquad d_\infty(f_1, f_2) = \sup_{0 < x < 1} |f_1(x) - f_2(x)|$$

and

$$(2.13) \qquad d_1(f_1, f_2) = \int_0^1 |f_1(x) - f_2(x)| \, dx.$$

Thus d_∞ is the largest deviation in ordinates between f_1 and f_2, whereas d_1 is the area between the curves f_1 and f_2. Although f_1 and f_2 are functions, d_1 and d_∞ are nonnegative real numbers. In Figure 6.2, Chapter 0, both $d_1(f_1, f_2)$ and $d_\infty(f_1, f_2)$ are small, whereas in Figure 6.3 $d_1(f_1, f_2)$ is small, but not $d_\infty(f_1, f_2)$. Now let f_1 and f_2 be two continuous functions satisfying $d_\infty(f_1, f_2) < \varepsilon$, and let u_1 and u_2 be the corresponding solutions of (2.8). Then

$$|u_1(x) - u_2(x)| = \left| \int_0^1 g(x, \xi) [f_1(\xi) - f_2(\xi)] \, d\xi \right| \leqslant \int_0^1 |g(x, \xi)| \, |f_1 - f_2| \, d\xi$$

$$\leqslant d_\infty(f_1, f_2) \int_0^1 |g(x, \xi)| \, d\xi.$$

Since $\sup_{0 \leqslant x, \xi \leqslant 1} |g| = \frac{1}{4}$, it follows that

$$d_\infty(u_1, u_2) = \sup_{0 < x < 1} |u_1(x) - u_2(x)| \leqslant \tfrac{1}{4} d_\infty(f_1, f_2) \leqslant \frac{\varepsilon}{4},$$

so that the solution of (2.8) depends continuously on the data. Similar calculations show that there is continuous dependence on the data if d_1 is used as a measure of the distance between functions. For (2.1) with nonzero α and β the situation is similar: if $d_\infty(f_1, f_2) < \varepsilon$, $|\alpha_1 - \alpha_2| < \varepsilon$, and $|\beta_1 - \beta_2| < \varepsilon$, then

$$d_\infty(u_1, u_2) \leqslant \tfrac{1}{4} d_\infty(f_1, f_2) + |\alpha_1 - \alpha_2| + |\beta_1 - \beta_2| \leqslant \tfrac{9}{4}\varepsilon,$$

and again there is continuous dependence on the data.

When dealing later with more general boundary value problems (or other equations such as integral equations), we shall still be faced with these three questions:

1. Is there at least one solution (*existence*)?
2. Is there at most one solution (*uniqueness*)?
3. Does the solution depend continuously on the data?

If the answer to this trio of questions is affirmative, the problem is said to be *well posed* (otherwise, *ill posed*). Until recently it was sound dogma to require that every "real" physical problem be well posed. However, it is now understood that ill-posed problems occur frequently in practice but that their physical interpretation and mathematical solution are somewhat more delicate.

Alternative Derivations for the Problem with Nonzero Boundary Data

There is no difficulty in visualizing the role of Green's function in solving the problem with data $\{f(x); 0, 0\}$. We proceed from (2.8) to (2.9) by straightforward, albeit intuitive, arguments. The extra terms in (2.10) corresponding to nonzero boundary data were obtained by a different procedure. Could we have used Green's function for this purpose as well? One way of doing this is by translating the problem with data $\{0; \alpha, \beta\}$ into a problem with nonzero f and vanishing boundary data. Consider the boundary value problem

$$(2.14) \qquad -u'' = 0, \quad 0 < x < 1; \quad u(0) = \alpha, \quad u(1) = \beta,$$

and let $h(x)$ be *any function* (not necessarily satisfying any related differential equation) such that $h(0) = \alpha, h(1) = \beta$. Setting

$$u = h + v,$$

we see that v satisfies

$$-v'' = h'', \quad 0 < x < 1; \quad v(0) = v(1) = 0,$$

whose solution by (2.9) is

$$v(x) = \int_0^1 g(x,\xi)h''(\xi)\,d\xi = \int_0^1 g(\xi,x)h''(\xi)\,d\xi,$$

where we have used the symmetry of g, that is, $g(x,\xi) = g(\xi,x)$. Splitting the interval of integration into $(0,x)$ and $(x,1)$, we obtain, after two integrations by parts,

$$v(x) = -g'(\xi,x)h(\xi)\big|_{\xi=0}^{\xi=x^-} - g'(\xi,x)h(\xi)\big|_{\xi=x+}^{\xi=1},$$

or, using the jump condition on dg/dx given in (2.5),

$$v(x) = -h(x) - g'(1,x)\beta + g'(0,x)\alpha.$$

Since $u = h + v$, we find that

$$(2.15) \qquad u(x) = \alpha g'(0,x) - \beta g'(1,x) = \alpha(1-x) + \beta x,$$

in accord with (2.10). Observe that $h(x)$ has disappeared from the final expression (2.15) for u.

Another way of arriving at (2.15) is to combine the differential equation of (2.14) with that for Green's function in the subintervals $(0,\xi)$ and $(\xi,1)$. Since $g'' = 0$ in each subinterval, we have $ug'' - gu'' = 0$ in $(0,\xi)$ and in $(\xi,1)$, so that

$$\int_0^\xi (ug'' - gu'')\,dx + \int_\xi^1 (ug'' - gu'')\,dx = 0.$$

The relation

$$(2.16) \qquad ug'' - gu'' = (ug' - gu')',$$

which is valid classically in each of the subintervals $(0,\xi)$ and $(\xi,1)$, and the jump condition on g' then yield

$$(2.17) \qquad u(\xi) = u(0)g'(0,\xi) - u(1)g'(1,\xi),$$

which is the same as (2.15).

Both methods used so far are rigorously based on (2.5). An alternative to the second of these methods is based formally on the symbolic characterization (2.7) and on the sifting property (2.6). Multiply (2.14) by g and (2.7) by u, subtract, and integrate from 0 to 1 to obtain

$$u(\xi) = - \int_0^1 (ug'' - gu'') \, dx.$$

We now use (2.16) over the whole interval from 0 to 1; we are entitled to do this because we have accounted for the jump in g' by including the term $\delta(x - \xi)$ in (2.7). Thus $u(\xi) = u(0)g'(0,\xi) - u(1)g'(1,\xi)$ as in (2.17).

There is a lesson worth remembering here. In the classical approach we use only the subintervals in which all functions are well behaved, the term $u(\xi)$ in (2.17) arising from the jump in g' at $x = \xi$. In the symbolic approach we deal with the whole interval at once, the term $u(\xi)$ in (2.17) now arising from the fact that there is a delta function on the right side of the differential equation. *Do not mix the two approaches!*

Eigenfunction Expansion

An apparently different approach to (2.1) is by way of the associated *eigenproblem*

(2.18) $-u'' = \lambda u, \quad 0 < x < 1; \qquad u(0) = u(1) = 0.$

Here λ is a complex number regarded as a parameter. Since we are dealing with a homogeneous equation of order 2 with two homogeneous boundary conditions, we might expect that (2.18) has only the trivial solution $u \equiv 0$, $0 \leqslant x \leqslant 1$. It turns out that this is true for most values of λ, but that there are exceptional values of λ, known as *eigenvalues*, for which the boundary value problem (2.18) has nontrivial solutions. These nontrivial solutions are called *eigenfunctions*. Observe that an eigenfunction corresponds to a definite eigenvalue but that to an eigenvalue may be associated more than one independent eigenfunction (it is clear of course that any constant multiple of an eigenfunction is again an eigenfunction corresponding to the same eigenvalue; if u_1 and u_2 are eigenfunctions corresponding to the same λ, then $Au_1 + Bu_2$ is also an eigenfunction corresponding to that λ).

For any complex λ we can easily solve the differential equation in (2.18); imposition of the boundary conditions then shows that nontrivial solutions are possible only for $\lambda_1 = \pi^2, \lambda_2 = 4\pi^2, \dots, \lambda_n = n^2\pi^2, \dots$. To the eigenvalue $\lambda_n = n^2\pi^2$ corresponds essentially one eigenfunction $u_n(x) = \sin n\pi x$ (what this means is that every eigenfunction corresponding to λ_n is necessarily of the form Au_n). We observe that eigenfunctions corresponding to different eigenvalues are orthogonal, that is,

(2.19) $\int_0^1 \sin m\pi x \sin n\pi x \, dx = 0, \qquad m \neq n.$

If we now multiply the differential equation in (2.1) by $u_n(x)$ and integrate from 0 to 1, we find that

$$-\int_0^1 u'' u_n \, dx = \int_0^1 f u_n \, dx,$$

or, after two integrations by parts and use of the boundary conditions,

$$\lambda_n \int_0^1 u u_n \, dx + n\pi(\beta \cos n\pi - \alpha) = \int_0^1 f u_n \, dx$$

or

$$\int_0^1 u u_n \, dx = \lambda_n^{-1} \left[\int_0^1 f u_n \, dx + n\pi(\alpha - \beta \cos n\pi) \right].$$

The number $\int_0^1 u u_n \, dx$ is just one-half the nth Fourier sine coefficient of $u(x)$, so that we can recover $u(x)$ through

$$(2.20) \qquad u(x) = \sum_{n=1}^{\infty} \frac{2}{n^2 \pi^2} \left[\int_0^1 f u_n \, dx + n\pi(\alpha - \beta \cos n\pi) \right] \sin n\pi x.$$

In particular, for problem (2.8) having vanishing boundary data, we find that

$$(2.21) \qquad u(x) = \sum_{n=1}^{\infty} \frac{2}{n^2 \pi^2} \left(\int_0^1 f u_n \, dx \right) \sin n\pi x,$$

which can be considered as an alternative representation to (2.9). Comparing the two forms, we deduce the *bilinear representation* of Green's function:

$$(2.22) \qquad g(x, \xi) = \sum_{n=1}^{\infty} \frac{2 \sin n\pi x \sin n\pi \xi}{n^2 \pi^2},$$

which we will study further in Chapters 6 and 7.

We may regard (2.18) as a problem of type (2.8) with forcing function $\lambda u(x)$; the "solution" is then given by (2.9), which becomes

$$(2.23) \qquad u(x) = \lambda \int_0^1 g(x, \xi) u(\xi) \, d\xi, \qquad 0 < x < 1.$$

Since u appears under the integral sign as well as outside, we have not really solved for $u(x)$. Instead we have shown that (2.18) is equivalent to the *integral equation* (2.23).

Exercises

2.1. Let $f_1(x)$ and $f_2(x)$ be piecewise continuous on $0 \leqslant x \leqslant 1$, and let $u_1(x)$ and $u_2(x)$ be the corresponding solutions of (2.8). Show that the following hold:

(a) If $d_\infty(f_1,f_2)<\varepsilon$, then $d_\infty(u_1,u_2)<\varepsilon/8$, which is a slight improvement of the result in the text.

(b) If $d_1(f_1,f_2)<\varepsilon$, then $d_1(u_1,u_2)<\varepsilon/6$.

(c) If $d_1(f_1,f_2)<\varepsilon$, then $d_\infty(u_1,u_2)<\varepsilon/4$.

The last result shows that, if f_1 is piecewise continuous and f_2 is a reasonable continuous approximation to f_1 (such as in the solid and dashed curves of Figure 4.2), the temperatures corresponding to these source functions are *uniformly* close over the entire interval. Thus the statement made in Section 1 about replacing certain continuous sources by idealized piecewise continuous ones has been substantiated.

2.2. Let $0<a<b<1$, and let

$$q(x)=\begin{cases} q, & a<x<b, \\ 0, & \text{otherwise.} \end{cases}$$

Determine the solution $u(x)$ of (1.1) with data $\{q(x);0,0\}$ by superposition from (1.8). Let $q=1/(b-a)$, and take the limit as $b\to a$. Show that $u(x)$ tends *uniformly* to $g(x,a)$ on $0\leqslant x\leqslant 1$. Thus it is reasonable in this case to approximate a unit concentrated source by a uniformly distributed density (of total strength 1) in a narrow neighborhood of a.

2.3. Let λ be an arbitrary complex number. We shall define the *principal value* of $\sqrt\lambda$ as follows. If $\sqrt\lambda=0$, then $\lambda=0$; if $\lambda\neq0$, then λ has a unique representation $\lambda=|\lambda|e^{i\theta}$, $0\leqslant\theta<2\pi$, and the principal value of $\sqrt\lambda$ is defined as $|\lambda|^{1/2}e^{i\theta/2}$, where $|\lambda|^{1/2}$ is the *positive square root* of the positive real number $|\lambda|$. Throughout this exercise $\sqrt\lambda$ will stand for the principal value just defined (note that as a function of a complex variable $\sqrt\lambda$ has a discontinuity on the positive real axis).

(a) The general solution of $-u''=\lambda u$ is $u(x)=A+Bx$ if $\lambda=0$; $u(x)=A\exp(i\sqrt\lambda\,x)+B\exp(-i\sqrt\lambda\,x)$ (or, alternatively, $u(x)=C\sin\sqrt\lambda\,x+D\cos\sqrt\lambda\,x$) if $\lambda\neq0$. Show that only the real values $\lambda_n=n^2\pi^2$ are eigenvalues of (2.18).

(b) Find the eigenvalues and eigenfunctions of

$$-u''=\lambda u, \quad 0<x<1; \qquad u'(0)=u'(1)=0.$$

2.4. Find Green's function $g(x,\xi)$ satisfying

(2.24)
$$\frac{d^4g}{dx^4} = \delta(x-\xi), \quad 0<x,\xi<1;$$

$$g(0,\xi)=g''(0,\xi)=g(1,\xi)=g''(1,\xi)=0$$

by a graphical method (see Exercise 1.2). Note that g is the deflection of a simply supported beam with a concentrated unit load at $x=\xi$. What is the equivalent classical formulation of (2.24)?

2.5. A simply supported beam $(0<x<l)$ is subject to the distributed transverse loading

$$f(x)=\begin{cases} 0, & \left|x-\dfrac{l}{2}\right|>\varepsilon, \\[2mm] p, & \dfrac{l}{2}<x<\dfrac{l}{2}+\varepsilon, \\[2mm] -p, & \dfrac{l}{2}-\varepsilon<x<\dfrac{l}{2}. \end{cases}$$

Find the reactions at the ends; plot shear and moment diagrams. Denote the moment in the beam by $M(x,\varepsilon)$, and calculate $\lim_{\varepsilon\to0} M(x,\varepsilon)$ in the following cases:

(a) p is fixed.
(b) $p=1/\varepsilon$.
(c) $p=1/\varepsilon^2$.

Calculate the limiting deflection corresponding to case (c). What is its physical significance? Formulate the limiting problem as a self-contained mathematical problem without any limiting process.

2.6. Let $\{f_n(x)\}$ be a sequence satisfying the following conditions: $f_n(x)\geqslant0$ for all x, $f_n(x)=0$ for $|x-t|>1/n$, and $\int_{t-1/n}^{t+1/n} f_n(x)\,dx=1$. Let $u_n(x)$ be the deflection of a string of unit length with fixed ends and unit tension subject to the transverse pressure $f_n(x)$. Draw a graph of $u_n'(x)$ for some large values of n. Show that C_n, the constant slope in $0<x<t-1/n$, approaches $(1-t)$ as $n\to\infty$. Next graph the corresponding $u_n(x)$, and prove that

$$\lim_{n\to\infty} u_n(x)=g(x,t) \qquad \text{uniformly in } 0\leqslant x\leqslant1,$$

where g is Green's function.

2.7. Suppose $f(x)$ is piecewise continuous on $0 \leqslant x \leqslant 1$. This means that f is continuous except at $a_1, \ldots, a_k$, where f has simple jumps of amounts $J_1, \ldots, J_k$, respectively. We can write

$$(2.25) \qquad f = [f] + \sum_{i=1}^{k} J_i H(x - a_i),$$

where $[f]$ is continuous on $0 \leqslant x \leqslant 1$ and all the jumps in f are accounted for in the sum of Heaviside functions.

We have already proved that, if f is continuous, (2.9) satisfies (2.8). To take care of the piecewise continuous case it is clear [in view of (2.25)] that it is enough to treat the special situation where the loading is $H(x - a)$. Show that in this case (2.9) satisfies (2.8) for all $x \neq a$ and that the deflection has a continuous derivative at $x = a$ (the requirements on a classical solution as defined in Section 1 will then be met).

3. MAXIMUM PRINCIPLE

If $f(x) < 0$ in (2.1), we have steady heat conduction with sinks. The temperature $u(x)$ satisfies the *differential inequality*

$$(3.1) \qquad -u'' < 0, \qquad 0 < x < 1.$$

Since heat is removed at every point of the rod, it is physically clear that the maximum temperature must occur on the boundary (which consists of the two points $x = 0$ and $x = 1$) and nowhere else. From the geometrical point of view, u, having a positive second derivative, is strictly convex. This again shows that the maximum is on the boundary. The proof is trivial: if u had even a relative maximum at the interior point x_0, then $u'(x_0) = 0$ and $u''(x_0) \leqslant 0$, contradicting (3.1).

If instead of the strict inequality (3.1) we know only that

$$(3.2) \qquad -u'' \leqslant 0, \qquad 0 < x < 1,$$

we can still conclude that the maximum of u occurs on the boundary, but now it is possible for the maximum to be also attained in the interior if u is identically constant. We have two versions of the *maximum principle*:

1. *Weak version.* Let u be continuous on $0 \leqslant x \leqslant 1$ and satisfy (3.2). Let the maximum of u on the boundary be M. Then $u(x) \leqslant M$, $0 \leqslant x \leqslant 1$.

2. *Strong version.* Let u be continuous on $0 \leqslant x \leqslant 1$ and satisfy (3.2). Suppose $u(x) \leqslant M$ in $0 \leqslant x \leqslant 1$ and $u(x_0) = M$ at an interior point x_0; then $u(x) \equiv M$ in $0 \leqslant x \leqslant 1$.

The first version makes no prediction as to whether the maximum can occur at interior points as well as on the boundary; in the strong version this is ruled out unless u is identically constant.

Proof of weak version. For $\varepsilon > 0$ set $v(x) = u(x) + \varepsilon x^2$; then v satisfies the strict inequality $-v'' < 0$, $0 < x < 1$, so that the maximum of v is on the boundary and

$$v(x) = u(x) + \varepsilon x^2 \leqslant M + \varepsilon.$$

Thus $u(x) \leqslant M + \varepsilon$ for every $\varepsilon > 0$, and hence $u(x) \leqslant M$.

Proof of strong version. Suppose $u(x_1) < M$, where with no loss of generality we can take $x_1 > x_0$. We will show that this leads to a contradiction by constructing a function v satisfying $-v'' < 0$ in $0 < x < x_1$, $v(0) < M$, $v(x_1) < M$, $v(x_0) = M$. Consider the function

$$z = e^{x - x_0} - 1,$$

which is positive for $x > x_0$, is negative for $x < x_0$, and vanishes at x_0. Now choose ε so that $0 < \varepsilon < [M - u(x_1)]/z(x_1)$, which is clearly possible since $M > u(x_1)$ and $z(x_1) > 0$. Then the function

$$v(x) = u(x) + \varepsilon z(x)$$

satisfies

$$-v'' = -u'' - \varepsilon z'' \leqslant -\varepsilon z'' = -\varepsilon e^{x - x_0} < 0,$$

which is the strict inequality (3.1). But $v(0) < M$, $v(x_1) < M$, and $v(x_0) = M$, contradicting the fact that v must have its maximum on the boundary of the interval $(0, x_1)$.

Remarks

1. If instead of (3.2) we had $-u'' \geqslant 0$, u would be concave and u would satisfy a *minimum* principle. The proof is obtained by noting that $w = -u$ satisfies the maximum principle (in either version).

2. If $u''=0$, then both the maximum and minimum principles apply (the result is trivial in one dimension but not in higher dimensions).

The weak version of these principles is easily extended to higher dimensions. Let Ω be a bounded domain with boundary Γ, and let u be continuous on $\overline{\Omega}$. If $-\Delta u \leqslant 0$ in Ω, then $\max u$ occurs on Γ; if $-\Delta u \geqslant 0$ in Ω, then $\min u$ occurs on Γ; if $\Delta u = 0$ in Ω, then $\max u$ and $\min u$ occur on Γ.

The strong version of these principles is extended to higher dimensions in Section 3, Chapter 7.

The weak version by itself leads to the following interesting consequences:

1.   The solution of the inhomogeneous problem (1.4) is unique. *Proof*: If u_1 and u_2 are two solutions, then $w = u_1 - u_2$ satisfies $-\Delta w = 0$ in Ω with $w = 0$ on Γ. Since the maximum and the minimum of w on the boundary are both 0, w must be identically 0 in the interior.

2.   The data $\{f_1(x); h_1(x)\}$ is said to *dominate* $\{f_2(x); h_2(x)\}$ if $f_1(x) \geqslant f_2(x)$ in Ω and $h_1(x) \geqslant h_2(x)$ on Γ. Suppose $\{f_1; h_1\}$ dominates $\{f_2; h_2\}$; then the corresponding solutions of (1.4) satisfy $u_1(x) \geqslant u_2(x)$. *Proof*: $w = u_2 - u_1$ satisfies $-\Delta w \leqslant 0$ in Ω and $w \leqslant 0$ on Γ. By the maximum principle, $\max w$ occurs on Γ; therefore $w(x) \leqslant 0$ in Ω.

3.   Exercise 3.2 shows that the solution of (1.4) depends continuously on the data (in the d_∞ sense).

For a comprehensive, yet accessible treatment of maximum principles, the reader should consult the book by Protter and Weinberger.

Exercises

3.1.   The equation $-(ku')' + qu = f$ governs steady diffusion in an absorbing medium. Here $u(x)$ is the *concentration* of the diffusing substance measured relative to some ambient value (so that u can be positive or negative), $-k(x) \operatorname{grad} u$ is the diffusion flux vector, $q(x)$ measures the absorption properties, and $f(x)$ is the source density. The effect of the term qu is to try to restore the concentration to its ambient value. The same equation also governs the transverse deflection of a string when there is a springlike resistance (the term qu) to such a deflection. In both cases it is natural to take $k(x) > 0$, $q(x) \geqslant 0$, and we shall do so.

   (a) Let $v(x)$ be continuous on $a \leqslant x \leqslant b$ and satisfy the strict inequality

   (3.3) $-(kv')' + qv < 0, \qquad a < x < b.$

Show that v cannot have a *positive* (or even nonnegative) relative maximum at an interior point. The example $v = -\cosh x - 1$ on $-1 < x < 1$ satisfies (3.3) with $k = q = 1$ and has a negative maximum at the interior point $x = 0$.

(b) Let $u(x)$ be continuous on $a \leqslant x \leqslant b$ and satisfy

(3.4) $-(ku')' + qu \leqslant 0, \qquad a \leqslant x \leqslant b.$

State and prove a weak version of the maximum principle for positive solutions of (3.4).

(c) If the inequality in (3.4) is reversed, a minimum principle is obtained for negative solutions of the inequality $-(ku')' + qu \geqslant 0$. State appropriate principles for solutions of the equation $-(ku')' + qu = 0$.

(d) Prove uniqueness and continuous dependence on data for $-(ku')' + qu = f;$ $u(0) = \alpha$, $u(1) = \beta$.

3.2. Let the continuous functions f_1 and f_2 satisfy $|f_1(x) - f_2(x)| < \varepsilon$ on a bounded domain Ω, while the continuous functions h_1 and h_2 satisfy $|h_1(x) - h_2(x)| < \varepsilon$ on the boundary Γ of Ω. Show that the corresponding solutions of (1.4) satisfy $|u_1(x) - u_2(x)| < \alpha\varepsilon$, where α is a constant which depends only on Ω.

3.3. Derive a strong maximum principle for solutions of

$$-(ku')' \leqslant 0, \qquad 0 < x < 1,$$

where $k(x) > 0$ in $0 \leqslant x \leqslant 1$.

3.4. (a) Derive a strong maximum principle for solutions of

$$-u'' + pu' \leqslant 0, \qquad 0 < x < 1,$$

where $p(x)$ is an arbitrary continuous function. *Hint:* First prove the result for solutions of the strict inequality, and then let $v = u + \varepsilon z$, where $z = \exp[\alpha(x - x_0)] - 1$ with α suitably chosen.

(b) Derive a strong minimum principle for solutions of $-u'' + pu' \geqslant 0$.

(c) State appropriate principles for solutions of $-u'' + pu' = 0$.

4.   EXAMPLES OF GREEN'S FUNCTIONS

Initial Value Problem

A particle of mass m moves along the u axis under the influence of a force $F(t)$ directed along the axis. The motion of the particle is determined by Newton's law with *initial conditions*:

(4.1) $$m\frac{d^2u}{dt^2} = F(t), \quad t>0; \quad u(0)=\alpha, \quad \frac{du}{dt}(0)=\beta.$$

If the problem were solved over a finite time interval $(0, T)$, T would play no role in the final result. Therefore we may as well consider the equation on the semi-infinite interval $t>0$.

Green's function $g(t,\tau)$ associated with (4.1) satisfies

(4.2) $$m\frac{d^2g}{dt^2} = \delta(t-\tau), \quad 0<t,\tau<\infty; \quad g(0,\tau)=0, \quad g'(0,\tau)=0.$$

The function $g(t,\tau)$ is the position of a particle initially at rest at the origin and subject to a unit impulse at time τ. We can regard the impulse as the limiting case of a very large force $X(t)$ acting over a very short period of time from τ to $\tau+\Delta\tau$ such that

$$\int_{\tau}^{\tau+\Delta\tau} X(t)\,dt = 1.$$

Such an impulse will cause an instantaneous unit change in the momentum $m(dg/dt)$ of the particle. Thus (4.2) can be written in the equivalent form

(4.3)
$$\begin{cases} m\dfrac{d^2g}{dt^2}=0, \quad 0<t<\tau, t>\tau; \quad g(0,\tau)=g'(0,\tau)=0, \\[2mm] g \text{ continuous at } t=\tau; \qquad m\dfrac{dg}{dt}\Big|_{t=\tau+} - m\dfrac{dg}{dt}\Big|_{t=\tau-} = 1. \end{cases}$$

Since both initial conditions apply to the interval $(0,\tau)$, we find that $g=0$ until $t=\tau$. The continuity of g and the jump condition on g' give

$$g(t,\tau)=\begin{cases} 0, & 0\leqslant t<\tau, \\[2mm] \dfrac{t-\tau}{m}, & t>\tau. \end{cases}$$

The superposition principle can then be applied to the problem with 0 initial data:

$$mu'' = F(t), \quad t > 0; \qquad u(0) = 0, \quad u'(0) = 0$$

with the result

(4.4) $$u(t) = \int_0^\infty g(t,\tau) F(\tau) \, d\tau = \int_0^t \frac{t - \tau}{m} F(\tau) \, d\tau.$$

Not surprisingly, the displacement $u(t)$ is independent of the force acting after time t. The solution of the problem with data $\{0; \alpha, \beta\}$ is $\alpha + \beta t$, so that the solution of (4.1) is the sum of $\alpha + \beta t$ and (4.4). Existence, uniqueness, and continuous dependence on data are easily proved.

Reverting to the x, ξ notation and setting $m = -1$, we see that the function

(4.5) $$h(x, \xi) = \begin{cases} 0, & 0 < x < \xi, \\ \xi - x, & x > \xi, \end{cases}$$

satisfies

(4.6) $$-\frac{d^2 h}{dx^2} = \delta(x - \xi), \quad 0 < x, \xi; \qquad h(0, \xi) = h'(0, \xi) = 0.$$

A Green's function for an initial value problem is sometimes called a *causal* Green's function. Green's function $g(x, \xi)$ given by (2.4) satisfies the same differential equation but with different side conditions. With ξ fixed, $h - g$ satisfies $(h - g)'' = 0$ for all x, and $h - g$ must coincide for all x with a solution of the homogeneous equation, which turns out to be $-(1 - \xi)x$. This suggests a method for constructing Green's function for a particular set of boundary conditions: first construct the causal Green's function for the same operator, and then add the appropriate solution of the homogeneous equation to satisfy the original boundary conditions (see the beam problem below, for instance).

Variable Conductivity

Let the thermal conductivity in a rod of unit length be a function $k(x)$ which is positive and continuously differentiable. The steady temperature $g(x, \xi)$ in a rod with a concentrated unit source at ξ, with its left end at 0 temperature, and with its right end insulated satisfies

(4.7) $$-\frac{d}{dx}\left(k(x)\frac{dg}{dx}\right) = \delta(x - \xi), \quad 0 < x, \xi < 1; \qquad g(0, \xi) = 0, \quad g'(1, \xi) = 0.$$

An equivalent formulation is

$$(4.8) \quad \begin{cases} -(kg')' = 0, & 0 < x < \xi, \, \xi < x < 1; \qquad g(0,\xi) = 0, \quad g'(1,\xi) = 0, \\ g \text{ continuous at } x = \xi; & k(\xi)[\, g'(\xi+,\xi) - g'(\xi-,\xi)\,] = -1, \end{cases}$$

the jump condition on g' stemming from a heat balance for a thin slice of the rod containing the source. The functions

$$u_1(x) = \int_0^x \frac{1}{k(y)}\, dy \qquad \text{and} \qquad u_2(x) = 1$$

are solutions of the homogeneous equation satisfying, respectively, the boundary conditions at the left and right endpoints. The matching conditions at $x = \xi$ give

$$g(x,\xi) = \begin{cases} \displaystyle\int_0^x \frac{1}{k(y)}\, dy, & 0 \leqslant x < \xi, \\[2mm] \displaystyle\int_0^\xi \frac{1}{k(y)}\, dy, & \xi < x \leqslant 1. \end{cases}$$

Simply Supported Beam

Consider a simply supported beam under a concentrated load at $x = \xi$. The deflection $g(x,\xi)$ satisfies

$$(4.9) \quad \frac{d^4 g}{dx^4} = \delta(x - \xi), \quad 0 < x, \xi < 1; \quad g(0,\xi) = g''(0,\xi) = g(1,\xi) = g''(1,\xi) = 0.$$

The shear force $V(x)$ experiences a jump discontinuity from $x = \xi-$ to $\xi+$ to balance the concentrated load:

$$V(\xi+) - V(\xi-) = -1.$$

The moment, the slope, and the deflection remain continuous even at $x = \xi$. Since $-V = d^3 g / dx^3$, we can write (4.9) as

$$(4.10) \quad \begin{cases} \dfrac{d^4 g}{dx^4} = 0, \quad 0 < x < \xi, \, \xi < x < 1; \\[2mm] g(0,\xi) = g''(0,\xi) = g(1,\xi) = g''(1,\xi) = 0, \\[2mm] g, g', g'' \text{ continuous at } x = \xi; \qquad g'''(\xi+,\xi) - g'''(\xi-,\xi) = 1. \end{cases}$$

Applying the boundary conditions, we find that the solution for $x<\xi$ is $Ax+Bx^3$, while for $x>\xi$ it is $C(1-x)+D(1-x)^3$. It remains to apply the matching conditions at ξ. The conditions on g'' and g''' should be used first to yield $g=Ax-(1-\xi)(x^3/6)$ for $x<\xi$ and $g=C(1-x)-\xi[(1-x)^3/6]$ for $x>\xi$. The continuity of g and g' then gives $A=\frac{1}{6}\xi(1-\xi)(2-\xi)$ and $C=\frac{1}{6}\xi(1-\xi)(1+\xi)$. The same result can of course be obtained (perhaps more intuitively) by using the shear and moment diagrams of Exercise 1.2.

We can also construct g by first finding the causal fundamental solution $h(x,\xi)$ satisfying

$$\frac{d^4h}{dx^4}=\delta(x-\xi),\quad 0<x,\xi;\qquad h(0,\xi)=h'(0,\xi)=h''(0,\xi)=h'''(0,\xi)=0.$$

An easy calculation gives

(4.11)
$$h(x,\xi)=\begin{cases} 0, & x<\xi, \\ \dfrac{(x-\xi)^3}{6}, & x>\xi. \end{cases}$$

Therefore g in (4.10) must be of the form $h+A+Bx+Cx^2+Dx^3$. The conditions at the end $x=0$ give $A=C=0$. At the right end we have $h''(1,\xi)+6D=0$ and $h(1,\xi)+B+D=0$, that is, $D=-(1-\xi)/6$ and $B=\xi(1-\xi)(2-\xi)/6$, which when substituted in $h+Bx+Dx^3$ confirm the earlier result.

The Infinite Rod with Absorption

The steady-state concentration $u(x)$ of a substance diffusing in a homogeneous absorbing medium satisfies

(4.12)
$$-\frac{d^2u}{dx^2}+q^2u=f(x),\qquad -\infty<x<\infty,$$

where q^2 is a positive constant, $f(x)$ is the source density of the substance, and the process can be considered as taking place in an infinitely long tube, $-\infty<x<\infty$. (The same equation governs the small transverse displacements of a string subject to an applied load and a springlike restoring mechanism.) Green's function corresponding to a steady unit input of the diffusing substance at $x=\xi$ satisfies

(4.13)
$$-\frac{d^2g}{dx^2}+q^2g=\delta(x-\xi),\qquad -\infty<x,\xi<\infty.$$

Since the coefficients of the differential equation are constants, it will suffice to find $g(x,0)$ and then set $g(x,\xi)=g(x-\xi,0)$. This argument obviously depends also on the fact that we are dealing with the infinite domain $-\infty<x<\infty$. Again we assume that $g(x,0)$ is continuous; conservation of matter gives $-g'(0+,0)+g'(0-,0)=1$. In keeping with the absorbing nature of the medium, we require that g *vanish* at $x=\pm\infty$, so that $g(x,0)=e^{-q|x|}/2q$, and

$$(4.14) \qquad\qquad g(x,\xi)=\frac{e^{-q|x-\xi|}}{2q}.$$

It is perhaps a little surprising that g has no limit as $q\to0$. The reason is that the nonabsorbing problem cannot obey the condition $g\to0$ as $|x|\to\infty$. On the other hand, the flux dg/dx obtained from (4.14) has the limits $-\frac{1}{2}$ for $x>\xi$ and $+\frac{1}{2}$ for $x<\xi$, so that, by integration, we might suspect that a solution of (4.13) for $q=0$ is $-(|x-\xi|/2)+C$, which is easily confirmed. Although there is no compelling physical argument for doing so, we often set $C=0$.

Method of Images

Consider (4.13) for the semi-infinite interval $0<x<\infty$. In addition to the condition $g\to0$ as $x\to\infty$, we now need a boundary condition at $x=0$, which we will take as $g(0,\xi)=0$. This means that any of the diffusing substance that reaches $x=0$ is removed [the boundary condition $g'(0,\xi)=0$ would model a reflecting wall at $x=0$]. Thus we wish to solve

$$(4.15) \qquad \begin{aligned} -\frac{d^2g}{dx^2}+q^2g&=\delta(x-\xi), \quad 0<x,\xi<\infty; \\ g(0,\xi)&=0, \quad g\to0 \text{ as } x\to\infty. \end{aligned}$$

Let us look instead at an infinite rod with a unit source at $x=\xi$ and a unit sink at $x=-\xi$. According to (4.14), the solution of this problem is

$$(4.16) \qquad\qquad \frac{e^{-q|x-\xi|}}{2q}-\frac{e^{-q|x+\xi|}}{2q}.$$

This function vanishes at $x=0$ and has only one source singularity in $0<x<\infty$ namely, the original source at $x=\xi$. The term $e^{-q|x+\xi|}/2q$ arising from the *image* source at $x=-\xi$ satisfies the homogeneous differential equation in $0<x<\infty$. Thus (4.16) is a solution of the boundary value problem (4.15).

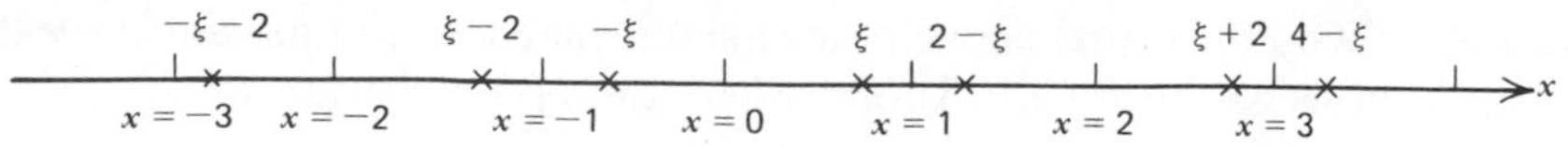

$$-\xi-2 \qquad \xi-2 \quad -\xi \qquad \xi \quad 2-\xi \qquad \xi+2 \;\; 4-\xi$$

$$x=-3 \quad x=-2 \qquad x=-1 \qquad x=0 \qquad x=1 \qquad x=2 \qquad x=3$$

Figure 4.1

For Green's function of a finite rod we use a similar idea. If the ends are both *reflecting*, say, the boundary condition is $dg/dx=0$ at both $x=0$ and $x=1$. We consider the related problem of an infinite rod with *positive* unit sources located at the set of points $\xi+2n$ and $-\xi+2n$, n ranging through the integers from $-\infty$ to ∞, as in Figure 4.1. The solution of this problem is

$$(4.17) \qquad \sum_{n=-\infty}^{\infty} \frac{e^{-q|x-(\xi+2n)|}}{2q} + \sum_{n=-\infty}^{\infty} \frac{e^{-q|x-(-\xi+2n)|}}{2q},$$

which has even symmetry about both $x=0$ and $x=1$. Thus this function has a vanishing derivative at $x=0$ and $x=1$. It is clear from the figure that of this array of sources the only one in the interval $(0,1)$ is the original source. Therefore (4.17) is a solution of the problem of the finite tube with reflecting walls.

Steady Diffusion in a Three-Dimensional Medium

Let Ω be a bounded or unbounded domain in R_3, and let x be the position vector in R_3. The concentration $u(x)$ of the diffusing substance satisfies the partial differential equation

$$(4.18) \qquad -\Delta u + q^2 u = f(x), \qquad x \in \Omega,$$

where the constant $q^2 \geqslant 0$ is a measure of the absorption of the medium and $f(x)$ is the density of the source. There will of course be boundary conditions on Γ, the boundary of Ω. The case $q=0$ corresponds to diffusion without absorption or to steady heat conduction.

Let us look at the case where Ω is the whole space and there is only a concentrated steady unit source at the origin. A mass balance (or heat balance) shows that the flux through a small sphere about the source must equal the input in the ball, that is,

$$(4.19) \qquad \lim_{\varepsilon \to 0} - \int_{|x|=\varepsilon} \frac{\partial u}{\partial n}\, dS = 1.$$

We also expect u to vanish at infinity. The concentration should clearly depend only on the radial coordinate r; since there are no sources for $r \neq 0$, we find, on using the spherical form of Δ, that $u(r)$ satisfies

$$(4.20) \qquad -\frac{1}{r^2}\frac{d}{dr}\left(r^2\frac{du}{dr}\right)+q^2u=0, \qquad r>0,$$

with (4.19) becoming

$$(4.21) \qquad -1=\lim_{\varepsilon\to 0}4\pi\varepsilon^2\left(\frac{du}{dr}\right)_{r=\varepsilon}.$$

The substitution $v=u/r$ transforms (4.20) into $-v''+q^2v=0$, whose general solution is a linear combination of e^{-qr} and e^{qr}. Taking account of the required behavior at $r=\infty$, we obtain $u=Ae^{-qr}/r$. Imposing (4.21) gives $A=1/4\pi$, and therefore

$$(4.22) \qquad u=\frac{e^{-qr}}{4\pi r}.$$

The effect of a source at ξ is obtained from (4.22) by translation. The concentration due to such a source is what we call the *free space* Green's function:

$$(4.23) \qquad g=\frac{e^{-q|x-\xi|}}{4\pi|x-\xi|}.$$

Note that, unlike the one-dimensional case, the limit as $q\to 0$ gives the solution $1/4\pi|x-\xi|$ for a nonabsorbing medium. A more important observation is that Green's function is now *singular* at $x=\xi$ (in one dimension g was continuous at $x=\xi$). This is of course the free space Green's function for the negative Laplacian:

$$(4.24) \qquad -\Delta\frac{1}{4\pi|x-\xi|}=\delta(x-\xi).$$

Green's functions for some simple domains (such as a half-space, a quarter-space, a slab, a rectangular parallelepiped) can be found by images when the boundary condition is that the function or its normal derivative vanishes on the boundary. Other methods for constructing Green's functions for partial differential equations will be discussed in later chapters, but suppose for the time being that Green's function $g(x,\xi)$ is known for the negative Laplacian in a domain Ω with $g=0$ on the boundary Γ. Then

g is the solution of

(4.25) $-\Delta g = \delta(x-\xi), \quad x,\xi \text{ in } \Omega; \qquad g=0, \quad x \text{ on } \Gamma.$

Let $u(x)$ be the solution of the problem with data $\{f;0\}$; that is, $u(x)$ satisfies

(4.26) $-\Delta u = f, \quad x \text{ in } \Omega; \qquad u=0, \quad x \text{ on } \Gamma.$

By the superposition principle we still expect the solution to be expressible as

(4.27) $u(x) = \int_\Omega g(x,\xi) f(\xi)\, d\xi.$

Clearly this function vanishes when x is on Γ because g does. If we formally calculate $-\Delta u$ by differentiating under the integral sign in (4.27) and use (4.25), we obtain $-\Delta u = f$ as required. The procedure is permissible if f obeys some very mild restrictions.

Next we express the solution of the problem with data $\{0;h\}$ in terms of Green's function. Let $v(x)$ satisfy

(4.28) $\Delta v = 0, \quad x \text{ in } \Omega; \qquad v=h(x), \quad x \text{ on } \Gamma,$

and multiply the differential equation by g, multiply (4.25) by v, subtract, and integrate over Ω to obtain

$$v(\xi) = \int_\Omega (g\,\Delta v - v\,\Delta g)\, dx.$$

By using Green's theorem and the fact that g vanishes for x on Γ, we find that

$$v(\xi) = -\int_\Gamma \frac{\partial g(x,\xi)}{\partial n_x} h(x)\, dS_x$$

or

(4.29) $v(x) = -\int_\Gamma \frac{\partial g(\xi,x)}{\partial n_\xi} h(\xi)\, dS_\xi,$

where the subscript indicates the variable of differentiation or integration.

The solution of the problem with data $\{f; h\}$ is then the sum of (4.27) and (4.29), as stated in (1.5), where the symmetry of $g(x, \xi)$ was also used.

Other problems of interest have sources spread on surfaces in R_3. The forcing function here stands somewhere between an ordinary volume density of sources and the most highly concentrated forcing function, $\delta(x - \xi)$, corresponding to a point source. Suppose, for instance, that a layer of sources whose total strength is unity is spread uniformly over the sphere $|x| = a$. The corresponding solution of (4.18) will then depend only on the radial coordinate r measured from the center of the sphere (the differential operator being invariant under rotation). Denoting the solution by $u(r)$, we see, by using the form of Δ appropriate for spherical coordinates, that

$$(4.30) \qquad -\frac{1}{r^2}\frac{d}{dr}\left(r^2\frac{du}{dr}\right) + q^2 u = 0, \qquad 0 < r < a, \quad r > a.$$

We search for a solution which is finite at $r = 0$, vanishes as $r \to \infty$, and represents the appropriate source at $r = a$. The total flux on the sphere $|x| = a + \varepsilon$ minus the flux on $|x| = a - \varepsilon$ must equal the input in the interior of the shell, that is,

$$\lim_{\varepsilon \to 0} -\left[\int_{|x| = a + \varepsilon}\frac{\partial u}{\partial n}\,dS + \int_{|x| = a - \varepsilon}\frac{\partial u}{\partial n}\,dS\right] = 1.$$

Since u depends only on r, this becomes

$$(4.31) \qquad -1 = 4\pi a^2\left[\left(\frac{du}{dr}\right)_{r = a+} - \left(\frac{du}{dr}\right)_{r = a-}\right].$$

The solution of (4.30) must therefore satisfy (4.31), vanish at infinity, and be bounded at $r = 0$. We find that

$$u = A\frac{\sinh qr}{r} \quad \text{for } r < a, \qquad u = B\frac{e^{-qr}}{r} \quad \text{for } r > a.$$

Since the problem has been reduced to a one-dimensional problem, it is appropriate to require that u be continuous at $r = a$; this condition, together with (4.31), then yields

$$u = \begin{cases} \dfrac{1}{4\pi q}\dfrac{e^{-qa}}{a}\dfrac{\sinh qr}{r}, & r < a, \\[3mm] \dfrac{1}{4\pi q}\dfrac{\sinh qa}{a}\dfrac{e^{-qr}}{r}, & r > a. \end{cases}$$

As $a \to 0$ we should recover the solution $g(x,0)$ for a unit source at the origin. Taking the limit in the expression valid for $r > a$, we find that

$$u = \frac{1}{4\pi} \frac{e^{-qr}}{r},$$

in agreement with (4.22).

In the case of a *line source* of uniform unit density, the response u is independent of the coordinate parallel to the line. It is therefore appropriate to use cylindrical polar coordinates (ρ, ϕ) with the source at $\rho = 0$; the axial symmetry of the problem suggests that u is independent of ϕ and (4.18) reduces to

$$-\frac{1}{\rho} \frac{d}{d\rho} \left(\rho \frac{du}{d\rho} \right) + q^2 u = 0, \qquad \rho > 0.$$

This is the modified Bessel equation whose independent solutions are $I_0(q\rho)$ and $K_0(q\rho)$. Since I_0 is exponentially large at ∞, we must have $u = AK_0(q\rho)$ with A determined from the unit source condition at $\rho = 0$. This condition has the form

$$-1 = \lim_{\varepsilon \to 0} 2\pi\varepsilon \left(\frac{du}{d\rho} \right)_{\rho = \varepsilon},$$

which in light of the logarithmic singularity of K_0 at the origin gives $A = \frac{1}{2}\pi$ and

(4.32) $$u = \frac{1}{2\pi} K_0(q\rho).$$

Note that in three dimensions a concentrated source gives rise to a singularity of order $1/|x|$, a line source to a logarithmic singularity, and a surface source to a simple discontinuity in the normal derivative. Only in the last case is the response continuous across the source.

Interface Problems

Consider steady one-dimensional heat conduction in a rod occupying the interval $-1 < x < 1$. The rod's thermal conductivity is the positive constant k_1 in $-1 < x < 0$ and another positive constant k_2 in $0 < x < 1$. Such a problem could arise in dealing with a *composite rod* constructed by joining end to end two rods of unit length and of different conductivities or in attempting to idealize a *heterogeneous rod* whose conductivity changes rapidly but continuously from k_1 to k_2. In both interpretations we want to reduce the problem to solving constant conductivity equations in the two halves of the rod, and the question that remains is how to match these solutions at the interface $x=0$.

For the heterogeneous rod we are considering the limiting case as $\varepsilon \to 0+$ of a problem with a continuously varying positive conductivity $k(x,\varepsilon)$ having the property

$$(4.33) \qquad \lim_{\varepsilon \to 0+} k(x,\varepsilon) = \begin{cases} k_1, & x<0, \\ k_2, & x>0. \end{cases}$$

The limiting conductivity will be denoted by $k(x)$; we have $k(x)=k_1+(k_2-k_1)H(x)$, where H is the Heaviside function. It turns out that it is not quite sufficient to ask that (4.33) hold pointwise. Instead we will need to require that $1/k(x,\varepsilon)$ tend to its limit in the L_1 sense (see Section 7, Chapter 0), that is,

$$(4.34) \qquad \lim_{\varepsilon \to 0} \int_{-1}^{1} \left| \frac{1}{k(x,\varepsilon)} - \frac{1}{k(x)} \right| dx = 0,$$

which means that the area between the curves $1/k(x,\varepsilon)$ and $1/k(x)$ must go to 0 as $\varepsilon \to 0$ (this does not follow from pointwise convergence alone). There are many ways of generating explicit expressions for $k(x,\varepsilon)$, for instance,

$$k(x,\varepsilon) = \frac{k_1+k_2}{2} + \frac{k_2-k_1}{\pi} \arctan \frac{x}{\varepsilon},$$

but our results are independent of the particular form of $k(x,\varepsilon)$. The resulting interface conditions are

$$(4.35) \qquad k_2 u'(0+) - k_1 u'(0-) = 0$$

and

$$(4.36) \qquad u(0+) - u(0-) = 0.$$

The first of these conditions is a consequence of the integral formulation of the law of heat conduction, which states in our case that the heat fluxes to the left and right of $x=0$ must be equal in the absence of concentrated sources at the interface (see Section 1, Chapter 0). The second condition is nearly obvious but does in fact require (4.34), as we shall see in the special case analyzed below.

For a *composite* rod made by joining two rods together, conditions (4.35) and (4.36) are appropriate only if the unit rods are joined perfectly at $x=0$ with no film or gap between them.

Let us now consider the explicitly solvable boundary value problem

$$(4.37) \quad -\frac{d}{dx}\left(k(x,\varepsilon)\frac{du}{dx}\right)=1, \quad -1<x<1; \quad u'(-1)=0=u(1).$$

Here we have a heterogeneous rod subject to a uniform density of sources with its left end insulated and its right end kept at 0 temperature. We are interested in the limiting case of ε tending to 0 for k satisfying (4.33) and (4.34); we take $k_2>k_1$ for the sake of definiteness. In view of (4.37) we have $-k(x,\varepsilon)u'=x+1$, so that, by (4.33), the pointwise limit of u' exists, and

$$(4.38) \qquad \lim_{\varepsilon\to 0} u'(x,\varepsilon)=\begin{cases} -\dfrac{x+1}{k_1}, & x<0, \\[2mm] -\dfrac{x+1}{k_2}, & x>0. \end{cases}$$

We shall denote the function on the right of (4.38) by $u'(x)$. We observe that $u'(x)$ is discontinuous at $x=0$ and satisfies (4.35) despite the fact that, for each $\varepsilon>0$, $u'(x,\varepsilon)$ is continuous at $x=0$. The situation is illustrated in Figure 4.2. For a fixed small value of ε there is a very sharp change in $u'(x,\varepsilon)$ in a thin transition layer around the interface; the continuity of $u'(x,\varepsilon)$ at $x=0$ is deceptive—the useful information is really contained in the discontinuous function $u'(x)$.

Since $u(x,\varepsilon)=-\int_x^1 u'(\eta,\varepsilon)d\eta$, we have

$$(4.39) \qquad u(x,\varepsilon)=\int_x^1 \frac{\eta+1}{k(\eta,\varepsilon)}\,d\eta,$$

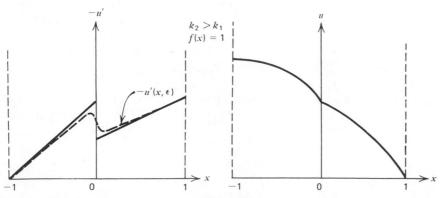

Figure 4.2

where, in view of (4.34), the limit as $\varepsilon \to 0$ may be taken under the integral sign, so that

$$(4.40) \quad u(x) = \begin{cases} \int_x^1 \dfrac{\eta+1}{k_2} \, d\eta = \dfrac{3}{2k_2} - \dfrac{1}{k_2}\left(x + \dfrac{x^2}{2}\right), & x > 0, \\[4mm] \int_x^0 \dfrac{\eta+1}{k_1} \, d\eta + \int_0^1 \dfrac{\eta+1}{k_2} \, d\eta = \dfrac{3}{2k_2} - \dfrac{1}{k_1}\left(x + \dfrac{x^2}{2}\right), & x < 0, \end{cases}$$

which is certainly continuous at $x = 0$. The continuity of $u(x)$ is guaranteed from the fact that we can pass to the limit under the integral sign in (4.39), making $u(x)$ the integral of a piecewise continuous function, and such an integral is of course continuous.

Since the interface conditions (4.35) and (4.36) are deduced by a limiting process from a continuously varying conductivity, we shall call them the *natural interface conditions*.

Let us now calculate Green's function $g(x, \xi)$ for a composite rod with natural interface conditions at $x = 0$ and the same boundary conditions as in (4.37). We first place the unit source at ξ in the left half of the rod, so that g satisfies

$$(4.41) \quad \begin{cases} -g'' = 0, \quad -1 < x < \xi, \, \xi < x < 0, \, 0 < x < 1; \\[2mm] g'(-1, \xi) = 0, \quad g(1, \xi) = 0; \\[2mm] g(\xi+, \xi) = g(\xi-, \xi), \qquad g'(\xi+, \xi) - g'(\xi-, \xi) = -\dfrac{1}{k_1}; \\[2mm] g(0-, \xi) = g(0+, \xi), \qquad k_1 g'(0-, \xi) = k_2 g'(0+, \xi). \end{cases}$$

If we solve the homogeneous equation in each of the three intervals and take into account the boundary conditions at $x = \pm 1$, we are left with four constants to be determined by the two interface conditions and the two matching conditions at the source. It is often preferable to begin with a solution which already satisfies the source conditions; an obvious candidate is the causal Green's function $h(x, \xi) = H(x - \xi)(\xi - x)/k_1$ for a rod of conductivity k_1. The desired Green's function $g(x, \xi)$ differs from h by a solution of the homogeneous equation in $-1 < x < 0$; in $0 < x < 1$, $g(x, \xi)$ is a solution of the homogeneous equation. In view of the boundary conditions we can write

$$(4.42) \qquad\qquad g = \begin{cases} h(x, \xi) + A, & x < 0, \\ B(1 - x), & x > 0. \end{cases}$$

It remains to apply the interface conditions to obtain

$$\frac{\xi}{k_1} + A = B, \qquad -1 = k_2 B,$$

so that

$$B = \frac{1}{k_2}, \qquad A = \frac{1}{k_2} - \frac{\xi}{k_1}.$$

We leave as an exercise the calculation of $g(x,\xi)$ when the source is in the right half of the rod.

What are the natural interface conditions for more complicated differential operators? We now present a method which avoids the limiting process described earlier. Suppose we want to solve the steady neutron diffusion problem

$$(4.43) \qquad\qquad -\frac{d}{dx}\left(k(x)\frac{du}{dx} \right) + q(x)u = f(x),$$

where the source density f is a given piecewise continuous function, $k(x)$ is the diffusion coefficient, and $q(x)$ is related to the collision cross section and the multiplication factor [see (5.23), Chapter 0]. Normally one assumes that q is continuous and k smooth and then searches for a classical solution u. However, the left side $-(ku')' + qu$ can be piecewise continuous under weaker conditions on k and q. Suppose, for instance, that k and q are only piecewise continuous; then $-(ku')' + qu$ will be piecewise continuous if (a) ku' is piecewise smooth so that $(ku')'$ is defined as a piecewise continuous function, and (b) qu is piecewise continuous. Now, if ku' is piecewise smooth, it is certainly continuous, and therefore u' is piecewise continuous, so that u is continuous, and the condition on qu is automatically satisfied. In applying these ideas to concrete problems, we usually solve (4.43) in the subintervals where both k and q are continuous; this leaves us with constants of integration that are explicitly found by applying interface or matching conditions at the ends of subintervals. Two conditions are needed at each interface x_i:

$$(4.44) \qquad\qquad \Delta(ku')_i = 0, \qquad \Delta u_i = 0,$$

where ΔF_i is the jump in F at x_i, that is, $F(x_i+) - F(x_i-)$. If only q is discontinuous at an interface, the matching conditions are

$$(4.45) \qquad\qquad \Delta u_i = 0, \qquad \Delta u_i' = 0.$$

In one-dimensional quantum mechanics, the Schrödinger equation has the form

(4.46) $$u'' + [E - V(x)]u = 0,$$

where E is a constant and $V(x)$ is the potential. Often V is only piecewise continuous (as in problems of a rectangular well or a rectangular potential barrier). One then solves for u in the various intervals of continuity of $V(x)$; at the points where V is discontinuous the matching conditions are (4.45).

For a more complicated problem such as

$$(r_2 u'')'' + (r_1 u')' + r_0 u = f,$$

where the coefficients may only be piecewise continuous, one writes the left side as

$$[(r_2 u'')' + r_1 u']' + r_0 u.$$

To make this piecewise continuous, we need the following: (a) $r_2 u''$ piecewise smooth (hence $r_2 u''$ continuous, u'' piecewise continuous, u' and u continuous); (b) $(r_2 u'')' + r_1 u'$ piecewise smooth (hence continuous); (c) $r_0 u$ piecewise continuous [follows automatically from (a)]. This gives us the four interface conditions:

(4.47) $\Delta u_i = 0$, $\Delta u_i' = 0$, $\Delta (r_2 u'')_i = 0$, $\Delta [(r_2 u'')' + r_1 u']_i = 0$.

As an illustration, consider the small transverse deflection of a beam of constant cross section whose stiffness changes abruptly at $x = 0$. If $E(x)$ is the stiffness of the beam, the deflection satisfies

$$(E(x)u'')'' = f(x),$$

so that, according to (4.47), the interface conditions at $x = 0$ are

(4.48) $\Delta u_0 = 0$, $\Delta u_0' = 0$, $\Delta (Eu'')_0 = 0$, $\Delta (Eu'')_0' = 0$.

These conditions have a very simple interpretation: the deflection, slope, moment, and shear are all continuous at $x = 0$. In particular, if E is the constant E_1 for $x < 0$ and the constant E_2 for $x > 0$, these conditions become

$$u(0+) = u(0-), \qquad u'(0+) = u'(0-),$$
$$E_2 u''(0+) = E_1 u''(0-), \qquad E_2 u'''(0+) = E_1 u'''(0-).$$

In Exercises 4.6 and 4.7 we give examples of different kinds of problems for composite beams that do *not* lead to natural interface conditions.

Exercises

4.1. Let q^2 be a positive constant. Find Green's function $g(x,\xi)$ satisfying

$$-g'' + q^2 g = \delta(x-\xi), \quad 0<x,\xi<1; \qquad g'(0,\xi)=g'(1,\xi)=0$$

by the direct method of Section 2, that is, by starting with two solutions u_1, u_2 of the homogeneous equation satisfying, respectively, the end conditions at $x=0$ and $x=1$ and then matching them under the load. Compare your result with the one obtained by images, (4.17). Do you notice anything strange as $q \to 0$?

4.2. (a) Show that the electrostatic potential for a line source of uniform unit density is $u=(1/2\pi)\log(1/\rho)$, where ρ is the cylindrical coordinate measured from the line source (which coincides with the z axis).

 (b) Consider the two-dimensional problem

$$(4.49) \qquad -\Delta u = -\frac{\partial^2 u}{\partial x_1^2} - \frac{\partial^2 u}{\partial x_2^2} = 0, \quad x_2>0, -\infty<x_1<\infty;$$

$$u(x_1,0)=h(x_1),$$

where $h(x_1)$ is a given function. First find Green's function $g(x,\xi)$ for 0 boundary data by the method of images [using part (a)]. Then write the solution of (4.49) by using (1.5) as

$$(4.50) \qquad u(x_1,x_2)=\frac{1}{\pi}\int_{-\infty}^{\infty}\frac{x_2}{x_2^2+(x_1-\xi_1)^2}h(\xi_1)\,d\xi_1.$$

4.3. An elastic beam is subject to a restoring force proportional to the local displacement and tending to oppose it. If a transverse distributed load $f(x)$ is applied, the appropriate differential equation satisfied by the deflection $u(x)$ is

$$\frac{d^4u}{dx^4} + k^4 u = f(x),$$

where k^4 is a positive constant regarded as known.

(a) Find the deflection in an infinite beam $-\infty < x < \infty$, when the applied load is a unit concentrated load at $x = \xi$. This deflection is the free space Green's function.

(b) Find the causal Green's function for the problem.

(c) For a beam simply supported at its ends $x = 0$ and $x = 1$, find Green's function first by using the causal Green's function of part (b) and then by using the method of images.

4.4. Consider neutron diffusion in all of three-dimensional space. The ball $|x| < R$ and its exterior are different homogeneous media that are in perfect contact. A uniform layer of sources of surface density ρ is located on the sphere $|x| = a$, where $a < R$. Set up and solve the boundary value problem for the neutron density. Consider also the limiting case of a point source at the origin.

4.5. *Quasi-derivatives.* Consider the equation of order $2p$:

$$(4.51) \qquad \left(r_p(x)u^{(p)}\right)^{(p)} + \cdots + (r_1 u')' + r_0 u = f.$$

We can write the left side as

$$u^{[2p]}(x),$$

where the *quasi-derivatives* $u^{[k]}(x)$ are defined by

$$u^{[0]} = u, \qquad u^{[1]} = u', \ldots, u^{[p-1]} = u^{(p-1)},$$
$$u^{[p]} = r_p u^{(p)}, \qquad u^{[p+1]} = (u^{[p]})' + r_{p-1} u^{[p-1]},$$
$$u^{[p+2]} = (u^{[p+1]})' + r_{p-2} u^{[p-2]}, \ldots.$$

Show that the differential equation (4.51) then becomes

$$u^{[2p]} = f,$$

and that the natural interface conditions take the simple form that $u, u^{[1]}, \ldots, u^{[2p-1]}$ be continuous.

4.6. A simply supported composite beam of constant cross section occupies the interval $0 < x < 2l$. The left half has $EI = 1$, whereas the right half is *rigid* and the two halves are welded together at $x = l$. A concentrated unit transverse force is applied at $x = \xi$, where $0 < \xi < l$. Draw the shear, moment, slope, and deflection diagrams. Express these analytically.

4.7. A homogeneous beam of constant cross section is attached to a string. The beam and string are stretched under tension H between the fixed points $x=0$ and $x=2l$, the beam occupying the interval $0<x<l$ and the string the interval $l<x<2l$. The left end of the beam is simply supported. A transverse concentrated unit force is applied at $x=l/2$. What are the interface conditions at $x=l$? Find the deflection.

4.8. Consider the case of a steep potential well or barrier in (4.46), which can be ideally represented by a potential $V(x)=\alpha\delta(x)$, where α is a real number. Although such a problem does not fall into the class studied in Section 4, it can nevertheless be solved. The principal interest is in the matching (connection) conditions at $x=0$. If we assume u continuous at the origin, a formal integration of (4.46) gives $u'(0+)-u'(0-)=\alpha u(0)$. Show that the same result is obtained by replacing the delta function in (4.46) by the sequence

$$f_n(x)=\begin{cases} n, & 0<x<\dfrac{1}{n}, \\ 0, & \text{otherwise,} \end{cases}$$

and then proceeding to the limit as $n \to \infty$.

REFERENCE AND ADDITIONAL READING

Protter, M. H. and Weinberger, H., *Maximum principles in differential equations*, Prentice-Hall, Englewood Cliffs, N. J., 1967.

2
The Theory of
Distributions

1. BASIC IDEAS, DEFINITIONS, EXAMPLES

Various examples of Chapter 1, Section 4 show that one frequently encounters sources that are nearly instantaneous (if time is the independent variable) or almost localized (if a space coordinate is the independent variable). To avoid the cumbersome study of the detailed functional dependence of such sources, we would like to replace them by idealized sources which are truly instantaneous or localized; such idealized sources are said to be *impulsive* or *concentrated* (as opposed to distributed sources). Typical instances of such sources are the concentrated forces and moments of solid mechanics; the heat sources and dipoles in heat conduction; the point masses in the theory of the gravitational potential; the impulsive forces in acoustics and in impact mechanics; the fluid sources and vortices of incompressible fluid mechanics; and the point charges, dipoles, multipoles, line charges, and surface layers in electrostatics.

What do we expect from a mathematical theory of concentrated sources? First, there should be a clear and unambiguous mathematical framework in which such sources have equal standing with distributed sources. Second, a method should be provided for calculating the response to a concentrated source, that is, a means of interpreting and solving a differential equation whose inhomogeneous term is a concentrated source. Third, if a concentrated source is "approximated" by a sequence of distributed sources, the response to the concentrated source should be a suitable limit of the sequence of responses to the distributed sources.

Functions as Linear Functionals

Consider a real-valued, continuous function on R_n. The function f is a rule which associates with each point x in R_n a real number $y = f(x)$, the value of f at x.

Another, indirect description of f is often useful. Instead of giving the value of f at each point x, we give the real number $\int_{R_n} f(x)\phi(x)\,dx$ for every ϕ belonging to a class K of accessory functions. The role of the class K is secondary; the functions ϕ are merely a vehicle for the description of f. In this alternative characterization we are viewing f as a *functional* on K [that is, f associates with each function ϕ in K the real number $\int_{R_n} f(x)\phi(x)\,dx$]. We have already encountered this point of view in ordinary Fourier series, for instance: if f is a differentiable real-valued function on $0 \leqslant x \leqslant \pi$, then f is equally well characterized by its Fourier sine coefficients $b_n = (2/\pi)\int_0^\pi f(x)\sin nx\,dx$. Here the class K of accessory functions is essentially the set $\sin x$, $\sin 2x, \dots$. We can of course find $f(x)$ from its Fourier sine coefficients by the usual simple Fourier series, but all we care about here is that $f(x)$ is unambiguously determined from the coefficients, with or without an explicit formula. Our purpose in introducing functionals on a space K of accessory functions is not merely to give an alternative description of ordinary functions f; rather, we shall see that the new point of view permits us to enlarge the kinds of functions f that can be described. By choosing the class K of accessory functions ϕ appropriately, we shall be able to characterize as functionals "singular" functions such as the delta function and the dipole function. The most suitable class K of accessory functions will be a class of very smooth functions known as test functions. We shall then be able to describe some rather "wild" functions as functionals on this space of very smooth functions. This point of view was introduced by L. Schwartz, and the comprehensive theory can be found in his book. (See also Gelfand and Shilov, Korevaar, and Zemanian).

Test Functions

Definition. A *test function* $\phi(x) = \phi(x_1, \dots, x_n)$ on R_n is a function which is infinitely differentiable on R_n (this means that every partial derivative to every order exists) and vanishes outside some bounded region (which may vary from test function to test function). The space of all test functions on R_n will be denoted by $C_0^\infty(R_n)$.

Remarks

1. It is not entirely evident that there exist nontrivial test functions. In R_1 a test function $\phi(x)$ would vanish identically outside an interval $a < x < b$ but would have to reach nonzero values in the interior, even though every derivative of ϕ is 0 at $x = a$ and $x = b$. At first it is hard to see how a function so "flat" at a and b can differ from the zero function. In fact, if $\phi(x)$ were expandable in a Taylor series of nonzero radius of

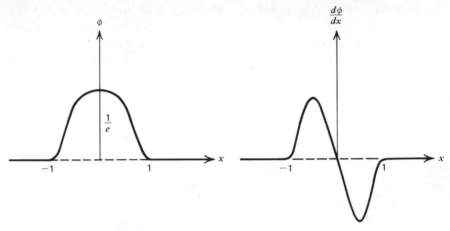

Figure 1.1

convergence about $x=a$, then ϕ would have to vanish identically, since all derivatives are 0 at $x=a$. The following example shows that test functions on R_1 do exist:

$$(1.1) \qquad \phi(x)=\begin{cases} \exp\left(\dfrac{1}{x^2-1}\right), & |x|<1, \\ 0, & |x|\geq 1. \end{cases}$$

This even function is sketched in Figure 1.1. There is no difficulty in seeing that ϕ is infinitely differentiable except possibly at $x=\pm 1$. Because of the evenness of ϕ it is enough to examine the point $x=1$. Since

$$\lim_{x\to 1-}\exp\left(\frac{1}{x^2-1}\right)=0,$$

$\phi(x)$ is continuous at $x=1$. To show that all derivatives of ϕ are 0 at $x=1$, it is sufficient to note that, for any m,

$$\lim_{x\to 1-}\frac{1}{(x^2-1)^m}\exp\left(\frac{1}{x^2-1}\right)=0.$$

A spherically symmetric test function on R_n is given by

$$(1.2) \qquad \phi(x)=\begin{cases} \exp\left(\dfrac{1}{|x|^2-1}\right), & |x|<1, \\ 0, & |x|\geq 1, \end{cases}$$

where $|x| = r$ is the radial coordinate (or distance from the origin).

2. If $\phi_1(x)$, $\phi_2(x)$ are test functions on R_n, so is $c_1\phi_1 + c_2\phi_2$ for any real numbers c_1 and c_2. The space $C_0^\infty(R_n)$ is therefore a real linear space (see Chapter 4).

3. Let $\phi(x)$ be the test function (1.2). Then

$$\phi\left(\frac{x - x_0}{\varepsilon}\right)$$

is also a test function on R_n which vanishes outside a ball of radius ε with center at x_0.

4. If $\phi(x) \in C_0^\infty(R_n)$, so does every partial derivative of $\phi(x)$.

5. If $\phi(x) \in C_0^\infty(R_n)$ and $a(x)$ is infinitely differentiable, then $a(x)\phi(x) \in C_0^\infty(R_n)$.

6. If $\phi(x_1, \ldots, x_m) \in C_0^\infty(R_m)$ and $\psi(x_{m+1}, \ldots, x_n) \in C_0^\infty(R_{n-m})$, then $\phi(x_1, \ldots, x_m)\psi(x_{m+1}, \ldots, x_n) \in C_0^\infty(R_n)$.

Convergence in the Space of Test Functions

We first introduce a concise notation for partial derivatives and differential operators in n independent variables. Let $k_1, \ldots, k_n$ be nonnegative integers; we shall call $k = (k_1, \ldots, k_n)$ a *multi-index of dimension n*. We define

$$|k| = k_1 + \cdots + k_n$$

and

$$D^k = \frac{\partial^{|k|}}{\partial x_1^{k_1} \cdots \partial x_n^{k_n}} = \frac{\partial^{k_1 + \cdots + k_n}}{\partial x_1^{k_1} \cdots \partial x_n^{k_n}},$$

with the understanding that if any component of k is 0, the differentiation with respect to the corresponding variable is omitted. As an example, if $n = 3$ and $k = (2, 0, 5)$, then

$$D^k = \frac{\partial^7}{\partial x_1^2 \partial x_3^5}.$$

An arbitrary linear differential operator L of order p in n variables can be written as

$$(1.3) \qquad\qquad L = \sum_{|k| \leqslant p} a_k(x) D^k,$$

where the coefficients $a_k(x) = a_{k_1, \ldots, k_n}(x_1, \ldots, x_n)$ are arbitrary functions. The most general differential operator of order 2 in two variables would then be

$$L = \sum_{|k| \leqslant 2} a_k(x) D^k = \sum_{|k|=0} a_k(x) D^k + \sum_{|k|=1} a_k(x) D^k + \sum_{|k|=2} a_k(x) D^k$$

$$= a_{0,0} + a_{1,0} \frac{\partial}{\partial x_1} + a_{0,1} \frac{\partial}{\partial x_2} + a_{2,0} \frac{\partial^2}{\partial x_1^2} + a_{1,1} \frac{\partial^2}{\partial x_1 \partial x_2} + a_{0,2} \frac{\partial^2}{\partial x_2^2}.$$

We also introduce the notion of support of a function $f(x)$.

Definition. The *support* of $f(x)$ is the closure of the set of points in R_n on which $f(x) \neq 0$. For instance, the support of the test function (1.2) is the closed ball $|x| \leqslant 1$, even though $\phi(x)$ vanishes on the boundary $|x| = 1$. The space of test functions $C_0^\infty(R_n)$ is also referred to as the space of infinitely differentiable functions with compact support.

We now introduce a very stringent form of convergence in $C_0^\infty(R_n)$. We say that the sequence of test functions $\{\phi_1(x), \ldots, \phi_m(x), \ldots\}$ is a *null sequence in* $C_0^\infty(R_n)$ if and only if the following conditions hold:

1. There exists a common bounded region outside of which all $\phi_m(x)$ vanish (that is, the support of all ϕ_m, $m = 1, 2, \ldots$, is contained within a single, sufficiently large ball).

2. For every multi-index k of dimension n,

$$\lim_{m \to \infty} \max_{x \in R_n} |D^k \phi_m(x)| = 0.$$

Thus $\{\phi_m(x)\}$ tends *uniformly* to 0 in R_n, and so does the sequence $\{D^k \phi_m(x)\}$. To say that $\{\phi_m\}$ is a null sequence in $C_0^\infty(R_n)$ therefore means that the approach to 0 is a very strong one: $\{\phi_m\}$ and all its derivatives tend to 0 uniformly in R_n. For example, (a) if $\phi(x)$ is a test function, then $\{(1/m)\phi(x)\}$ is a null sequence in $C_0^\infty(R_n)$; (b) if $\phi(x)$ is the test function (1.1), the sequence of test functions $\{(1/m)\phi(x/m)\}$ fails to meet criterion (1) for a null sequence in $C_0^\infty(R_1)$, whereas $\{(1/m)\phi(mx)\}$

fails to meet criterion (2). The latter sequence tends to 0 uniformly in R_1, but the differentiated sequence tends to 0 only pointwise, not uniformly.

Distributions

Definition. We say that f is a linear functional on $C_0^\infty(R_n)$ if there exists a rule which assigns to each $\phi(x)$ in $C_0^\infty(R_n)$ a real number [denoted by $\langle f, \phi \rangle$ rather than $f(\phi)$] such that

$$\langle f, \alpha_1\phi_1 + \alpha_2\phi_2 \rangle = \alpha_1 \langle f, \phi_1 \rangle + \alpha_2 \langle f, \phi_2 \rangle$$

for all real numbers α_1, α_2 and all ϕ_1, ϕ_2 in $C_0^\infty(R_n)$.

Note that for any linear functional we have $\langle f, 0 \rangle = 0$ and

$$\langle f, \sum_{k=1}^{m} \alpha_k\phi_k \rangle = \sum_{k=1}^{m} \alpha_k \langle f, \phi_k \rangle.$$

Definition. A linear functional on $C_0^\infty(R_n)$ is said to be *continuous* if, whenever $\{\psi_m(x)\}$ is a null sequence in $C_0^\infty(R_n)$, the numerical sequence $\langle f, \phi_m \rangle$ tends to 0 as $m \to \infty$. A continuous linear functional on $C_0^\infty(R_n)$ is said to be a *distribution* (or an *n-dimensional distribution*). The number $\langle f, \phi \rangle$ is the value of f at ϕ or, perhaps more picturesquely, the *action* of f on ϕ. If f_1, f_2 are any distributions and if c_1, c_2 are any real numbers, the distribution $f = c_1 f_1 + c_2 f_2$ is defined in the obvious manner:

$$\langle f, \phi \rangle = c_1 \langle f_1, \phi \rangle + c_2 \langle f_2, \phi \rangle.$$

It is easy to verify that all the conditions for a distribution are satisfied. The space $\mathcal{D}_n$ of n-dimensional distributions is therefore a linear space (see Chapter 4).

Our hope is that the framework of distributions will enable us to incorporate such extraordinary "functions" as the delta function, but first we must make sure that run-of-the-mill functions can be viewed as distributions.

Definition. A function $f(x)$ on R_n is said to be *locally integrable* if $\int_\Omega |f| \, dx$ exists for every bounded domain Ω in R_n.

Thus behavior at infinity does not affect local integrability; however, we cannot allow singularities that are too large at any finite point. For instance $1/|x|^\alpha$ is locally integrable in R_n if $\alpha < n$ but not if $\alpha \geq n$. Of course continuous functions and piecewise continuous functions are locally integrable.

Theorem 1.   A locally integrable function $f(x)$ in R_n defines (generates) an n-dimensional distribution f through the rule

(1.4)

$$\langle f,\phi\rangle = \int_{R_n} f(x)\phi(x)\,dx = \int_{-\infty}^{\infty} \cdots \int_{-\infty}^{\infty} f(x_1,\ldots,x_n)\phi(x_1,\ldots,x_n)\,dx_1 \cdots dx_n.$$

Proof.   It is clear that a linear functional on $C_0^\infty(R_n)$ has been defined. To prove continuity, let $\{\phi_m(x)\}$ be a null sequence all of whose elements vanish outside the finite ball Ω. Then

$$|\langle f,\phi_m\rangle| < \left(\max_{x\in\Omega} |\phi_m(x)| \right) \int_\Omega |f(x)|\,dx.$$

Since $\{\phi_m\}$ is a null sequence, it certainly follows that $\lim_{m\to\infty}\max|\phi_m(x)| = 0$. The local integrability of f guarantees that $\int_\Omega |f(x)|\,dx$ is finite, so that $\lim_{m\to\infty}\langle f,\phi_m\rangle = 0$ and the functional (1.4) is continuous and is therefore a distribution.

Remark.   By means of (1.4) every locally integrable function f can also be regarded as a distribution. As a point function, f has a value $f(x)$ at a point x in R_n; as a distribution, f has an action (value) $\langle f,\phi\rangle$ on a test function $\phi(x)$ belonging to $C_0^\infty(R_n)$. To what extent does the distribution f determine the point function f? We answer this question below.

If $f_1(x)$, $f_2(x)$ are different continuous functions, they generate different distributions. We must show that, for some ϕ in $C_0^\infty(R_n)$, $\langle f_2,\phi\rangle \neq \langle f_1,\phi\rangle$ or $\langle f_2 - f_1,\phi\rangle \neq 0$. Since $f_1(x)$ and $f_2(x)$ are different, there exists x_0 such that $f_2(x_0)\neq f_1(x_0)$, and we may suppose $f_2(x_0)>f_1(x_0)$. In view of the continuity of these functions, x_0 must have a neighborhood $|x - x_0|<\varepsilon$ in which $f_2(x) - f_1(x)>0$. We have seen that there is a test function ϕ which is positive in $|x - x_0|<\varepsilon$ and vanishes elsewhere. For this ϕ, rule (1.4) shows that $\langle f_2,\phi\rangle$ is larger than $\langle f_1,\phi\rangle$, so that the distributions f_1 and f_2 are different.

If $f_1(x)$ and $f_2(x)$ coincide except at a finite number of points, they generate the same distribution. More generally, two functions are said to be equal almost everywhere if $\int_\Omega |f_1 - f_2|\,dx = 0$ for every bounded domain Ω. Two locally integrable functions that are equal almost everywhere generate the same distribution. If they are not equal almost everywhere, they generate different distributions (proof omitted). *We shall regard functions that are equal almost everywhere as the same function.* (To make all this entirely consistent it is necessary to introduce the notion of equivalence classes, but we shall not do so.)

Definition. A distribution is *regular* if it can be written in form (1.4) with $f(x)$ locally integrable. All other distributions are *singular* (and we shall see that the idea is not vacuous). Even in the latter case we sometimes use formula (1.4) symbolically: given a distribution f, we assign to it a *generalized function* $f(x)$ and write symbolically

$$(1.5) \qquad \langle f, \phi \rangle = \int_{R_n} f(x)\phi(x)\,dx.$$

The symbols f, $f(x)$, and $\langle f, \phi \rangle$ are used interchangeably.

Example 1. Let c be a constant, and consider the functional on $C_0^\infty(R_n)$ defined by

$$\int_{R_n} c\phi(x)\,dx.$$

This is clearly a continuous linear functional on $C_0^\infty(R_n)$, so it is a distribution; in fact, it is a regular distribution generated via (1.4) by the constant function $f(x) = c$. We do not distinguish between the number c, the constant point function c, and the distribution c whose value $\langle c, \phi \rangle$ at ϕ is given by $\int_{R_n} c\phi(x)\,dx$.

Example 2. Let Ω be a domain in R_n, and consider the functional on $C_0^\infty(R_n)$ defined by

$$\langle I_\Omega, \phi \rangle = \int_\Omega \phi(x)\,dx.$$

This functional is clearly linear and can be written in the form (1.4): $\int_{R_n} I_\Omega(x)\phi(x)\,dx$, where $I_\Omega(x)$ is the *indicator* function of Ω; that is, $I_\Omega(x) = 1$, $x \in \Omega$, and $I_\Omega(x) = 0$, otherwise. Since $I_\Omega(x)$ is piecewise continuous, Theorem 1 tells us that $\langle I_\Omega, \phi \rangle$ is a regular distribution. A particular case of importance occurs in R_1 with Ω the interval $(0, \infty)$; then $I_\Omega(x)$ is usually denoted by $H(x)$, the *Heaviside function*.

Example 3. Let ξ be a fixed point in R_n. Consider the linear functional δ_ξ, defined from $\langle \delta_\xi, \phi \rangle = \phi(\xi)$, which thus assigns to each test function its value at ξ. If $\{\phi_m(x)\}$ is a null sequence in $C_0^\infty(R_n)$, then surely the numerical sequence $\{\phi_m(\xi)\}$ tends to 0. Therefore δ_ξ is a continuous linear functional on $C_0^\infty(R_n)$, that is, a distribution, known as the *n-dimensional Dirac distribution* (with *pole* at ξ). Let us show that δ_0 (denoted simply by δ) is a *singular* distribution. If δ were regular, there would exist a locally integrable function $f(x)$ such that

$$(1.6) \qquad \int_{R_n} f(x)\phi(x)\,dx = \phi(0) \qquad \text{for every } \phi \in C_0^\infty(R_n).$$

Consider the test functions $\psi_a(x) = \phi(x/a)$, where $\phi(x)$ is given by (1.2). Thus

$$\psi_a(x) = \begin{cases} \exp\left(\dfrac{a^2}{|x|^2 - a^2} \right), & |x| < a, \\ 0, & |x| \geqslant a, \end{cases}$$

and

$$\psi_a(0) = \frac{1}{e}, \qquad |\psi_a(x)| \leqslant \frac{1}{e}.$$

Hence

$$\left| \int_{R_n} f(x)\psi_a(x)\,dx \right| = \left| \int_{|x|<a} f(x)\exp\left(\frac{a^2}{|x|^2 - a^2} \right) dx \right| \leqslant \frac{1}{e} \int_{|x|<a} |f(x)|\,dx.$$

If $f(x)$ is locally integrable, we must have $\lim_{a\to 0} \int_{|x|<a} |f(x)|\,dx = 0$, and hence $\lim_{a\to 0} \int_{R_n} f(x)\psi_a(x)\,dx = 0$. On the other hand, if (1.6) holds, $\int_{R_n} f(x)\psi_a(x)\,dx = 1/e$, independently of a. This is a contradiction, so that the Dirac distribution δ is singular (the same proof holds if the pole is at ξ instead of at the origin). We visualize $\delta_\xi(x)$ as the symbolic source density for a concentrated unit source at ξ.

Example 4.   Translation of a distribution.   If the locally integrable function $f(x)$ on R_n is translated through the vector a, we obtain the locally integrable function $f(x-a)$, which defines a distribution of its own:

$$\langle f(x-a), \phi(x) \rangle = \int_{R_n} f(x-a)\phi(x)\,dx$$

$$= \int_{R_n} f(x)\phi(x+a)\,dx = \langle f(x), \phi(x+a) \rangle.$$

We are therefore led to use the formula

(1.7) $\langle f(x-a), \phi(x) \rangle = \langle f(x), \phi(x+a) \rangle$

to define the translation through a of any distribution f [note that in (1.7) we have used the symbolic notation for generalized functions]. To dot the i's one must show that $\langle f(x), \phi(x+a) \rangle$ really defines a distribution. Since $\phi(x)$ is a test function, so is $\phi(x+a)$; f being a distribution, its action on $\phi(x+a)$ is well defined and linearity is obviously preserved; moreover, if $\{\phi_m(x)\}$ is a null sequence in $C_0^\infty(R_n)$, so is $\{\phi_m(x+a)\}$, and we have therefore defined a continuous functional.

As a particular case

$$\langle \delta(x-\xi), \phi(x) \rangle = \langle \delta(x), \phi(x+\xi) \rangle = \phi(\xi),$$

so that we can write

$$\delta_\xi(x) = \delta(x-\xi).$$

In the rest of the work we shall use the notation $\delta(x-\xi)$ for the generalized function corresponding to the Dirac distribution with pole at ξ.

Example 5.   Scale expansion and contraction (Similarity transformation). If $f(x)$ is locally integrable, so is $f(\alpha x)$ for any real $\alpha \neq 0$. The distribution corresponding to $f(\alpha x)$ is $\int_{R_n} f(\alpha x)\phi(x)\,dx$. Setting $\alpha x = y$ and observing that the limits of integration are reversed if α is negative, we find that

(1.8) $\langle f(\alpha x), \phi(x) \rangle = \dfrac{1}{|\alpha|^n} \left\langle f(x), \phi\left(\dfrac{x}{\alpha}\right) \right\rangle.$

Even if f is a singular distribution, we use (1.8) to define the distribution $f(\alpha x)$.

Example 6.   The dipole.   Consider in R_3 an electrostatic source configuration consisting of a positive source $1/\varepsilon$ at $\xi + (\varepsilon/2)l$ and a negative source of the same strength at $\xi - (\varepsilon/2)l$, where ξ is a point in R_3 and l is a unit vector (see Figure 1.2). The distribution corresponding to these sources has the action

$$\frac{1}{\varepsilon}\phi\left(\xi + \frac{\varepsilon}{2}l\right) - \frac{1}{\varepsilon}\phi\left(\xi - \frac{\varepsilon}{2}l\right)$$

on any test function in $C_0^\infty(R_n)$. The *unit dipole with axis l* is obtained as $\varepsilon \to 0$ (note that this brings the charges together, at the same time increasing

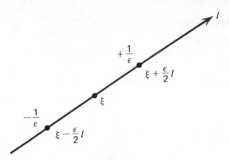

Figure 1.2

the magnitude of the charges; otherwise there would be no effect). The corresponding distribution is

$$(1.9) \qquad\qquad \langle f, \phi \rangle = \frac{d\phi}{dl}(\xi),$$

which can also be used to define a dipole distribution even if $n \neq 3$. We shall see later that (1.9) can be regarded as a distributional derivative of the delta distribution. In any event it is clear that (1.9) defines a *singular* distribution.

Example 7. Let Γ be a surface in R_3, and let dS be an area element on Γ. Consider the following functional on $C_0^\infty(R_3)$:

$$\langle f, \phi \rangle = \int_\Gamma a(x)\phi(x)\,dS,$$

where $a(x)$ is a given, locally integrable function on Γ. This distribution is singular and corresponds to a surface layer of charges (*simple layer*) on Γ with surface density $a(x)$. A special case occurs when Γ is the sphere $|x| = r$ and $a(x) = 1/4\pi r^2$. Our distribution f when acting on a test function $\phi(x)$ averages that test function on a sphere of radius r:

$$(1.10) \qquad\qquad \langle f, \phi \rangle = \frac{1}{4\pi r^2} \int_{|x|=r} \phi(x)\,dS.$$

Example 8. If $f(x)$ and $a(x)$ are both locally integrable, the product $a(x)f(x)$ need not be locally integrable; and even if af is locally integrable, its action on ϕ may not be related to the individual actions of a and f. If, however, $a(x)$ is infinitely differentiable (but not necessarily of compact support), then $a\phi$ is a test function and we have

$$(1.11) \qquad \langle af, \phi \rangle = \int_{R_n} a(x)f(x)\phi(x)\,dx = \langle f, a\phi \rangle.$$

Thus $a(x)f(x)$ generates a distribution through (1.11). Obviously the definition can be used to define the multiplication of any distribution f by an infinitely differentiable function. As a particular case, note that

$$(1.11a) \qquad a(x)\delta(x) = a(0)\delta(x).$$

Differentiation of Distributions

If $f(x)$ is a differentiable function in R_1 whose first derivative $f'(x)$ is locally integrable, f' defines its own distribution from (1.4):

$$\langle f', \phi \rangle = \int_{-\infty}^{\infty} f'(x)\phi(x)\,dx = -\int_{-\infty}^{\infty} f(x)\phi'(x)\,dx = \langle f, -\phi' \rangle,$$

the middle step being a result of integration by parts combined with the fact that $\phi \equiv 0$ outside a bounded interval. This suggests defining the derivative f' of any distribution f from

$$(1.12) \qquad \langle f', \phi \rangle = \langle f, -\phi' \rangle.$$

We must check that a distribution is really defined in this way. Since ϕ is a test function, so is $-\phi'$; and f being a distribution, the action of f on $-\phi'$ is defined. Linearity is no problem, but what about continuity? If $\{\phi_m\}$ is a null sequence in $C_0^\infty(R_1)$, so is $\{-\phi'_m\}$; therefore $\langle f, -\phi'_m \rangle$ tends to 0 as $m \to \infty$, and (1.12) defines a distribution f'.

Similar arguments apply to the partial derivatives of an n-dimensional distribution f. We define

$$(1.13) \qquad \left\langle \frac{\partial f}{\partial x_i}, \phi \right\rangle = \left\langle f, -\frac{\partial \phi}{\partial x_i} \right\rangle,$$

and by repeated application of this definition we find that

$$(1.14) \qquad \langle D^k f, \phi \rangle = (-1)^{|k|} \langle f, D^k \phi \rangle.$$

Thus we have the remarkable conclusion that every distribution can be differentiated as often as desired. Distributions can of course be generated by functions which are not differentiable in the ordinary sense, but we now have a way to differentiate such functions in a distributional sense.

Example 9. We can now attach a meaning to the derivative of the Heaviside function $H(x)$ which vanishes for $x < 0$ and has the value 1 for $x > 0$. Classically this function of a real variable is not differentiable at the origin, regardless of how we define $H(0)$. Since H is piecewise continuous, it defines a distribution from (1.4):

$$\langle H, \phi \rangle = \int_0^\infty \phi(x)\, dx,$$

so that, by (1.12), H' is defined as a distribution

$$\langle H', \phi \rangle = \langle H, -\phi' \rangle = -\int_0^\infty \phi'(x)\, dx = \phi(0).$$

Therefore

(1.15) $H'(x) = \delta(x)$

in the sense of distributions.

An alternative way of defining $H'(x)$ is as the limit of the difference quotient $(1/\varepsilon)[H(x + \varepsilon) - H(x)]$. As a distribution this difference quotient is well defined:

$$\left\langle \frac{1}{\varepsilon}[H(x+\varepsilon) - H(x)], \phi \right\rangle = \frac{1}{\varepsilon}\left(\int_{-\varepsilon}^\infty \phi\, dx - \int_0^\infty \phi\, dx \right) = \frac{1}{\varepsilon}\int_{-\varepsilon}^0 \phi\, dx.$$

Since ϕ is continuous at $x = 0$, the right side has the limit $\phi(0)$ as ε tends to 0, confirming the earlier result. We take up the question of sequences of distributions in Section 2.

In ordinary analysis $H(x)$ has derivative 0 except at the origin, where the derivative fails to exist. One might be tempted to ignore the origin and say that $H'(x) = 0$; such a definition would be fruitless, for it would not make it possible to calculate $H(x)$ as the integral of its derivative. On the other hand, (1.15) can at least be formally integrated to give

$$H(x) = \int_{-\infty}^x \delta(t)\, dt = \begin{cases} 0, & x < 0, \\ 1, & x > 0. \end{cases}$$

Example 10. If $\delta(x)$ is the Dirac distribution in n dimensions, we can calculate $\partial \delta / \partial x_i$ as follows:

$$\left\langle \frac{\partial \delta}{\partial x_i}, \phi \right\rangle = \left\langle \delta, -\frac{\partial \phi}{\partial x_i} \right\rangle = -\frac{\partial \phi}{\partial x_i}(0),$$

so that $\partial \delta / \partial x_i$ is a unit dipole at the origin with axis in the negative x_i direction [see (1.9)].

Example 11. A calculus for functions with jumps. We saw in Example 9 that $H' = \delta$ in the sense of distributions. More generally, if $f(x)$ has simple jumps, its distributional derivative will contain delta contributions from the jumps. Suppose $f(x)$ is infinitely differentiable on the real line *except* at the points $a_1, \ldots, a_k$, where, however, all left- and right-hand derivatives exist. We denote by $[f'], [f''], \ldots$ the functions obtained by differentiating f without regard to the jumps; for instance, $[H'] = 0$. It is clear that the piecewise continuous function $[f']$ will not usually coincide with the distributional derivative of f. If f has jumps of amounts $\Delta f_1, \ldots, \Delta f_k$ at $a_1, \ldots, a_k$, respectively, the function

$$(1.16) \qquad g = f - \sum_{i=1}^{k} \Delta f_i H(x - a_i)$$

is continuous and has a piecewise continuous derivative g' which coincides with its distributional derivative. Thus, differentiating (1.16) in the sense of distributions, we find that

$$g' = f' - \sum_{i=1}^{k} \Delta f_i \delta(x - a_i),$$

and since $g' = [f']$,

$$(1.17) \qquad f' = [f'] + \sum_{i=1}^{k} \Delta f_i \delta(x - a_i).$$

Therefore each simple jump discontinuity in f contributes a term $\Delta f_i \delta(x - a_i)$ to its distributional derivative.

We can now calculate higher derivatives in this way. Suppose for simplicity that there is only one point, $a_1 = 0$, where discontinuities occur. Let the jump in $[f^{(m)}]$ at $x = 0$ be $\Delta f^{(m)}$. Then

$$f' = [f'] + \Delta f^{(0)} \delta(x),$$

$$f'' = [f''] + \Delta f^{(1)} \delta(x) + \Delta f^{(0)} \delta'(x),$$

$$f^{(m)} = [f^{(m)}] + \Delta f^{(m-1)} \delta + \Delta f^{(m-2)} \delta' + \cdots + \Delta f^{(0)} \delta^{(m-1)}.$$

As an illustration let us calculate the successive derivatives of $f = e^{-|x|}$:

$$f' = [f'] = \begin{cases} -e^{-x}, & x > 0, \\ e^{x}, & x < 0, \end{cases}$$

$$f'' = [f''] - 2\delta(x) = f - 2\delta(x),$$

$$f''' = f' - 2\delta'(x),$$

etc.

In particular, we have shown that $f = e^{-|x|}/2$ satisfies $-f'' + f = \delta(x)$ in the sense of distributions, confirming the more intuitive treatment that led to (4.14), Chapter 1.

Example 12. Let Ω be a bounded domain in R_n with smooth boundary Γ, and let $I_\Omega(x)$ be the indicator function of Ω (see Example 2). Obviously I_Ω has a simple jump on Γ, but we are now dealing with a function of more than one variable, so that $\partial I_\Omega/\partial x_i$ cannot be calculated from Example 11. We can, however, go back to definition (1.13), which gives

$$(1.18) \qquad \left\langle \frac{\partial I_\Omega}{\partial x_i}, \phi \right\rangle = \left\langle I_\Omega, -\frac{\partial \phi}{\partial x_i} \right\rangle = -\int_\Omega \frac{\partial \phi}{\partial x_i}\, dx = -\int_\Gamma \phi \cos \Theta_i\, dS,$$

where Θ_i is the angle between the outward normal to Γ and the x_i axis. Therefore $\partial I_\Omega/\partial x_i$ is a simple layer of density $-\cos \Theta_i$ on the boundary Γ.

Example 13. Let L be the general linear differential operator of order p in n variables $x_1, \ldots, x_n$ [see (1.3)]. Assume that the coefficients $\{a_k(x)\}$ are infinitely differentiable. Then Lf is defined for any distribution f by using (1.11) and (1.14). These give

$$(1.19) \qquad \langle Lf, \phi \rangle = \left\langle \sum_{|k| \leqslant p} a_k D^k f, \phi \right\rangle = \left\langle f, \sum_{|k| \leqslant p} (-1)^{|k|} D^k (a_k \phi) \right\rangle.$$

The pth-order operator appearing on the right side of (1.19) is known as the *formal adjoint* of L and is denoted by L^*. We can then write (1.19) as

$$(1.20) \qquad \langle Lf, \phi \rangle = \langle f, L^* \phi \rangle,$$

which defines the distribution Lf in terms of the action of the distribution f on the test function $L^*\phi$. Note that the operator L^* is the one that would appear if we integrated by parts the left side of (1.20), treating f as an ordinary p-times-differentiable function. We always have $(L^*)^* = L$. If $L = L^*$, we say that L is *formally self-adjoint*.

Let $L = a_2(x)D^2 + a_1(x)D + a_0(x)$ be the most general second-order operator in one variable, x. Then L^* is defined from

(1.21) $$L^*\phi = a_2\phi'' + (2a_2' - a_1)\phi' + (a_2'' - a_1' + a_0)\phi,$$

and the necessary and sufficient condition for L^* to be formally self-adjoint is

(1.22) $$a_2' = a_1.$$

Note that an operator L of any order in any number of variables will be formally self-adjoint if it has constant coefficients and only partial derivatives of even order.

Example 14.   We now consider the differentiation of some locally integrable functions whose discontinuities are more severe than the simple jumps covered by (1.17).

In R_1 the locally integrable function $f(x) = H(x)\log x$, where H is the Heaviside function, defines a distribution

$$\langle f, \phi \rangle = \int_0^\infty \phi(x)\log x\, dx.$$

By definition (1.12) we have

(1.23) $$\langle f', \phi \rangle = -\langle f, \phi' \rangle = -\int_0^\infty \phi'(x)\log x\, dx,$$

which we would like to transform so as to exhibit the function $H(x)(1/x)$, which must somehow be related to $f'(x)$. Since $H(x)/x$ is not locally integrable, it does not define a distribution, so that some qualification is needed. Since the right side of (1.23) is a convergent integral, we can write

$$\langle f', \phi \rangle = -\lim_{\varepsilon \to 0} \int_\varepsilon^\infty \phi'(x)\log x\, dx,$$

where, performing the integration by parts, we find that

(1.24)

$$\langle f', \phi \rangle = \lim_{\varepsilon \to 0}\left[\int_\varepsilon^\infty \frac{\phi(x)}{x}\, dx + \phi(\varepsilon)\log\varepsilon \right] = \lim_{\varepsilon \to 0}\left[\int_\varepsilon^\infty \frac{\phi(x)}{x}\, dx + \phi(0)\log\varepsilon \right].$$

Although the individual terms in the brackets will have no limit as $\varepsilon \to 0$ (unless by chance ϕ vanishes at 0), each bracket as a whole has a limit, and this limit defines a distribution (no proof is needed since this is merely a restatement, in the present context, of the definition of a distributional derivative). The right side of (1.24) can be regarded as giving a meaning (*regularization, finite part*) to the divergent integral $\int_0^\infty [\phi(x)/x]dx$. In this spirit we write the right side of (1.24) as $\langle \mathrm{pf}[H(x)/x], \phi \rangle$, where pf $[H(x)/x]$ is the *pseudo function* $H(x)/x$. Of course this means no more or less than (1.24), but in this notation we can write

$$(1.25) \qquad\qquad \frac{d}{dx}H(x)\log x = \mathrm{pf}\,\frac{H(x)}{x}.$$

In a similar vein $H(-x)\log(-x)$ coincides with $\log|x|$ for $x<0$ and vanishes for $x>0$; with appropriate definition mimicking the ones just given, we find that

$$(1.26) \qquad\qquad \frac{d}{dx}H(-x)\log(-x) = \mathrm{pf}\,\frac{H(-x)}{x}.$$

Combining the two formulas, we have

$$(1.27) \qquad\qquad \frac{d}{dx}\log|x| = \mathrm{pf}\,\frac{H(x)}{x} + \mathrm{pf}\,\frac{H(-x)}{x} = \mathrm{pf}\,\frac{1}{x},$$

where the pseudo function $1/x$ is defined as the distribution

$$(1.28) \qquad \left\langle \mathrm{pf}\,\frac{1}{x}, \phi \right\rangle = \lim_{\varepsilon \to 0}\left[\int_\varepsilon^\infty \frac{\phi(x)}{x}dx + \int_{-\infty}^{-\varepsilon} \frac{\phi(x)}{x}dx \right],$$

which stems from (1.24) and a similar formula for the derivative of $H(-x)\log(-x)$, the combination of the integrated terms vanishing as $\varepsilon \to 0$. The right side of (1.28) converges as $\varepsilon \to 0$ and defines a distribution, but the individual terms diverge unless $\phi(0)=0$. In this way we have assigned a value to the usually divergent integral $\int_{-\infty}^\infty [\phi(x)/x]dx$; this value is known as the *Cauchy principal value*.

Example 15. Consider the locally integrable function $1/|x|$ in R_3. Obviously this function is not differentiable at the origin in the classical sense. Let us calculate the Laplacian of $1/|x|$ in the distributional sense. Since Δ is formally self-adjoint, we have from (1.20)

$$\left\langle \Delta \frac{1}{|x|}, \phi \right\rangle = \left\langle \frac{1}{|x|}, \Delta\phi \right\rangle = \int_{R_3} \frac{\Delta\phi}{|x|}dx.$$

The integral on the right side is convergent because the singularity of $1/|x|$ at the origin is rather weak, so that we may write

$$\int_{R_3} \frac{\Delta\phi}{|x|}\, dx = \lim_{\varepsilon \to 0} \int_{|x|>\varepsilon} \frac{\Delta\phi}{|x|}\, dx.$$

We calculate the integral on the right by using Green's theorem and the fact that $\phi \equiv 0$ outside a bounded region:

$$\int_{|x|>\varepsilon} \frac{\Delta\phi}{|x|}\, dx = \int_{|x|>\varepsilon} \phi \Delta\left(\frac{1}{|x|}\right) dx + \int_{|x|=\varepsilon} \left[\frac{1}{|x|}\frac{\partial\phi}{\partial n} - \phi \frac{\partial}{\partial n}\left(\frac{1}{|x|}\right)\right] dS,$$

where n is the outward normal to the domain $|x|>\varepsilon$. Setting $|x|=r$, noting that $\partial/\partial n = -(\partial/\partial r)$, and using the fact that $\Delta(1/r)=0$, $r\neq 0$, we have

$$\int_{|x|>\varepsilon} \frac{\Delta\phi}{|x|}\, dx = -\int_{|x|=\varepsilon}\left(\frac{1}{r}\frac{\partial\phi}{\partial r} + \frac{\phi}{r^2}\right) dS.$$

Since the derivatives of a test function are bounded,

$$\left|\frac{\partial\phi}{\partial r}\right| < M \qquad \text{for all } x,$$

and

$$\left|\int_{|x|=\varepsilon} \frac{1}{r}\frac{\partial\phi}{\partial r}\, dS\right| \leqslant \frac{M}{\varepsilon}(4\pi\varepsilon^2) = 4\pi\varepsilon M,$$

which tends to 0 as $\varepsilon \to 0$. We also find that

$$\int_{|x|=\varepsilon} \frac{\phi}{r^2}\, dS = \int_{|x|=\varepsilon} \frac{\phi(0) + [\phi(x)-\phi(0)]}{r^2}\, dS$$

$$= 4\pi\phi(0) + \int_{|x|=\varepsilon} \frac{\phi(x)-\phi(0)}{r^2}\, dS.$$

Since $\phi(x)$ is continuous at $x=0$, it is easy to show that the last integral tends to 0 as $\varepsilon \to 0$. It therefore follows that

$$\left\langle \Delta\frac{1}{|x|}, \phi \right\rangle = -4\pi\phi(0),$$

so that, in the distributional sense,

$$(1.29) \qquad\qquad \Delta \frac{1}{|x|} = -4\pi \delta(x).$$

Exercises

1.1. Consider the test function on R_n:

$$(1.30) \qquad \phi(x) = \begin{cases} k_n \exp\left(\dfrac{1}{|x|^2 - 1}\right), & |x| < 1, \\ 0, & |x| \geqslant 1. \end{cases}$$

where k_n has been chosen so that the integral of ϕ over the unit ball is 1. Thus k_n is unambiguously determined from $\int_{|x|<1}\phi(x)\,dx = 1$ or, equivalently, from

$$k_n S_n \int_0^1 r^{n-1} \exp\left(\frac{1}{r^2 - 1}\right) dr = 1,$$

where S_n is the area of the unit sphere (that is, the surface area of the unit ball in R_n). The function $\phi_\varepsilon(x - \xi) = (1/\varepsilon)^n \phi[(x - \xi)/\varepsilon]$ is then a test function that vanishes for $|x - \xi| \geqslant \varepsilon$ and has the property $\int_{R_n} \phi_\varepsilon(x - \xi)\,dx = 1$.

Now let $u(x)$ be a continuous function vanishing outside some bounded domain in R_n. Show that the *convolution*

$$\psi_\varepsilon(x) = \int_{R_n} \phi_\varepsilon(x - \xi) u(\xi)\,d\xi$$

is a test function and that

$$\lim_{\varepsilon \to 0} \psi_\varepsilon(x) = u(x) \qquad \textit{uniformly on } R_n.$$

Thus we have the remarkable result that over a bounded domain any continuous function can be uniformly approximated by test functions. At the same time we have also shown how to recover the continuous point function $u(x)$ from knowledge of its actions as a distribution. Treating x as a parameter, we see that

$$\int_{R_n} u(\xi)\phi_\varepsilon(x - \xi)\,d\xi = \langle u(\xi), \phi_\varepsilon(x - \xi)\rangle$$

is well defined, since u is a distribution and $\phi_\varepsilon(x-\xi)$ is a test function in ξ for each x and each $\varepsilon > 0$. By taking the limit as $\varepsilon \to 0$ of $\langle u(\xi), \phi_\varepsilon(x-\xi) \rangle$, we obtain $u(x)$.

1.2. (a) On R_1 construct a test function that has the value 1 in $-1 \leqslant x \leqslant 1$. *Hint:* Look at functions of the type $\int_{-\infty}^{x} \phi(\xi) d\xi$, where $\phi(x)$ is the test function on R_1, given in Exercise 1.1.

(b) If Ω is a bounded domain in R_n and $\phi_\varepsilon(x)$ is the test function of Exercise 1.1, show that the function

$$\psi(x) = \int_\Omega \phi_\varepsilon(x-\xi) d\xi$$

is itself a test function and has the value 1 in the part of the interior of Ω lying at a distance greater than ε from the boundary.

1.3. We have defined the derivative f' of a one-dimensional distribution f from $\langle f', \phi \rangle = \langle f, -\phi' \rangle$. It is equally reasonable to try to define f' from $\lim_{\varepsilon \to 0} [f(x+\varepsilon) - f(x)]/\varepsilon$, where $f(x+\varepsilon)$ is a translated distribution. Adopting this point of view, we find that f' has to satisfy

$$\langle f', \phi \rangle = \langle f, -\phi' \rangle - \lim_{\varepsilon \to 0} \left\langle f(x), \frac{\phi(x) - \phi(x-\varepsilon)}{\varepsilon} - \phi'(x) \right\rangle,$$

which will be in agreement with the earlier definition if

$$\lim_{\varepsilon \to 0} \left\langle f(x), \frac{\phi(x) - \phi(x-\varepsilon)}{\varepsilon} - \phi'(x) \right\rangle = 0$$

for every ϕ in $C_0^\infty(R_1)$. Since f is a distribution, all we have to show is that $\{[\phi(x) - \phi(x-\varepsilon)]/\varepsilon\} - \phi'(x)$ is a null sequence in $C_0^\infty(R_1)$. Prove this last statement, thereby reconciling the two definitions.

1.4. Show that $g = -(1/2\pi)\log|x|$ satisfies $-\Delta g = \delta(x)$ in the distributional sense (in two dimensions).

1.5. Show that $g = e^{-q|x|}/4\pi|x|$ satisfies $-\Delta g + q^2 g = \delta(x)$ in the distributional sense (in three dimensions).

1.6. Show that $g = -\frac{1}{2}|x-\xi|$, where ξ is fixed, satisfies $-g'' = \delta(x-\xi)$ in the distributional sense (in one dimension). Show that the same is true for Green's functions defined by (1.3) and (4.5), Chapter 1 (these functions having been suitably extended to the whole real line). Use the method of Example 11 or definition (1.20). Also show that $u = H(x)x^3/6$ satisfies $u^{(4)} = \delta(x)$.

1.7. The function $H(x)x^{-1/2}$ is locally integrable on R_1. Express its distributional derivative in terms of a suitably defined pseudo function: pf $H(x)x^{-3/2}$.

1.8. In R_2 the function $1/r^2$, where $r=|x|$, is *not* locally integrable because the singularity at the origin is too strong. We can, however, find a locally integrable function F such that $\Delta F = 1/r^2$ for $r \neq 0$. In fact, it is easy to see that $F = \frac{1}{2}\log^2 r$ has this property. We then *define* the pseudo function $1/r^2$ as a distribution from the formula

$$\Delta \tfrac{1}{2}\log^2 r = \mathrm{pf}\,\frac{1}{r^2}.$$

Carry out the calculations to show that

$$\left\langle \mathrm{pf}\,\frac{1}{r^2},\phi\right\rangle = \lim_{\varepsilon\to 0}\left[\int_{r\geqslant\varepsilon}(\phi/r^2)\,dx + 2\pi\phi(0)\log\varepsilon\right].$$

1.9. In R_1 let $a(x)$ be an infinitely differentiable function, and let $f(x)$ be an arbitrary generalized function. Show that

$$[a(x)f(x)]' = a(x)f' + a'(x)f,$$

and therefore

$$a(x)\delta'(x) = a(0)\delta'(x) - a'(0)\delta(x).$$

2. CONVERGENCE OF SEQUENCES AND SERIES OF DISTRIBUTIONS

One of the most natural ways of looking at the delta function is as a limit of ordinary functions (say, locally integrable). For instance, in one dimension we might think of a concentrated unit source at $x=0$ as the limit as $k\to\infty$ of the sequence $\{s_k(x)\}$ of uniformly distributed sources defined by

$$s_k(x) = \begin{cases} 0, & |x| > \dfrac{1}{2k}, \\[2mm] k, & |x| < \dfrac{1}{2k}. \end{cases}$$

In the string problem we would regard a concentrated unit load at $x=0$ as the limit of distributed loads of increasing density over narrower intervals.

The way in which we approximate the concentrated load is not unique; we could, for instance, use distributed loads $s_k(x)$ that are bell-shaped with increasing peak and decreasing base in such a way that the total load is unity. Some one-dimensional examples of this type are

$$s_k(x) = \frac{1}{\pi} \frac{k}{1 + k^2 x^2} \quad \text{and} \quad s_k(x) = \frac{k}{\sqrt{\pi}} e^{-k^2 x^2}.$$

In any event we realize that we cannot have $\lim_{k\to\infty} s_k(x) = \delta(x)$ in any classical sense; on the other hand, both $s_k(x)$ and $\delta(x)$ are well defined as one-dimensional distributions, and it would be reasonable to say that $\lim_{k\to\infty} s_k = \delta$ in the distributional sense if

$$\lim_{k\to\infty} \langle s_k, \phi \rangle = \langle \delta, \phi \rangle \quad \text{for every } \phi \text{ in } C_0^\infty(R_1)$$

or, equivalently,

$$\lim_{k\to\infty} \int_{R_n} s_k(x)\phi(x)\, dx = \phi(0) \quad \text{for every } \phi \text{ in } C_0^\infty(R_1).$$

Consider a family $\{f_\alpha(x)\}$ of distributions depending on a parameter α that belongs to an index set I (often this index set will be the set of positive integers, and we then let $\alpha = k$ so that we would be talking about a sequence of distributions); thus, for each α in I, a distribution f_α is defined.

Definition 1. Let $\{f_\alpha\}$ be a family of n-dimensional distributions. We say that $\{f_\alpha\}$ *converges distributionally* to the distribution f as $\alpha \to \alpha_0$, and write $f_\alpha \to f$ (as $\alpha \to \alpha_0$) if

(2.1) $$\lim_{\alpha \to \alpha_0} \langle f_\alpha, \phi \rangle = \langle f, \phi \rangle \quad \text{for each } \phi \text{ in } C_0^\infty(R_n).$$

Observe that convergence of distributions has been reduced to convergence of numbers.

Often one can ascertain that for each ϕ the left side of (2.1) has a limit as $\alpha \to \alpha_0$, but one is not sure that the limiting values are the values of a distribution f. These limiting values clearly are the values of a linear functional f on $C_0^\infty(R_n)$; it is also true (proof omitted) but not obvious that the functional is continuous. We therefore have the following theorem.

Theorem 1. If $\lim_{\alpha \to \alpha_0} \langle f_\alpha, \phi \rangle$ exists for each ϕ in $C_0^\infty(R_n)$, there exists one and only one distribution f such that $f_\alpha \to f$ as $\alpha \to \alpha_0$, that is,

$$\lim_{\alpha \to \alpha_0} \langle f_\alpha, \phi \rangle = \langle f, \phi \rangle \quad \text{for each } \phi \in C_0^\infty(R_n).$$

As a particular case of the theorem, consider a sequence of distributions $\{f_k\}$ such that $\lim_{k\to\infty}\langle f_k,\phi\rangle$ exists for each ϕ; it then follows that there exists a unique distribution f such that $f_k\to f$ as $k\to\infty$ (i.e., $\langle f_k,\phi\rangle\to\langle f,\phi\rangle$ for each ϕ).

We also have an immediate definition for the convergence of a *series* of distributions. Let $u_1,\dots,u_k,\dots$ be n-dimensional distributions; we write $\sum_{k=1}^{\infty}u_k=f$ if and only if the sequence of distributions $f_k=\sum_{j=1}^{k}u_j$ converges to f.

When dealing with a family $\{f_\alpha\}$ of locally integrable functions on R_n, we may wish to inquire about the relation between distributional convergence and more classical types of convergence. The following theorem will serve our purposes.

Theorem 2. Let $\{f_\alpha(x)\}$ be a family of locally integrable functions on R_n. If $\lim_{\alpha\to\alpha_0}f_\alpha(x)=f(x)$ uniformly over every bounded ball in R_n, then $f_\alpha\to f$ distributionally as $\alpha\to\alpha_0$.

Proof. It follows from uniformity that $f(x)$ is locally integrable, so that

$$\langle f_\alpha,\phi\rangle=\int_{R_n}f_\alpha(x)\phi(x)\,dx\to\int_{R_n}f\phi\,dx=\langle f,\phi\rangle.$$

The middle step is a consequence of the uniform convergence of $f_\alpha\phi$ to $f\phi$ and of the fact that ϕ vanishes outside a bounded ball. It is possible to have distributional convergence without even pointwise convergence. For instance, the sequence $\{\sin kx\}$ in R_1 does not converge pointwise to a limiting function on R_1; yet we can show using integration by parts that

$$\lim_{k\to\infty}\int_{-\infty}^{\infty}\sin kx\,\phi(x)\,dx=0 \qquad \text{for each test function } \phi,$$

so that $\sin kx\to 0$ distributionally as $k\to\infty$.

Let $\{f_\alpha\}$ be a family of arbitrary distributions with the property $f_\alpha\to f$ distributionally as $\alpha\to\alpha_0$. What can be said about the derivatives (which necessarily exist, since everything in sight is a distribution and distributions are differentiable)? We claim that

$$\frac{\partial f_\alpha}{\partial x_i}\to\frac{\partial f}{\partial x_i} \qquad \text{distributionally as } \alpha\to\alpha_0.$$

The proof is simple:

$$\left\langle\frac{\partial f_\alpha}{\partial x_i},\phi\right\rangle=\left\langle f_\alpha,-\frac{\partial\phi}{\partial x_i}\right\rangle\to\left\langle f,-\frac{\partial\phi}{\partial x_i}\right\rangle=\left\langle\frac{\partial f}{\partial x_i},\phi\right\rangle.$$

It follows immediately that, as $\alpha \to \alpha_0$, $D^k f_\alpha \to D^k f$ distributionally for any multi-index k. *Thus every convergent sequence or series of distributions can be differentiated term by term as often as required.* How unlike classical convergence, where term-by-term differentiation is possible only under rather severe restrictions! As a simple one-dimensional example note that, since $\lim_{m \to \infty} \sin mx = 0$ and $\lim_{m \to \infty} \cos mx = 0$, we must also have

$$0 = \lim_{m \to \infty} m^p \sin mx = \lim_{m \to \infty} m^p \cos mx \qquad \text{for any integer } p \geq 0.$$

In mathematical physics one frequently encounters sequences of functions which fail to converge uniformly on R_n. Some of these sequences converge pointwise; others fail altogether to converge. In either case a distributional interpretation is often possible. Suppose $\{f_k(x)\}$ is the sequence in question (say on R_1, for simplicity) and that a positive integer m can be found such that $\{f_k(x)\}$ is the mth derivative of a sequence $\{g_k(x)\}$ which converges uniformly to $g(x)$. Then $g_k \to g$ distributionally; and, since we are permitted to differentiate term by term, $g_k^{(m)} \to g^{(m)}$, that is, $f_k \to g^{(m)}$, where $g^{(m)}$ is the mth derivative (in the sense of distributions) of g. This idea will be applied to Fourier series in Section 3.

Example 1. In R_1 consider the sequence $\{s_k(x)\}$ of piecewise continuous functions

$$(2.2) \qquad s_k(x) = \begin{cases} 0, & |x| > \dfrac{1}{2k}, \\[2mm] k, & |x| < \dfrac{1}{2k}. \end{cases}$$

Then, denoting the interval $|x| < 1/2k$ by I_k, we have

$$\langle s_k, \phi \rangle = \int_{I_k} k\phi(x)\, dx = \phi(0) + \int_{I_k} k[\phi(x) - \phi(0)]\, dx$$

and

$$\left| k \int_{I_k} [\phi(x) - \phi(0)]\, dx \right| \leq k \int_{I_k} |\phi(x) - \phi(0)|\, dx \leq \max_{x \in I_k} |\phi(x) - \phi(0)|.$$

If ϕ is continuous at $x = 0$, the maximum tends to 0 as $k \to \infty$. Therefore

$$(2.3) \qquad \lim_{k \to \infty} \langle s_k, \phi \rangle = \phi(0)$$

for every ϕ continuous at the origin. This means that (2.3) is certainly true for all test functions (with plenty to spare). In the distributional sense we therefore have

$$\lim_{k \to \infty} s_k(x) = \delta(x),$$

but it is important in other applications to know that (2.3) holds under much less severe restrictions on ϕ.

Example 2. Let $\{f_\alpha(x)\}$ be a family of locally integrable functions in R_n with the property

$$(2.4) \qquad \lim_{\alpha \to \alpha_0} \int_{R_n} f_\alpha(x)\phi(x)\,dx = \phi(0) \qquad \text{for each } \phi \text{ in } C_0^\infty(R_n).$$

We call $\{f_\alpha\}$ an *n-dimensional delta family* (as $\alpha \to \alpha_0$). (Often the index α runs through the positive integers k; we then use the name *delta sequence* for $\{f_k\}$.)

In Example 1, $\{s_k(x)\}$ was a one-dimensional delta sequence. The following theorem shows how easy it is to construct delta sequences by starting with any nonnegative function $f(x)$ and suitably compressing it while at the same time increasing its peak.

Theorem 3. Let $f(x) = f(x_1, \ldots, x_n)$ be a *nonnegative* locally integrable function on R_n for which $\int_{R_n} f(x)\,dx = 1$. With $\alpha > 0$ define

$$(2.5) \qquad f_\alpha(x) = \frac{1}{\alpha^n} f\left(\frac{x}{\alpha}\right) = \frac{1}{\alpha^n} f\left(\frac{x_1}{\alpha}, \ldots, \frac{x_n}{\alpha}\right);$$

then $\{f_\alpha(x)\}$ is a delta family as $\alpha \to 0$ [and, setting $\alpha = 1/k$, the sequence $\{s_k(x) \doteq k^n f(kx_1, \ldots, kx_n)\}$ is a delta sequence as $k \to \infty$].

Proof. The substitution $y = x/\alpha$ yields these three properties:

(a) $\displaystyle \int_{R_n} f_\alpha(x)\,dx = 1,$

(b) $\displaystyle \lim_{\alpha \to 0} \int_{|x| > A} f_\alpha(x)\,dx = 0 \qquad \text{for each } A > 0,$

(c) $\displaystyle \lim_{\alpha \to 0} \int_{|x| < A} f_\alpha(x)\,dx = 1 \qquad \text{for each } A > 0,$

so that, for small positive α, $f_\alpha(x)$ is highly peaked about $x = 0$ in such a way that the total strength of this distributed source is unity, with most of it near the origin. We must show that, for each test function ϕ, $\lim_{\alpha \to 0} \int_{R_n} f_\alpha \phi \, dx = \phi(0)$; in view of (a) it suffices to show that

$$(2.6) \qquad \lim_{\alpha \to 0} \int_{R_n} f_\alpha(x) \eta(x) \, dx = 0,$$

where $\eta(x) = \phi(x) - \phi(0)$.

We divide the region of integration R_n into the two parts $|x| \leqslant B$, $|x| > B$, so that

$$\left| \int_{R_n} f_\alpha(x) \eta(x) \, dx \right| \leqslant \left| \int_{|x| \leqslant B} f_\alpha \eta \, dx \right| + \left| \int_{|x| > B} f_\alpha \eta \, dx \right|.$$

Since $\phi(x)$ is bounded, so is $\eta(x)$; thus $|\eta| \leqslant M$ for all x. Setting $p(B) = \max_{|x| \leqslant B} |\eta|$ and using the *nonnegativity* of f and property (a), we have

$$\left| \int_{R_n} f_\alpha \eta \, dx \right| \leqslant p(B) + M \int_{|x| > B} f_\alpha \, dx.$$

To prove (2.6) we must show that, for each $\varepsilon > 0$, there exists $\gamma > 0$ such that

$$\left| \int_{R_n} f_\alpha \eta \, dx \right| < \varepsilon \qquad \text{for all } \alpha, \quad 0 < \alpha < \gamma.$$

Since $\eta(x)$ is continuous and $\eta(0) = 0$, $\lim_{B \to 0} p(B) = 0$. Therefore we may choose B (independently of α) such that $p(B) < \varepsilon/2$. With B so chosen, we use property (b) to select γ such that

$$\int_{|x| > B} f_\alpha(x) \, dx < \frac{\varepsilon}{2M} \qquad \text{whenever } 0 < \alpha < \gamma.$$

This completes the proof of Theorem 3.

Extensions are possible in the following directions:

1. We retain $\phi(x)$ as a test function, but we relax the conditions on $\{f_\alpha\}$. Exercise 2.4 still has $\{f_\alpha(x)\}$ nonnegative, but the construction is more general than (2.5). Exercise 2.5 considers the possibility that $\{f_\alpha(x)\}$ may change sign; in that case additional restrictions are needed for a sequence to be a delta sequence.

2. We want (2.4) to hold for a more general class of functions $\phi(x)$. If we only require $\phi(x)$ continuous at $x = 0$ and $\int_{R_n} |\phi(x)| \, dx < \infty$, then (2.4)

will follow if $\{f_\alpha(x)\}$ is a family of nonnegative functions satisfying the requirements of Exercise 2.4 [which includes families of type (2.5)]. If we want to use families $\{f_\alpha\}$ which change sign, additional conditions will have to be placed on ϕ (see Exercise 2.5).

3. If $\{f_\alpha(x)\}$ is a delta family, it follows that

$$\lim_{\alpha \to \alpha_0} \int_{R_n} f_\alpha(x-\xi)\phi(\xi)\,d\xi = \phi(x) \qquad \text{pointwise for } x \text{ in } R_n,$$

but it is often more important to know whether the convergence is *uniform* (see Exercises 2.6 and 2.7).

Example 3. We now look at some special cases of delta families of the nonnegative type (2.5). The first four examples are in R_1:

(a) $f(x) = \dfrac{1}{\pi(1+x^2)},$ $\qquad \left\{ f_y(x) = \dfrac{y}{\pi(y^2 + x^2)} \right\}$

(delta family as $y \to 0+$),

(b) $f(x) = \dfrac{e^{-x^2/4}}{\sqrt{4\pi}},$ $\qquad \left\{ f_t(x) = \dfrac{e^{-x^2/4t}}{\sqrt{4\pi t}} \right\}$

(delta family as $t \to 0+$; we have set $\alpha^2 = t$),

(c) $f(t) = H(t)\dfrac{e^{-1/4t}}{\sqrt{4\pi}\ t^{3/2}},$ $\qquad \left\{ f_x(t) = \dfrac{xe^{-x^2/4t}}{\sqrt{4\pi}\ t^{3/2}} \right\}$

(delta family as $x \to 0+$; we have set $\alpha^2 = x$),

(d) $f(x) = \dfrac{\sin^2 x}{\pi x^2},$ $\qquad \left\{ f_R(x) = \dfrac{\sin^2 Rx}{\pi Rx^2} \right\}$

(delta family as $R \to \infty$).

Case (a) occurs in potential theory; see Exercise 4.2, Chapter 1. Cases (b) and (c) both occur for the time-dependent equation of heat conduction (see Chapter 8). Case (d) is the Féjer kernel as it occurs in the theory of Fourier integrals.

Let us look at some cases in R_n:

(e) $f(x) = \psi(x)$, the test function (1.30) renamed ψ; $f_\alpha(x) = \psi(x/\alpha)$ is a delta family as $\alpha \to 0+$. Thus

$$\lim_{\alpha \to 0+} \int_{R_n} \psi\left(\frac{x}{\alpha}\right)\phi(x)\,dx = \phi(0) \qquad \text{for every } \phi \text{ in } C_0^\infty(R_n).$$

Actually we showed much more in Exercise 1.1, where the present ϕ was denoted by u and was merely assumed continuous.

(f) $f(x)$ depends only on $r=|x|$. The requirement $\int_R f(x)\,dx = 1$ becomes $\int_0^\infty r^{n-1} f(x)\,dr = 1/S_n$, where S_n is the area of the unit sphere. Case (e) was of this type; another one is the Gauss kernel

$$f(x) = (4\pi)^{-n/2} e^{-r^2/4},$$

$$f_t(x) = (4\pi t)^{-n/2} e^{-r^2/4t} \qquad \text{(delta family as } t \to 0+\text{)}.$$

This family occurs in n-dimensional heat conduction and in probability.

(g) In R_3 take $f(x) = 1/\pi^2(r^2+1)^2$; then $f_\alpha(x) = \alpha/\pi^2(r^2+\alpha^2)^2$ is a delta family as $\alpha \to 0+$.

(h) In R_2 take $f(x) = 1/2\pi(r^2+1)^{3/2}$; then $f_\alpha(x) = \alpha/2\pi(r^2+\alpha^2)^{3/2}$ is a delta family as $\alpha \to 0+$.

Both (g) and (h) have applications in potential theory.

Example 4. We now give some examples of delta families in R_1 that are not of type (2.5).

(i) The Dirichlet kernel for Fourier integrals,

$$\frac{1}{2\pi} \int_{-R}^{R} e^{i\omega x}\,d\omega = \frac{\sin Rx}{\pi x},$$

is a delta family as $R \to \infty$. Note that we can write $(\sin Rx)/\pi x = Rf(Rx)$, where $f(x) = (\sin x)/\pi x$, $\int_{-\infty}^{\infty} f(x)\,dx = 1$; but $f(x)$ is not $\geqslant 0$.

(j) Let $I_k = \int_{-1}^{1}(1-x^2)^k\,dx$. The sequence

$$f_k(x) = \begin{cases} (1-x^2)^k/I_k, & |x| < 1, \\ 0, & |x| \geqslant 1, \end{cases}$$

is a delta sequence as $k \to \infty$ and is useful in proving the Weierstrass approximation theorem.

(k) The family

$$f_r(\theta) = \begin{cases} \dfrac{1}{2\pi}\left(\dfrac{1-r^2}{1+r^2-2r\cos\theta}\right), & |\theta| \leqslant \pi, \\ 0, & |\theta| > \pi, \end{cases}$$

is a delta family as $r \to 1-$. It is the so-called Poisson kernel that comes up in potential theory (Dirichlet problem for the unit disk).

(1)   The sequence

$$f_k(x) = \begin{cases} \displaystyle\sum_{m=-k}^{k} \frac{1}{2\pi} e^{imx} = \frac{\sin\left(k+\frac{1}{2}\right)x}{2\pi\sin\frac{1}{2}x}, & |x| \leqslant \pi, \\ 0, & |x| > \pi, \end{cases}$$

is a delta sequence as $k \to \infty$. This sequence is the Dirichlet kernel for Fourier series and plays a central role in proving convergence theorems for Fourier series.

Example 5.   Consider the family $g_t(x) = -x(e^{-x^2/4t}/4\sqrt{\pi}\ t^{3/2})$, which is the x derivative of case (b), Example 3 (see also Figure 2.1). In view of the fact that convergent families of distributions can be differentiated term-wise, it follows that

$$\lim_{t \to 0+} \int_{-\infty}^{\infty} g_t(x)\phi(x)\,dx = -\phi'(0) \qquad \text{for each } \phi \in C_0^\infty(R_1).$$

If we examine $g_t(x)$ under the rather dim light of pointwise convergence, we see that $\lim_{t \to 0} g_t(x) = 0$ for every x in $-\infty < x < \infty$. Thus the very large bumps in $g_t(x)$ do not show under pointwise convergence as $t \to 0+$. It is

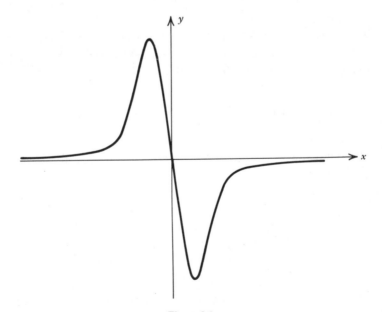

Figure 2.1

also worth noting that the family

$$f_t(x) + g_t(x) = \frac{e^{-x^2/4t}}{2\sqrt{\pi t}}\left(1 - \frac{x}{2t}\right)$$

satisfies all conditions on a delta family (Exercise 2.4) except nonnegativity, yet it is far from a delta family since

$$\lim_{t\to 0+} \int_{-\infty}^{\infty} [f_t(x) + g_t(x)]\phi(x)\,dx = \phi(0) - \phi'(0).$$

Example 6. Differentiation with respect to a parameter. Let $f_\alpha(x)$ be an n-dimensional distribution depending on the continuous parameter α. With α fixed, $[f_{\alpha+h}(x) - f_\alpha(x)]/h$ is a distribution depending on the parameter h. The limit as $h \to 0$ (*if it exists*) gives us the distribution $df_\alpha/d\alpha$:

(2.7)
$$\left\langle \frac{df_\alpha}{d\alpha}, \phi \right\rangle = \lim_{h\to 0} \frac{1}{h}\left[\langle f_{\alpha+h}, \phi \rangle - \langle f_\alpha, \phi \rangle\right].$$

Whereas derivatives of f_α with respect to the action variables $x_1, \ldots, x_n$ always exist, $df_\alpha/d\alpha$ may fail to exist. For instance, if $f_\alpha(x) = |\alpha| f(x)$, where $f(x)$ is a distribution, then $f_\alpha(x)$ is a distribution whose derivative with respect to the parameter α does not exist at $\alpha = 0$.

(a)   As our first illustration of differentiation with respect to a parameter, let $f(x)$ be an n-dimensional distribution, and let $\xi = (\xi_1, \ldots, \xi_n)$ be a vector parameter. For each fixed ξ the distribution $f(x - \xi)$ is defined from (1.7) as

$$\langle f(x - \xi), \phi(x) \rangle = \langle f(x), \phi(x + \xi) \rangle,$$

so that $f(x - \xi)$ can be regarded as a distribution depending on the parameters $\xi_1, \ldots, \xi_n$. We can now calculate the derivative of this distribution with respect to ξ_i:

$$\left\langle \frac{\partial}{\partial \xi_i} f(x - \xi), \phi(x) \right\rangle = \left\langle f(x), \frac{\partial}{\partial \xi_i}\phi(x + \xi) \right\rangle = \left\langle f(x), \frac{\partial}{\partial x_i}\phi(x + \xi) \right\rangle$$

$$= -\left\langle \frac{\partial}{\partial x_i} f(x), \phi(x + \xi) \right\rangle = -\left\langle \frac{\partial}{\partial x_i} f(x - \xi), \phi(x) \right\rangle.$$

Thus, as expected,

(2.8)
$$\frac{\partial}{\partial \xi_i} f(x - \xi) = -\frac{\partial}{\partial x_i} f(x - \xi),$$

and, in particular,

$$(2.9) \qquad \frac{\partial}{\partial \xi_i} \delta(x - \xi) = - \frac{\partial}{\partial x_i} \delta(x - \xi),$$

so that $(\partial/\partial \xi_i)\delta(x - \xi)$ is the volume density arising from a unit dipole at ξ with axis in the positive x_i direction.

(b)   Next, let Γ be a surface in R_3 carrying a normally oriented dipole layer of *surface density* $b(x)$. The surface element dS_ξ at ξ carries a dipole whose equivalent volume source density can be expressed by (2.9) as

$$b(\xi) \, dS_\xi \frac{\partial}{\partial n_\xi} \delta(x - \xi)$$

with action

$$b(\xi) \frac{\partial \phi}{\partial n}(\xi) \, dS_\xi.$$

The action of the whole layer is therefore given by

$$(2.10) \qquad \int_\Gamma b(\xi) \frac{\partial \phi}{\partial n}(\xi) \, dS_\xi,$$

which we interpret as the value of a functional on $C_0^\infty(R_3)$ corresponding to a *dipole* (or *double*) *layer* of density $b(x)$ on Γ.

(c)   If t is regarded as a parameter, the function $f(x,t) = \frac{1}{2}H(t - |x|)$ defines a one-dimensional distribution. Here we have written $f(x, t)$ instead of $f_t(x)$. Clearly, for $t > 0$,

$$(2.11) \qquad \langle f(x,t), \phi(x) \rangle = \frac{1}{2} \int_{-t}^{t} \phi(x) \, dx,$$

whereas, for $t \leq 0$, f is the zero distribution. Let us calculate $\partial f/\partial t$ and $\partial^2 f/\partial t^2$ for $t > 0$:

$$(2.12) \qquad \begin{aligned} \left\langle \frac{\partial f}{\partial t}, \phi(x) \right\rangle &= \tfrac{1}{2}\phi(t) + \tfrac{1}{2}\phi(-t), \\ \left\langle \frac{\partial^2 f}{\partial t^2}, \phi(x) \right\rangle &= \tfrac{1}{2}\phi'(t) - \tfrac{1}{2}\phi'(-t). \end{aligned}$$

Since we also have

$$\left\langle \frac{\partial^2 f}{\partial x^2}, \phi \right\rangle = \left\langle f, \frac{\partial^2 \phi}{\partial x^2} \right\rangle = \frac{1}{2} \int_{-t}^{t} \frac{\partial^2 \phi}{\partial x^2} \, dx = \tfrac{1}{2}\phi'(t) - \tfrac{1}{2}\phi'(-t),$$

we can say that f satisfies the wave equation

(2.13)
$$\frac{\partial^2 f}{\partial t^2} - \frac{\partial^2 f}{\partial x^2} = 0 \qquad \text{for } t > 0$$

with the initial conditions, obtained from (2.11) and (2.12),

(2.14)
$$\lim_{t \to 0+} f = 0, \qquad \lim_{t \to 0+} \frac{\partial f}{\partial t} = \delta(x).$$

In (2.13) the distributions are one-dimensional, $\partial^2 f / \partial t^2$ representing differentiation with respect to a parameter. If we gave x, t equal status in $f(x, t)$ and treated f as a two-dimensional distribution, we would still find that (2.13) is satisfied for $t > 0$ (and of course for $t < 0$), but the equation satisfied for $-\infty < t < \infty$ is then $\partial^2 f / \partial t^2 - \partial^2 f / \partial x^2 = \delta(x, t)$. (See Exercise 5.4.)

Other interesting examples can be found in the books by Korevaar and by Gelfand and Shilov.

Exercises

2.1. Let $V_n(r)$ and $S_n(r)$ denote, respectively, the volume and surface area of the n-dimensional ball of radius r. To obtain a formula for $V_n(r)$ we first relate it to the volume of an $(n-1)$-dimensional ball as follows. Let z denote a coordinate on an axis through the origin. If A is a constant smaller than r, the intersection of the hyperplane $z = A$ and the n-dimensional ball is an $(n-1)$-dimensional ball of radius $(r^2 - z^2)^{1/2}$. (The reader should draw appropriate figures for both the three-dimensional and two-dimensional cases.) Thus

$$V_n(r) = \int_{-r}^{r} V_{n-1}\left(\sqrt{r^2 - z^2}\right) dz.$$

By induction, show that $V_n(r) = C_n r^n$, where

$$C_{n+1} = 2 C_n \int_0^1 (1 - u^2)^{n/2} du.$$

Show that

$$\int_0^1 (1 - u^2)^{n/2} du = \frac{\sqrt{\pi}}{2} \frac{(n/2)!}{[(n+1)/2]!}.$$

and hence

(2.15)
$$V_n(r) = \frac{\pi^{n/2} r^n}{(n/2)!},$$

where $z!$ is defined as $\Gamma(z+1)$.

By considering the volume of a thin spherical shell, show that

(2.16)
$$S_n(r) = \frac{n\pi^{n/2} r^{n-1}}{(n/2)!} = \frac{2\pi^{n/2} r^{n-1}}{[(n/2)-1]!}.$$

Using $\frac{1}{2}! = \sqrt{\pi}/2$, check that (2.15) and (2.16) are correct for $n=2$ and $n=3$.

2.2. Show that cases (a) through (d) in Example 3 are in fact delta families as indicated.

2.3. Show that the Gauss kernel of Example 3(f) and the kernels of Examples 3(g) and 3(h) all generate delta families.

2.4. Let $\{f_\alpha(x)\}$ be a family of *nonnegative* locally integrable functions on R_n. Make the following assumptions:

(a) For some $A>0$, $\lim_{\alpha\to\alpha_0}\int_{|x|<A} f_\alpha(x)\,dx = 1$.

(b) For every $A>0$, $f_\alpha(x)\to 0$ as $\alpha\to\alpha_0$ uniformly on $|x|\geqslant A$.

It follows immediately that (a) holds for every $A>0$.

Show that, if $\phi(x)$ is any function continuous at $x=0$ and satisfying $\int_{R_n}|\phi(x)|\,dx < \infty$, then

$$\lim_{\alpha\to\alpha_0}\int_{R_n} f_\alpha(x)\phi(x)\,dx = \phi(0).$$

This certainly proves that $\{f_\alpha\}$ is a delta family as $\alpha\to\alpha_0$, but of course also proves a great deal more, since ϕ is not restricted to be a test function.

Hint: Write

$$\int_{R_n} f_\alpha\phi\,dx - \phi(0) = \int_{|x|<B} f_\alpha[\phi-\phi(0)]\,dx$$

$$+ \phi(0)\left[\int_{|x|<B} f_\alpha\,dx - 1\right] + \int_{|x|>B} f_\alpha\phi\,dx,$$

and first choose B so that $\max_{|x| \leqslant B} |\phi(x) - \phi(0)|$ is small. Then pick α close enough to α_0 so that the magnitude of the three terms on the right is small.

2.5. (a) A sequence $\{r_k(x)\}$ of locally integrable functions in R_1 will be said to be an *H sequence* if

(2.17)

$$\lim_{k \to \infty} \int_{-\infty}^{\infty} r_k(x)\phi(x)\,dx = \int_0^{\infty} \phi(x)\,dx \qquad \text{for each } \phi \text{ in } C_0^{\infty}(R_1).$$

Thus r_k tends distributionally to the Heaviside distribution. A sufficient condition for $\{r_k(x)\}$ to be an H sequence is that $\lim_{k \to \infty} r_k(x) = H(x)$, *pointwise* for $x \neq 0$, *and* $|r_k(x)| \leqslant C$, where C is a constant independent of k and x. (This sufficient condition follows directly from the Lebesgue convergence theorem, Chapter 0.) If $\{r_k\}$ satisfies this sufficient condition, show that (2.17) holds for any ϕ such that $\int_{-\infty}^{\infty} |\phi(x)|\,dx < \infty$.

(b) Prove that $r_k(x) = (1/\pi)\int_{-\infty}^{x} y^{-1} \sin ky\,dy$ is an H sequence. *Hint*: Show that $|\int_a^b v^{-1} \sin v\,dv| \leqslant \int_{-\pi}^{\pi} v^{-1} \sin v\,dv$.

(c) Show that

(2.18)

$$\lim_{k \to \infty} \int_{-\infty}^{\infty} \frac{\sin kx}{\pi x}\phi(x)\,dx = \phi(0)$$

for any ϕ which is differentiable with $|\phi'|$ integrable from $-\infty$ to ∞ and $\lim_{|x| \to \infty} \phi = 0$. The sequence $\{(\sin kx)/\pi x)\}$ is typical of *oscillatory* delta sequences.

(d) Show that (2.18) still holds if ϕ is differentiable in $|x| < A$ and vanishes for $|x| > A$. Note that ϕ is allowed to be discontinuous at $|x| = A$. (See also Section 4.)

2.6. Let $\{f_\alpha(x)\}$ be a delta family of the type of Exercise 2.4. Show that if in addition to the conditions already imposed there on ϕ we also require that $\phi(x)$ be bounded and *uniformly continuous* on R_n, then

$$\lim_{\alpha \to \alpha_0} \int_{R_n} f_\alpha(x - \xi)\phi(\xi)\,d\xi = \lim_{\alpha \to \alpha_0} \int_{R_n} \phi(x - \xi)f_\alpha(\xi)\,d\xi = \phi(x)$$

uniformly on Ω.

The proof follows that of Exercise 2.4. Write

$$\left| \int_{R_n} \phi(x-\xi) f_\alpha(\xi)\, d\xi - \phi(x) \right| \leqslant \left| \int_{|\xi|<B} f_\alpha(\xi) \left[\phi(x-\xi) - \phi(x) \right] d\xi \right|$$

$$+ \left| \phi(x) \left(\int_{|\xi|<B} f_\alpha(\xi)\, d\xi - 1 \right) \right|$$

$$+ \left| \int_{|\xi|>B} f_\alpha(\xi) \phi(x-\xi)\, d\xi \right|.$$

Since $\phi(x)$ is uniformly continuous on R_n, we can choose B so small that $\max_{|\xi| \leqslant B} |\phi(x-\xi) - \phi(x)| < \varepsilon$ for all x in R_n. After that it is easy to estimate the remaining terms to obtain the desired result.

2.7.   *The Weierstrass approximation theorem.* Consider the sequence $\{f_k(x)\}$ of Example 4(j).

(a) Show from Exercise 2.1 that $I_k = 2^{2k+1}(k!)^2/(2k+1)!$; by using Stirling's formula, $k! \sim k^k e^{-k} \sqrt{2\pi k}$, valid for large k, show that $\lim_{k\to\infty} \sqrt{k}\, I_k = \sqrt{\pi}$.

(b) Show that the sequence $\{f_k(x)\}$ satisfies the conditions of Exercise 2.4. Then let $\phi(x)$ be a continuous function on R_1 which vanishes for $|x| > \frac{1}{2}$ [this implies $\phi(\frac{1}{2}) = \phi(-\frac{1}{2}) = 0$ and the uniform continuity of $\phi(x)$ in $-\infty < x < \infty$]. Show that $\int_{-\infty}^{\infty} f_n (x - \xi)\phi(\xi)\, d\xi$ reduces to a *polynomial* in x when $-\frac{1}{2} \leqslant x \leqslant \frac{1}{2}$. In view of Exercise 2.6 we have therefore approximated $\phi(x)$ uniformly by polynomials on $-\frac{1}{2} \leqslant x \leqslant \frac{1}{2}$.

(c) By a suitable change of variable show that any function $\phi(x)$ continuous on the finite interval $a \leqslant x \leqslant b$ can be uniformly approximated by polynomials over this interval.

(d) Consider the function $|x|$ on $-1 \leqslant x \leqslant 1$. According to (c), we can approximate this function as closely as we desire by polynomials, yet we cannot expand the function in a power series. Explain the apparent paradox.

2.8.   Let $f(x)$ be a nonnegative function such that $\int_0^\infty f(x)\, dx = A$. Show that, if $\phi(x)$ is an arbitrary, bounded, locally integrable function for which $\phi(0+)$ exists, then

(2.19) $$\lim_{\alpha \to 0+} \int_0^\infty \frac{1}{\alpha} f\left(\frac{x}{\alpha}\right) \phi(x)\, dx = A\phi(0+).$$

2.9. (a) By setting $e^{i\theta} = z$, transform the real integral

$$J = \int_{-\pi}^{\pi} \frac{1-r^2}{1+r^2-2r\cos\theta} \, d\theta$$

to a counterclockwise integral on the boundary of the unit circle in the complex z plane, and obtain

$$J = \frac{i(1-r^2)}{r} \int \frac{dz}{(z-r)(z-1/r)}.$$

If $0 < r < 1$, there is only the simple pole at $z = r$ in the interior of the circle, so that

$$J = 2\pi.$$

For $r > 1$ show that

$$J = -2\pi.$$

(b) Show that, for $0 \leqslant r < 1$,

$$\frac{1-r^2}{1+r^2-2r\cos\theta} = 2\left(\tfrac{1}{2} + r\cos\theta + r^2\cos 2\theta + \cdots\right),$$

and hence derive the first result of part (a) by term-by-term integration.

(c) Show that case (k) in Example 4 is a delta family by using the result of Exercise 2.4.

2.10. Let $\{f_k(x)\}$ be the Dirichlet kernel of case (l) of Example 4. By using (2.18), show that

$$(2.20) \qquad \lim_{k\to\infty} \int_{-\pi}^{\pi} \frac{\sin\left(k+\tfrac{1}{2}\right)x}{2\pi\sin\tfrac{1}{2}x} \phi(x)\, dx = \phi(0)$$

if ϕ is differentiable, $-\pi < x < \pi$. For a stronger theorem, see Section 3.

2.11. In R_1 show that, distributionally,

$$\lim_{R\to\infty} \frac{1-\cos Rx}{x} = \mathrm{pf}\,\frac{1}{x}.$$

Hint: Since $(1 - \cos Rx)/x$ is locally integrable, we can write

$$\int_{-\infty}^{\infty} \frac{1 - \cos Rx}{x} \phi(x)\,dx = \lim_{\varepsilon \to 0} \int_{|x| > \varepsilon} \frac{1 - \cos Rx}{x} \phi(x)\,dx$$

$$= \left\langle \operatorname{pf}\frac{1}{x}, \varphi \right\rangle - \lim_{\varepsilon \to 0} \int_{|x| > \varepsilon} \frac{\cos Rx}{x} \phi(x)\,dx.$$

Show that as $\varepsilon \to 0$ the limiting value of this last integral is

$$\int_{-A}^{A} \cos Rx \left[\frac{\phi(x) - \phi(0)}{x} \right] dx,$$

where $(-A, A)$ contains the support of ϕ. The Riemann-Lebesgue lemma then puts the finishing touch on the problem.

Alternatively, as pointed out by R. P. Boas, one can write

$$\log|x| = \lim_{R \to \infty} \int_{1}^{|x|} \frac{1 - \cos Rt}{t}\,dt,$$

differentiate this equation with respect to x, and use (1.27).

2.12. It is fortuitous that we were able to calculate the exact value of the integral of the Poisson kernel. All that is necessary in the proof of the delta behavior of the family are the properties of Exercise 2.4. Let us now try to establish these properties for the Poisson kernel without reference to the exact value of the integral.

Hint: Given $\varepsilon > 0$, choose δ such that

$$1 - (1+\varepsilon)\frac{\theta^2}{2} \leqslant \cos\theta \leqslant 1 - (1-\varepsilon)\frac{\theta^2}{2}, \qquad |\theta| \leqslant \delta.$$

In this way obtain bounds for $\int_{-\delta}^{\delta} P(r,\theta)\,d\theta$, where $P(r,\theta)$ is the Poisson kernel. Now choose r sufficiently near 1 so that $\int_{-\delta}^{\delta} P(r,\theta)\,d\theta$ is close to 1 and $\int_{|\theta| \geqslant \delta} P(r,\theta)\,d\theta$ is close to 0.

2.13. *Direct product of distributions.* Let $f_1(x_1)$ and $f_2(x_2)$ be locally integrable functions of one variable. Then $f_1(x_1)f_2(x_2)$ is locally integrable in the plane and defines a two-dimensional distribution whose action can be expressed in the equivalent forms

(2.21) $\langle f_1 f_2, \phi(x_1, x_2) \rangle = \langle f_1(x_1), \psi(x_1) \rangle = \langle f_2(x_2), \chi(x_2) \rangle,$

where

$$(2.22) \quad \psi(x_1) \doteq \langle f_2(x_2), \phi(x_1, x_2) \rangle, \qquad \chi(x_2) \doteq \langle f_1(x_1), \phi(x_1, x_2) \rangle.$$

Clearly $\psi(x_1)$ and $\chi(x_2)$ belong to $C_0^\infty(R_1)$, so that the second and third terms of (2.21) are well-defined actions of one-dimensional distributions.

If f_1 and f_2 are arbitrary one-dimensional distributions, we use (2.21) and (2.22) to define the *direct product* $f_1 f_2$, which is a two-dimensional distribution since the variables are different (no attempt is made to define the product of two arbitrary distributions in the same variable). Obviously, the same idea can be used to define an n-dimensional distribution as the product of n one-dimensional distributions in n different variables.

(a) Show that the n-dimensional Dirac distribution $\delta(x) = \delta(x_1, \ldots, x_n)$ can be written as the direct product of n one-dimensional distributions: $\delta(x_1)\delta(x_2)\ldots\delta(x_n)$.

(b) Define $\delta(x_1)f(x_2, x_3)$ as the direct product of a one-dimensional Dirac distribution with a two-dimensional distribution, and show that the resulting three-dimensional distribution corresponds to a simple layer of sources of surface density $f(x_2, x_3)$ spread on the plane $x_1 = 0$ in R_3.

(c) Show that $f(x_1)\delta(x_2)\delta(x_3)$ represents a line source along the x_1 axis in R_3 with line density $f(x_1)$.

2.14. *Coordinate transformations.* A locally integrable function $f(x) = f(x_1, \ldots, x_n)$ defines a distribution through $\langle f, \phi \rangle = \int_{R_n} f(x)\phi(x)\,dx$. It is sometimes useful to express this distribution in terms of a different coordinate system. Proceeding in a purely formal manner, we let $u_1, \ldots, u_n$ be new coordinates that can be obtained from $x_1, \ldots, x_n$ by the transformation law $u = u(x)$, that is, $u_1 = u_1(x_1, \ldots, x_n), \ldots, u_n = u_n(x_1, \ldots, x_n)$. If the Jacobian J of x with respect to u is *positive* everywhere,

$$\langle f, \phi \rangle = \int_{R_n} f(x)\phi(x)\,dx = \int_{u\text{-space}} f(x(u))\phi(x(u))J\,du$$

(2.23)

$$= \int_{u\text{-space}} \tilde{f}(u)J\tilde{\phi}(u)\,du,$$

where

$$\tilde{\phi}(u) \doteq \phi(x(u)), \qquad \tilde{f}(u) \doteq f(x(u)).$$

Equation (2.23) can be used to interpret coordinate changes for arbitrary distributions. For instance, consider $f = \delta(x - x') = \delta(x_1 - x'_1) \cdots \delta(x_n - x'_n)$, then $\langle f, \phi \rangle = \phi(x') = \tilde{\phi}(u')$, where $u' = u(x')$. Thus, for (2.23) to be consistent, we need

$$\tilde{f}(u)J = \delta(u_1 - u'_1) \cdots \delta(u_n - u'_n)$$

or, since

$$\tilde{f}(u) = \delta[x(u) - x'],$$

we find that

(2.24) $$\delta(x - x') = \frac{\delta(u_1 - u'_1) \cdots \delta(u_n - u'_n)}{J}.$$

If we know only that $J \geqslant 0$, (2.24) can still be used if J does not vanish at u', since only the value of J at u' affects the right side of (2.24).

(a) Show that, if u_1, u_2, u_3 are the spherical coordinates in R_3 usually denoted by r, θ, ϕ, respectively, then $J = r^2 \sin \theta$ and

(2.25)
$$\delta(x - x') = \frac{1}{r^2 \sin \theta} \delta(r - r') \delta(\theta - \theta') \delta(\phi - \phi'),$$

$$r' \neq 0, \quad \theta' \neq 0, \pi.$$

If $r' = 0$ or $\theta' = 0$ or π, the coordinate transformation is singular and J vanishes. For instance, if the source is on the positive x_3 axis ($\theta' = 0, r' > 0$), (2.25) must be modified. One method of attacking the problem is to replace the point source by a uniform ring of sources, of unit total strength, located on the intersection of the sphere $r = r'$ and the cone $\theta = \theta'$, and then to let $\theta' \to 0$. For $\theta' > 0$ the intersection is a circle traced as ϕ ranges from 0 to 2π, say. Since the portion of the ring between ϕ' and $\phi' + \Delta\phi'$ carries a source of strength $\Delta\phi'/2\pi$, its distributional representation is given by (2.25) multiplied by $\Delta\phi'/2\pi$. The representation of the whole ring is then obtained by integrating from $\phi' = 0$ to $\phi' = 2\pi$ with the result

$$\frac{\delta(r - r')\delta(\theta - \theta')}{2\pi r^2 \sin \theta}.$$

As $\theta' \to 0$, the configuration tends to a unit point source at $r = r'$, $\theta = 0$, whose distributional representation is therefore

$$\frac{\delta(r-r')\,\delta(\theta)}{2\pi r^2 \sin\theta}.$$

(b) For the spherical coordinates of part (a), show that a point source at the origin can be represented as

$$\frac{\delta(r)}{4\pi r^2}$$

by considering the limit as $r' \to 0$ of a uniform simple layer of unit strength on the sphere $r = r'$.

(c) For cylindrical coordinates ρ, θ, z, show that

$$(2.26) \quad \delta(x-x') = \frac{\delta(\rho-\rho')\,\delta(\theta-\theta')\,\delta(z-z')}{\rho}, \qquad \rho' > 0.$$

Show that a uniform ring of sources of unit strength on $\rho = \rho', z = 0$ has the representation

$$(2.27) \qquad\qquad \frac{\delta(z)\,\delta(\rho-\rho')}{2\pi\rho},$$

and that for a unit source at the origin

$$(2.28) \qquad\qquad \delta(x) = \frac{\delta(z)\,\delta(\rho)}{2\pi\rho}.$$

Expressions (2.26) and (2.28) without the z factor remain valid for two-dimensional Dirac distributions in plane polar coordinates.

3. FOURIER SERIES

Review of Classical Results

The term "Fourier series" is used in Chapter 4 to describe an expansion in a general orthonormal set of functions, but in the present section the term is reserved for the special case of expansion in *trigonometric* functions.

Let $f(x)$ be a complex-valued function on the real line $-\infty < x < \infty$, and let f have *period* 2π, that is, $f(x+2\pi)=f(x)$ for all x. If $f(x)$ is given only on some 2π interval, say $-\pi \leqslant x < \pi$, we can always extend f periodically for all x by the rule $f(x+2\pi)=f(x)$. Of course this extension procedure may introduce additional discontinuities into the function; for instance, if we start with $f(x)=x$ in $-\pi \leqslant x < \pi$, the periodic extension will have jump discontinuities at odd integral multiples of π. Instead of looking at a periodic f on $(-\infty, \infty)$ it is sometimes more helpful to view the interval $(-\pi, \pi)$ as wrapped around the unit circle S (whose circumference has length 2π, of course), and in this way the endpoints $-\pi$, π are identified as a single point on the unit circle; for the function x on $-\pi \leqslant x < \pi$, this construction gives a function on S with a single point of discontinuity.

The class of functions f which we are trying to expand will always be a subset of $L_1(-\pi, \pi)$, the set of functions absolutely integrable on $(-\pi, \pi)$, that is,

$$\int_{-\pi}^{\pi} |f(x)|\,dx < \infty.$$

Usually we shall require $f(x)$ to belong to $L_2(-\pi, \pi)$, the set of square integrable functions. This means that

$$\int_{-\pi}^{\pi} |f^2(x)|\,dx < \infty,$$

and it can be seen that $L_2(-\pi, \pi)$ is a subset of $L_1(-\pi, \pi)$. For certain theorems we will also demand some continuity or smoothness for $f(x)$.

A *trigonometric* series (or, more properly, a trigonometric series in complex-exponential form) is of the type

(3.1)
$$\sum_{k=-\infty}^{\infty} c_k e^{ikx},$$

where the coefficients c_k are complex numbers. The sequence $\{s_n(x)\}$ of partial sums of (3.1) is defined symmetrically by

(3.2)
$$s_n(x) = \sum_{k=-n}^{n} c_k e^{ikx},$$

where the index n runs from 0 to infinity.

A series of the form

(3.3)
$$\frac{a_0}{2} + \sum_{k=1}^{\infty} (a_k \cos kx + b_k \sin kx)$$

is completely equivalent to (3.1) if the coefficients $\{a_k\}$ and $\{b_k\}$ are related to the $\{c_k\}$ by the formulas

(3.4) $a_0 = 2c_0;$ $a_k = c_k + c_{-k}$ and $b_k = i(c_k - c_{-k})$ for $k \geqslant 1$.

If (3.4) holds, it is easy to see that the nth partial sum $(a_0/2) + \sum_{k=1}^{n}(a_k \cos kx + b_k \sin kx)$ of (3.3) is exactly equal to $s_n(x)$ as given by (3.2). For most of our purposes it is more convenient to deal with (3.1) than with (3.3).

All the complex exponentials in (3.1) have period 2π, so that the series, if it converges, will also have period 2π. Therefore the relation between (3.1) and any 2π-periodic function it purports to represent needs to be studied only in the fundamental interval $(-\pi, \pi)$ or on the unit circle.

A set $\{e_k(x)\}$ of complex-valued functions is said to be orthonormal over (a, b) if

(3.5) $$\int_a^b e_k(x)\bar{e}_j(x)\,dx = \begin{cases} 0, & k \neq j, \\ 1, & k = j, \end{cases}$$

where the overbar denotes complex conjugation. The set

(3.6) $$\left\{ e_k(x) \doteq \frac{e^{ikx}}{\sqrt{2\pi}} \right\},$$

where the index k ranges over all integers from $-\infty$ to ∞, is an orthonormal set over any interval of length 2π.

Suppose now that the series (3.1) converges uniformly on $-\pi \leqslant x \leqslant \pi$, and let us call the sum $f(x)$. We multiply the equality $f(x) = \sum_{k=-\infty}^{\infty} c_k e^{ikx}$ by e^{-imx} and integrate from $-\pi$ to π. The multiplication by e^{-imx} does not affect the uniform convergence, so that we can perform the integration term by term to obtain

(3.7) $$c_m = \frac{1}{2\pi} \int_{-\pi}^{\pi} f(x) e^{-imx}\,dx, \quad m = 0, \pm 1, \pm 2, \ldots.$$

Thus, if the series (3.1) converges uniformly on $-\pi \leqslant x \leqslant \pi$ to a function f (which is necessarily continuous, since a uniformly convergent series of continuous functions is continuous), then the coefficients in (3.1) are given in terms of f by (3.7). If the form (3.3) is used instead, the coefficients are

(3.8)
$$a_m = \frac{1}{\pi} \int_{-\pi}^{\pi} f(x) \cos mx\,dx, \quad m = 0, 1, 2, \ldots;$$
$$b_m = \frac{1}{\pi} \int_{-\pi}^{\pi} f(x) \sin mx\,dx, \quad m = 1, 2, \ldots.$$

Note that $c_0 (= a_0/2)$ is the average value of f over a period.

Let us now reverse the process and look at the possible expansion of a given function $f(x)$ on $(-\pi, \pi)$ in the series (3.1). Formula (3.7) still defines coefficients c_m as long as f is in $L_1(-\pi, \pi)$; these coefficients are the *Fourier coefficients* of f, and the corresponding series (3.1) is the *Fourier series* of f. *We no longer know that the series converges to f.* For what functions f does the Fourier series represent f, and in what sense? Obviously, if f_1 and f_2 are equal almost everywhere, they define the same $\{c_m\}$ and hence the same series, so that we cannot in general expect pointwise convergence. Probably the simplest results, and in many ways the most useful, are set in the framework of the Hilbert space $L_2(-\pi, \pi)$. The main theorem (see Chapter 4) states that for each f in $L_2(-\pi, \pi)$ the Fourier series defined by (3.1) and (3.7) always converges to f in the L_2 sense, that is,

$$(3.9) \qquad \lim_{n \to \infty} \int_{-\pi}^{\pi} \left| f(x) - \sum_{k=-n}^{n} c_k e^{ikx} \right|^2 dx = 0.$$

We shall not use result (3.9) here. Instead we shall discuss some aspects of pointwise and uniform convergence of Fourier series. We first note the following lemma.

Lemma. Let $f \in L_2(-\pi, \pi)$, and let $c_k = (1/2\pi)\int_{-\pi}^{\pi} f(x)e^{-ikx} dx$; then

$$(3.10) \qquad \lim_{|k| \to \infty} c_k = 0,$$

$$(3.11) \qquad \sum_{k=-\infty}^{\infty} |c_k|^2 \quad \text{converges},$$

and

$$(3.12) \qquad \sum_{k=-\infty}^{\infty} |c_k|^2 \leqslant \frac{1}{2\pi} \int_{-\pi}^{\pi} |f(x)|^2 dx \qquad (\textit{Bessel's inequality}).$$

Proof. We have

$$0 \leqslant \int_{-\pi}^{\pi} \left| f - \sum_{k=-n}^{n} c_k e^{ikx} \right|^2 dx = \int_{-\pi}^{\pi} \left(f - \sum_{k=-n}^{n} c_k e^{ikx} \right) \overline{\left(f - \sum_{k=-n}^{n} c_k e^{ikx} \right)} dx$$

$$= \int_{-\pi}^{\pi} |f|^2 dx - 2\pi \sum_{k=-n}^{n} |c_k|^2,$$

so that, for every n,

$$\sum_{k=-n}^{n} |c_k|^2 \leqslant \frac{1}{2\pi} \int_{-\pi}^{\pi} |f|^2 dx.$$

If we let $\{t_n\}$ be the sequence of symmetric partial sums

$$t_n = \sum_{k=-n}^{n} |c_k|^2,$$

then t_n is monotonically increasing with n and is bounded above by the constant $(1/2\pi)\int_{-\pi}^{\pi} |f|^2 dx$. Thus t_n converges, so that (3.11) and Bessel's inequality obviously hold. It follows that $|c_k|^2$ tends to 0 as $|k| \to \infty$, implying (3.10).

Remarks

1.   Result (3.10) holds even if $f \subset L_1(-\pi, \pi)$ and is then known as the *Riemann-Lebesgue lemma*. Suppose $f \in L_1$; then for any $\varepsilon > 0$ we can write $f = g + h$, where g is bounded and $\int_{-\pi}^{\pi} |h(x)| dx < \varepsilon$. It follows that

$$\left| \frac{1}{2\pi} \int_{-\pi}^{\pi} h(x) e^{-ikx} dx \right| < \frac{\varepsilon}{2\pi} \qquad \text{for all } k,$$

and, by (3.10) applied to g, $(1/2\pi)\int_{-\pi}^{\pi} g e^{-ikx} dx \to 0$. Therefore $(1/2\pi)\int_{-\pi}^{\pi} f e^{-ikx} dx \to 0$, as required. An easy consequence is that for $f \in L_1(a, b)$, where the interval may be finite or infinite,

$$(3.13) \qquad \int_a^b f(x) e^{-ikx} dx, \qquad \int_a^b f(x) \cos kx \, dx, \qquad \int_a^b f(x) \sin kx \, dx,$$

all tend to 0 as $k \to \infty$ (k is not restricted to integral values).

2.   Inequality (3.12) is actually an equality known as Parseval's identity, which holds for all $f \in L_2(-\pi, \pi)$. We shall give a proof in this section for smooth functions and the general proof in Chapter 4. However, (3.11) and (3.12) do not necessarily hold for L_1 functions.

We shall now try to improve on (3.10) for the special classes of functions defined below. It turns out that, the smoother f is (as a function on the unit circle S), the faster the Fourier coefficients vanish.

A function f belongs to $C^k(S)$ if it is continuous and has continuous derivatives of orders up to and including k on the unit circle. [If we think of f as a function initially defined on a 2π interval on the real axis, we are requiring that its periodic extension have continuous derivatives of orders up to and including k on the real line. Thus $C^0(S) = C(S)$ is the class of continuous 2π-periodic functions.]

A function f belongs to $D(S)$ (obeys *Dirichlet conditions*) if it is piecewise continuous and if there exists a finite subdivision of S such that f has a continuous derivative within each subinterval and such that the left- and right-hand derivatives exist at the end of the subintervals.

The class D is both worse and better than C: D includes some discontinuous but rather simple functions; D does not include continuous functions which are nowhere differentiable.

Theorem 1. If f belongs to $C^p(S)$, its Fourier coefficients $\{c_m\}$ satisfy

$$\lim_{m \to \infty} c_m m^p = 0.$$

Proof. Integration by parts p times gives

$$c_m = \frac{1}{2\pi} \int_{-\pi}^{\pi} f(x) e^{-imx} \, dx = \frac{1}{2\pi(im)^p} \int_{-\pi}^{\pi} f^{(p)}(x) e^{-imx} \, dx,$$

the integrated terms dropping out because of periodicity. Since $f^{(p)}(x)$ is continuous, its Fourier coefficients tend to zero as $m \to \infty$. Therefore $c_m m^p \to 0$ as $m \to \infty$.

Theorem 2. If $f \in C^{p-1}$ and $f^{(p)} \in D$, the Fourier coefficients $\{c_m\}$ of f satisfy

$$|c_m m^{p+1}| \leq M, \qquad \text{for all } m.$$

Proof. Suppose that $p = 0$ (that is, f belongs to D), and that f has a simple jump at only one point, say ξ. Then

$$c_m = \frac{1}{2\pi} \int_{-\pi}^{\xi} f e^{-imx} \, dx + \frac{1}{2\pi} \int_{\xi}^{\pi} f e^{-imx} \, dx,$$

where in each interval integration by parts is permitted since f is differentiable. It follows that

$$2\pi i m c_m = \left(\int_{-\pi}^{\xi} f' e^{-imx} \, dx + \int_{\xi}^{\pi} f' e^{-imx} \, dx \right) + e^{-im\xi} [f(\xi+) - f(\xi-)].$$

The sum of integrals, being proportional to the mth Fourier coefficient of an L_2 function, certainly tends to 0 as $m \to \infty$. The other term on the right side has nonzero modulus independently of m. Thus mc_m remains bounded as $m \to \infty$ while, for $\varepsilon > 0$, $m^{1+\varepsilon} c_m$ becomes unbounded as $m \to \infty$.

The proof for $p > 0$ proceeds along similar lines. We have thus established a relation between the smoothness properties of f and the behavior of its Fourier coefficients for large index.

Example 1. Let $f(x) = x$ in $-\pi \leqslant x < \pi$ and $f(x + 2\pi) = f(x)$. It is clear from (3.8) that an odd function such as this has $a_m = 0$ and $b_m = (2/\pi)\int_0^\pi f(x) \sin mx \, dx$. In our case we find $b_m = (-1)^{m+1} 2/m$, which has the behavior for large m expected of a discontinuous function in D.

Example 2. Let $f(x) = |x|$ in $-\pi \leqslant x < \pi$ and $f(x + 2\pi) = f(x)$. This is an even function, so that $b_m = 0$ and $a_m = (2/\pi)\int_0^\pi f(x) \cos mx \, dx$. We find $a_0 = \pi, a_m = 0$ for m even, $a_m = -(4/\pi m^2)$ for m odd. The coefficients behave like $1/m^2$ for large m as befits a differentiable function whose discontinuous derivative is in D.

Example 3. For a given real α let $f(x) = \cos \alpha x$, $-\pi \leqslant x < \pi$, and $f(x + 2\pi) = f(x)$. Note that the 2π-periodic function constructed in this way is *not* the same as $\cos \alpha x$ on the whole real line (unless α is an integer). If α is not an integer, a straightforward calculation gives

$$a_m = (-1)^{m+1} \frac{2\alpha}{\pi} \frac{\sin \alpha \pi}{m^2 - \alpha^2}, \qquad m = 0, 1, 2, \ldots,$$

so that $m^2 a_m$ is bounded as $m \to \infty$. The function f is continuous but has a discontinuous derivative at odd multiples of π. If α is an integer, the calculation is even simpler: $a_m = 0$ except for $m = \alpha$, and $a_\alpha = 1$. Thus, as expected, the Fourier series of $\cos \alpha x$ is just the single term $\cos \alpha x$ *when α is an integer*. Note that in this case $f(x)$ is infinitely differentiable, so that $m^p a_m$ should tend to 0 with m for every p, a condition which is certainly satisfied here since the coefficients are identically 0 past the index α. There are, however, other 2π-periodic, infinitely differentiable functions whose Fourier coefficients do not vanish identically past a certain index. For instance, the temperature $u_t(x)$ at time t in an infinite rod whose initial temperature is the 2π-periodic function of Example 1 is given by (Chapter 8)

$$\sum_{m=1}^{\infty} \frac{2(-1)^{m+1}}{m} e^{-m^2 t} \sin mx, \qquad -\infty < x < \infty.$$

The coefficients are exponentially small as $m \to \infty$, so that they satisfy Theorem 1 for every p. This implies that $u_t(x)$ is an infinitely differentiable 2π-periodic function.

We now give a version of the principal theorem dealing with pointwise convergence of Fourier series.

Theorem 3. Let $f(x)$ belong to $L_1(S)$, and let x_0 be a fixed point on S at which $f(x_0+)$ and $f(x_0-)$ exist and at which the right- and left-hand derivatives of f also exist. Then the Fourier series (3.1)-(3.7) of f converges at that point to the value $[f(x_0+)+f(x_0-)]/2$.

Remarks

1. The local behavior of f at x_0 determines the convergence of the Fourier series at x_0, even though the coefficients (unlike Taylor series coefficients) are given by the global formulas (3.7).

2. If $f(x)\in D(S)$, the conditions of the theorem are met at every point x_0 on the circle (or, equivalently, at every point x on the real line for the periodic extension of f). In particular, if $f\in D(S)$ *and also* $f\in C(S)$, the Fourier series converges to f at every point. If $f\in D(S)$ and is discontinuous, we may as well redefine f at points of discontinuity by the rule $f(x)=[f(x+)+f(x-)]/2$, so that the Fourier series converges everywhere to this redefined f (of course redefining f at isolated points does not change its Fourier coefficients, since integrals are insensitive to the value of the integrand at isolated points).

Proof. Let $s_n(x_0)=\sum_{k=-n}^{n}c_k e^{ikx_0}$, where $c_k=(1/2\pi)\int_{-\pi}^{\pi}f(x)e^{-ikx}\,dx$. We must show that $\lim_{n\to\infty}s_n(x_0)=[f(x_0+)+f(x_0-)]/2$. Since rotations leave the circle invariant, it suffices to prove the theorem for $x_0=0$, that is,

$$(3.14) \qquad \lim_{n\to\infty} s_n(0)=\frac{f(0+)+f(0-)}{2}.$$

We note first that

$$\sum_{k=-n}^{n} e^{ikx}=\sum_{k=-n}^{n} e^{-ikx}=-1+\sum_{k=0}^{n} e^{ikx}+\sum_{k=0}^{n} e^{-ikx}.$$

Since each sum on the right side is a geometric series, we find, after some simplifications, that

$$(3.15) \qquad \frac{1}{2\pi}\sum_{k=-n}^{n} e^{ikx}=\frac{1}{2\pi}\sum_{k=-n}^{n} e^{-ikx}=\frac{\sin\left(n+\frac{1}{2}\right)x}{2\pi\sin(x/2)}\doteq D_n(x),$$

an immediate consequence of which is

$$(3.16) \qquad \int_{-\pi}^{\pi} D_n(x)\, dx = 1, \qquad \int_{0}^{\pi} D_n(x)\, dx = \tfrac{1}{2}.$$

The function $D_n(x)$ is known as the *Dirichlet kernel* and has period 2π. The value of D_n at $x=0$ (and therefore at all even multiples of 2π) is understood to be its limiting value as $x \to 0$, that is, $(2n+1)/2\pi$. This understanding is consistent with (3.15) at $x=0$. We have

$$s_n(0) = \sum_{k=-n}^{n} c_k = \sum_{k=-n}^{n} \frac{1}{2\pi} \int_{-\pi}^{\pi} f(x) e^{-ikx}\, dx = \int_{-\pi}^{\pi} D_n(x) f(x)\, dx.$$

To prove (3.14) we must essentially show that $D_n(x)$ is a delta sequence on $-\pi < x < \pi$. Let us split the interval $(-\pi, \pi)$ into $(-\pi, 0)$ and $(0, \pi)$. Looking at the latter, we find, on using (3.16), that

$$(3.17) \qquad \int_{0}^{\pi} D_n(x) f(x)\, dx = \frac{f(0+)}{2} + \int_{0}^{\pi} D_n(x) [\, f(x) - f(0+) \,]\, dx.$$

Writing

$$(3.18) \qquad \frac{f(x) - f(0+)}{\sin(x/2)} = \left[\frac{f(x) - f(0+)}{x} \right] \left[\frac{x}{\sin(x/2)} \right],$$

we observe that the first factor has a limit as $x \to 0+$ by the hypothesis that the right-hand derivative of f exists at $x=0$. It follows that $[f(x) - f(0+)]/x$ belongs to $L_1(0, \pi)$. The second factor, $x/\sin(x/2)$, is bounded on $(0, \pi)$, so that the product (3.18) is in $L_1(0, \pi)$. From the Riemann-Lebesgue lemma (3.13), we conclude that (3.17) tends to $f(0+)/2$ as $n \to \infty$. Similar considerations for the integral from $-\pi$ to 0 then yield (3.14) as required.

We can now prove uniform convergence of the Fourier series for smooth functions.

Theorem. If $f(x) \in C^1(S)$, its Fourier series converges uniformly to f.

Proof. Let $\{c_m\}$ be the Fourier coefficients of f, and $\{d_m\}$ those of f'. Starting with (3.7) for d_m and integrating by parts, we find that $d_m = imc_m$, or $c_m = (1/im)d_m$ for $m \neq 0$. Using the Schwartz inequality for sequences, we obtain

$$\sum_{m=-\infty}^{\infty} |c_m| = |c_0| + \sum_{m\neq0} \frac{|d_m|}{|m|} \leqslant |c_0| + \left(\sum_{m\neq0} \frac{1}{m^2}\right)^{1/2} \left(\sum_{m\neq0} |d_m|^2\right)^{1/2}.$$

Since f' is continuous, it is in L_2, so that $\sum |d_m|^2 < \infty$ and, of course $\sum (1/m^2) < \infty$. Thus $\sum |c_m| < \infty$, and $\sum c_m e^{imx}$ converges uniformly (by the Weierstrass M-test) to a function, which, according to Theorem 3, must be f.

Corollary. If $f(x) \in C^1(S)$, its Fourier series converges in the L_2 sense to f.

Proof. We have

$$(3.19) \qquad \int_{-\pi}^{\pi} |f|^2 dx - 2\pi \sum_{k=-m}^{m} |c_k|^2 = \int_{-\pi}^{\pi} \left| f(x) - \sum_{k=-m}^{m} c_k e^{ikx} \right|^2 dx,$$

and in view of the uniform convergence of the Fourier series to f, we can pick M so large that

$$\left| f - \sum_{k=-m}^{m} c_k e^{ikx} \right| < \epsilon, \quad m > M.$$

Hence the right side of (3.19) tends to 0 as m approaches infinity, and Parseval's identity holds for $f \in C^1(S)$.

It is now relatively easy to prove that Parseval's identity holds for any L_2 function f by approximating it in the L_2 sense by a function g in $C^1(S)$. Thus, if $f \in L_2(-\pi, \pi)$, we have

$$(3.20) \qquad \sum_{m=-\infty}^{\infty} |c_m|^2 = \frac{1}{2\pi} \int_{-\pi}^{\pi} |f|^2 dx \qquad \text{(Parseval's identity)},$$

where $c_m = (1/2\pi) \int_{-\pi}^{\pi} f(x) e^{-imx} dx$. The proof is given in Chapter 4.

The Gibbs Phenomenon

The Fourier series of a discontinuous function cannot converge uniformly on an interval enclosing the discontinuity, since a uniformly convergent series of continuous functions is continuous. Let us examine more closely the details of the nonuniformity in the simple case of the function $f(x) =$

$\operatorname{sgn} x [= 2H(x) - 1]$ for $-\pi \leqslant x < \pi$ and $f(x + 2\pi) = f(x)$. Its Fourier series is a sine series, and we find that

$$b_n = 0, \quad n \text{ even}; \qquad b_n = \frac{4}{\pi n}, \quad n \text{ odd}.$$

Thus we can write the partial sums as

$$s_{2n+1}(x) = \frac{4}{\pi} \sum_{k=0}^{n} \frac{\sin(2k+1)x}{2k+1} = \frac{4}{\pi} \sum_{k=0}^{n} \int_0^x \cos(2k+1)y\, dy$$

$$= \frac{2}{\pi} \int_0^x \left[\sum_{k=0}^{n} e^{i(2k+1)y} + \sum_{k=0}^{n} e^{-i(2k+1)y} \right] dy$$

$$= \frac{2}{\pi} \int_0^x \left[e^{iy} \frac{1 - e^{i2(n+1)y}}{1 - e^{i2y}} + e^{-iy} \frac{1 - e^{-i2(n+1)y}}{1 - e^{-i2y}} \right] dy$$

$$= \frac{2}{\pi} \int_0^x \left[\frac{1 - e^{i2(n+1)y}}{-2i\sin y} + \frac{1 - e^{-i2(n+1)y}}{2i\sin y} \right] dy$$

$$= \frac{2}{\pi} \int_0^x \frac{\sin 2(n+1)y}{\sin y}\, dy$$

$$= \frac{2}{\pi} \int_0^x \frac{\sin 2(n+1)y}{y} \frac{y}{\sin y}\, dy.$$

We shall investigate the behavior near the jump at the origin. Let us choose $x > 0$ and small. Then $y/\sin y$ is close to 1 in the entire interval from 0 to x, and $s_{2n+1}(x)$ is closely approximated by

$$\frac{2}{\pi} \int_0^x \frac{\sin 2(n+1)y}{y}\, dy = \frac{2}{\pi} \int_0^{2(n+1)x} \frac{\sin u}{u}\, du \doteq Si[2(n+1)x].$$

We have already encountered $Si(z)$ in Exercise 2.5 and in Example 4(i) of Section 2. Since $Si(\infty) = 1$, it is easy to see that $Si(\pi)$ must exceed 1, with an actual value that can be estimated as 1.18. Thus, for large n, there is always a value of x_0 near 0 $[x_0 \doteq \pi/(2n+2)]$ where the partial sum s_{2n+1} takes on the value 1.18 (Figure 3.1). This does not contradict the fact that the Fourier series converges pointwise to 1 in $0 < x < \pi$, but it does show clearly that the convergence is not uniform in an interval including $x = 0$, either in its interior or as an endpoint. At the jump discontinuity the partial sums of high order *overshoot* the limiting function by 9% of the jump. This is known as the *Gibbs phenomenon* and holds for the Fourier series of any function in D with simple jump discontinuities; indeed any such function can be written as the sum of a continuous function in D and

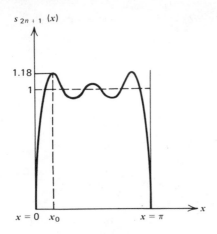

Figure 3.1

terms of the form $J_k H(x - x_k)$, where the x_k's are the points of discontinuity and the J_k's are the numerical values of the jumps. Since $2H(x - x_k) = \text{sgn}(x - x_k) + 1$, the analysis of this section applies almost directly to the terms $J_k H(x - x_k)$; on the other hand, the continuous function in D has Fourier coefficients of order $1/n^2$, and its Fourier series therefore converges uniformly. Thus the entire overshoot is due to the terms $J_k H(x - x_k)$.

In electrical engineering a typical low-pass filter can be regarded as a transformation which cuts off frequencies above a critical frequency ω_c. If the periodic square wave just studied is the input to a low-pass filter, the output will be a *partial sum* of the Fourier series. Any such partial sum exhibits an overshoot at the points where the square wave has jumps, so that the output to the filter looks like Figure 3.1. The phenomenon was once considered mystifying. The overshoot can be avoided by using the Cesaro partial sums (see Exercise 3.6).

Change of Scale

Suppose $f(x)$ is periodic with period T instead of 2π. We may think of f as defined on a circle whose circumference has length T instead of 2π. Since

$$g(x) \doteq f\left(\frac{2\pi}{T} x\right)$$

has period 2π, the previous results on 2π-periodic functions are easily modified. The Fourier series (in complex exponential form) associated with a T-periodic function $f(x)$ is

$$\sum_{k=-\infty}^{\infty} c_k e^{ik2\pi x/T},$$

where

$$c_k = \frac{1}{T} \int_{-T/2}^{T/2} f(x) e^{-ik2\pi x/T} dx = \frac{1}{T} \int_0^T f(x) e^{-ik2\pi x/T} dx.$$

In view of the T-periodicity, the integration can be over any interval of the form $(a, a+T)$.

Fourier Series as Distributions

We first extend slightly the notion of distributions to include generalizations of complex-valued functions (whereas in our earlier treatment we generalized only real-valued functions). We now understand by $C_0^\infty(R_1)$ the set of all *complex-valued*, infinitely differentiable functions $\phi(x)$ vanishing outside a bounded interval. A null sequence in $C_0^\infty(R_1)$ is defined just as in Section 1. A distribution f becomes a continuous, complex-valued, linear functional on $C_0^\infty(R_1)$; that is, to each test function ϕ in $C_0^\infty(R_1)$ there is assigned a complex number $\langle f, \phi \rangle$ such that

$$\langle f, \alpha_1 \phi_1 + \alpha_2 \phi_2 \rangle = \alpha_1 \langle f, \phi_1 \rangle + \alpha_2 \langle f, \phi_2 \rangle$$

for all *complex* numbers α_1, α_2 and all ϕ_1, ϕ_2 in $C_0^\infty(R_1)$, and

$$\lim_{n \to \infty} \langle f, \phi_n \rangle = 0 \qquad \text{whenever } \{\phi_n\} \text{ is a null sequence in } C_0^\infty(R_1).$$

Obviously any complex-valued function f that is locally integrable defines a distribution by (1.4). All our earlier work survives with only trivial modifications.

A Fourier series $\sum_{n=-\infty}^{\infty} c_n e^{inx}$ will certainly converge uniformly if $|c_n| \leqslant M/n^2$ for large $|n|$. It therefore also converges in the sense of distributions. How should we interpret $\sum_{n=-\infty}^{\infty} c_n e^{inx}$ if all we know is that, for large $|n|$, $|c_n| \leqslant M n^\alpha$ for some integer α? This class includes series that diverge badly in the pointwise sense. Consider the related series

$$\sum_{n \neq 0} (in)^{-\alpha-2} c_n e^{inx},$$

which clearly converges uniformly, and hence distributionally, to $g(x)$, say. Here the qualification $n \neq 0$ on the summation symbol means that the sum runs from $-\infty$ to ∞ *excluding the term* $n=0$. On taking the $(\alpha+2)$th distributional derivative of the equality

$$g(x) = \sum_{n \neq 0} (in)^{-\alpha-2} c_n e^{inx},$$

we find that

$$g^{(\alpha+2)}(x) = \sum_{n \neq 0} c_n e^{inx}$$

and

$$\sum_{n=-\infty}^{\infty} c_n e^{inx} = c_0 + g^{(\alpha+2)}(x),$$

where $g^{(\alpha+2)}$ is the $(\alpha+2)$th derivative of g in the sense of distributions.

Example 4. Consider the function $f(x) = x - \pi$ in $0 \leqslant x < 2\pi$, with $f(x+2\pi) = f(x)$ for other values of x. This function is in D and has the Fourier series

$$(3.21) \qquad f(x) = -2 \sum_{n=1}^{\infty} \frac{\sin nx}{n} = i \frac{\sum\limits_{n \neq 0} e^{inx}}{|n|}.$$

Series (3.21) does not converge uniformly to f but nevertheless converges distributionally to f. This can be seen by looking at the "integrated" series $2\sum_{n=1}^{\infty}(\cos nx)/n^2$, which converges uniformly, and hence distributionally, to a function $F(x)$ whose derivative is $f(x)$. We can therefore differentiate this integrated series termwise to recover (3.21) in the distributional sense. Another differentiation then gives

$$f'(x) = -2 \sum_{n=1}^{\infty} \cos nx,$$

where by (1.17) $f' = 1 - \sum_{k=-\infty}^{\infty} 2\pi \delta(x-2k\pi)$, so that

$$(3.22) \qquad \sum_{k=-\infty}^{\infty} \delta(x-2k\pi) = \frac{1}{2\pi} + \frac{1}{\pi} \sum_{n=1}^{\infty} \cos nx.$$

In a classical sense the series on the right does not converge, and the left side is totally meaningless. Within the theory of distributions, (3.22) means that the action of both sides on a test function ϕ in $C_0^\infty(R_1)$ is the same. Thus (3.22) states that, for every test function ϕ,

$$\sum_{k=-\infty}^{\infty} \phi(2k\pi) = \frac{1}{2\pi} \int_{-\infty}^{\infty} \phi(x)\,dx + \frac{1}{\pi} \sum_{n=1}^{\infty} \int_{-\infty}^{\infty} (\cos nx)\phi(x)\,dx.$$

The series on the left is really a finite sum, since ϕ has bounded support. It is less obvious that the series on the right converges, but this is guaranteed by the fact that (3.22) has been correctly derived within the theory of distributions.

We can rewrite (3.22) as

$$(3.23) \qquad \sum_{k=-\infty}^{\infty} \delta(x-2k\pi) = \frac{1}{2\pi} \sum_{n=-\infty}^{\infty} e^{inx} = \frac{1}{2\pi} \sum_{n=-\infty}^{\infty} e^{-inx}$$

or, replacing x by $x-\xi$, as

$$\sum_{k=-\infty}^{\infty} \delta(x-\xi-2k\pi) = \frac{1}{2\pi} \sum_{n=-\infty}^{\infty} e^{in(\xi-x)}$$

$$(3.24)$$

$$= \frac{1}{2\pi} + \frac{1}{\pi} \sum_{n=1}^{\infty} (\cos n\xi \cos nx + \sin n\xi \sin nx).$$

Thinking of the parameter ξ as fixed in $(-\pi,\pi)$, let us apply (3.24) to a test function whose support is in $(-\pi,\pi)$. Only the delta function with $k=0$ is effective, and we obtain

$$(3.25) \qquad \phi(\xi) = \frac{1}{2\pi} \sum_{n=-\infty}^{\infty} e^{in\xi} \int_{-\pi}^{\pi} e^{-inx}\phi(x)\,dx, \qquad -\pi<\xi<\pi,$$

which is the just the Fourier series expansion formula for ϕ. Of course in (3.25) ϕ has to be a test function; nevertheless it appears that (3.23) contains in a nutshell much of the information needed to generate Fourier series expansions. Equation (3.23) is often called a *completeness relation*.

Example 5. The Poisson summation formula. Let $\phi(x)$ be a complex-valued test function on R_1; if λ, t are real parameters, then $\phi(\lambda x/2\pi)e^{ixt/2\pi}$ is also a test function, to which we apply (3.23) to obtain

$$\sum_{k=-\infty}^{\infty} \phi(\lambda k)e^{ikt} = \frac{1}{2\pi} \sum_{n=-\infty}^{\infty} \int_{-\infty}^{\infty} e^{inx}\phi\left(\frac{\lambda x}{2\pi}\right)e^{ixt/2\pi}\,dx.$$

Setting $y = \lambda x / 2\pi$, we find that

$$(3.26) \qquad \sum_{k=-\infty}^{\infty} \phi(\lambda k) e^{ikt} = \frac{1}{|\lambda|} \sum_{n=-\infty}^{\infty} \phi^{\wedge} \left(\frac{t + 2\pi n}{\lambda} \right),$$

where

$$\phi^{\wedge}(\omega) = \int_{-\infty}^{\infty} e^{i\omega x} \phi(x) \, dx$$

is the *Fourier transform* of $\phi(x)$, to be discussed at greater length in the following section.

If we take successively $t = 0$ and then $\lambda = 1$ in (3.26), we obtain

$$(3.27) \qquad \sum_{k=-\infty}^{\infty} \phi(\lambda k) = \frac{1}{|\lambda|} \sum_{n=-\infty}^{\infty} \phi^{\wedge} \left(\frac{2\pi n}{\lambda} \right)$$

and

$$(3.28) \qquad \sum_{k=-\infty}^{\infty} \phi(k) = \sum_{n=-\infty}^{\infty} \phi^{\wedge}(2\pi n).$$

Identities (3.26), (3.27), and (3.28) are all given the name *Poisson's summation formula*. They can be regarded as transforming one series to another for the purpose of improving the rate of convergence or, in rare instances, carrying out the summation in closed form. Although our formulas have been derived only for test functions, they hold for a much larger class of functions: it suffices for ϕ to be a piecewise smooth function, vanishing at $|x| = \infty$, with $\int_{-\infty}^{\infty} |\phi| \, dx$ finite.

As an illustration of the use of (3.27), let $\phi(x) = e^{-x^2}$. An easy calculation gives $\phi^{\wedge}(\omega) = \sqrt{\pi} \, e^{-\omega^2/4}$ (see Example 3, Section 4). Thus we find that

$$(3.29) \qquad \sum_{k=-\infty}^{\infty} e^{-\lambda^2 k^2} = \frac{\sqrt{\pi}}{|\lambda|} \sum_{n=-\infty}^{\infty} e^{-n^2 \pi^2 / \lambda^2}.$$

By examining the ratio of successive terms for the two series in (3.29), we see that the series on the left converges rapidly for large $|\lambda|$, and the one on the right for small $|\lambda|$. A similar procedure is often possible when we want to improve the convergence of an infinite series. In fact, (3.29) expresses a duality between two forms of the solution of the heat equation with λ^2 proportional to the time [see (2.33) and (2.35), Chapter 8].

Exercises

3.1. Let

$$f(x) = x(\pi - x), \qquad 0 \leqslant x \leqslant \pi.$$

(a) Graph $f(x)$ in the interval $(0, \pi)$. Extend $f(x)$ as an odd function in $-\pi \leqslant x \leqslant 0$. What is the formula for this extended function in $(-\pi, 0)$? Extend this new f as a periodic function of period 2π.

(b) Expand the periodic f of (a) into a Fourier series.

(c) Justify the fact that the coefficients behave as $1/n^3$ for large n.

(d) Show that

$$\frac{\pi^3}{32} = 1 - \frac{1}{3^3} + \frac{1}{5^3} - \frac{1}{7^3} + \cdots.$$

(e) Write Parseval's formula for the present case.

3.2. We wish to approximate $\sin x$ in $-\pi \leqslant x \leqslant \pi$ by a polynomial $s_n(x) = a_0 + a_1 x + \cdots + a_n x^n$.

(a) Plot $s_1(x)$ and $s_3(x)$, when the Maclaurin expansion for $\sin x$ is used $[s_1(x) = x, s_3(x) = x - (x^3/6)]$.

(b) Find the $s_1(x)$ and $s_3(x)$ which give the best L_2 approximation. Plot $s_1(x)$ and $s_3(x)$.

(c) Find the $s_1(x)$ which minimizes the maximum absolute deviation between s_1 and $\sin x$ from $-\pi$ to π, that is, choose $s_1(x)$ so that

$$\max_{-\pi \leqslant x \leqslant \pi} |s_1(x) - \sin x|$$

is as small as possible.

3.3. Let $f(x)$ be an unknown, odd, continuous function with period 1. We observe the values $f_1, \ldots, f_N$ of f at the N points $x_k = k/2(N+1)$, $k = 1, 2, \ldots, N$. Show that there is one and only one interpolating function $g(x)$ of the form $\sum_{m=1}^{N} c_m \sin 2\pi m x$ whose value at x_k coincides with f_k, for $k = 1, \ldots, N$. The coefficients c_m are the ones obtained by numerical integration, that is,

(3.30)
$$c_m = \frac{2}{N+1} \sum_{k=1}^{N} f_k \sin 2\pi m x_k.$$

3.4. Use the Poisson formula (3.27) with $\lambda > 0$ and $\phi(x) = e^{-|x|}$ to show that

(3.31)
$$\sum_{k=-\infty}^{\infty} e^{-\lambda|x|} = \sum_{n=-\infty}^{\infty} \frac{2\lambda}{\lambda^2 + 4\pi^2 n^2}.$$

By noting that the left side is essentially a geometric series, obtain the identity

(3.32)
$$\sum_{n=-\infty}^{\infty} \frac{2\lambda}{\lambda^2 + 4\pi^2 n^2} = \frac{1 + e^{-\lambda}}{1 - e^{-\lambda}}.$$

3.5. (a) Consider a 2π-periodic function $f(x)$ whose definition on the fundamental interval $|x| < \pi$ is $f = 1$ for $|x| < \xi$ and $f = 0$ for $\xi < |x| < \pi$ (here ξ is fixed, $0 < \xi < \pi$). Expand $f(x)$ in a cosine series, and differentiate term by term to obtain

(3.33)
$$\frac{2}{\pi} \sum_{n=1}^{\infty} \sin n\xi \sin nx = \sum_{k=-\infty}^{\infty} \delta(x - \xi - 2\pi k) - \sum_{k=-\infty}^{\infty} \delta(x + \xi - 2\pi k).$$

Apply (3.33) to a test function $\phi(x)$ whose support lies in $(0, \pi)$ to derive the sine series expansion

$$\phi(\xi) = \frac{2}{\pi} \sum_{n=1}^{\infty} \sin n\xi \int_0^{\pi} \phi(x) \sin nx \, dx.$$

(b) Combine (3.33) and (3.24) to find

$$\frac{1}{\pi} + \frac{2}{\pi} \sum_{n=1}^{\infty} \cos n\xi \cos nx = \sum_{k=-\infty}^{\infty} \delta(x - \xi - 2\pi k) + \sum_{k=-\infty}^{\infty} \delta(x + \xi - 2\pi k).$$

3.6. Let $\{s_n(x)\}$ be the sequence of partial sums of the Fourier series for $f(x)$, and define σ_n as the mean of the first n partial sums:

$$\sigma_n(x) = \frac{s_0(x) + \cdots + s_{n-1}(x)}{n}.$$

Using (3.15), we can write

$$\sigma_n(0) = \int_{-\pi}^{\pi} \frac{D_0(x) + \cdots + D_{n-1}(x)}{n} f(x) \, dx \doteq \int_{-\pi}^{\pi} F_n(x) f(x) \, dx,$$

where $F_n(x)$ is the *Féjer kernel*. Show that

$$F_n(x) = \frac{1}{2\pi n} \frac{\sin^2(nx/2)}{\sin^2(x/2)},$$

and that $\{F_n(x)\}$ is a delta sequence. Prove that, for any $f(x)$ in $C(S)$, $\sigma_n(x)$ converges to $f(x)$. We say that the Fourier series is Cesaro summable to $f(x)$.

3.7. The function $f(x) = \log(1 - \cos x)$ has period 2π and belongs to $L_1(-\pi, \pi)$ despite its singularity at $x = 0$. Show that

$$\operatorname{Re}\log(1 - e^{ix}) = \log 2 + \log(1 - \cos x);$$

use the power series expansion of $\log(1 - z)$ to show that

$$(3.34) \quad \log(1 - \cos x) = -\log 2 - 2 \sum_{n=1}^{\infty} \frac{\cos nx}{n}, \qquad x \neq 2k\pi.$$

By differentiating term by term in the distributional sense, obtain a series for $\cot(x/2)$.

3.8. Let $f(x)$ and $g(x)$ have period 2π, and let their Fourier coefficients be $\{f_n\}$ and $\{g_n\}$, respectively. Express the Fourier coefficients of

$$h(x) \doteq \int_{-\pi}^{\pi} f(x - \xi)g(\xi)\,d\xi = \int_{-\pi}^{\pi} g(x - \xi)f(\xi)\,d\xi$$

in terms of f_n and g_n. The function $h(x)$ is known as the (Fourier series) *convolution* of f and g.

4. FOURIER TRANSFORMS AND INTEGRALS

Classical Results

The Fourier integral expansion is the analog for functions small at $|x| = \infty$ (hence nonperiodic) of the Fourier series expansion of periodic functions. Whereas a periodic function is expanded into harmonics of the same period, the nonperiodic function is a continuous superposition over all frequencies.

Let $f(x)$ be a function vanishing sufficiently fast as $|x| \to \infty$, and let $f_T(x)$ be the T-periodic function coinciding with f on $-T/2 \leqslant x < T/2$. Then, if

f_T obeys Dirichlet conditions, we can write, for all x,

$$f_T(x) = \sum_{k=-\infty}^{\infty} c_k e^{2\pi i k x/T}, \qquad c_k = \frac{1}{T} \int_{-T/2}^{T/2} f(\xi) e^{-2\pi i k \xi/T} \, d\xi,$$

so that, for $|x| < T/2$,

$$(4.1) \qquad f(x) = \frac{1}{2\pi} \sum_{k=-\infty}^{\infty} \left[\frac{2\pi}{T} \int_{-T/2}^{T/2} f(\xi) e^{i 2\pi k (x-\xi)/T} \, d\xi \right].$$

Setting $\omega_k = 2\pi k/T$, $\Delta\omega = 2\pi/T$, and $g(\omega, x, T) = \int_{-T/2}^{T/2} f(\xi) e^{i\omega(x-\xi)} \, d\xi$, we find that (4.1) becomes

$$(4.2) \qquad f(x) = \frac{1}{2\pi} \sum_{k=-\infty}^{\infty} \Delta\omega \, g(\omega_k, x, T), \qquad |x| < T/2,$$

which, for large T, can be regarded as a Riemann sum over a partition of spacing $2\pi/T$ of the whole ω axis. The sum in (4.2) is therefore essentially an integral from $-\infty$ to ∞ in ω. At the same time, since $T \to \infty$, the expression for g also becomes an integral from $-\infty$ to ∞ in ξ (assuming that $\int_{-\infty}^{\infty} f(\xi) e^{-i\omega\xi} \, d\xi$ exists). Replacing ω by $-\omega$, we are led to the following *Fourier integral expansion* for a function $f(x)$ in $L_1(-\infty, \infty)$:

$$(4.3) \qquad f(x) = \frac{1}{2\pi} \int_{-\infty}^{\infty} f^{\wedge}(\omega) e^{-i\omega x} \, d\omega,$$

where

$$(4.4) \qquad f^{\wedge}(\omega) \doteq \int_{-\infty}^{\infty} f(\xi) e^{i\omega\xi} \, d\xi \left[= \int_{-\infty}^{\infty} f(x) e^{i\omega x} \, dx \right].$$

Formula (4.4) defines $f^{\wedge}$ as the *Fourier transform of* f, and (4.3) shows how to reconstruct $f(x)$ from its continuous spectrum. It turns out that the integral in (4.3) has to be interpreted as

$$(4.5) \qquad \lim_{R \to \infty} \frac{1}{2\pi} \int_{-R}^{R} f^{\wedge}(\omega) e^{-i\omega x} \, d\omega.$$

[Some authors prefer to define the transform with $e^{-i\omega\xi}$ instead of $e^{i\omega\xi}$; this requires a change in the exponential in (4.3) to $e^{i\omega x}$. The factor $1/2\pi$ is sometimes apportioned differently between (4.3) and (4.4). In some ways the best solution is to use the exponential $e^{i2\pi\omega\xi}$ in definition (4.4); this then eliminates all factors of 2π except in the exponent, the drawback being that the differentiation formulas are slightly more cumbersome.]

To prove (4.5) we first note that $f^\wedge$ exists in view of the L_1 assumption on f. Substituting the definition of $f^\wedge$, we find that

$$\frac{1}{2\pi} \int_{-R}^{R} f^\wedge(\omega) e^{-i\omega x} d\omega = \frac{1}{2\pi} \int_{-R}^{R} d\omega \, e^{-i\omega x} \int_{-\infty}^{\infty} f(\xi) e^{i\omega\xi} d\xi$$

$$= \int_{-\infty}^{\infty} f(\xi) \frac{\sin R(\xi - x)}{\pi(\xi - x)} d\xi = \int_{-\infty}^{\infty} f(x+y) \frac{\sin Ry}{\pi y} dy.$$

To calculate the limit as $R \to \infty$ we recall from Exercise 2.5 that $(\sin Ry)/\pi y$ is a delta sequence, so that the limit is $f(x)$ for a fairly large class of functions. Alternatively, we can use a procedure similar to that for Theorem 3, Section 3, on Fourier series. We state the result as follows.

Fourier Integral Theorem. Let $f \in L_1(-\infty, \infty)$, and let f satisfy Dirichlet conditions; then $f^\wedge(\omega)$ exists, and the inversion formula (4.3) holds in the sense

$$(4.6) \qquad \frac{f(x+) + f(x-)}{2} = \lim_{R \to \infty} \frac{1}{2\pi} \int_{-R}^{R} f^\wedge(\omega) e^{-i\omega x} d\omega.$$

For further reference we mention two relations, both of which go under the name *Parseval's formula*:

$$(4.7) \qquad \int_{-\infty}^{\infty} |f(x)|^2 dx = \frac{1}{2\pi} \int_{-\infty}^{\infty} |f^\wedge(\omega)|^2 d\omega,$$

$$(4.8) \qquad \int_{-\infty}^{\infty} f^\wedge(x) g(x) dx = \int_{-\infty}^{\infty} f(x) g^\wedge(x) dx.$$

The Parseval formulas are easily derived from the definition of the Fourier transform by formally interchanging orders of integration. We shall not, for the present, state precise conditions for the validity of these formulas.

Band-Limited Functions

The interrelation between Fourier series and integrals is nicely illustrated by so-called band-limited functions. A function $\phi(x)$ defined on the real axis is *band limited* if its Fourier transform $\phi^\wedge(\omega)$ vanishes for $|\omega|$ larger than some critical frequency, that is, if $\phi^\wedge$ has bounded support. Suppose, for instance, that the support of $\phi^\wedge$ is contained in $|\omega| < 2\pi$. Then only the term corresponding to $n = 0$ remains on the right side of (3.28), which becomes

$$(4.9) \qquad \sum_{k=-\infty}^{\infty} \phi(k) = \phi^\wedge(0) = \int_{-\infty}^{\infty} \phi(x) dx,$$

where the last equality stems from the definition of the Fourier transform. Note that the left side of (4.9) is just the trapezoidal approximation, with unit subdivision size, of the integral on the right side. Thus this trapezoidal approximation is perfect! We can also interpret the right side of (4.9) as the continuous analog of the series on the left (see Boas and Pollard).

An important property of band-limited functions is described by the *sampling formula*. Set $\lambda = 1$ in (3.26), multiply both sides by e^{-ixt}, and integrate from $t = -\pi$ to π to obtain

$$\sum_{k=-\infty}^{\infty} \phi(k) \frac{2\sin(x-k)\pi}{x-k} = \sum_{n=-\infty}^{\infty} \int_{-\pi}^{\pi} e^{-ixt}\phi^{\wedge}(t+2\pi n)\,dt,$$

where term-by-term integration is permitted.

If $\phi^{\wedge}$ vanishes for $|\omega| \geqslant \pi$, only one term remains on the right side: $\int_{-\pi}^{\pi} e^{-ixt}\phi^{\wedge}(t)\,dt$, which by the inversion formula (4.3) is just $2\pi\phi(x)$. We conclude that

$$(4.10) \qquad \phi(x) = \sum_{k=-\infty}^{\infty} \phi(k) \frac{\sin\pi(x-k)}{\pi(x-k)}, \qquad -\infty < x < \infty,$$

so that a band-limited function is completely determined by "sampling" at integral values.

Heisenberg's Uncertainty Principle

In its original quantum-mechanical setting the principle gives a quantitative bound on the possible precision of simultaneous measurements of the position and momentum of a particle. Mathematically the principle reflects a relationship between a function and its Fourier transform. It will be convenient to use the language of electrical engineering and of probability.

Let $f(t)$ be a complex-valued time signal of unit energy: $\int_{-\infty}^{\infty} |f(t)|^2 dt = 1$. The energy spectrum of the signal is given by $|f^{\wedge}(\omega)|^2/2\pi$. By Parseval's identity (4.7) the total energy in the signal and that in the spectrum are equal, so that

$$\int_{-\infty}^{\infty} |f(t)|^2 dt = \frac{1}{2\pi} \int_{-\infty}^{\infty} |f^{\wedge}(\omega)|^2 d\omega = 1.$$

Both $|f(t)|^2$ and $(1/2\pi)|f^{\wedge}(\omega)|^2$ can be regarded as densities of probability distribution on their respective axes. A translation in either $f(t)$ or $f^{\wedge}(\omega)$ merely leads to a phase shift in the other, not affecting its energy distribution. We may therefore assume that the mean of each distribution is 0. The variances of these distributions are then given by

$$\sigma_t^2 \doteq \int_{-\infty}^{\infty} t^2 |f(t)|^2 \, dt, \qquad \sigma_\omega^2 \doteq \frac{1}{2\pi} \int_{-\infty}^{\infty} \omega^2 |f^\wedge(\omega)|^2 \, d\omega,$$

where σ_t is a measure of the duration of the time signal, and σ_ω a measure of the frequency spread.

Theorem.   If $|t|^{1/2}f$ and $|t|^{1/2}f'$ both tend to 0 as $t \to \infty$, then

(4.11) $\sigma_t \sigma_\omega \geqslant \frac{1}{2},$

and the equality holds if and only if f is the Gaussian signal $Ce^{-\alpha t^2}$, where α is an arbitrary positive number and C is chosen so that its energy is unity.

Proof.   For simplicity we shall assume that $f(t)$ is *real*; the reader is invited to supply the appropriate changes for f complex. Since the Fourier transform of $f'(t)$ is $-i\omega f^\wedge(\omega)$, Parseval's identity gives

$$\frac{1}{2\pi} \int_{-\infty}^{\infty} \omega^2 |f^\wedge(\omega)|^2 \, d\omega = \int_{-\infty}^{\infty} [f'(t)]^2 \, dt.$$

On the other hand, we infer from the Schwarz inequality [(5.14), Chapter 4] that

$$\left(\int_{-\infty}^{\infty} \frac{1}{2} t \frac{d}{dt} f^2 \, dt \right)^2 = \left(\int_{-\infty}^{\infty} tff' \, dt \right)^2 \leqslant \int_{-\infty}^{\infty} t^2 f^2 \, dt \int_{-\infty}^{\infty} f'^2 \, dt = \sigma_t^2 \sigma_\omega^2,$$

with equality only if f' is a multiple of tf. Integrating the left side by parts and using the condition on f at infinity, we obtain

$$\sigma_t^2 \sigma_\omega^2 \geqslant \frac{1}{4} \left(\int_{-\infty}^{\infty} f^2 \, dt \right)^2 = \frac{1}{4},$$

with equality only if $f' = cft$, that is, only if f is the predicted Gaussian signal.

Extensions and further applications can be found in Dym and McKean and in Papoulis.

Fourier Transform in the Complex Plane

Equations (4.4) and (4.3) can be generalized to some extent by allowing the transform variable ω to take on complex values. Writing $\omega = u + iv$, where u and v are real, we obtain

$$(4.12) \quad f^\wedge(\omega) = f^\wedge(u + iv) = \int_{-\infty}^{\infty} e^{i\omega x} f(x)\, dx = \int_{-\infty}^{\infty} e^{iux} e^{-vx} f(x)\, dx.$$

Thus $f^\wedge(\omega)$ is just the "old" Fourier transform (4.4) with transform variable u of the function $e^{-vx} f(x)$. If v is chosen so that $e^{-vx} f(x)$ is in $L_1(-\infty, \infty)$, we deduce from (4.5) that

$$e^{-vx} f(x) = \lim_{R \to \infty} \frac{1}{2\pi} \int_{-R}^{R} f^\wedge(u + iv) e^{-iux}\, du.$$

The last integral can be interpreted as an integral in the complex ω plane along a line parallel to the real axis. In fact, we have

$$f(x) = \lim_{R \to \infty} \frac{1}{2\pi} \int_{iv-R}^{iv+R} f^\wedge(\omega) e^{-i\omega x}\, d\omega,$$

or, by a slight abuse of notation,

$$(4.13) \qquad\qquad f(x) = \frac{1}{2\pi} \int_{iv-\infty}^{iv+\infty} f^\wedge(\omega) e^{-i\omega x}\, d\omega,$$

where v is any real number for which

$$(4.14) \qquad\qquad \int_{-\infty}^{\infty} |e^{-vx} f(x)|\, dx < \infty.$$

One should observe that the factor e^{-vx} which occurs in (4.14) is not necessarily helpful; if $v > 0$, the factor improves convergence at the upper limit but impairs it at the lower limit, and conversely for $v < 0$. Even in the simple case $f(x) = 1$, there is no value of v for which (4.14) holds, so that our approach will have to be modified. On the other hand, there are cases where inequality (4.14) is satisfied in an entire strip $v_1 < v < v_2$; it can then be shown that $f^\wedge(\omega)$ is an analytic function in that strip.

Example 1. $f(x) = e^{-|x|}$. Then $f(x) e^{-vx}$ is in $L_1(-\infty, \infty)$ for all v such that $-1 < v < 1$. We find that

$$f^\wedge(\omega) = \int_{-\infty}^{0} e^{i\omega x} e^x\, dx + \int_{0}^{\infty} e^{i\omega x} e^{-x}\, dx$$

$$= \frac{1}{1 + i\omega} + \frac{1}{1 - i\omega} = \frac{2}{1 + \omega^2}, \qquad -1 < v < 1.$$

We now illustrate how (4.13), with $v=0$, can be used to recover $f(x)$ from $f\hat{\,}(\omega)$ by the method of contour integration. For $x>0$, $2e^{-i\omega x}/(1+\omega^2)$ is exponentially small in the lower half of the ω plane ($v<0$). Consider the contour C_R, consisting of the entire boundary of a large semicircle of radius R. in the lower half-plane (with the diameter on the real axis between $-R$ and R). Then, by Cauchy's theorem,

$$\lim_{R\to\infty} \int_{C_R} \frac{1}{2\pi} \frac{2}{1+\omega^2} e^{-i\omega x}\, d\omega = 2\pi i r,$$

where r is the sum of the residues of $e^{-i\omega x}/\pi(1+\omega^2)$ within C_R. The only singularity is a simple pole at $\omega=-i$, and the corresponding residue is $ie^{-x}/2\pi$. Now, as $R\to\infty$, the contribution from the curved portion of C_R disappears because of the behavior of the integrand. Taking into account the contribution from the diameter, we find that

$$\frac{1}{2\pi} \int_{-\infty}^{\infty} \frac{2}{1+\omega^2} e^{-i\omega x}\, d\omega = e^{-x}, \qquad x>0.$$

Similarly, by considering a semicircle in the upper half-plane, one can easily show that the value of the inversion integral is e^x for $x<0$.

Example 2. $f(x)=-2H(x)\sinh x$, where $H(x)$ is the Heaviside function. Thus

$$f(x)=\begin{cases} -2\sinh x, & x>0, \\ 0, & x<0. \end{cases}$$

Since $2\sinh x = e^x - e^{-x}$, we see that $e^{-vx}\sinh x$ is in $L_1(0,\infty)$ for $v>1$, and therefore $e^{-vx}f(x)$ is in $L_1(-\infty,\infty)$ for $v>1$. A simple calculation yields

$$f\hat{\,}(\omega) = \int_0^{\infty} (e^{-x}-e^x)e^{i\omega x}\, dx = \frac{2}{1+\omega^2}, \qquad v>1.$$

When compared with the result of Example 1, this may at first seem mystifying. Do two different functions have the same Fourier transforms? The answer is that transforms of different functions may have the same functional form valid in *different, nonoverlapping* regions of the ω plane. We can still use the inversion formula (4.13), this time with $v>1$, to recover the original $f(x)$ of our present example. First, for $x<0$ we use a semicircle in the region $v>1$ to find $f=0$ for $x<0$. Then, for $x>0$ we use a lower semicircle, which includes both poles of the integrand (at $\omega=+i$ and $\omega=-i$), and obtain $f=-2\sinh x$.

Example 3. $f(x) = e^{-x^2}$. Then fe^{-vx} is in $L_1(-\infty, \infty)$ for all real v, so that $f^\wedge(\omega)$ is analytic in the whole ω plane. We find for $\omega = iv$, v real, that

$$f^\wedge(iv) = \int_{-\infty}^{\infty} e^{-vx} e^{-x^2} dx = e^{v^2/4} \int_{-\infty}^{\infty} e^{-(x+v/2)^2} dx = \sqrt{\pi} \, e^{v^2/4},$$

so that, by analytic continuation,

$$f^\wedge(\omega) = \sqrt{\pi} \, e^{-\omega^2/4}.$$

Example 4. Let θ be a given positive constant; consider

$$f_\theta(x) = \begin{cases} \dfrac{1}{2\theta}, & |x| < \theta \\ 0, & |x| > \theta. \end{cases}$$

Clearly $f_\theta e^{-vx}$ is in $L_1(-\infty, \infty)$ for all real v, so that $f_\theta^\wedge(\omega)$ will be analytic in the entire ω plane. We have

$$f_\theta^\wedge(\omega) = \frac{1}{2\pi} \int_{-\theta}^{\theta} e^{i\omega x} dx = \frac{\sin \omega \theta}{\omega \theta},$$

which remains correct at $\omega = 0$ with the obvious interpretation of the right side. The function so defined is, in fact, analytic in the entire ω plane, including $\omega = 0$. The inversion formula can be used on the real axis of the ω plane, so that

$$f_\theta(x) = \frac{1}{2\pi} \int_{-\infty}^{\infty} \frac{\sin u\theta}{u\theta} e^{-iux} du.$$

Without attention to rigor, let us examine the limit as $\theta \to 0$. Then $f_\theta(x) \to \delta(x)$ and $(\sin u\theta)/u\theta \to 1$. Thus we surmise that in some appropriate sense

$$\delta^\wedge = 1, \qquad 1^\wedge = 2\pi\delta.$$

These formulas will be shown to hold rigorously in the distributional sense.

One-Sided Functions

A function which vanishes for $x < 0$ is said to be *right-sided* (or causal) and will be denoted by $f_+(x)$; a function which vanishes for $x > 0$ is said to be *left-sided* and will be denoted by $f_-(x)$.

Consider a right-sided function f_+ which is $0(e^{\alpha x})$ at $x = +\infty$, that is, such that there exists a constant C with the property

$$|f_+(x)| < Ce^{\alpha x} \qquad \text{for } x \text{ sufficiently large.}$$

Then $f_+(x)e^{-vx}$ is in $L_1(-\infty,\infty)$ for $v>\alpha$, and therefore the Fourier transform $\hat{f}_+(\omega)$ of $f_+(x)$ is analytic in the upper half-plane $v>\alpha$. Formulas (4.12) and (4.13) then become

$$(4.15) \qquad \hat{f}_+(\omega)=\int_0^\infty f_+(x)e^{i\omega x}\,dx, \qquad v>\alpha$$

and

$$(4.16) \qquad f_+(x)=\frac{1}{2\pi}\int_{iv-\infty}^{iv+\infty}\hat{f}_+(\omega)e^{-i\omega x}\,d\omega, \qquad v>\alpha,$$

respectively. In particular, this implies that the integral in (4.16) vanishes identically for $x<0$.

For a left-sided function f_-, which is $0(e^{\beta x})$ at $x=-\infty$, $f_-(x)e^{-vx}$ belongs to $L_1(-\infty,\infty)$ for $v<\beta$, and the Fourier transform $\hat{f}_-(\omega)$ of $f_-(x)$ is analytic in the lower half-plane $v<\beta$. We therefore have the transform relations

$$(4.17) \qquad \hat{f}_-(\omega)=\int_{-\infty}^0 f_-(x)e^{i\omega x}\,dx, \qquad v<\beta$$

and

$$(4.18) \qquad f_-(x)=\frac{1}{2\pi}\int_{iv-\infty}^{iv+\infty}\hat{f}_-(\omega)e^{-i\omega x}\,d\omega, \qquad v<\beta.$$

The integral in (4.18) vanishes identically for $x>0$.

Now, if $f(x)$ is an arbitrary function on the real line, we can write

$$f(x)=f_+(x)+f_-(x),$$

where

$$f_+(x)=\begin{cases} f(x), & x>0, \\ 0, & x<0, \end{cases} \qquad f_-(x)=\begin{cases} 0, & x>0, \\ f(x), & x<0. \end{cases}$$

If $f(x)$ is $0(e^{\alpha x})$ at $x=+\infty$ and $0(e^{\beta x})$ at $x=-\infty$, we have, by combining our previous results,

$$(4.19) \qquad \hat{f}_+(\omega)=\int_0^\infty f(x)e^{i\omega x}\,dx, \qquad v>\alpha,$$

$$(4.20) \qquad \hat{f}_-(\omega)=\int_{-\infty}^0 f(x)e^{i\omega x}\,dx, \qquad v<\beta,$$

$$(4.21) \qquad f(x)=\frac{1}{2\pi}\int_{ia-\infty}^{ia+\infty}\hat{f}_+(\omega)e^{-i\omega x}\,d\omega+\frac{1}{2\pi}\int_{ib-\infty}^{ib+\infty}\hat{f}_-(\omega)e^{-i\omega x}\,d\omega,$$

where $a>\alpha$ and $b<\beta$. These formulas provide a useful generalization of (4.12) and (4.13). If it happens that $\beta>\alpha$, the Fourier transform $f\hat{}(\omega)$ exists for $\alpha<v<\beta$, and we can choose $a=b$ in this strip to reduce (4.21) to (4.13).

Functions of Slow Growth

Definition. A function $f(x)$ on the real line is said to be of *slow growth* if the following conditions hold:

1. f is locally integrable, that is, $\int_I |f(x)|\,dx$ is finite for each bounded interval I.

2. There exist constants C, n, and R such that

$$|f(x)|<C|x|^n \qquad \text{for } |x|>R.$$

Thus a function of slow growth is one which grows at infinity more slowly than some polynomial. Of course a function $f(x)$ of slow growth does not usually have a Fourier transform in the sense of (4.4) or (4.12); but $f_+(x)e^{-vx}$ is in $L_1(-\infty,\infty)$ for each $v>0$, and $f_-(x)e^{-vx}$ is in $L_1(-\infty,\infty)$ for each $v<0$. Hence we can use (4.21) for any $a>0$ and $b<0$, and therefore

$$(4.22) \quad f(x)=\frac{1}{2\pi}\lim_{\varepsilon\to 0+}\left[\int_{i\varepsilon-\infty}^{i\varepsilon+\infty}f\hat{}_+(\omega)e^{-i\omega x}\,d\omega+\int_{-i\varepsilon-\infty}^{-i\varepsilon+\infty}f\hat{}_-(\omega)e^{-i\omega x}\,d\omega\right]$$

or

$$(4.23) \quad f(x)=\frac{1}{2\pi}\lim_{\varepsilon\to 0+}\int_{-\infty}^{\infty}e^{-iux}\left[e^{\varepsilon x}f\hat{}_+(u+i\varepsilon)+e^{-\varepsilon x}f\hat{}_-(u-i\varepsilon)\right]du.$$

One would hope that in some appropriate sense it could be said that $f\hat{}(u)$ exists and that

$$(4.24)$$

$$f\hat{}(u)=\lim_{\varepsilon\to 0+}\left[f\hat{}_+(u+i\varepsilon)+f\hat{}_-(u-i\varepsilon)\right]=\lim_{\varepsilon\to 0+}\int_{-\infty}^{\infty}f(x)e^{iux}e^{-\varepsilon|x|}\,dx.$$

Such an interpretation will be shown to be possible in the theory of distributions (see Exercise 4.2). At present we content ourselves with a simple example. Let $f(x)=1$, $-\infty<x<\infty$; then $f(x)$ is clearly of slow growth, and

$$f\hat{}_+(u+i\varepsilon)=\int_0^{\infty}e^{i(u+i\varepsilon)x}\,dx=\frac{i}{u+i\varepsilon},$$

$$f\hat{}_-(u-i\varepsilon)=\int_{-\infty}^0 e^{i(u-i\varepsilon)x}\,dx=-\frac{i}{u-i\varepsilon}.$$

Thus

$$f^{\hat{}}_+(u+i\varepsilon)+f^{\hat{}}_-(u-i\varepsilon)=\frac{2\varepsilon}{u^2+\varepsilon^2},$$

and, as was shown in Section 2,

$$\lim_{\varepsilon\to 0}\frac{2\varepsilon}{u^2+\varepsilon^2}=2\pi\,\delta(u).$$

We are therefore led from (4.24) to state that

$$1^{\hat{}}(u)=2\pi\,\delta(u),$$

that is, the Fourier transform of 1 is $2\pi\,\delta(u)$. This confirms the result conjectured in Example 4.

Transforms of Distributions on the Line

In attempting to define the Fourier transform of a distribution f, we would like to use (4.4), with the real transform variable u, but unfortunately e^{iux} is not a test function in $C_0^\infty(R_1)$, so that the action of f on e^{iux} is not defined. Instead we try to use Parseval's formula (4.8) to define $f^{\hat{}}$ from

$$\langle f^{\hat{}},\phi\rangle=\langle f,\phi^{\hat{}}\rangle.$$

Again the right side is not defined because $\phi^{\hat{}}$ is not a test function, even though ϕ is. The remedy is to introduce a more suitable class of test functions and correspondingly a new class of distributions. We begin with the one-dimensional case.

Definition.   A complex-valued function $\phi(x)$ of a single real variable is said to belong to $C_\downarrow^\infty(R_1)$, the space of *test functions of rapid decay*, if the following conditions hold:

1. $\phi(x)$, is infinitely differentiable.
2. $\phi(x)$, together with all its derivatives, vanishes at $|x|=\infty$ faster than any negative power of x. Thus, for every pair of nonnegative integers k and l,

(4.25) $$\lim_{|x|\to\infty}\left|x^k\frac{d^l\phi}{dx^l}\right|=0.$$

This class of test functions is larger than the class $C_0^\infty(R_1)$ introduced in Section 1. The test functions in $C_0^\infty(R_1)$ vanish identically outside a finite interval, whereas those in $C_\downarrow^\infty(R_1)$ merely decrease rapidly at infinity.

Every test function in $C_0^\infty(R_1)$ also belongs to $C_\downarrow^\infty(R_1)$, but e^{-x^2} belongs to $C_\downarrow^\infty(R_1)$ and not to $C_0^\infty(R_1)$. The test functions in $C_\downarrow^\infty(R_1)$ form a linear space; moreover, if ϕ is in $C_\downarrow^\infty(R_1)$, so is $x^k\phi^{(l)}(x)$ for any nonnegative integers k and l.

Convergence in $C_\downarrow^\infty(R_1)$

A sequence $\{\phi_m(x)\}$ of functions in $C_\downarrow^\infty(R_1)$ is said to be a *null sequence* in $C_\downarrow^\infty(R_1)$ if, for each pair of nonnegative integers k and l,

$$\lim_{m\to\infty} \quad \max_{-\infty<x<\infty} \left| x^k \frac{d^l\phi_m}{dx^l} \right| = 0.$$

Definition. A *distribution of slow growth* is a continuous linear functional on $C_\downarrow^\infty(R_1)$. Thus to each ϕ in $C_\downarrow^\infty(R_1)$ there is assigned a complex number $\langle f,\phi \rangle$ with the properties

$$\langle f, \alpha_1\phi_1 + \alpha_2\phi_2 \rangle = \alpha_1\langle f,\phi_1 \rangle + \alpha_2\langle f,\phi_2 \rangle,$$

$$\lim_{m\to\infty} \langle f,\phi_m \rangle = 0 \qquad \text{for every null sequence in } C_\downarrow^\infty(R_1).$$

Theorem. Every function $f(x)$ of slow growth generates a distribution of slow growth by the formula

$$\langle f,\phi \rangle = \int_{-\infty}^{\infty} f(x)\phi(x)\,dx.$$

Proof. The integral converges absolutely by the assumptions on f and ϕ. It is also clear that the functional is linear; we must still prove continuity. Let $\phi_n \to 0$ in $C_\downarrow^\infty(R_1)$; then

$$\left| \int_{-\infty}^{\infty} f(x)\phi_n(x)\,dx \right| = \left| \int_{-\infty}^{\infty} \frac{f(x)}{(1+x^2)^p} (1+x^2)^p\phi_n(x)\,dx \right|.$$

For p sufficiently large, $f(x)/(1+x^2)^p$ is absolutely integrable from $-\infty$ to ∞, since f is a function of slow growth. With such a value of p, the integral on the right is dominated by

$$\max_{-\infty<x<\infty} \left[(1+x^2)^p|\phi_n(x)| \right] \int_{-\infty}^{\infty} \frac{|f(x)|}{(1+x^2)^p}\,dx.$$

Since $\phi_n \to 0$ in $C_\downarrow^\infty(R_1)$, the maximum which appears also approaches 0. Therefore $\langle f,\phi_n \rangle \to 0$ whenever $\phi_n \to 0$ in $C_\downarrow^\infty(R_1)$, and $\langle f,\phi \rangle$ is a distribution of slow growth on $C_\downarrow^\infty(R_1)$.

Nearly all important distributions on $C_0^\infty(R_1)$ are also distributions on $C_\downarrow^\infty(R_1)$. Only the distributions on $C_0^\infty(R_1)$ which grow too rapidly at infinity cannot be extended to $C_\downarrow^\infty(R_1)$. Much of the theory of Sections 1 and 2 can be applied to distributions on $C_\downarrow^\infty(R_1)$ with only slight modifications. We shall accept this statement and proceed with the new aspects of the theory.

Theorem. If ϕ is in $C_\downarrow^\infty(R_1)$, then $\phi^\hat{}(u)$ exists and is also in $C_\downarrow^\infty(R_1)$.

Proof. The rapid decay of $\phi(x)$ at $|x| = \infty$ implies the absolute convergence of

$$\int_{-\infty}^\infty (ix)^k e^{iux}\phi(x)\,dx, \qquad k = 0, 1, 2, \dots .$$

Since this integral is the result of differentiating k times, under the integral sign, the expression for $\phi^\hat{}$, it must represent the kth derivative of $\phi^\hat{}$. Thus

$$\frac{d^k\phi^\hat{}}{du^k}(u) = \int_{-\infty}^\infty (ix)^k e^{iux}\phi(x)\,dx,$$

$$\left| \frac{d^k\phi^\hat{}}{du^k}(u) \right| \leqslant \int_{-\infty}^\infty |x^k\phi|\,dx,$$

so that the quantity on the left is bounded for all u. Moreover,

$$(iu)^p \frac{d^k\phi^\hat{}}{du^k} = \int_{-\infty}^\infty (ix)^k\phi(x)\frac{d^p}{dx^p}e^{iux}\,dx,$$

and integration by parts converts the right side to

$$(-1)^p \int_{-\infty}^\infty \left[\frac{d^p}{dx^p}(ix)^k\phi(x) \right] e^{iux}\,dx.$$

Since the term multiplying e^{iux} is in $C_\downarrow^\infty(R_1)$, the integrand is absolutely integrable, and therefore

$$\left| u^p \frac{d^k\phi^\hat{}}{du^k} \right|$$

is bounded for all u. Since p and k are arbitrary, it follows that $\phi^\hat{}(u)$ is in $C_\downarrow^\infty(R_1)$.

The same considerations also apply to the inverse transformation, so that we conclude that every function $\psi(u)$ in $C_\downarrow^\infty(R_1)$ is the transform of a function $\phi(x)$ in $C_\downarrow^\infty(R_1)$. Before proceeding with the principal task of defining the transform of a distribution of slow growth, we list some properties of the transforms of test functions in $C_\downarrow^\infty(R_1)$. Consider the transform of $d^k\phi/dx^k$; then

$$\int_{-\infty}^{\infty} \frac{d^k\phi}{dx^k} e^{iux}\,dx = (-iu)^k \int_{-\infty}^{\infty} \phi e^{iux}\,dx,$$

by integration by parts. In more compact notation,

(4.26) $$[\phi^{(k)}]\hat{}(u) = (-iu)^k \phi\hat{}(u).$$

Also we have

$$\int_{-\infty}^{\infty} (ix)^k \phi(x) e^{iux}\,dx = \frac{d^k}{du^k} \int_{-\infty}^{\infty} \phi(x) e^{iux}\,dx,$$

that is,

(4.27) $$[(ix)^k\phi]\hat{}(u) = \frac{d^k}{du^k} \phi\hat{}(u).$$

By the inversion formula for Fourier transforms,

$$\int_{-\infty}^{\infty} e^{ixz} \phi\hat{}(z)\,dz = 2\pi\phi(-x),$$

and therefore

(4.28) $$\phi\hat{}\hat{}(x) = 2\pi\phi(-x).$$

The change of variable $x - a = y$ shows that

$$\int_{-\infty}^{\infty} \phi(x-a) e^{iux}\,dx = e^{iau} \int_{-\infty}^{\infty} \phi(y) e^{iay}\,dy;$$

hence

(4.29) $$[\phi(x-a)]\hat{}(u) = e^{iau}\phi\hat{}(u).$$

Definition. Let f be any distribution of slow growth. Its Fourier transform $f\hat{}$ is the distribution of slow growth defined from

(4.30) $$\langle f\hat{}, \phi \rangle = \langle f, \phi\hat{} \rangle.$$

We must show that we have in fact defined a distribution. Since $\phi^{\wedge}$ is in $C_{\downarrow}^{\infty}(R_1)$, the action of f on $\phi^{\wedge}$ is defined, so that $f^{\wedge}$ is a functional on $C_{\downarrow}^{\infty}(R_1)$. The functional is clearly linear; moreover, it is continuous, since, whenever $\phi_m \to 0$ in $C_{\downarrow}^{\infty}(R_1)$, then $\phi_m^{\wedge} \to 0$ in $C_{\downarrow}^{\infty}(R_1)$ and therefore $\langle f, \phi_m^{\wedge} \rangle \to 0$.

To show that definition (4.30) really provides an extension of the ordinary Fourier transform, we must show that it coincides with the usual definition when f is an L_1 function with Fourier transform F. Suppose this to be the case; then by (4.30)

$$\langle f^{\wedge}, \phi \rangle = \langle f, \phi^{\wedge} \rangle = \int_{-\infty}^{\infty} f(x)\,dx \int_{-\infty}^{\infty} e^{ixy} \phi(y)\,dy$$

$$= \int_{-\infty}^{\infty} \phi(y)\,dy \int_{-\infty}^{\infty} f(x) e^{ixy}\,dx$$

$$= \int_{-\infty}^{\infty} F(y)\phi(y)\,dy.$$

Thus the action of $f^{\wedge}$ on ϕ is the same as the action of F on ϕ. Hence $f^{\wedge} = F$, and our definition is consistent.

We now show that properties (4.26) to (4.29) hold for the transform $f^{\wedge}$ of any distribution. First, consider the Fourier transform of the kth derivative of a distribution f. Then, by definition (4.30),

$$\langle [\,f^{(k)}]^{\wedge}, \phi \rangle = \langle f^{(k)}(u), \phi^{\wedge}(u) \rangle = (-1)^k \langle f(u), (\phi^{\wedge})^{(k)}(u) \rangle.$$

Now by (4.27) we have

$$(-1)^k \langle f, (\phi^{\wedge})^{(k)} \rangle = \langle f, [(-ix)^k \phi]^{\wedge} \rangle = \langle f^{\wedge}(x), (-ix)^k \phi(x) \rangle$$

$$= \langle (-ix)^k f^{\wedge}(x), \phi(x) \rangle,$$

where for the last step we have used the definition of multiplication of a distribution by an infinitely differentiable function of slow growth. Thus we find that

(4.31) $$[\,f^{(k)}]^{\wedge}(x) = (-ix)^k f^{\wedge}(x),$$

which is just (4.26) with a relabeling of the variables.

Turning next to the transform of $(ix)^k f$, we have

$$\langle [(ix)^k f]^{\wedge}, \phi \rangle = \langle (ix)^k f(x), \phi^{\wedge}(x) \rangle = (-1)^k \langle f(x), (-ix)^k \phi^{\wedge}(x) \rangle,$$

and, by using (4.26),

$$\langle f(x),(-ix)^k\phi^\wedge(x)\rangle=\langle f(x),[\phi^{(k)}]^\wedge(x)\rangle=\langle f^\wedge(u),\phi^{(k)}(u)\rangle$$

$$=(-1)^k\langle\frac{d^kf^\wedge}{du^k}(u),\phi(u)\rangle.$$

Consequently,

$$(4.32)\qquad\qquad\qquad [(ix)^kf]^\wedge(u)=\frac{d^kf^\wedge}{du^k}.$$

We leave the proofs of the following properties to the reader:

$$(4.33)\qquad\qquad\qquad f^{\wedge\wedge}(x)=2\pi f(-x),$$
$$(4.34)\qquad\qquad\qquad [f(x-a)]^\wedge(u)=e^{iau}f^\wedge(u),$$
$$(4.35)\qquad\qquad\qquad [f(-x)]^\wedge(u)=f^\wedge(-u).$$

Example 5.   Consider the transform of $\delta(x)$. Then

$$\langle\delta^\wedge,\phi\rangle=\langle\delta(x),\phi^\wedge(x)\rangle=\langle\delta(x),\int_{-\infty}^{\infty}\phi(y)e^{ixy}\,dy\rangle$$

$$=\int_{-\infty}^{\infty}\phi(y)\,dy=\langle 1,\phi\rangle.$$

Therefore

$$(4.36)\qquad\qquad\qquad\qquad \delta^\wedge=1.$$

Example 6.   To find the transform of $f(x)=1$,

$$\langle 1^\wedge,\phi\rangle=\langle 1,\phi^\wedge(x)\rangle=\int_{-\infty}^{\infty}\phi^\wedge(x)\,dx$$

$$=\left[\int_{-\infty}^{\infty}\phi^\wedge(x)e^{ixy}\,dx\right]_{y=0}.$$

By the inversion formula for $\phi^\wedge$, the last integral is just $2\pi\phi(0)$. Thus

$$(4.37)\qquad\qquad\qquad\qquad 1^\wedge=2\pi\delta.$$

The same result can also be obtained by using (4.36) and (4.33). In fact, from (4.36)

$$1^\wedge=\delta^{\wedge\wedge},$$

and from (4.33)

$$\delta^{\wedge\wedge}(x) = 2\pi\,\delta(-x) = 2\pi\,\delta(x).$$

Example 7. The transform of $\delta(x-a)$ is e^{iau}.

Example 8. We now calculate the transform of the Heaviside function $H(x)$ in three different ways.

(a) $$\langle H^{\wedge}, \phi \rangle = \langle H, \phi^{\wedge} \rangle = \int_0^\infty \phi^{\wedge}(x)\,dx$$

$$= \int_0^\infty dx \int_{-\infty}^\infty \phi(y) e^{ixy}\,dy$$

$$= \lim_{R\to\infty} \int_{-\infty}^\infty \phi(y)\,dy \int_0^R e^{ixy}\,dx$$

$$= \lim_{R\to\infty} \int_{-\infty}^\infty \frac{e^{iRy}-1}{iy}\phi(y)\,dy.$$

Now, by Exercise 2.11, $\lim_{R\to\infty}(1-\cos Ry)/y = \mathrm{pf}(1/y)$. Moreover $\lim_{R\to\infty}\sin Ry/y = \pi\delta(y)$. Therefore

$$\langle H^{\wedge}, \phi \rangle = \langle i\mathrm{pf}\frac{1}{y} + \pi\,\delta(y), \phi(y)\rangle,$$

(4.38) $$H^{\wedge}(y) = i\mathrm{pf}\frac{1}{y} + \pi\,\delta(y).$$

(b) We have $H'(x) = \delta(x)$. By (4.31)

$$(H')^{\wedge}(x) = -ixH^{\wedge}(x),$$

and, using (4.36),

$$1 = -ixH^{\wedge}(x).$$

Thus $H^{\wedge}$ satisfies the distributional equation

(4.39) $$1 = -ixf(x),$$

a particular solution of which is

$$f(x) = i\mathrm{pf}\frac{1}{x}.$$

In fact, substituting in (4.39), we find, using (1.28),

$$\langle 1,\phi\rangle = \langle x\mathrm{pf}\frac{1}{x},\phi\rangle = \langle \mathrm{pf}\frac{1}{x},x\phi\rangle$$

$$= \lim_{\varepsilon\to 0}\int_{-\infty}^{-\varepsilon}+\int_{\varepsilon}^{\infty}\frac{1}{x}x\phi\,dx = \int_{-\infty}^{\infty}\phi\,dx,$$

which is an identity.

To find the general solution of (4.39) we must add to the particular solution just obtained the general solution of the homogeneous equation $-ixf=0$, which, by Exercise 5.11(c), is $C\delta(x)$. Therefore

(4.40) $$H^{\hat{}}(x)=i\mathrm{pf}\frac{1}{x}+C\delta(x),$$

where C is a constant to be determined. The following trick enables us to find C. Consider the equation

$$H(x)+H(-x)=1,$$

whose transform by (4.35) and (4.37) yields

$$H^{\hat{}}(u)+H^{\hat{}}(-u)=2\pi\delta(u).$$

Comparing with (4.40) we find that $C=\pi$. Thus

$$H^{\hat{}}(x)=i\mathrm{pf}\frac{1}{x}+\pi\delta(x).$$

(c) The distribution $H(x)$ may be considered as the limit as $\varepsilon\to 0+$ of $H(x)e^{-\varepsilon x}$ (see Exercise 4.2). Therefore, by Exercise 4.1, $H^{\hat{}}= \lim_{\varepsilon\to 0+}(He^{-\varepsilon x})^{\hat{}}$. But $He^{-\varepsilon x}$ has a Fourier transform in the ordinary sense:

$$(He^{-\varepsilon x})^{\hat{}}=\int_{0}^{\infty}e^{-\varepsilon x}e^{iux}\,dx=\frac{1}{\varepsilon-iu}=\frac{\varepsilon+iu}{\varepsilon^2+u^2}.$$

The distributional limit as $\varepsilon\to 0+$ is easily calculated. We have

$$\lim_{\varepsilon\to 0+}\frac{\varepsilon}{\varepsilon^2+u^2}=\pi\delta(u),$$

$$\lim_{\varepsilon\to 0+}\frac{iu}{\varepsilon^2+u^2}=i\mathrm{pf}\frac{1}{u},$$

which lead again to (4.38).

Transforms in More Than One Variable

Definition. A complex-valued function $\phi(x_1,\ldots,x_n)=\phi(x)$ is said to belong to $C_\downarrow^\infty(R_n)$, the space of n-*dimensional test functions of rapid decay*, if the following conditions hold:

1. $\phi(x)$ is infinitely differentiable, that is, $D^l\phi$ exists for any multi-index l of dimension n.

2. For each pair of multi-indices k and l of dimension n,

$$\lim_{|x|\to\infty} |x^k D^l\phi| = 0,$$

where

$$x^k = x_1^{k_1}\ldots x_n^{k_n}$$

and

$$D^l\phi = \frac{\partial^{l_1+\cdots+l_n}}{\partial x_1^{l_1}\ldots x_n^{l_n}}.$$

A sequence $\{\phi_m(x)\}$ of functions in $C_\downarrow^\infty(R_n)$ is said to be a *null sequence* in $C_\downarrow^\infty(R_n)$ if, for each pair of multi-indices k and l of dimension n,

$$\lim_{m\to\infty}\ \max_{x\ \text{in}\ R_n} |x^k D^l\phi_m| = 0.$$

Definition. An n-*dimensional distribution of slow growth* is a continuous linear functional on $C_\downarrow^\infty(R_n)$. To each test function ϕ in $C_\downarrow^\infty(R_n)$ there is assigned a complex number $\langle f,\phi \rangle$ with the properties

$$\langle f,\alpha_1\phi_1 + \alpha_2\phi_2 \rangle = \alpha_1\langle f,\phi_1 \rangle + \alpha_2\langle f,\phi_2 \rangle,$$

$$\lim_{m\to\infty} \langle f,\phi_m \rangle = 0 \qquad \text{for every null sequence in } C_\downarrow^\infty(R_n).$$

We can now define n-dimensional transforms. First, if ϕ is in $C_\downarrow^\infty(R_n)$, we define, for real u,

$$\phi^\wedge(u) = \phi^\wedge(u_1,\ldots,u_n) = \int_{-\infty}^{\infty}\cdots\int_{-\infty}^{\infty} \phi(x_1,\ldots,x_n)e^{iu_1x_1}\ldots e^{iu_nx_n}\,dx_1\ldots dx_n$$

$$= \int_{R_n} \phi(x)e^{iu\cdot x}\,dx.$$

It is easily seen that $\phi^\wedge(u)$ is an n-dimensional test function of rapid decay.

The inversion formula is

$$\phi(x) = \frac{1}{(2\pi)^n} \int_{R_n} \phi^{\wedge}(u) e^{-iu \cdot x} \, du.$$

The transform of a distribution of slow growth is then defined by use of Parseval's formula:

$$\langle f^{\wedge}, \phi \rangle = \langle f, \phi^{\wedge} \rangle.$$

As examples we observe that

$$1^{\wedge} = (2\pi)^n, \qquad \delta^{\wedge} = 1,$$

where 1 is the constant function equal to 1 everywhere in R_n, and δ is the n-dimensional Dirac distribution. Formulas similar to (4.31) to (4.35) are easily derived and are left to the reader.

Exercises

4.1. Let $\{f_n\}$ be a sequence of distributions on $C_{\downarrow}^{\infty}(R_1)$ such that $f_n \to f$, where f is a distribution on $C_{\downarrow}^{\infty}(R_1)$. Show that $f_n^{\wedge} \to f^{\wedge}$.

4.2. Show that, if $f(x)$ is a function of slow growth on the real line,

$$\lim_{\varepsilon \to 0+} \langle f(x) e^{-\varepsilon|x|}, \phi(x) \rangle = \langle f, \phi \rangle.$$

Thus in the distributional sense

$$f(x) = \lim_{\varepsilon \to 0+} f(x) e^{-\varepsilon|x|} = \lim_{\varepsilon \to 0+} \left[f_+(x) e^{-\varepsilon x} + f_-(x) e^{\varepsilon x} \right].$$

Therefore, by Exercise 4.1,

$$f^{\wedge}(u) = \lim_{\varepsilon \to 0+} \left[\int_0^{\infty} f(x) e^{-\varepsilon x} e^{iux} \, dx + \int_{-\infty}^0 f(x) e^{\varepsilon x} e^{iux} \, dx \right]$$

or

$$f^{\wedge}(u) = \lim_{\varepsilon \to 0+} \left[f_+^{\wedge}(u + i\varepsilon) + f_-^{\wedge}(u - i\varepsilon) \right],$$

where $f_+^{\wedge}$ and $f_-^{\wedge}$ are the one-sided transforms defined in (4.19) and (4.20). Thus we have established (4.24). In particular, if f vanishes for $x < 0$, we have

$$f^{\wedge}(u) = \lim_{\varepsilon \to 0+} f_+^{\wedge}(u + i\varepsilon).$$

4.3. Find the Fourier transforms of the following distributions on $C_\downarrow^\infty(R_1)$:

(a) $\mathrm{pf}(1/x)$.

(b) $\mathrm{sgn}\, x = \begin{cases} 1, & x > 0, \\ -1, & x < 0. \end{cases}$

(c) $\log|x|$.

It will help to recall that $(d/dx)\log|x| = \mathrm{pf}(1/x)$.

4.4. Find the Fourier transform of x^k, where k is a positive integer and x is a single real variable.

4.5. The *convolution* of f and g is defined as

$$(4.41) \qquad h(x) \doteq \int_{-\infty}^{\infty} f(x-\xi)g(\xi)\, d\xi.$$

Show in a purely formal manner that

$$(4.42) \qquad h\hat{\ }(\omega) = f\hat{\ }(\omega)g\hat{\ }(\omega)$$

and that

$$\int_{-\infty}^{\infty} f(x-\xi)g(\xi)\, d\xi = \int_{-\infty}^{\infty} g(x-\xi)f(\xi)\, d\xi.$$

4.6 (a) Let $f(t)$ be defined on $0 \leqslant t < \infty$ and be $0(e^{\alpha t})$ at $t = +\infty$. Its *Laplace transform* is defined as

$$(4.43) \qquad \tilde{f}(s) \doteq \int_0^{\infty} e^{-st}f(t)\, dt,$$

where s is a complex variable. Clearly the integral converges for $\mathrm{Re}\, s > \alpha$ and is an analytic function of s in that right half-plane. This definition is the same as that of the Fourier transform of a right-sided function (4.15), with $s \doteq -i\omega, t \doteq x$. Show that the inversion (4.16) then becomes

$$(4.44) \qquad \frac{1}{2\pi i} \int_{a-i\infty}^{a+i\infty} e^{st}\tilde{f}(s)\, ds = \begin{cases} f(t), & t > 0; \\ 0, & t < 0; \end{cases} \quad \mathrm{Re}\, a > \alpha.$$

The inversion integral is therefore being taken on a vertical line in the right half-plane of analycity.

(b) By specializing Exercise 4.5 to right-sided functions, show that the convolution of f and g becomes

$$(4.45) \qquad h(t) = \int_0^t f(t-\tau)g(\tau)\,d\tau = \int_0^t g(t-\tau)f(\tau)\,d\tau$$

and that

$$(4.46) \qquad\qquad\qquad \tilde{h}(s) = \tilde{f}(s)\tilde{g}(s).$$

5. DIFFERENTIAL EQUATIONS IN DISTRIBUTIONS

Local Properties of Distributions

When a distribution f is generated by a continuous function, one can recover the point values of f from a knowledge of its actions on test functions (see Exercise 1.1). If all we know is that the distribution is generated by a locally integrable function f, we cannot obtain complete information about $f(x)$ from $\langle f,\phi \rangle$, since two functions f_1 and f_2 which are equal almost everywhere generate the same distribution. It is possible, however, to determine $f(x)$ up to equality almost everywhere. Suppose, for instance, that $f(x)$ is equal to 0 almost everywhere in an open set Ω; then $\langle f,\phi \rangle = 0$ for all test functions ϕ *whose support is contained in* Ω; and, vice versa, if a distribution is generated by a locally integrable function f and if $\langle f,\phi \rangle = 0$ for all test functions ϕ whose support is contained in Ω, then $f(x)$ is equal to 0 almost everywhere in Ω. The same ideas are used for an arbitrary distribution.

Definition.   The distribution f is said to *vanish on the open set* Ω if $\langle f,\phi \rangle = 0$ for every test function ϕ with support in Ω. Two distributions f_1 and f_2 are said to be *equal in* Ω if $f_1 - f_2$ vanishes on Ω.

Examples

(a) Let Ω be the open set consisting of all R_n with the origin removed. Then δ vanishes on Ω. Indeed, if ϕ has its support in Ω (remember that the support is a closed set), then ϕ must vanish in some neighborhood of the origin and $\langle \delta,\phi \rangle = \phi(0) = 0$ for any such ϕ.

(b) In (3.22) we considered the distribution $f(x) = \sum_{k=-\infty}^{\infty} \delta(x-2k\pi)$, whose action on a test function $\phi(x)$ in $C_0^\infty(R_1)$ is $\sum_{k=-\infty}^{\infty} \phi(2k\pi)$. Let Ω be the open interval $-\pi < x < \pi$; then, if ϕ has its support in Ω, $\langle f,\phi \rangle = \phi(0)$, so that we can say that $f(x)$ coincides with $\delta(x)$ in $-\pi < x < \pi$.

(c) Let $a > 0$. The distributions

$$f(x) = \operatorname{pf} H(x)\frac{1}{x} \qquad \text{and} \qquad f(x) = \frac{H(x-a)}{x}$$

coincide for $x > a$.

The Differential Equation $u' = f$ in R_1

Consider first the homogeneous equation

$$(5.1) \hspace{4cm} u' = 0,$$

regarded as an equation for distributions on the real line. By definition this means that we are looking for distributions u such that

$$(5.2) \hspace{3cm} \langle u, \phi' \rangle = 0 \qquad \text{for every } \phi \in C_0^\infty(R_1).$$

Equation (5.2) tells us that the action of u is 0 on any test function which is the *derivative* of some other test function. Of course not every test function has this property. For instance, test function (1.1) is not the derivative of a test function. Indeed any antiderivative $F(x)$ of (1.1) will have $F(\infty) \neq F(-\infty)$.

Let M be the subset of $C_0^\infty(R_1)$ consisting of the elements which are the first derivatives of elements of $C_0^\infty(R_1)$. Then we have the following lemmas.

Lemma 1. Let $\phi \in C_0^\infty(R_1)$. Then $\phi \in M$ if and only if

$$(5.3) \hspace{4cm} \int_{-\infty}^{\infty} \phi \, dx = 0.$$

Proof. (a) If $\phi \in M$, then $\phi = \chi'$, where χ belongs to $C_0^\infty(R_1)$. It follows that $\int_{-\infty}^{\infty} \phi(x) \, dx = \chi]_{-\infty}^{\infty} = 0$, so that (5.3) is satisfied.

(b) Let $\phi \in C_0^\infty(R_1)$, and let $\int_{-\infty}^{\infty} \phi(x) \, dx = 0$. Define $\chi(x) = \int_{-\infty}^{x} \phi(s) \, ds$; then χ is infinitely differentiable, and χ vanishes outside a bounded interval by (5.3). Thus χ is a test function; and since $\chi' = \phi$, χ also belongs to M.

Lemma 2. Let $\phi_0(x)$ be a fixed (but arbitrary) test function such that $\int_{-\infty}^{\infty} \phi_0(x) \, dx = 1$. Then for each $\phi(x) \in C_0^\infty(R_1)$ there are a unique constant a and a unique element ψ in M such that

$$(5.4) \hspace{4cm} \phi(x) = a\phi_0(x) + \psi(x).$$

Proof. Choose $a = \langle 1, \phi \rangle = \int_{-\infty}^{\infty} \phi(x) dx$, and define $\psi = \phi - a\phi_0$ so that (5.4) is clearly satisfied. From the definition of ψ, we see that it is a test function and that $\int_{-\infty}^{\infty} \psi \, dx = 0$. Therefore $\psi \in M$. The proof of uniqueness is left to the reader.

We are now ready to solve (5.1). If u is any distribution, then, from (5.4),

$$\langle u, \phi \rangle = a \langle u, \phi_0 \rangle + \langle u, \psi \rangle,$$

where $\psi \in M$. If u is a solution of (5.1), then $\langle u, \psi \rangle = 0$ whenever $\psi \in M$, so that, for every $\phi \in C_0^{\infty}(R_1)$,

$$\langle u, \phi \rangle = a \langle u, \phi_0 \rangle = \langle u, \phi_0 \rangle \int_{-\infty}^{\infty} \phi \, dx = \langle c, \phi \rangle,$$

where c is the constant $\langle u, \phi_0 \rangle$. We have thus shown that only constant distributions can be solutions of (5.1), and it is easy to check that constant distributions do in fact satisfy the equation.

Next we turn to the inhomogeneous equation

$$(5.5) \qquad\qquad\qquad u' = f,$$

where f is a given arbitrary distribution. By definition a distribution u satisfies (5.5) if and only if

$$\langle u, \phi' \rangle = -\langle f, \phi \rangle \qquad \text{for every } \phi \in C_0^{\infty}(R_1).$$

To find the general solution of (5.5) we use the decomposition (5.4) to write

$$\langle u, \phi \rangle = a \langle u, \phi_0 \rangle + \langle u, \psi \rangle,$$

where $\psi \in M$, say $\psi = \chi'$, $\chi \in C_0^{\infty}(R_1)$. The explicit expression for χ in terms of ϕ is

$$\chi = \int_{-\infty}^{x} \psi(s) \, ds = \int_{-\infty}^{x} \phi(s) \, ds - \langle 1, \phi \rangle \int_{-\infty}^{x} \phi_0(s) \, ds.$$

Since u is a solution of (5.5),

$$\langle u, \psi \rangle = \langle u, \chi' \rangle = -\langle f, \chi \rangle,$$

and therefore

$$\langle u, \phi \rangle = \langle u, \phi_0 \rangle \langle 1, \phi \rangle - \langle f, \chi \rangle.$$

We claim that it is legitimate to define a distribution u_p from

(5.6) $$\langle u_p, \phi \rangle = -\langle f, \chi \rangle.$$

Indeed χ is a test function depending linearly on ϕ, so that (5.6) defines a linear functional on the space of test functions $\phi(x)$. If $\{\phi_m\}$ is a null sequence in $C_0^\infty(R_1)$, so is $\{\psi_m\}$ and hence $\{\chi_m\}$; therefore the functional defined by (5.6) is continuous—in other words, a distribution. Thus every solution of (5.5) must be of the form

$$\langle u, \phi \rangle = c \langle 1, \phi \rangle + \langle u_p, \phi \rangle,$$

and it is easily verified that every distribution of this form is indeed a solution.

Green's Formula and Lagrange's Identity

We interrupt the distributional treatment to collect some results on linear differential operators.

To fix ideas consider the ordinary differential operator of order 2 given by

$$L = a_2(x)D^2 + a_1(x)D + a_0(x),$$

where $D = d/dx$ and the coefficients $a_k(x)$ are in $C^2(R_1)$. Starting from

$$\int_a^b vLu \, dx = \int_a^b (va_2u'' + va_1u' + va_0u) \, dx,$$

where u, v are arbitrary functions in $C^2(R_1)$, we integrate by parts until all the differentiations are transferred to v. In this way we find that

(5.7) $$\int_a^b vLu \, dx - \int_a^b uL^*v \, dx = J(u,v)\big]_a^b,$$

where the operator L^*, known as the *formal adjoint* of L, is given by

(5.8) $$L^* = a_2D^2 + (2a_2' - a_1)D + (a_2'' - a_1' + a_0),$$

and the bilinear form J, the *conjunct* of u and v, is

(5.9) $$J(u,v) = a_2(vu' - uv') + (a_1 - a_2')uv.$$

Since (5.7) is valid for any upper limit b, we can differentiate with respect to b and then set $b = x$ to obtain

(5.10) $$vLu - uL^*v = \frac{d}{dx} J(u,v),$$

which is known as *Lagrange's identity*. The integrated form (5.7) is called *Green's formula*.

If the operators L and L^* coincide, we say that L is *formally self-adjoint*; in our case of a second-order ordinary differential operator, L is self-adjoint if and only if

(5.11) $a_2' = a_1$ so that $Lu = D(a_2 Du) + a_0 u.$

For a formally self-adjoint operator [that is, one satisfying (5.11)], we have the simplifications

(5.12) $$J(u,v) = a_2(vu' - uv'),$$

(5.13) $$vLu - uLv = \frac{d}{dx} J(u,v),$$

(5.14) $$\int_a^b (vLu - uLv)\,dx = J(u,v) \Big]_a^b.$$

Of course if L does not satisfy (5.11), we must use the earlier formulas involving L^*.

Let us turn next to an ordinary differential operator of order p,

$$L = a_p(x)D^p + \cdots + a_1(x)D + a_0(x),$$

where the coefficients are in $C^p(R_1)$. If u and v are in $C^p(R_1)$, we have

$$D\left[vD^{m-1}u - v'D^{m-2}u + \cdots + (-1)^{m-1}(D^{m-1}v)u \right]$$
$$= vD^m u + (-1)^{m-1} uD^m v,$$

the result following from the observation that most of the terms on the left side cancel out by telescopic action. Thus we have

$$vD^m u = (-1)^m uD^m v + D \sum_{j+k=m-1} (-1)^k (D^k v)(D^j u),$$

where the sum ranges over the $j \geqslant 0$ and $k \geqslant 0$ satisfying $j + k = m - 1$. For $m = 0$ the summation disappears.

Substituting $a_m v$ for v and summing from $m = 0$ to p, we obtain

(5.15) $$vLu - uL^*v = \frac{d}{dx} J(u,v),$$

where

$$(5.16) \qquad L^*v = \sum_{m=0}^{p} (-1)^m D^m(a_m v)$$

and

$$(5.17) \qquad J(u,v) = \sum_{m=1}^{p} \sum_{j+k=m-1} (-1)^k D^k(a_m v) D^j u.$$

The term "Lagrange identity" is used for (5.15), while the integrated form

$$(5.18) \qquad \int_a^b (vLu - uL^*v)\,dx = J(u,v)\big]_a^b$$

is known as Green's formula. Observe that J contains only derivatives of order up to $p-1$.

If $L = L^*$, we say that L is formally self-adjoint. This is possible if and only if L is of even order and can be written in the form

$$(5.19) \qquad D^r(b_r D^r) + D^{r-1}(b_{r-1} D^{r-1}) + \cdots + D(b_1 D) + b_0,$$

where $p = 2r$ and $b_0, \ldots, b_r$ are arbitrary functions.

Partial differential operators pose more of a problem. We take our cue from the Laplacian Δ in R_n, which satisfies

$$(5.20) \qquad v\Delta u - u\Delta v = \operatorname{div}(v\operatorname{grad} u - u\operatorname{grad} v)$$

or, in integrated form,

$$(5.21) \qquad \int_\Omega (v\Delta u - u\Delta v)\,dx = \int_\Gamma n\cdot(v\operatorname{grad} u - u\operatorname{grad} v)\,dS,$$

which is of course the classical Green's formula. For an arbitrary linear operator of order p, as given by (1.3), the corresponding forms of (5.20) and (5.21) are

$$(5.22) \qquad vLu - uL^*v = \operatorname{div} J(u,v)$$

and

$$(5.23) \qquad \int_\Omega (vLu - uL^*v)\,dx = \int_\Gamma n\cdot J(u,v)\,dS.$$

The operator L and its formal adjoint L^* are given by

$$(5.24) \qquad Lu = \sum_{|k| \leqslant p} a_k(x) D^k u, \qquad L^* v = \sum_{|k| \leqslant p} (-1)^{|k|} D^k(a_k v),$$

the expression for L^* being obtained by integration by parts as in (1.19). If $L = L^*$, we say that L is formally self-adjoint. The expression for the vector bilinear form J is somewhat complicated in general, so that we will merely present explicit expressions for J in specific examples.

Example 1. $x = (x_1, \ldots, x_n)$, $p = 2$, $L = \Delta = \partial^2 / \partial x_1^2 + \cdots + \partial^2 / \partial x_n^2$. Since the coefficients are constants and there are only terms of even order, Δ is formally self-adjoint. The appropriate formulas are (5.20) and (5.21).

Example 2. $x = (x_1, \ldots, x_n)$, $p = 4$, $L = \Delta\Delta$. Again L is formally self-adjoint, and either by direct calculation or by using Exercise 5.10 we find that (5.22) becomes

$$(5.25)$$
$$v\Delta\Delta u - u\Delta\Delta v = \operatorname{div}[\,v \operatorname{grad} \Delta u - u \operatorname{grad} \Delta v + (\Delta v)\operatorname{grad} u - (\Delta u)\operatorname{grad} v\,].$$

Example 3. The diffusion operator is

$$(5.26) \qquad L = \frac{\partial}{\partial t} - \left(\frac{\partial^2}{\partial x_1^2} + \cdots + \frac{\partial^2}{\partial x_n^2} \right),$$

where the time variable t has been distinguished from the space variables $x_1, \ldots, x_n$. We find that

$$(5.27) \qquad L^* = -\frac{\partial}{\partial t} - \left(\frac{\partial^2}{\partial x_1^2} + \cdots + \frac{\partial^2}{\partial x_n^2} \right)$$

and

$$vLu - uL^*v = \operatorname{div} J,$$

with

$$J = e_t uv - (v \operatorname{grad}_x u - u \operatorname{grad}_x v),$$

where e_t is a unit vector in the t direction, and the subscript x indicates that differentiation is only with respect to the space variables. Green's formula becomes

$$(5.28) \quad \int_\Omega (vLu - uL^*v)\, dx\, dt = \int_\Gamma n \cdot (e_t uv + u\operatorname{grad}_x v - v\operatorname{grad}_x u)\, dS,$$

where Ω is a domain in space-time, Γ its boundary, $dx\, dt$ an element of volume in space-time, and dS a hypersurface element on Γ. In most applications of (5.28) Ω is a cylinder in space-time having base Ω_x and bounded by the parallel hyperplanes $t = t_1, t = t_2 (> t_1)$. The boundary Γ consists of (a) the two bases: $x \in \Omega_x$, $t = t_1$, and $x \in \Omega_x$, $t = t_2$, and (b) the lateral surface: $x \in \Gamma_x$, $t_1 < t < t_2$, where Γ_x is the ordinary bounding surface of the space domain Ω_x. On the base at t_2 the outward normal n is e_t, while on the other base $n = -e_t$; in both cases n is orthogonal to the space directions. On the lateral surface, n is just the outward normal n_x to Ω_x, and n_x is orthogonal to e_t. Thus (5.28) reduces to

(5.29)

$$\int_{t_1}^{t_2} dt \int_{\Omega_x} dx (vLu - uL^*v) = \int_{\Omega_x} [uv]_{t_1}^{t_2}\, dx + \int_{t_1}^{t_2} dt \int_{\Omega_x} dS_x u\left(\frac{\partial v}{\partial n_x} - v\frac{\partial u}{\partial n_x}\right).$$

Example 4. The wave operator

$$(5.30) \qquad \Box^2 \doteq \frac{\partial^2}{\partial t^2} - \left(\frac{\partial^2}{\partial x_1^2} + \cdots + \frac{\partial^2}{\partial x_n^2}\right)$$

is formally self-adjoint, and (5.22) becomes

$$(5.31) \quad v\Box^2 u - u\Box^2 v = \operatorname{div}\left[e_t\left(v\frac{\partial u}{\partial t} - u\frac{\partial v}{\partial t}\right) + u\operatorname{grad}_x v - v\operatorname{grad}_x u\right].$$

Applying Green's formula to a cylindrical domain as in Example 3, we obtain

$$(5.32) \quad \int_{t_1}^{t_2} dt \int_{\Omega_x} dx (v\Box^2 u - u\Box^2 v)$$

$$= \int_{\Omega_x}\left(v\frac{\partial u}{\partial t} - u\frac{\partial v}{\partial t}\right)_{t_1}^{t_2} dx + \int_{t_1}^{t_2} dt \int_{\Gamma_x} dS_x\left(u\frac{\partial v}{\partial n_x} - v\frac{\partial u}{\partial n_x}\right).$$

Classical, Weak, and Distributional Solutions

Let us look at the ordinary differential equation

$$(5.33) \qquad \frac{du}{dx} = f(x) \quad \text{on the interval } \Omega: \quad a < x < b.$$

If $f(x)$ is a continuous function, we can define the notion of solution in the classical sense: $u(x)$ is a *classical (or strict) solution* if it has a continuous derivative which satisfies (5.33) pointwise on Ω. (An almost trivial extension to the case where f is piecewise continuous was made in Section 1, Chapter 1.) Denoting the class of test functions with support in Ω by $C_0^\infty(\Omega)$, we find that, for any classical solution of (5.33),

$$(5.34) \qquad \int_\Omega f\phi \, dx = -\int_\Omega u \frac{d\phi}{dx} \, dx \qquad \text{for each } \phi \text{ in } C_0^\infty(\Omega),$$

the result stemming from integration by parts and the fact that $\phi \equiv 0$ in a neighborhood of the boundary. The two sides of (5.34) make sense even if f and u are only locally integrable. This leads to the definition of a *weak solution* of (5.33): If f is locally integrable, a locally integrable function u is a weak solution of (5.33) if and only if it satisfies (5.34) for each ϕ in $C_0^\infty(\Omega)$. If u is a weak solution of (5.33), we also say that $du/dx = f$ in the *weak sense*.

Equation (5.33) can also be interpreted distributionally. If f is a distribution, we say that a distribution u is a solution of (5.33) if and only if

$$(5.35) \qquad -\left\langle u, \frac{d\phi}{dx} \right\rangle = \langle f, \phi \rangle \qquad \text{for each } \phi \text{ in } C_0^\infty(\Omega).$$

Note that the left side is the definition of the distribution u'. If f is a distribution generated by a locally integrable function and if we are looking for solutions u that are functions, (5.35) reduces to (5.34), that is, to the concept of a weak solution.

We can now extend these ideas to more general operators. Let L be an arbitrary linear differential operator of order p in the n variables $x_1, \ldots, x_n$:

$$L = \sum_{|k| \leqslant p} a_k(x) D^k, \qquad \begin{cases} k = (k_1, \ldots, k_n), \\ |k| = k_1 + k_2 + \cdots + k_n. \end{cases}$$

Assuming that the $a_k(x)$ are infinitely differentiable, the distribution Lu always exists for any distribution u and [see (1.19)] is defined by

$$\langle Lu, \phi \rangle = \langle u, L^*\phi \rangle,$$

where L^* is the formal adjoint of L, introduced in (5.24),

$$(5.36) \qquad L^*\phi = \sum_{|k| \leqslant p} (-1)^{|k|} D^k(a_k \phi).$$

We are therefore in a position to give a distributional meaning to the differential equation

(5.37) $Lu = f$, x in Ω,

where f is a given distribution.

What we require is that the distributions Lu and f coincide in Ω. (See the beginning of Section 5.)

Definition.   A distribution u is a solution of (5.37) *on* Ω if

(5.38) $\langle u, L^*\phi \rangle = \langle f, \phi \rangle$ for each ϕ in $C_0^\infty(\Omega)$.

Definition.   Let f be locally integrable. A locally integrable function u which satisfies (5.38) is said to be a *weak solution* of (5.37) on Ω.

Remark.   If $f(x)$ is a continuous function, we can give a classical interpretation to (5.37). The function $u(x)$ is a *classical* solution of (5.37) if it belongs to $C^p(\Omega)$, the class of functions with continuous derivatives of order up to p, and if it satisfies (5.37) at every point of Ω. Since functions can also be interpreted as distributions, we would like to compare the two notions of solutions: weak solutions and classical solutions.

Theorem.   Let $f(x)$ be continuous on Ω. Then (a) a classical solution of (5.37) on Ω is also a weak solution, and (b) any weak solution on Ω which has p continuous derivatives is a classical solution.

Proof.   (a) Let u be a classical solution on Ω, and let $\phi(x)$ be a test function with support in Ω. Then

$$\langle u, L^*\phi \rangle = \int_\Omega uL^*\phi\,dx$$

can be integrated by Green's theorem. The support of ϕ being in Ω, ϕ vanishes in a neighborhood of Γ so that ϕ and all its derivatives vanish on Γ. This guarantees that $J(u,\phi) = 0$ on Γ, and therefore, by (5.23),

$$\langle u, L^*\phi \rangle = \int_\Omega \phi Lu\,dx.$$

Since $Lu = f$ at every point of Ω, we obtain

$$\langle u, L^*\phi \rangle = \langle f, \phi \rangle \qquad \text{for } \phi \text{ in } C_0^\infty(\Omega).$$

(b) By assumption $\langle u, L^*\phi \rangle = \langle f, \phi \rangle$ for ϕ with support in Ω. Since u has continuous derivatives of order p, we can use Green's theorem on Ω. Again we have $J=0$ on Γ, so that

$$(5.39) \qquad\qquad 0 = \int_\Omega \phi q \, dx \qquad \text{for } \phi \text{ in } C_0^\infty(\Omega),$$

where q is the continuous function $Lu-f$. We wish to show that (5.39) implies that $q \equiv 0$ in Ω. Suppose $q(x_0) \neq 0$ at x_0 in Ω. There is no loss of generality in assuming $q(x_0) > 0$. Since q is continuous, we can find a small ball with center at x_0, lying wholly in Ω, such that $q > 0$ in the ball. We can also construct a test function ϕ positive in the ball and 0 outside. For this ϕ we would have $\int_\Omega \phi q \, dx > 0$, contradicting the hypothesis. Therefore $q = Lu - f = 0$ at every point in Ω, and u is a classical solution of (5.37) in Ω.

The question naturally arises as to whether (5.37) can have weak solutions that are not classical solutions. By part (b) of the theorem just proved, any such weak solution could not belong to $C^p(\Omega)$. A simple example with $\Omega = R_1$ is the first-order equation

$$x \frac{du}{dx} = 0,$$

which has the weak solution $u(x) = H(x)$. Indeed, $\langle xH', \phi \rangle = \langle H', x\phi \rangle = \langle \delta, x\phi \rangle = 0$, so that $u = H$ is a weak solution which is clearly not a classical solution. In R_1 a weak solution is associated with the fact that the differential equation has a singular point (in our particular case, at $x = 0$), that is, a point at which the coefficient $a_p(x)$ of the highest derivative $D^p u$ vanishes. The situation is more complicated for partial differential equations. The role formerly played by the coefficient of the highest order term is now taken over by a matrix of coefficients of the terms of order p. (See also Section 1, Chapter 8.) Let us look at a few examples.

Example 5. In R_2, with $x = (x_1, x_2)$, consider the equation

$$(5.40) \qquad\qquad \frac{\partial u}{\partial x_1} = 0,$$

whose classical solutions are $u = f(x_2)$ with f differentiable. Since no differentiation with respect to x_2 is involved in (5.40), the requirement that $f(x_2)$ be differentiable can be dispensed with; for instance, $u = H(x_2)$ is a weak solution of (5.40) since

$$\left\langle \frac{\partial H(x_2)}{\partial x_1}, \phi(x_1, x_2) \right\rangle = \left\langle H(x_2), -\frac{\partial \phi}{\partial x_1} \right\rangle = -\int_0^\infty dx_2 \int_{-\infty}^\infty dx_1 \frac{\partial \phi}{\partial x_1}$$

$$= -\int_0^\infty dx_2 [\phi(+\infty, x_2) - \phi(-\infty, x_2)] = 0.$$

Example 6. In R_2, with $x = (x_1, x_2)$, the equation

$$(5.41) \qquad \frac{\partial^2 u}{\partial x_1 \partial x_2} = 0$$

has classical solutions of the form $u = f(x_1) + g(x_2)$, where f and g are twice differentiable. Even if f and g are not differentiable, u of this type is still a solution. Indeed, $u = H(x_1) + H(x_2)$ is a weak solution of (5.41) for

$$\left\langle \frac{\partial^2 [H(x_1) + H(x_2)]}{\partial x_1 \partial x_2}, \phi \right\rangle = \left\langle H(x_1) + H(x_2), \frac{\partial^2 \phi}{\partial x_1 \partial x_2} \right\rangle$$

$$= \int_{-\infty}^{\infty} dx_2 \int_0^{\infty} \frac{\partial^2 \phi}{\partial x_1 \partial x_2} dx_1 + \int_{-\infty}^{\infty} dx_1 \int_0^{\infty} \frac{\partial^2 \phi}{\partial x_1 \partial x_2} dx_2 = 0.$$

Example 7. Consider the homogeneous wave equation in one space dimension x:

$$(5.42) \qquad \frac{\partial^2 u}{\partial t^2} - \frac{\partial^2 u}{\partial x^2} = \square^2 u = 0.$$

If f is any function of a real variable with a continuous second derivative, then $u = f(x - t)$ is easily seen to be a solution of (5.42). This solution represents a wave traveling to the right with velocity 1. Snapshots taken at times t_1 and t_2 show the same wave form, the one at the later time t_2 being displaced an amount $t_2 - t_1$ to the right with respect to the one taken at time t_1. There seems to be no reason physically to restrict oneself to wave forms that are twice differentiable. Let us show that $u = H(x - t)$ is a weak solution of (5.42). Since $\square^2$ is formally self-adjoint, we must show, according to (5.38), that

$$\langle H(x - t), \square^2 \phi(x, t) \rangle = 0 \qquad \text{for each test function } \phi(x, t),$$

or, equivalently, that

$$(5.43) \qquad \iint_{x > t} \left(\frac{\partial^2 \phi}{\partial t^2} - \frac{\partial^2 \phi}{\partial x^2} \right) dx\, dt = 0 \qquad \text{for } \phi \text{ in } C_0^{\infty}(R_2).$$

By changing variables to $x_1 = x - t$, $x_2 = x + t$, we find that the last integral reduces to an integration of $\partial^2 \phi / \partial x_1 \partial x_2$ over the half-plane $x_1 > 0$. Since

$$\int_{-\infty}^{\infty} \frac{\partial^2 \phi}{\partial x_1 \partial x_2} dx_2 = 0,$$

we have verified (5.43). Weak solutions of (5.42) can have jump discontinuities on the *characteristics* $x = t$ and $x = -t$ [the latter arising from solutions of the type $f(x + t)$]. Further discussion will be found in Section 1,
Chapter 8.

Example 8. *Weak solutions of Laplace's equation are necessarily classical.*
We shall only show that a solution of $\Delta u = 0$ in Ω cannot have a simple
jump across a hypersurface σ. Suppose that σ divides Ω into the two
domains Ω_+ and Ω_-. If u is a weak solution of $\Delta u = 0$ on R_n, we have, for
ϕ with support in Ω,

$$0 = \int_\Omega u\Delta\phi\, dx = \int_{\Omega_+} u\Delta\phi\, dx + \int_{\Omega_-} u\Delta\phi\, dx.$$

If u and its first derivatives have limiting values on both sides of σ, we find
by Green's theorem that

$$\int_{\Omega_+} u\Delta\phi\, dx = \int_{\Omega_+} \phi\Delta u + \int_{\sigma_+} \left(u\frac{\partial\phi}{\partial n} - \phi\frac{\partial u}{\partial n} \right) dS,$$

where σ_+ is the side of σ bounding Ω_+. There is no contribution from the
boundary Γ of Ω, since ϕ vanishes in a neighborhood of Γ. Combining the
last equation with a similar one for Ω_- and using the assumption that u is
a classical solution within Ω_+ and Ω_-, we obtain

$$(5.44) \quad 0 = \int_\sigma \left[(u_+ - u_-)\frac{\partial\phi}{\partial n} - \left(\frac{\partial u_+}{\partial n} - \frac{\partial u_-}{\partial n} \right)\phi \right] dS, \qquad \phi \in C_0^\infty(\Omega),$$

where n is the outward normal to σ from Ω_+. The factors multiplying
$\partial\phi/\partial n$ and ϕ must each vanish on σ. The proof can be adapted from the
special case where $\Omega = R_2$ and σ is the line $x_2 = 0$. Consider the test
function $\phi(x_1, x_2) = \phi_1(x_1)\phi_2(x_2)$, where $\phi_2(0) = 0$, $\phi_2'(0) \neq 0$, and $\phi_1(x_1) \geqslant 0$
has its support in a small neighborhood of the point $x_1 = \xi_1$. It then follows
from (5.44) that $u(\xi_1, 0+) = u(\xi_1, 0-)$; since ξ_1 is arbitrary, we have
$u(x_1, 0+) = u(x_1, 0-)$ for any point on σ. Similarly, it can be shown that
$(\partial u/\partial x_2)(x_1, 0+) = (\partial u/\partial x_2)(x_1, 0-)$. In the general case the conclusion is
$u_+ = u_-$ and $\partial u_+/\partial n = \partial u_-/\partial n$. The first of these relations also implies
that tangential derivatives are continuous on crossing σ. Thus u and all its
first derivatives must be continuous on crossing σ. More can be shown: u
is infinitely differentiable in Ω. We already know this in two dimensions,
where a solution of Laplace's equation is the real part of an analytic
function of the complex variable $x_1 + ix_2$.

Fundamental Solutions

We consider differential equations on the whole of R_n (that is, $\Omega = R_n$). A powerful method for studying such equations is based on the fundamental solution.

Definition. A *fundamental solution for L with pole at ξ* is a solution of the equation

$$(5.45) \qquad Lu = \delta_\xi(x) = \delta(x - \xi),$$

where ξ is regarded as a parameter.

Remarks

1. Equation (5.45) is to be interpreted in the sense of distributions. A solution of (5.45) is denoted by $E(x, \xi)$. It is a distribution in x depending parametrically on ξ. Often (but not always) E will correspond to a locally integrable function of x. In any event, according to (5.38), E satisfies (5.45) if and only if

$$(5.46) \qquad \langle E, L^*\phi \rangle = \phi(\xi) \qquad \text{for each test function } \phi \text{ in } C_0^\infty(R_n).$$

2. Equation (5.45) will usually have many solutions differing from one another by a solution of the homogeneous equation. For problems in which R_n can be interpreted as an isotropic geometrical space we often select a particular solution on grounds of symmetry and behavior at infinity; if, however, one of the coordinates is timelike, we may use causality as the appropriate criterion with respect to that coordinate.

3. If L has *constant coefficients*, it suffices to find the fundamental solution with pole at 0 [that is, $E(x, 0)$] and then translate to obtain the solution with pole at ξ:

$$E(x, \xi) = E(x - \xi, 0).$$

The fundamental solution $E(x, 0)$ will also be denoted by $E(x)$.

There are two parts to determining a fundamental solution. First we must construct, often by intuitive means, a likely candidate, and then we must check that (5.46) is in fact satisfied. The first part can itself be divided into two steps: solve the homogeneous equation for $x \neq \xi$ with proper regard to physical considerations, and then build in the right singularity at $x = \xi$ by using the integrated form of (5.45) near $x = \xi$ [for one dimension this involves appropriate matching at ξ of the solutions on either side; for higher dimensions the condition will be a modification of (4.19), Chapter 1, which is valid for $-\Delta$]. We have already acquired some experience in verifying that a given distribution satisfies a differential equation; see, for instance, Example 15, Section 1, and Exercises 1.4, 1.5, and 1.6.

Example 9. Find a fundamental solution E for $-(d^2/dx^2) + q^2$. Since the coefficients are constant, it suffices to find $E(x)$ satisfying

$$- \frac{d^2 E}{dx^2} + q^2 E = \delta(x), \qquad -\infty < x < \infty.$$

On intuitive grounds based on our experience in Chapter 1 we require the continuity of E at $x = 0$ and $E'(0+, 0) - E'(0-, 0) = -1$. If E is to represent the concentration in an absorbing medium, we also demand that E vanish at $|x| = \infty$. This leads to the candidate [see (4.14), Chapter 1]

$$E(x) = \frac{e^{-q|x|}}{2q}, \qquad E(x, \xi) = \frac{e^{-q|x - \xi|}}{2q}.$$

To check that $E(x)$ is a fundamental solution with pole at 0, we must show that (5.46) holds. We have

$$\langle E, L^* \phi \rangle = \int_{-\infty}^{0} \frac{e^{qx}}{2q} L^* \phi \, dx + \int_{0}^{\infty} \frac{e^{-qx}}{2q} L^* \phi \, dx.$$

Using Green's theorem in each interval, we find easily that

$$\langle E, L^* \phi \rangle = \phi(0) \qquad \text{for each } \phi \text{ in } C_0^\infty(R_1).$$

Alternatively, we can use the result of Example 11, Section 1, to differentiate the function $E(x)$ in the sense of distributions.

Example 10. Consider the general ordinary differential operator L of order p. Let us find the *causal* fundamental solution $E(t, \tau)$ which *vanishes* for $t < \tau$ and satisfies

$$(5.47) \quad LE = a_p \frac{d^p E}{dt^p} + \cdots + a_1 \frac{dE}{dt} + a_0 E = \delta(t - \tau), \qquad -\infty < t, \tau < \infty.$$

Proceeding intuitively, we have $E \equiv 0$ for $t < \tau$ and $E, E', \dots, E^{(p-2)}$ continuous at $t = \tau$ (therefore all are 0 at $\tau +$), and $a_p E^{(p-1)}$ has a unit jump at $t = \tau$, that is,

$$E^{(p-1)}(\tau +, \tau) = \frac{1}{a_p(\tau)}.$$

The last condition was obtained by integrating (5.47) from $\tau -$ to $\tau +$. This suggests that, for $t > \tau$, $E(t, \tau)$ will coincide with the solution $u_\tau(t)$ of the initial value problem

(5.48)

$$Lu_\tau(t) = 0, \qquad u_\tau(\tau) = u_\tau'(\tau) = \cdots = u_\tau^{(p-2)}(\tau) = 0, \qquad u_\tau^{(p-1)}(\tau) = \frac{1}{a_p(\tau)}.$$

Since (5.48) has one and only one solution by the existence and uniqueness theorem for initial value problems (see Chapter 3), we tentatively set

(5.49) $$E(t, \tau) = H(t - \tau)u_\tau(t).$$

Let us now check that (5.49) satisfies (5.46) with $x = t, \xi = \tau$. We have

$$\langle E, L^* \phi \rangle = \int_\tau^\infty u_\tau(t) L^* \phi \, dt,$$

and, using Green's formula (5.18),

$$\langle E, L^* \phi \rangle = \int_\tau^\infty \phi L u_\tau(t) \, dt - J(u_\tau, \phi) \big]_{t = \tau}^{t = \infty}.$$

Since $Lu_\tau = 0$, the integral vanishes. The fact that $\phi \equiv 0$ outside a bounded interval shows that $J = 0$ at the upper limit. At the lower limit τ we must look a little more closely at expression (5.17) for J. All terms involving derivatives of u_τ of order $p - 2$ or less are 0 by (5.48); this leaves only a single term with $j = p - 1, k = 0, m = p$. Thus, at $t = \tau$, $J(u_\tau, \phi) = u_\tau^{(p-1)}(\tau)a_p(\tau)\phi(\tau) = \phi(\tau)$, which gives $\langle E, L^* \phi \rangle = \phi(\tau)$ as required. Note that causality removes all arbitrariness from fundamental solutions.

If the coefficients in (5.47) are all *constants*, then

$$u_\tau(t) = v(t - \tau),$$

where $v(t)$ is the solution of the initial value problem (5.48) with $\tau = 0$. There is a corresponding simplification in (5.49).

Example 11. We have already shown in Example 15, Section 1, that $E(x) = 1/4\pi|x|$ is a fundamental solution for $-\Delta$ with pole at the origin.

Example 12. Let us find the causal fundamental solution for the equation of diffusion in one space dimension (the term "causal" applies to the time coordinate). With the source at $x=0$, $t=0$, we are searching for the solution $E(x,t;0,0)$ of

(5.50) $$LE = \frac{\partial E}{\partial t} - \frac{\partial^2 E}{\partial x^2} = \delta(x,t), \qquad -\infty < t, x < \infty,$$

with $E \equiv 0$, $t<0$, and $E \to 0$ as $|x| \to \infty$. Proceeding without regard to rigor, we take a Fourier transform on the space coordinate; setting

$$E^{\wedge} = \int_{-\infty}^{\infty} e^{i\omega x} E \, dx,$$

we obtain, by multiplying (5.50) by $e^{i\omega x}$ and integrating,

$$\frac{dE^{\wedge}}{dt} + \omega^2 E^{\wedge} = \delta(t), \qquad E^{\wedge} = 0, \quad t < 0.$$

Thus $E^{\wedge}$ jumps by unity at $t=0$, and

$$E^{\wedge} = e^{-\omega^2 t}, \quad t > 0 \qquad (\text{of course } E^{\wedge} = 0 \quad \text{for } t < 0).$$

The inversion is easily performed:

$$E = \frac{1}{2\pi} \int_{-\infty}^{\infty} e^{-i\omega x} E^{\wedge} \, d\omega = \frac{e^{-x^2/4t}}{\sqrt{4\pi t}}, \qquad t > 0.$$

We are therefore led to believe that

(5.51) $$E = H(t) \frac{e^{-x^2/4t}}{\sqrt{4\pi t}}$$

is a fundamental solution of (5.50). To verify this we must show that (5.51) satisfies

$$\langle E, L^*\phi \rangle = \phi(0,0) \qquad \text{for each test function } \phi(x,t).$$

The left side is the convergent integral

$$\int_0^{\infty} dt \int_{-\infty}^{\infty} dx \frac{e^{-x^2/4t}}{\sqrt{4\pi t}} L^*\phi,$$

which can be written as

$$\lim_{\varepsilon \to 0} \int_\varepsilon^\infty dt \int_{-\infty}^\infty dx \frac{e^{-x^2/4t}}{\sqrt{4\pi t}} L^*\phi.$$

Applying Green's formula (5.29) to this integral, and noting that $e^{-x^2/4t}/\sqrt{4\pi t}$ satisfies the homogeneous equation for $t>0$ (that is how it was constructed!), we find that

$$\int_\varepsilon^\infty dt \int_{-\infty}^\infty dx \frac{e^{-x^2/4t}}{\sqrt{4\pi t}} L^*\phi = \int_{-\infty}^\infty \frac{e^{-x^2/4\varepsilon}}{\sqrt{4\pi\varepsilon}} \phi(x,\varepsilon)\, dx.$$

It is now a simple modification of the argument in case (b), Example 3, Section 2, to show that the limit as $\varepsilon \to 0$ is $\phi(0,0)$.

Exercises

5.1. Consider the operator in R_2 defined by

$$Lu = \frac{\partial^2 u}{\partial x\, \partial y} + \gamma^2 u,$$

where γ^2 is a positive constant.
(a) Write Green's formula for L.
(b) Show that $J_0(2\gamma\sqrt{xy}\,)$ is a classical solution of $Lu=0$ for $x,y>0$.
(c) Show that $E = H(x)H(y)J_0(2\gamma\sqrt{xy}\,)$ is a fundamental solution for L with pole at the origin.

5.2. Consider the wave operator in one space dimension:

$$\Box^2 \doteq \frac{\partial^2}{\partial t^2} - \frac{\partial^2}{\partial x^2}.$$

Write Green's formula for an arbitrary domain Ω with boundary Γ. On Γ introduce a vector q, known as the *transversal* to Γ, defined by

$$q \cdot e_t = n \cdot e_t, \qquad q \cdot e_x = -n \cdot e_x,$$

where e_x and e_t are unit vectors in the x and t directions, respectively. Clearly q is a unit vector. Show that Green's formula takes the form

$$(5.52) \qquad \int_\Omega (v \Box^2 u - u \Box^2 v)\, dx\, dt = \int_\Gamma \left(v\frac{\partial u}{\partial q} - u\frac{\partial v}{\partial q} \right) dl,$$

where dl is an element of arclength along the bounding curve Γ. Note that q is tangent to Γ if and only if Γ is one of the two families of straight lines $x - t = \text{constant}$ and $x + t = \text{constant}$. These are the *characteristics* of the differential equation. Suppose we are looking for a curve C across which solutions of $\square^2 u = 0$ can suffer jump discontinuities. On either side of C we assume that u is a classical solution. Since u is to be a generalized solution, we have $0 = \langle u, \square^2 \phi \rangle$; using Green's formula (5.31) first on one side of C and then on the other, and adding the results, we find that

$$(5.53) \qquad 0 = \int_C \left[\frac{\partial \phi}{\partial q} (u_+ - u_-) - \phi \left(\frac{\partial u_+}{\partial q} - \frac{\partial u_-}{\partial q} \right) \right] dl,$$

where q is the transversal on the positive side of C. If C is nowhere tangent to a characteristic, show that $u_+ - u_- = 0$ and $\partial u_+ / \partial q - \partial u_- / \partial q = 0$. If C is a characteristic, then $\partial / \partial q$ is proportional to $\partial / \partial l$ and (5.53) reduces to

$$0 = \int_C \left\{ \frac{\partial}{\partial l} [\phi(u_+ - u_-)] - 2\phi \frac{\partial}{\partial l} (u_+ - u_-) \right\} dl.$$

Since ϕ vanishes outside a finite segment of C, the integral of the first term on the right is automatically 0. From the arbitrariness of ϕ, we can, however, conclude that

$$(5.54) \qquad \frac{\partial}{\partial l} (u_+ - u_-) = 0 \qquad \text{or} \qquad u_+ - u_- = \text{constant on } C.$$

Thus the only curves that can propagate discontinuities are the characteristics, and across a characteristic the jump in the solution remains constant.

5.3. Show that $E = (1/2\pi) K_0(q|x|)$ is a fundamental solution in R_2 for $L = -\Delta + q^2$. Since E is radially symmetric and vanishes at infinity, we can interpret E as the steady concentration in an absorbing medium due to a source placed at the origin of R_2.

5.4. Consider the wave operator of Exercise 5.2. Show that $E = \frac{1}{2} H (t - |x|)$ is a causal fundamental solution with pole at $x = 0$, $t = 0$. Find a fundamental solution which vanishes for $t > 0$.

5.5. Show that $E = H(t)(4\pi a t)^{-n/2} e^{-r^2/4at}$ is a causal fundamental solution for the diffusion operator $L = (\partial / \partial t) - a\Delta$ in n space dimensions. [See case (f), Example 3, Section 2.]

5.6. Find a spherically symmetric fundamental solution (with pole at the origin) for the biharmonic operator $\Delta\Delta$ in two dimensions and in three dimensions. Verify your results distributionally.

5.7. (a) Consider one-dimensional unsteady diffusion in an absorbing medium. The causal fundamental solution E with pole at $x=0$, $t=0$ satisfies

$$\frac{\partial E}{\partial t} - \frac{\partial^2 E}{\partial x^2} + q^2 E = \delta(x,t), \qquad E \equiv 0, \quad t<0.$$

Reduce the problem to ordinary diffusion by the transformation $E = e^{-q^2 t}F$ and find E.

(b) What would be the significance of the problem in which $q^2 E$ is replaced by $-q^2 E$? What about the fundamental solution?

5.8. In diffusion with *drift*, the causal fundamental solution E satisfies

$$\frac{\partial E}{\partial t} - \frac{\partial^2 E}{\partial x^2} + v\frac{\partial E}{\partial x} = \delta(x,t), \qquad E \equiv 0, \quad t<0,$$

where v is a real constant. Find E.

5.9. Let L be the ordinary differential operator of order p:

$$a_p D^p + \cdots + a_1 D + a_0,$$

where the coefficients are infinitely differentiable. Let $E(x,\xi)$ be a fundamental solution for L with pole at ξ. Then of course $E, E', \ldots, E^{(p-2)}$ are continuous at ξ, and $E^{(p-1)}$ has a jump $1/a_p(\xi)$. The functions $E^{(p)}, E^{(p-1)}, \ldots,$ are well defined for $x>\xi$ and $x<\xi$. Find the jumps $E^{(p)}(\xi+,\xi) - E^{(p)}(\xi-,\xi)$ and $E^{(p+1)}(\xi+,\xi) - E^{(p+1)}(\xi-,\xi)$, and indicate how jumps in higher derivatives could be found (note that it suffices to consider a causal fundamental solution).

5.10. Let L and M be two linear differential operators on R_n, and let $P = LM$. Show that

$$P^* = M^* L^*,$$

and express the right-hand side of Green's theorem for P in terms of the conjuncts J and K corresponding to L and M, respectively.

5.11. Let $\phi_0(x)$ be a fixed element of $C_0^\infty(R_1)$ with $\phi_0(0) = 1$, and let M be the subset of $C_0^\infty(R_1)$ consisting of elements $\psi(x)$ of the form $\psi = x\chi$, where χ is itself in $C_0^\infty(R_1)$. Show that the following hold:

(a) A test function ψ belongs to M if and only if $\psi(0) = 0$.

(b) For each ϕ in $C_0^\infty(R_1)$ there are a unique constant a and a unique element ψ in M such that

$$\phi = a\phi_0 + \psi.$$

(c) The general distributional solution of $xu = 0$ is $C\delta(x)$, where C is an arbitrary constant.
What is the general solution of $xu = f$, where f is a given distribution?

REFERENCES AND ADDITIONAL READING

Beals, R., *Advanced mathematical analysis,* Springer Verlag, New York, 1973.

Boas, R. P., Jr., and Pollard, H., Continuous analogues of series, *Am. Math. Mon.* **80** (1973), 18.

Dym, H. and McKean, H. P., *Fourier series and integrals,* Academic Press, New York, 1972.

Gelfand, I. M. and Shilov, G. E., *Generalized functions,* Vol. I, Academic Press, New York, 1964.

Korevaar, J., *Mathematical methods,* Vol. I, Academic Press, New York, 1968.

Papoulis, A., *The Fourier integral and its applications,* McGraw-Hill, New York, 1962.

Schwartz, L., *Théorie des distributions,* Hermann, Paris, 1966.

Zemanian, A. H., *Distribution theory and transform analysis,* McGraw-Hill, New York, 1965.

3

One-Dimensional
Boundary Value
Problems

1. REVIEW

In this section we deal with classical solutions of the differential equation $Lu = f$ on an open interval I on the x axis. Here L is the general, linear, ordinary differential operator of order p;

$$(1.1) \qquad L = a_p(x)\frac{d^p}{dx^p} + \cdots + a_1(x)\frac{d}{dx} + a_0(x),$$

where the coefficients $\{a_k(x)\}$ are continuous on the closure $\bar{I}$ of I, and *the leading coefficient $a_p(x)$ does not vanish anywhere on $\bar{I}$.* A point x at which $a_p(x)$ vanishes is a *singular point*, and the only purpose of introducing I is to avoid such points. If $a_p(x)$ does not vanish anywhere on the real line, we can take I to be the whole line. The inhomogeneous term $f(x)$ is assumed to be piecewise continuous on $\bar{I}$. We recall that a classical solution of $Lu = f$ is a function $u(x)$ with $p - 1$ continuous derivatives and a piecewise continuous pth derivative such that the differential equation is satisfied at all points of continuity of f.

From elementary differential equations we know that the general solution of the equation $Lu = f$ involves p arbitrary constants. By imposing p additional conditions, known generally as boundary conditions, one might hope to single out a specific solution of the differential equation. This conclusion is warranted in the special case of initial conditions, when $u, u', \ldots, D^{p-1}u$ are all given at the same point x_0 in $\bar{I}$, but is *not* always correct for general boundary conditions. We now state the precise form of the existence and uniqueness theorem for initial value problems, with the proof postponed until Chapter 4.

Theorem 1.   Let L be the operator (1.1), and let $a_p(x) \neq 0$ in $\bar{I}$. Let x_0 be a fixed point in $\bar{I}$; let $\gamma_1, \ldots, \gamma_p$ be given numbers, and let $f(x)$ be a given piecewise continuous function on $\bar{I}$. The initial value problem (IVP)

$$(1.2) \quad Lu = f, \quad x \text{ in } I; \qquad u(x_0) = \gamma_1, \quad u'(x_0) = \gamma_2, \ldots, u^{(p-1)}(x_0) = \gamma_p$$

has one and only one classical solution.

Remarks

1.   We refer to (1.2) as the IVP with data $\{f(x); \gamma_1, \ldots, \gamma_p\}_{x_0}$. The subscript x_0 serves to remind us of the point where the initial values are specified.

2.   The only solution of the completely homogeneous problem

$$(1.3) \qquad Lu = 0, \quad x \text{ in } I; \qquad u(x_0) = u'(x_0) = \cdots = u^{(p-1)}(x_0) = 0$$

is $u \equiv 0$, that is, the only solution of the IVP with data $\{0; 0, \ldots, 0\}_{x_0}$ is $u \equiv 0$.

3.   The essential nature of the assumption $a_p(x) \neq 0$ is illustrated by the example

$$xu' - 2u = 0 \quad \text{on } -\infty < x < \infty, \qquad u(0) = 0.$$

The function $u = Ax^2$ satisfies the condition $u(0) = 0$ and the differential equation for all x, so we clearly do not have uniqueness (in fact, it is even possible to use a function defined as Ax^2 for $x \geqslant 0$ and Bx^2 for $x < 0$ without $A = B$).

Another difficulty is illustrated by the example

$$xu' + u = 0, \quad -\infty < x < \infty, \qquad u(0) = \gamma \neq 0.$$

The general solution of the differential equation for $x \neq 0$ is $u = A/x$, and no solution can satisfy the initial condition (in fact, the solution is not even continuous at $x = 0$).

Linear Dependence; Wronskians

Consider n continuous functions $f_1(x), \ldots, f_n(x)$ defined on the interval I. These functions are said to be *dependent* over I if there exist constants

$c_1, \ldots, c_n$, not all 0, such that

$$(1.4) \qquad c_1 f_1(x) + \cdots + c_n f_n(x) \equiv 0 \qquad \text{on } I.$$

A set of n functions is therefore dependent if any one of them can be expressed as a linear combination of the others. A set of n functions is independent over I if (1.4) can be satisfied only if $c_1 = \cdots = c_n = 0$.

Let $f_1, \ldots, f_n$ have continuous derivatives of orders up to $n-1$. The *Wronskian* of $f_1, \ldots, f_n$ is the function of x defined as the n by n determinant

$$(1.5) \qquad W(f_1, \ldots, f_n; x) = \begin{vmatrix} f_1 & f_2 & \cdots & f_n \\ f_1' & f_2' & \cdots & f_n' \\ \vdots & & & \\ f_1^{(n-1)} & f_2^{(n-1)} & \cdots & f_n^{(n-1)} \end{vmatrix}.$$

In particular, the Wronskian of f_1 and f_2 is

$$(1.6) \qquad W(f_1, f_2; x) = f_1(x) f_2'(x) - f_1'(x) f_2(x).$$

If $f_1, \ldots, f_n$ are dependent over I, their Wronskian vanishes identically over I. The converse is not true, however, as the following example shows. Let I be interval $-1 < x < 1$, and let $f_1 = x^2$, $f_2 = x|x|$; then f_1 and f_2 are independent over I, but their Wronskian vanishes identically.

Abel's Formula for the Wronskian

Let $u_1, \ldots, u_p$ be p solutions (independent or not) of the *homogeneous* equation $Lu = 0$. It can then be shown (see Exercise 1.1) that there exists a constant C such that

$$(1.7) \qquad W(u_1, \ldots, u_p; x) = Ce^{-m(x)}, \qquad x \in I,$$

where $m(x)$ is a particular solution of $m' = a_{p-1}/a_p$. Formula (1.7) is known as *Abel's formula* for the Wronskian. If $p = 2$ and L is formally self-adjoint, then $a_2' = a_1$, so that (1.7) takes the simple form

$$(1.8) \qquad W(u_1, u_2; x) = \frac{C}{a_2(x)}, \qquad x \in I.$$

In both (1.7) and (1.8) the constant C is determined by the solutions $\{u_k\}$ used in forming the Wronskian. Different sets of solutions may yield different constants C. In particular, if $u_1, \ldots, u_p$ are dependent, $W \equiv 0$ so that $C=0$. An immediate conclusion of (1.7) is that, if $W=0$ at a single point x_0, then $C=0$ and hence $W\equiv 0$ on I. We are led to the following theorem.

Theorem 2. Let $u_1,\ldots,u_p$ be solutions of $Lu=0$. The necessary and sufficient condition for these p functions to be dependent is that their Wronskian vanishes at a single point x_0 in I.

Proof. If $u_1,\ldots,u_p$ are dependent, it is clear that $W\equiv 0$ on I and hence surely vanishes at x_0. If $W(u_1,\ldots,u_p;x_0)=0$, we have precisely the necessary and sufficient condition for the existence of a nontrivial solution $(c_1,\ldots,c_p)$ to the set of algebraic equations

$$c_1 u_1(x_0)+c_2 u_2(x_0)+\cdots+c_p u_p(x_0)=0,$$

$$\vdots$$

$$c_1 u_1^{(p-1)}(x_0)+c_2 u_2^{(p-1)}(x_0)+\cdots+c_p u_p^{(p-1)}(x_0)=0.$$

Let $(c_1,\ldots,c_p)$ be a nontrivial solution to this set of equations, and let $U(x)=c_1 u_1(x)+\cdots+c_p u_p(x)$. Clearly U is a solution of the homogeneous differential equation satisfying the initial conditions $U(x_0)=U'(x_0)=\cdots=U^{(p-1)}(x_0)=0$. By Theorem 1, $U\equiv 0$ and hence the set $\{u_1(x),\ldots,u_p(x)\}$ is dependent.

Theorem 3. Let $u_1(x),u_2(x),\ldots,u_p(x)$ be the respective solutions of the IVPs with data $\{0;1,0,\ldots,0\}_{x_0}$, $\{0;0,1,0,\ldots,0\}_{x_0},\ldots,\{0;0,0,\ldots,0,1\}_{x_0}$, that is, u_k is the solution of the homogeneous equation with 0 initial data at x_0 except that $u^{(k-1)}(x_0)=1$. Then the set $(u_1,\ldots,u_p)$ is independent over I, and each solution of $Lu=0$ can be written in the form $u=c_1 u_1+\cdots+c_p u_p$ for some constants $c_1,\ldots,c_p$.

Proof. The set $(u_1,\ldots,u_p)$ is independent since its Wronskian is equal to 1 at x_0. If $u(x)$ is any solution of $Lu=0$, we can write $u(x)=u(x_0)u_1(x)+u'(x_0)u_2(x)+\cdots+u^{(p-1)}(x_0)u_p(x)$, which is just the desired form.

Remark. Any set of p independent solutions of $Lu=0$ is called a *basis* for this equation. In Theorem 3 we exhibited a special, simple basis, but there are others. Let us look at the equation $u''+u=0$, which has the basis $\{\cos x,\sin x\}$ (which is of the type in Theorem 3 with $x_0=0$, $u_1=\cos x,u_2=\sin x$); but the functions $v_1(x)=\cos x+\sin x$, $v_2(x)=2\cos x-3\sin x$ also form a basis. Every solution of $u''+u=0$ can be expressed either as Au_1+Bu_2 or as Cv_1+Dv_2.

The Inhomogeneous Equation

Let $a_p(x) \neq 0$ on $-\infty < x < \infty$, and let $u_\xi(x)$ be the solution of the homogeneous equation satisfying the initial conditions $u_\xi(\xi) = 0$, $u'_\xi(\xi) = 0, \ldots, u_\xi^{(p-2)}(\xi) = 0$, $u_\xi^{(p-1)}(\xi) = 1/a_p(\xi)$. We have already seen (at least in the case where the coefficients are infinitely differentiable) that $E(x, \xi) = H(x - \xi)u_\xi(x)$ is the causal fundamental solution satisfying, in the distributional sense, $LE = \delta(x - \xi)$ with $E \equiv 0$, $x < \xi$. Even when the coefficients in L are only continuous, we still define the causal fundamental solution as $H(x - \xi)u_\xi(x)$. By superposition one would then expect that the solution of

$$Lu = f, \quad x > a; \quad u(a) = 0, \ldots, u^{(p-1)}(a) = 0$$

would be given by

$$u(x) = \int_a^\infty H(x - \xi)u_\xi(x)f(\xi)\,d\xi = \int_a^x u_\xi(x)f(\xi)\,d\xi.$$

Let us verify that this is true. We have $u(a) = 0$ and $u'(x) = u_x(x)f(x) + \int_a^x u'_\xi(x)f(\xi)\,d\xi$, so that $u'(a) = 0$. Since $u_x(x) = \cdots = u_x^{(p-2)}(x) = 0$, we find that

$$u^{(k)}(x) = u_x^{(k-1)}(x)f(x) + \int_a^x u_\xi^{(k)}(x)f(\xi)\,d\zeta$$

$$= \int_a^x u_\xi^{(k)}(x)f(\xi)\,d\xi, \quad k = 1, 2, \ldots, p - 1.$$

Thus $u^{(k)}(a) = 0$ for $k = 1, 2, \ldots, p - 1$. We also note that

$$u^{(p)}(x) = u_x^{(p-1)}(x)f(x) + \int_a^x u_\xi^{(p)}(x)f(\xi)\,d\xi = \frac{f(x)}{a_p(x)} + \int_a^x u_\xi^{(p)}(x)f(\xi)\,d\xi.$$

Hence $Lu = a_p(x)u^{(p)} + \cdots + a_0(x)u = f(x) + \int_a^x Lu_\xi(x)f(\xi)\,d\xi = f(x)$, since $Lu_\xi = 0$. We observe that we have not used the fact that $x > a$. Therefore we have proved the following theorem.

Theorem 4. If $a_p(x) \neq 0$ in $-\infty < x < \infty$, the one and only solution of the IVP with data $\{f; 0, \ldots, 0\}_a$ is

(1.9)
$$u(x) = \int_a^x u_\xi(x)f(\xi)\,d\xi,$$

where $u_\xi(x)$ is the solution of the IVP for the homogeneous equation

(1.10)
$$Lu = 0, \quad -\infty < x < \infty; \quad u(\xi) = 0, \ldots, u^{(p-2)}(\xi) = 0, u^{(p-1)}(\xi) = \frac{1}{a_p(\xi)}.$$

Remarks

1. The solution of the problem

(1.11) $Lv = f; \qquad v(a) = \gamma_1, \ldots, v^{(p-1)}(a) = \gamma_p$

can be written as

(1.12) $v(x) = \int_a^x u_\xi(x) f(\xi) \, d\xi + \gamma_1 u_1(x) + \cdots + \gamma_p u_p(x),$

where $(u_1, \ldots, u_p)$ is the basis of Theorem 3 (with $x_0 = a$).

2. We have reduced the problem of solving the inhomogeneous differential equation to that of finding the general solution of the homogeneous differential equation. Except for first-order equations and equations with constant coefficients, there is no systematic method for finding explicitly the solutions of the homogeneous equation. If one solution of the homogeneous equation is known, it is possible to reduce the order of the equation by 1. Thus, if we know a solution to a second-order equation, we can find all its solutions.

Exercises

1.1. Prove Abel's formula for the Wronskian. *Hint:* First show that the derivative of a p by p determinant is the sum of p determinants, each of which has only one row differentiated.

1.2. Consider the forced wave equation

$$\frac{\partial^2 u}{\partial t^2} - \frac{\partial^2 u}{\partial x^2} = f(x,t), \qquad 0 < x < 1, \quad t > 0,$$

with the boundary conditions $u(0,t) = u(1,t) = 0$, $t > 0$, and the initial conditions $u(x,0) = (\partial u / \partial t)(x,0) = 0$, $0 < x < 1$. Write $u(x,t) = \sum_{k=1}^{\infty} u_k(t) \sin k\pi x$, and show that u_k satisfies an IVP for an ordinary differential equation. Using (1.9), find $u_k(t)$ and hence $u(x,t)$. Consider the particular case $f(x,t) = \sin \omega t$, and discuss the behavior of the solution for large t (the case where $\omega = k\pi$ has to be treated separately).

1.3. The causal fundamental solution $E(t,\tau)$ is known as the *impulse response* in electrical engineering. E satisfies

$$LE = \delta(t - \tau), \qquad E \equiv 0, \quad t < \tau.$$

The *step response* $F(t,\tau)$ is the solution of

$$(1.13) \qquad LF = H(t-\tau), \qquad F \equiv 0, \quad t < \tau.$$

(a) Show that $-\partial F/\partial \tau$ is a fundamental solution. Since $\partial F/\partial \tau = 0$, $t < \tau$, we must have $-\partial F/\partial \tau = E(t,\tau)$.

(b) Show that the solution of

$$Lu = f(t), \qquad u(a) = u'(a) = \cdots = u^{(p-1)}(a) = 0$$

can be written in either of the forms

$$(1.14) \qquad \begin{aligned} u(t) &= \int_a^t E(t,\tau)f(\tau)\,d\tau, \\ u(t) &= F(t,a)f(a) + \int_a^t F(t,\tau)f'(\tau)\,d\tau. \end{aligned}$$

(c) What simplifications are possible when the coefficients in L are constant?

2. BOUNDARY VALUE PROBLEMS FOR SECOND-ORDER EQUATIONS

Formulation

In the initial value problem for a second-order equation we specify as accessory conditions the values of u and u' at some point x_0. In a boundary value problem we impose two conditions involving the values of u and u' at the points a and b; we are then interested in solving the differential equation subject to these boundary conditions in the interval $a < x < b$ (which we take to be bounded). Unlike the IVP, the general boundary value problem may not have a solution, or may have more than one solution.

We shall consider the differential equation

$$(2.1) \qquad Lu \doteq a_2(x)u'' + a_1(x)u' + a_0(x)u = f(x), \qquad a < x < b,$$

where the coefficients are continuous in $a \leqslant x \leqslant b$, $a_2(x) \neq 0$ in $a \leqslant x \leqslant b$, and $f(x)$ is piecewise continuous in $a \leqslant x \leqslant b$. The solution $u(x)$ is required to satisfy the two boundary conditions

$$(2.2) \qquad \begin{aligned} B_1 u &\doteq \alpha_{11} u(a) + \alpha_{12} u'(a) + \beta_{11} u(b) + \beta_{12} u'(b) = \gamma_1, \\ B_2 u &\doteq \alpha_{21} u(a) + \alpha_{22} u'(a) + \beta_{21} u(b) + \beta_{22} u'(b) = \gamma_2, \end{aligned}$$

where the row vectors $(\alpha_{11}, \alpha_{12}, \beta_{11}, \beta_{12})$ and $(\alpha_{21}, \alpha_{22}, \beta_{21}, \beta_{22})$ are independent (neither row is a multiple of the other).

We refer to B_1 and B_2 as *boundary functionals*, since they assign to each sufficiently smooth function $u(x)$ the numbers $B_1 u$ and $B_2 u$, respectively. The differential equation (2.1), together with the boundary conditions (2.2), forms a *boundary value problem* (BVP).

Remarks

1.   We regard the coefficients $a_i(x), \alpha_{ij}, \beta_{ij}$ as fixed (that is, L, B_1, B_2 are fixed), and we are interested in studying the dependence of the solution on $f(x), \gamma_1, \gamma_2$. We therefore refer to $\{f; \gamma_1, \gamma_2\}$ as the *data* for the problem. The coefficients and the data consist of *real* functions and *real* numbers.

2.   The independence of the row vectors $(\alpha_{11}, \alpha_{12}, \beta_{11}, \beta_{12})$ and $(\alpha_{21}, \alpha_{22}, \beta_{21}, \beta_{22})$ guarantees that we really have two distinct boundary conditions. (If the row vectors were dependent, one boundary functional would be a multiple of the other, and the two boundary conditions would be either identical or inconsistent.)

3.   If $\beta_{11} = \beta_{12} = \alpha_{21} = \alpha_{22} = 0$, the boundary conditions are said to be *unmixed* (one condition per endpoint):

$$
(2.3) \qquad
\begin{aligned}
B_1 u &= \alpha_{11} u(a) + \alpha_{12} u'(a) = \gamma_1, \\
B_2 u &= \beta_{21} u(b) + \beta_{22} u'(b) = \gamma_2.
\end{aligned}
$$

If $\alpha_{12} = \beta_{11} = \beta_{12} = \alpha_{21} = \beta_{21} = \beta_{22} = 0$, $\alpha_{11} = 1$, $\alpha_{22} = 1$, we have the *initial conditions* $u(a) = \gamma_1$, $u'(a) = \gamma_2$.

4.   Why do we confine ourselves to boundary conditions of type (2.2)? We do so partly because these occur most frequently in applications and partly because a general mathematical theory is most easily developed for conditions of this type. The reason for having two conditions instead of some other number is rather obvious; fewer conditions than the order of the equation would not determine the solution, and more conditions would usually prevent us from having any solution at all. We also exclude in (2.2) all derivatives of order greater than or equal to the order of the differential equation. The reason for this can perhaps be seen by studying the example $u'' = f(x)$; clearly u'' is already determined at the endpoints by the differential equation, so that it would not make sense to specify u'' independently; higher derivatives of u are already known at the endpoints by differentiating the differential equation, so that it would be inappropriate to prescribe such a derivative.

The *superposition principle* applies to solutions of (2.1)-(2.2). If u satisfies the BVP with data $\{f; \gamma_1, \gamma_2\}$ and U the one with data $\{F; \Gamma_1, \Gamma_2\}$, then $Au + Bu$ satisfies the BVP with data $\{Af + BF; A\gamma_1 + B\Gamma_1, A\gamma_2 + B\Gamma_2\}$. If u and v both satisfy the BVP with data $\{f; \gamma_1, \gamma_2\}$, then $u - v$ satisfies the BVP with data $\{0; 0, 0\}$, that is, the completely homogeneous problem. This leads to the important *uniqueness* condition: *If the BVP with data $\{0; 0, 0\}$ has only the solution $u \equiv 0$, then the BVP with data $\{f; \gamma_1, \gamma_2\}$ has at most one solution; if the BVP with data $\{0; 0, 0\}$ has a nontrivial solution, then the BVP with data $\{f; \gamma_1, \gamma_2\}$ either has no solution or has more than one solution.*

Example 1. The difficulties that can arise are illustrated by the simple BVP

$$(2.4) \qquad -u'' = f(x), \quad 0 < x < 1, \qquad u'(0) = \gamma_1, \quad -u'(1) = \gamma_2,$$

which represents steady one-dimensional heat flow in a rod with pre-scribed source density along the rod and prescribed heat flux at the ends. By integrating the differential equation from 0 to 1, we see immediately that $f(x), \gamma_1, \gamma_2$ must satisfy the relation

$$(2.5) \qquad \int_0^1 f(x)\, dx = \gamma_2 + \gamma_1,$$

which merely states that a steady state is possible only if the heat supplied along the rod is removed at the ends. For the data $\{1; 0, 0\}$, (2.4) has *no* solution. However, for the data $\{\sin 2\pi x; 0, 0\}$, which satisfies (2.5), there are many solutions: $u(x) = A - (x/2\pi) + (1/4\pi^2)\sin 2\pi x$, with A arbitrary.

Example 2. As another illustration consider the BVP

$$(2.6) \qquad u'' + a_0 u = f(x), \quad 0 < x < 1, \qquad u(0) = \gamma_1, \quad u(1) = \gamma_2,$$

where a_0 is a fixed constant.

(a) If a_0 is *not* one of the numbers $\pi^2, 4\pi^2, 9\pi^2, \ldots$, the BVP with data $\{0; 0, 0\}$ has only the trivial solution and therefore the BVP with data $\{f; \gamma_1, \gamma_2\}$ has at most one solution. We shall show later that there does in fact exist a solution for any data.

(b) If $a_0 = \pi^2$, say, the BVP with data $\{0; 0, 0\}$ has the nontrivial solutions $A \sin \pi x$, where A is an arbitrary constant. The BVP with data $\{1; 0, 0\}$ has *no solution*: indeed the general solution of the differential equation is then $u = (1/\pi^2) + A \sin \pi x + B \cos \pi x$, and the boundary conditions give $(1/\pi^2) + B = 0$ and $(1/\pi^2) - B = 0$, which are inconsistent. The

BVP with data $\{x - \frac{1}{2}; 0, 0\}$ has *many solutions*: the general solution of the differential equation is $u = (x/\pi^2) - (1/2\pi^2) + A \sin \pi x + B \cos \pi x$, and the boundary conditions give $-(1/2\pi^2) + B = 0$ and $(1/2\pi^2) - B = 0$, so that $B = (1/2\pi^2)$ and A remains arbitrary.

Another way of stating what is happening is in terms of eigenvalues [see (2.16), Chapter 1]. If a_0 is not an eigenvalue λ_k of $u'' + \lambda u = 0$, $0 < x < 1$, $u(0) = 0$, $u(1) = 0$, then (2.6) has at most one solution for any data $\{f; \gamma_1, \gamma_2\}$. If a_0 is an eigenvalue, then we either have no solution of (2.6) or have more than one solution.

Green's Function and Its Uses

We assume that the completely homogeneous problem (that is, with data $\{0; 0, 0\}$*) has only the trivial solution.* We shall then show that problem (2.1)-(2.2) with data $\{f; \gamma_1, \gamma_2\}$ has one and only one solution. This will be done by an explicit construction using Green's function $g(x, \xi)$, which is the solution corresponding to the data $\{\delta(x - \xi); 0, 0\}$, where ξ is a fixed point in $a < x < b$. Of course the forcing function $\delta(x - \xi)$ is not a piecewise continuous function, so that this problem does not fall directly in the classical category being studied. We regard $g(x, \xi)$ as the fundamental solution that satisfies the boundary conditions $B_1 g = 0$, $B_2 g = 0$. When the coefficients in L are infinitely differentiable, we found in Chapter 2 that a fundamental solution E satisfies the homogeneous equation for $x < \xi$ and $x > \xi$, that E is continuous at $x = \xi$, and that its first derivative has a jump $1/a_2(\xi)$ at $x = \xi$. Using this experience and the physical examples of Chapter 1 as a guide, it is natural to *define* Green's function $g(x, \xi)$ associated with BVP (2.1)-(2.2) as the solution of

$$(2.7) \quad \begin{cases} Lg = 0, \quad a < x < \xi, \xi < x < b; \quad B_1 g = 0, \quad B_2 g = 0, \\ g \text{ continuous at } x = \xi; \quad \left. \dfrac{dg}{dx} \right|_{x = \xi +} - \left. \dfrac{dg}{dx} \right|_{x = \xi -} = \dfrac{1}{a_2(\xi)}. \end{cases}$$

Boundary value problem (2.7) can be written more succinctly in the delta function notation:

$$(2.8) \qquad Lg = \delta(x - \xi), \quad a < x, \xi < b; \qquad B_1 g = 0, \quad B_2 g = 0.$$

Thus Green's function is the response, under homogeneous boundary conditions, to a forcing function consisting of a concentrated unit of inhomogeneity at $x = \xi$. Since the difference between two solutions of (2.7) has a continuous derivative everywhere, it is a classical solution of the completely homogeneous problem which, by our assumption, necessarily vanishes identically. Therefore (2.7) has at most one solution, which we now construct explicitly in different cases.

1. *Unmixed conditions.* Green's function is to satisfy the boundary conditions $B_1 g = \alpha_{11} g(a, \xi) + \alpha_{12} g'(a, \xi) = 0$ and $B_2 g = \beta_{21} g(b, \xi) + \beta_{22} g'(b, \xi) = 0$. Let $u_1(x)$ be a nontrivial solution of $Lu = 0$ satisfying $B_1 u = 0$. Such a solution must exist. One can, for instance, choose u_1 to be the solution of $Lu = 0$ with the initial conditions $u(a) = \alpha_{12}$, $u'(a) = -\alpha_{11}$; since α_{12} and α_{11} are not both 0, $u_1(x)$ is not identically 0. Similarly let $u_2(x)$ be a nontrivial solution of $Lu = 0$ satisfying the boundary condition at $x = b$ ($B_2 u = 0$). *We note that u_1 and u_2 are independent,* since by assumption the completely homogeneous problem has only the trivial solution. Green's function is therefore of the form

$$g(x, \xi) = A u_1(x), \quad a < x < \xi; \qquad g = B u_2(x), \quad \xi < x < b,$$

where the constants A and B may of course depend on ξ. The continuity of g and the jump condition on g' at $x = \xi$ yield

$$A u_1(\xi) - B u_2(\xi) = 0,$$

$$-A u_1'(\xi) + B u_2'(\xi) = \frac{1}{a_2(\xi)}.$$

This inhomogeneous system of two algebraic equations in A and B has one and only one solution if and only if the determinant of the coefficients does not vanish. This determinant is just the Wronskian of u_1 and u_2 evaluated at ξ. Since u_1 and u_2 are independent, their Wronskian does not vanish anywhere, so that we can solve for A and B to obtain

$$A = \frac{u_2(\xi)}{a_2(\xi) W(u_1, u_2; \xi)}, \qquad B = \frac{u_1(\xi)}{a_2(\xi) W(u_1, u_2; \xi)},$$

and, hence Green's function is given by

(2.9) $$g(x, \xi) = \frac{u_1(x_<) u_2(x_>)}{a_2(\xi) W(u_1, u_2; \; \xi)}, \qquad a < x, \xi < b,$$

where $x_< = \min(x, \xi)$, $x_> = \max(x, \xi)$. It is easy to verify that g as given by (2.9) actually satisfies (2.7). One can use Abel's formula (1.7) for the Wronskian to cast (2.9) in a somewhat simpler form. In the self-adjoint case, C is defined from (1.8) and

(2.10) $$g(x, \xi) = \frac{1}{C} u_1(x_<) u_2(x_>).$$

2. *Initial conditions.* We have already seen that in this case Green's function (known as the causal fundamental solution) is given by

$$(2.11) \qquad\qquad g(x,\xi) = H(x-\xi)u_\xi(x),$$

where $u_\xi(x)$ is the solution of $Lu=0$ satisfying the conditions $u(\xi)=0$, $u'(\xi)=1/a_2(\xi)$. If the coefficients are constant, we have $g(x,\xi) = H(x-\xi)u_0(x-\xi)$.

3. *General boundary conditions.* If g is to satisfy mixed conditions $B_1 g = B_2 g = 0$, where B_1 and B_2 are the most general boundary functionals (2.2), we write g as the sum of the causal fundamental solution $H(x-\xi)u_\xi(x)$ and of a solution of the homogeneous equation to be determined. We already know that the causal solution satisfies (2.7) except for the boundary conditions. Let $u_1(x)$ be a nontrivial solution of $Lu=0$ satisfying $B_1 u = 0$ (such a solution must exist!), and let $u_2(x)$ be a nontrivial solution of $Lu=0$ with $B_2 u = 0$. Setting

$$(2.12) \quad g(x,\xi) = H(x-\xi)u_\xi(x) + Au_1(x) + Bu_2(x), \qquad a<x, \xi<b,$$

we impose the boundary conditions to find that

$$0 = \beta_{11} u_\xi(b) + \beta_{12} u_\xi'(b) + B(B_1 u_2),$$
$$0 = \beta_{21} u_\xi(b) + \beta_{22} u_\xi'(b) + A(B_2 u_1),$$

from which we can solve for A and B (since neither $B_1 u_2$ nor $B_2 u_1$ can vanish).

Green's function just constructed enables us to solve (2.1)-(2.2). We begin with the case of homogeneous boundary conditions, that is, the data is $\{f(x); 0, 0\}$. We claim that the one and only solution of the BVP

$$(2.13) \qquad\qquad Lu = f, \quad a<x<b; \qquad B_1 u = 0, \quad B_2 u = 0$$

is given by

$$(2.14) \qquad\qquad u(x) = \int_a^b g(x,\xi)f(\xi)\, d\xi.$$

The proof is easy. We note first that u satisfies the homogeneous boundary conditions because g does. Since g can be written in the form (2.12), where A and B depend only on ξ, it follows that the right side of

(2.14) reduces to

$$\int_a^x u_\xi(x)f(\xi)\,d\xi + \theta_1 u_1(x) + \theta_2 u_2(x),$$

where θ_1, θ_2 are constants. The first term satisfies the inhomogeneous differential equation (Theorem 4, Section 1), while the remaining terms are solutions of the homogeneous equation. The sum therefore satisfies $Lu = f$. Uniqueness follows from the assumption that the BVP with data $\{0;0,0\}$ has only the trivial solution.

Next we observe that the problem with data $\{0; \gamma_1, \gamma_2\}$ has the unique solution

$$v(x) = \frac{\gamma_2}{B_2 u_1} u_1(x) + \frac{\gamma_1}{B_1 u_2} u_2(x),$$

where u_1 is a nontrivial solution of the homogeneous equation satisfying $B_1 u_1 = 0$, and u_2 is a nontrivial solution of the same equation with $B_2 u_2 = 0$. Therefore, by superposition, the problem

$$(2.15) \qquad Lu = f, \quad a < x < b; \qquad B_1 u = \gamma_1, \quad B_2 u = \gamma_2$$

has the unique solution

$$(2.16) \qquad u = \int_a^b g(x,\xi)f(\xi)\,d\xi + \frac{\gamma_2}{B_2 u_1} u_1(x) + \frac{\gamma_1}{B_1 u_2} u_2(x).$$

To recapitulate, we have the following theorem.

Theorem 1. If the completely homogeneous BVP has only the trivial solution, then the BVP with data $\{f; \gamma_1, \gamma_2\}$ has one and only one solution.

Remark. The theorem holds because we have exactly two boundary conditions of type (2.2). If we imposed a third condition of the same type, the completely homogeneous BVP would still have only the trivial solution, but then the inhomogeneous problem would usually have no solution.

The Adjoint Problem

As we saw in Section 5, Chapter 2, each second-order operator with coefficients in $C^2(a,b)$,

$$(2.17) \qquad L = a_2(x)D^2 + a_1(x)D + a_0(x),$$

has a well-defined formal adjoint

(2.18) $$L^* = a_2 D^2 + (2a_2' - a_1)D + (a_2'' - a_1' + a_0)$$

such that, for any pair of functions u, v in $C^2(a, b)$,

(2.19) $$\int_a^b (vLu - uL^*v)\, dx = J(u, v)\big]_a^b,$$

where

(2.20) $$J(u, v) = a_2(vu' - uv') + (a_1 - a_2')uv.$$

A formally self-adjoint operator is one for which $L = L^*$, which is equivalent to $a_2' = a_1$, so that

(2.21) $$L = L^* = D(a_2 D) + a_0, \qquad J = a_2(vu' - uv'),$$

leading to a simplification of (2.19).

We now wish to introduce the notion of adjoint boundary conditions. Given the operator (2.17) and a function u satisfying two homogeneous conditions $B_1 u = 0$ and $B_2 u = 0$ of type (2.2), the right side of (2.19) will vanish only if v itself satisfies a pair of homogeneous boundary conditions —the so-called adjoint boundary conditions. To make matters precise, let M be the set of functions $u(x)$ in $C^2(a, b)$ satisfying $B_1 u = 0$ and $B_2 u = 0$. M is a linear space, that is, if $u_1(x)$ and $u_2(x)$ are in M, so is $\alpha u_1(x) + \beta u_2(x)$ for any constants α and β. The set M^* is defined as the set of functions $v(x)$ in $C^2(a, b)$ such that $J(u, v)]_a^b = 0$ *for all* $u \in M$. It is clear that M^* is a linear space. Functions in M^* can be characterized by two homogeneous boundary conditions $B_1^* v = 0$ and $B_2^* v = 0$, where B_1^* and B_2^* are boundary functionals of type (2.2) but usually with different coefficients from those of B_1 and B_2. Although M^* is unambiguously determined from M, the specific functionals B_1^*, B_2^* are not (for instance, we can exchange the order of the boundary functionals, or we could consider other equivalent functionals such as $B_1^{*\prime} = B_1^* + B_2^*$, $B_2^{*\prime} = B_1^* - B_2^*$). This leads to the definition of a self-adjoint problem.

Definition. We say that a boundary value problem (L, B_1, B_2) is *self-adjoint* if $L^* = L$ and $M^* = M$ (that is, the adjoint boundary conditions define the same set of functions as the homogeneous boundary conditions $B_1 u = B_2 u = 0$, so that it is possible to choose B_1^* and B_2^* with $B_1^* = B_1$, $B_2^* = B_2$).

We now illustrate these ideas through some examples.

Example 3. $L = D^2$, $a = 0$, $b = 1$, $B_1 u = u'(0) - u(1)$, $B_2 u = u'(1)$. Then $L^* = L$, and Green's formula becomes

$$(2.22) \quad \int_0^1 (vu'' - uv'')\,dx = v(1)u'(1) - u(1)v'(1) - v(0)u'(0) + u(0)v'(0).$$

The set M consists of all functions u such that $u(1) = u'(0)$ and $u'(1) = 0$. We want to find the set M^* of all functions v such that the right side of (2.22) vanishes whenever u is in M. For u in M the right side of (2.22) simplifies to

$$- u(1)[v(0) + v'(1)] + u(0)v'(0),$$

where $u(0)$ and $u(1)$ are arbitrary. Thus the set M^* consists of the functions v for which $v'(0) = 0$ and $v(0) + v'(1) = 0$. The adjoint boundary conditions can be taken as $B_1^* v = v'(0) = 0$, $B_2^* v = v(0) + v'(1) = 0$. [One could just as well characterize the same set M^* by the conditions $v'(0) + v(0) + v'(1) = 0$, $2v'(0) - v(0) - v'(1) = 0$, or by any pair of conditions obtained by taking independent linear combinations of B_1^* and B_2^*.]

Example 4. Let L be formally self-adjoint, $L = D(a_2 D) + a_0$, with *unmixed* boundary functionals, $B_1 u = \alpha_{11} u(a) + \alpha_{12} u'(a)$, $B_2 u = \beta_{21} u(b) + \beta_{22} u'(b)$. Then $M^* = M$ (see Exercise 2.2). This is a typical self-adjoint BVP.

Example 5. $L = a_2 D^2 + a_1 D + a_0$, $a = 0$, $b = 1$, $B_1 u = u(0)$, $B_2 u = u'(1)$. Then M is the set of functions u with $u(0) = u'(1) = 0$. For u in M, $J(u,v)]_0^1 = \{[a_1(1) - a_2'(1)]v(1) - a_2(1)v'(1)\}u(1) - a_2(0)v(0)u'(0)$. Since $u(1)$ and $u'(0)$ are arbitrary, M^* is characterized by $v(0) = 0$, $[a_1(1) - a_2'(1)]v(1) - a_2(1)v'(1) = 0$. We see that M^* coincides with M only if $a_2'(1) = a_1(1)$, which will certainly be the case if $L = L^*$.

Example 6. Initial value problem. Whether or not L is formally self-adjoint, we always have $M^* \neq M$. Take $L = a_2 D^2 + a_1 D + a_0$ on an interval $a < x < b$ with $B_1 u = u(a)$, $B_2(u) = u'(a)$. Then M is the set of functions u with $u(a) = u'(a) = 0$. For u in M,

$$J(u,v)]_a^b = \{[a_1(b) - a_2'(b)]v(b) - a_2(b)v'(b)\}u(b) + a_2(b)v(b)u'(b),$$

and, since $u(b)$ and $u'(b)$ are arbitrary, M^* consists of functions v such that $v(b) = v'(b) = 0$. Thus the adjoint boundary conditions never coincide with the original boundary conditions.

A slightly different approach to adjoints is taken in Section 3. Let us now look at some consequences of the ideas just introduced.

Consider the BVP

(2.23) $Lg(x,\xi) = \delta(x-\xi), \quad a<x,\xi<b; \qquad B_1 g = 0, \quad B_2 g = 0,$

whose solution will be called the *direct Green's function*. Let $g^*(x,\xi)$ be the *adjoint Green's function* satisfying

(2.24) $L^* g^*(x,\xi) = \delta(x-\xi), \quad a<x,\xi<b; \qquad B_1^* g^* = 0, \qquad B_2^* g^* = 0.$

As usual all differentiations in (2.23) and (2.24) are with respect to x. We shall show that g and g^* are related by

(2.25) $g^*(x,\xi) = g(\xi,x).$

Once (2.25) has been established, there is no further need to solve (2.24). Its solution is just the solution of (2.23) with variables interchanged. Another way of saying this is that $g(x,\xi)$ satisfies the direct problem in its first variable and the adjoint problem in its second variable; thus $g(x,\xi)$ satisfies the adjoint boundary conditions in the variable ξ.

To prove (2.25) set $\xi = \eta$ in (2.24), multiply (2.23) by $g^*(x,\eta)$ and (2.24) by $g(x,\xi)$, subtract, and integrate from $x=a$ to $x=b$. Using the formal properties of the delta function, we find that

$$\int_a^b \left[g^*(x,\eta) Lg(x,\xi) - g(x,\xi) L^* g^*(x,\eta) \right] dx = g^*(\xi,\eta) - g(\eta,\xi),$$

or, by Green's formula,

$$J(g^*,g)\big]_a^b = g^*(\xi,\eta) - g(\eta,\xi).$$

Since g^* satisfies the adjoint homogeneous boundary conditions, $J(g^*,g)\big]_a^b = 0$ (in fact, that is exactly how adjoint boundary conditions were defined!). This then proves (2.25). As an important corollary we note that, if (L, B_1, B_2) is *self-adjoint*, then $g(x,\xi)$ is symmetric, that is,

(2.26) $g(x,\xi) = g(\xi,x).$

Relation (2.26) is also known as the *reciprocity principle*: The response at x caused by a unit source at ξ is the same as the response at ξ due to a unit source at x.

One can use the fact that $g(\xi,x)$ satisfies (2.24) to write the solution of a problem with nonzero boundary data in terms of Green's function. Consider first

(2.27) $Lu = 0, \quad a<x<b; \qquad B_1 u = \gamma_1, \quad B_2 u = \gamma_2.$

Multiply (2.27) by $g(\xi, x)$ and (2.24) by u, subtract, and integrate to obtain

$$-u(\xi) = \int_a^b \left[g(\xi, x) Lu - u L^* g(\xi, x) \right] dx = J(u(x), g(\xi, x))]_{x=a}^{x=b},$$

or, on relabeling variables,

(2.28) $$u(x) = -J(u(\xi), g(x, \xi))]_{\xi=a}^{\xi=b}.$$

The term on the right side of (2.28) involves the direct Green's function $g(x, \xi)$, *its derivative with respect to* ξ, and the given boundary data γ_1, γ_2. The last part of this statement is not entirely obvious but is illustrated in the following example and perhaps more convincingly in Section 3.

Example 7. Consider the problem

$$u'' = 0, \quad 0 < x < 1; \quad B_1 u = u'(0) - u(1) = \gamma_1, \quad B_2 u = u'(1) = \gamma_2,$$

which is based on Example 3. We must first calculate $g(x, \xi)$, the solution of $Lg = \delta(x - \xi)$, $B_1 g = 0$, $B_2 g = 0$. By (2.12) we have

$$g(x, \xi) = (x - \xi) H(x - \xi) - x + \xi - 1,$$

so that $g(x, 0) = -1$, $g(x, 1) = -x$, $(dg/d\xi)(x, 0) = 0$, $(dg/d\xi)(x, 1) = 1$, and, by (2.28),

$$u(x) = -\left[g(x, \xi) u'(\xi) - u(\xi) \frac{dg}{d\xi}(x, \xi) \right]_{\xi=0}^{\xi=1} = x\gamma_2 - \gamma_1.$$

Now, by superposition of (2.13) and (2.27), we can express the solution of the completely inhomogeneous problem (2.15) as the sum of (2.14) and (2.28):

(2.29) $$u(x) = \int_a^b g(x, \xi) f(\xi) d\xi - J(u(\xi), g(x, \xi))]_{\xi=a}^{\xi=b},$$

which should be compared with the earlier form (2.16). For ordinary differential equations (2.16) might be regarded as simpler than (2.29), but when dealing with partial differential equations nothing like (2.16) is available, whereas (2.29) survives with some obvious modifications.

Exercises

2.1. *Extension of the superposition principle.* Suppose $Lu = 0$, $a < x < b$, $B_1 u = B_2 u = 0$ has only the trivial solution.

(a) Let θ be a real parameter. The BVP

$$Lu = f(x, \theta), \quad a < x < b; \qquad B_1 u = \gamma_1(\theta), \quad B_2 u = \gamma_2(\theta)$$

has one and only one solution $u(x, \theta)$. Show that the solution of

$$Lv = \frac{\partial f}{\partial \theta}, \quad a < x < b; \qquad B_1 v = \frac{d\gamma_1}{d\theta}, \quad B_2 v = \frac{d\gamma_2}{d\theta}$$

is given by $v(x, \theta) = (\partial / \partial \theta) u(x, \theta)$.

(b) Show that the solution $h(x, \xi)$ of

$$Lh = \delta'(x - \xi), \quad a < x, \xi < b; \qquad B_1 h = B_2 h = 0,$$

is given by $h = -\partial g / \partial \xi$, where g is Green's function. Find the solution of

$$-h'' = \delta'(x - \xi), \quad 0 < x, \xi < 1, \qquad h(0, \xi) = h(1, \xi) = 0.$$

2.2. Let $L = L^*$, and let the boundary functionals be of type (2.2). Show that the necessary and sufficient condition for self-adjointness is $a_2(a) P_{34} = a_2(b) P_{12}$, where

$$P_{12} = \begin{vmatrix} \alpha_{11} & \alpha_{12} \\ \alpha_{21} & \alpha_{22} \end{vmatrix}, \qquad P_{34} = \begin{vmatrix} \beta_{11} & \beta_{12} \\ \beta_{21} & \beta_{22} \end{vmatrix}.$$

This condition is satisfied for unmixed boundary conditions and for the "periodic" boundary conditions $u(a) = u(b)$, $a_2(a) u'(a) = a_2(b) u'(b)$.

2.3. Let $L = D^2 + 4D - 3$. Find L^* and J. If the boundary functionals are $B_1 u = u'(a) + 4u(a)$ and $B_2 u = u'(b) + 4u(b)$, find B_1^* and B_2^*.

2.4. Let $0 < a < b$; consider the BVP

$$(xg')' + \left(x - \frac{\nu^2}{x} \right) g = \delta(x - \xi), \quad a < x, \xi < b; \qquad g(a, \xi) = g(b, \xi) = 0.$$

Using information about Bessel functions, determine $g(x, \xi)$. Do the same for the problem

$$(xg')' - \left(x + \frac{\nu^2}{x} \right) g = \delta(x - \xi), \quad a < x, \xi < b; \qquad g(a, \xi) = g(b, \xi) = 0.$$

Are there any values of ν^2 for which either construction fails?

2.5. Find Green's function $g(x,\xi)$ of $L=D^2$ subject to the conditions $u'(0)=u(1)$, $u'(1)=0$. Then $L^*=L$, and the adjoint boundary conditions were found in Example 3. Find directly Green's function $g^*(x,\xi)$ for the adjoint problem, and show, by comparing explicit formulas, that $g^*(x,\xi)=g(\xi,x)$.

2.6. Let $L=D^2$ on $0<x<1$, and consider a problem with only one boundary functional $B_1(u)=u(0)$. Show that the adjoint homogeneous problem has three boundary conditions.

2.7. A boundary condition of type (2.2) specifies the value of a linear functional of a particular type. One can of course consider other kinds of linear functionals such as $\int_a^b u(x)h(x)\,dx$, where $h(x)$ is a given function.

Consider the BVP

$$(2.30) \qquad \begin{array}{l} -u''=f, \quad -1<x<1; \\[2mm] B_1 u \doteq \displaystyle\int_{-1}^{1} xu(x)\,dx = \gamma_1, B_2 u \doteq \int_{-1}^{1} u(x)\,dx = \gamma_2. \end{array}$$

Find the corresponding Green's function, and then solve (2.30).

3. BOUNDARY VALUE PROBLEMS FOR EQUATIONS OF ORDER p

Again we let $a \leqslant x \leqslant b$ be a bounded interval on the x axis. The linear operator $L=a_p(x)D^p+\cdots+a_1(x)D+a_0(x)$ has coefficients that belong to C^p on the closed interval $a \leqslant x \leqslant b$, and $a_p(x)\neq 0$ on $a \leqslant x \leqslant b$. We shall consider the boundary value problem

$$(3.1) \qquad Lu \doteq a_p u^{(p)} + \cdots + a_1 u' + a_0 u = f, \qquad a<x<b,$$

with the p boundary conditions

$$(3.2)$$

$$B_1 u \doteq \alpha_{11} u(a) + \cdots + \alpha_{1p} u^{(p-1)}(a) + \beta_{11} u(b) + \cdots + \beta_{1p} u^{(p-1)}(b) = \gamma_1,$$

$$\vdots$$

$$B_p u \doteq \alpha_{p1} u(a) + \cdots + \alpha_{pp} u^{(p-1)}(a) + \beta_{p1} u(b) + \cdots + \beta_{pp} u^{(p-1)}(b) = \gamma_p.$$

The p row vectors $(\alpha_{11},\ldots,\alpha_{1p},\beta_{11},\ldots,\beta_{1p}),\ldots,(\alpha_{p1},\ldots,\alpha_{pp},\beta_{p1},\ldots,\beta_{pp})$ are assumed independent (no one of them is a linear combination of the others; in particular, no row vector has all entries $=0$). The differential operator L and the boundary functionals $B_1,\ldots,B_p$ are fixed, and we are interested in studying the dependence of the solution u of (3.1)-(3.2) on the data $\{f;\gamma_1,\ldots,\gamma_p\}$. The problem with $f=0, \gamma_1=\cdots=\gamma_p=0$ is called the completely homogeneous problem. If the completely homogeneous problem has only the trivial solution, (3.1)-(3.2) has *at most one solution* (and it turns out that a solution actually exists for any data). If the completely homogeneous problem has nontrivial solutions, (3.1)-(3.2) either has *no solution* or has *more than one solution*.

Assume next that the completely homogeneous problem

$$(3.3) \qquad Lu=0, \quad a<x<b; \qquad B_1u=\cdots=B_pu=0$$

has *only the trivial solution*. We can then construct Green's function $g(x,\xi)$, necessarily unique, associated with $(L,B_1,\ldots,B_p)$. This Green's function satisfies

$$(3.4) \qquad Lg=\delta(x-\xi), \quad a<x,\xi<b; \qquad B_1g=\cdots=B_pg=0,$$

or, equivalently,

$$(3.5) \quad \begin{cases} Lg=0, \quad a<x<\xi, \xi<x<b; \qquad B_1g=\cdots=B_pg=0, \\[2mm] g,\ldots,g^{(p-2)} \text{ continuous}; \qquad g^{(p-1)}(\xi+,\xi)-g^{(p-1)}(\xi-,\xi)=\dfrac{1}{a_p(\xi)}. \end{cases}$$

If u and v are any functions in C^p, Green's formula gives [see (5.18), Chapter 2]

$$(3.6) \quad vLu-uL^*v=J(u,v)]_a^b, \quad J=\sum_{m=1}^{p}\sum_{j+k=m-1}(-1)^kD^k(a_mv)D^ju.$$

We observe that $J(u,v)]_a^b$ can be written as the sum of $2p$ terms

$$(3.7) \quad u(a)A_{2p}v+\cdots+u^{(p-1)}(a)A_{p+1}v+u(b)A_pv+\cdots+u^{(p-1)}(b)A_1v,$$

where each A_k is a linear combination of the $2p$ quantities $v(a),\ldots,v^{(p-1)}(b)$. If we are given p independent boundary functionals $B_1,\ldots,B_p$, we can rewrite (3.7) so as to feature the p quantities $B_1u,\ldots,B_pu$ instead of the $2p$ quantities $u(a),\ldots,u^{(p-1)}(b)$. This suggests introducing p additional boundary (complementary) functionals $B_{p+1},\ldots,B_{2p}$, so that $B_1,\ldots,B_{2p}$ is a set of $2p$ independent boundary functionals; then (3.7) can be rewritten as

$$(3.8) \qquad J(u,v)]_a^b = (B_1 u)(B_{2p}^* v) + \cdots + (B_p u)(B_{p+1}^* v)$$
$$+ (B_{p+1} u)(B_p^* v) + \cdots + (B_{2p} u)(B_1^* v).$$

Although $B_{p+1}, \ldots, B_{2p}$ can be introduced in many different ways, all turn out to be ultimately equivalent for our purposes. Thus we regard $B_{p+1}, \ldots, B_{2p}$ as fixed, and therefore $B_1^*, \ldots, B_{2p}^*$ are unambiguously determined in (3.8). The p boundary functionals $B_1^*, \ldots, B_p^*$ are said to be *adjoint* to $B_1, \ldots, B_p$. Given p homogeneous boundary conditions $B_1 u = \cdots = B_p u = 0$, we see from (3.8) that $J(u,v)]_a^b$ will vanish if and only if v satisfies the p adjoint conditions $B_1^* v = \cdots = B_p^* v = 0$. If $L = L^*$ and if the adjoint conditions define the same set of functions as $B_1 u = \cdots = B_p u = 0$, then $(L, B_1, \ldots, B_p)$ is said to be *self-adjoint*.

Example 1. Consider the first-order operator $L = D$. A typical problem of form (3.1)-(3.2) would be

$$\frac{du}{dx} = f, \quad a < x < b; \qquad B_1 u \doteq u(a) + \beta u(b) = \gamma.$$

Here $L^* = -D$, and Green's formula for any u, v in C^1 becomes

$$(3.9) \qquad \int_a^b \left(v \frac{du}{dx} + u \frac{dv}{dx} \right) dx = uv \,]_a^b = J(u,v)]_a^b = u(b)v(b) - u(a)v(a).$$

With $B_1 u$ defined above, we may introduce the complementary boundary functional $B_2 u \doteq u(b)$, in terms of which we can write

$$u(b)v(b) - u(a)v(a) = [u(a) + \beta u(b)][-v(a)] + u(b)[v(b) + \beta v(a)].$$

Thus $B_2^* v = -v(a)$, $B_1^* v = v(b) + \beta v(a)$. Therefore the adjoint condition is $v(b) + \beta v(a) = 0$. Thus, if u satisfies $B_1 u = 0$, the necessary and sufficient condition for $J(u,v)]_a^b = 0$ is that v satisfies $B_1^* v = 0$.

Example 2. Consider the fourth-order operator $L = D^2(b_2 D^2) + D(b_1 D) + b_0$, where $b_0(x), b_1(x), b_2(x)$ are arbitrary functions. Then $L = L^*$, and

$$\int_a^b (vLu - uL^*v) \, dx = v(b_2 u'')' - v' b_2 u'' + b_2 v'' u' - u(b_2 v'')' + b_1 (vu' - uv')]_a^b.$$

If $B_1 u = u(a)$, $B_2 u = u''(a)$, $B_3 u = u(b)$, $B_4 u = u''(b)$, then (L, B_1, B_2, B_3, B_4) is self-adjoint. This is also true for the case $B_1u = u(a)$, $B_2u = u'(a)$, $B_3u = u(b)$, $B_4u = u'(b)$. But if we have three conditions at a and one condition at b, the problem is not self-adjoint.

The adjoint Green's function $g^*(x, \xi)$ satisfies

$$(3.10) \quad L^*g^* = \delta(x - \xi), \quad a < x, \xi < b; \qquad B_1^*g^* = \cdots = B_p^*g^* = 0,$$

and we can once more show that $g^*(x, \xi) = g(\xi, x)$. To find the solution of (3.1)-(3.2), multiply (3.1) by $g(\xi, x)$ and (3.10) by $u(x)$, subtract, and integrate to obtain

$$\int_a^b \left[g(\xi, x)Lu - uL^*g(\xi, x) \right] dx = -u(\xi) + \int_a^b g(\xi, x)f(x)\,dx$$

or, changing the roles of x and ξ,

$$(3.11) \qquad u(x) = \int_a^b g(x, \xi)f(\xi)\,d\xi - J(u(\xi), g(x, \xi))]_{\xi = a}^{\xi = b},$$

where [say, by using (3.8)] J can be expressed in terms of $\gamma_1, \ldots, \gamma_p$ and $g, \ldots, d^{p-1}g/d\xi^{p-1}$.

A complete treatment of BVPs for equations of any order can be found in Coddington and Levinson.

Exercises

3.1. *Adjoints for unbalanced problems*

For an operator of order p there are at most $2p$ independent conditions of type (3.2). Consider the problem

$$Lu = f, \quad a < x < b; \qquad B_1u = \gamma_1, \ldots, B_mu = \gamma_m,$$

where L is the general differential operator of order p, $B_1, \ldots, B_m$ are independent boundary functionals of type (3.2), and $0 \leqslant m \leqslant 2p$. If $m \neq p$, the problem $(L, B_1, \ldots, B_m)$ is unbalanced. We can still define the adjoint of $(L, B_1, \ldots, B_m)$ as $(L^*, B_1^*, \ldots, B_{p-m}^*)$; equivalently, if u satisfies $B_1u = 0, \ldots, B_mu = 0$, the conditions $B_1^*v = 0, \ldots, B_{p-m}^*v = 0$ are just what is required to make $J(u, v)]_a^b$ vanish. As an example, if $L = D$ on $0 < x < 1$ and B_1, B_2 are defined by $B_1u = u(0)$, $B_2u = u(1)$, we clearly have "too many" boundary functionals; the adjoint consists of $L^* = -D$ and *no* boundary functional. If we consider $L = D^2$ on $0 < x < 1$ with only one boundary functional, defined by $B_1u = u(0)$, the adjoint consists of $L^* = D^2$ with B_1^*, B_2^*, B_3^* defined by $B_1^*v = v(1)$, $B_2^*v = v'(1)$, $B_3^*v = v(0)$.

Using these ideas find the adjoint of $L = D^4$ on $0 < x < 1$, with B_1, B_2, B_3, B_4, B_5 defined from $B_1 u = u(0)$, $B_2 u = u(1)$, $B_3 u = u''(0)$, $B_4 u = u''(1)$, $B_5 u = u'(1) - u'(0)$.

3.2. Consider the fourth-order operator $L = D^4$ on $0 < x < 1$, and let the boundary functionals be $B_1 u = u(0)$, $B_2 u = u^{(3)}(0)$, $B_3(u) = u(1)$, $B_4(u) = u^{(3)}(1)$. Find the adjoint boundary conditions, and show by explicit calculation that the direct Green's function is not symmetric.

4. ALTERNATIVE THEOREMS

Consider the boundary value problem (3.1)-(3.2) for a pth-order operator:

$$(4.1) \qquad Lu = f, \quad a < x < b; \qquad B_1 u = \gamma_1, \quad \dots, \quad B_p u = \gamma_p.$$

For boundary conditions of the type (3.2), we saw that (4.1) has one and only one solution if the corresponding completely homogeneous problem

$$(4.2) \qquad Lu = 0, \quad a < x < b; \qquad B_1 u = 0, \quad \dots, \quad B_p u = 0$$

has only the trivial solution. If (4.2) has nontrivial solutions, (4.1) usually has no solution unless the data $\{f; \gamma_1, \dots, \gamma_p\}$ is of a particular type. To characterize the data for which solutions to (4.1) exist we need to consider the related homogeneous adjoint problem

$$(4.3) \qquad L^* v = 0, \quad a < x < b; \qquad B_1^* v = 0, \quad \dots, \quad B_p^* v = 0.$$

The situation bears considerable resemblance to that encountered for systems of algebraic equations, and we begin our discussion with such systems.

Alternative Theorem for a System of n Equations in n Unknowns

Consider the set of n linear equations in n unknowns

$$\sum_{j=1}^{n} a_{ij} u_j = f_i, \qquad i = 1, \dots, n,$$

where $a_{11}, \dots, a_{nn}, f_1, \dots, f_n$ are real numbers, and we are looking for a real solution $u_1, \dots, u_n$. We write (4.4) as the single equation

$$(4.4) \qquad \qquad Au = f,$$

where A is the n by n matrix with entries a_{ij}, and f and u are vectors (that is, n-tuples of real numbers). Together with (4.4), we consider the related problems

(4.5) $Au = 0$ (direct homogeneous equation),

(4.6) $A^*v = 0$ (adjoint homogeneous equation).

The matrix A^* is the *transposed* or *adjoint* matrix of A. The element a_{ij}^* in the ith row and jth column of A^* is equal to a_{ji}, the element in the jth row and ith column of A.

We denote the *inner product* of two vectors u and v by $\langle u,v \rangle$. By definition $\langle u,v \rangle = \sum_{j=1}^{n} u_j v_j$. The inner product is also known as the dot product or scalar product between two vectors. We say that u and v are *orthogonal* if $\langle u,v \rangle = 0$.

The following results stem from elementary linear algebra (although they will be proved in a more general setting in Chapters 4 and 5).

1. If (4.5) has only the trivial solution, then so does (4.6). This occurs if and only if $\det A \neq 0$ (and then automatically $\det A^* \neq 0$). In this case (4.4) has one and only one solution for each f, the unique solution being written as $u = A^{-1}f$, where A^{-1} is the inverse matrix of A. This then is the normal situation, the one where A is nonsingular.

2. If (4.5) has nontrivial solutions, then $\det A = 0$. The number of independent solutions of (4.5) depends on how "badly" $\det A$ vanishes. If $\det A$ vanishes, this means that there are interdependencies among the rows of the matrix A. Therefore we cannot expect solutions of (4.4) unless the interdependencies in A are reflected in f. (For instance, if the second row of A is the sum of the first and third rows, we will need $f_2 = f_1 + f_3$ in order to have any hope of a solution.) If there are k ($\leq n$) independent solutions of (4.5), A is said to have a k-*dimensional null space*. It then turns out that A^* also has a k-dimensional null space, but the solutions of (4.6) will in general be different from those of (4.5). *A necessary and sufficient condition for* (4.4) *to have solutions is that f be orthogonal to all the solutions of* (4.6). This condition will be known as a *solvability condition* (also, consistency or compatibility condition).

The proof of the necessity of this condition is easy. Let u be a solution of (4.4), and v a solution of (4.6). Take the inner product of (4.4) with v and the inner product of (4.6) with u, and subtract the resulting equations to obtain

$$\langle Au,v \rangle - \langle u,A^*v \rangle = \langle f,v \rangle.$$

Now $\langle Au, v \rangle = \sum_{i,j} a_{ij} u_j v_i$, whereas $\langle u, A^* v \rangle = \sum_{k,l} u_k a_{kl}^* v_l$. Since $a_{kl}^* = a_{lk}$, we have $\langle Au, v \rangle = \langle u, A^* v \rangle$ and therefore $\langle f, v \rangle = 0$.

The sufficiency proof is a little more subtle. We want to show that, if f is orthogonal to the solutions of (4.6), then (4.4) is solvable, that is, f is in the range R_A of A. The proof will be given in a more general context in Chapter 5.

Remarks

1.   If $\det A \neq 0$, the solvability conditions are automatically satisfied for any f, since the only solution of (4.6) is $v \equiv 0$. Thus (4.4) has one and only one solution.

2.   If $\det A = 0$ and the null space of A is k-dimensional, we can find k independent solutions $u^{(1)}, \ldots, u^{(k)}$ of (4.5) and k independent solutions $v^{(1)}, \ldots, v^{(k)}$ of (4.6). If f satisfies the solvability conditions $\langle f, v^{(1)} \rangle = \cdots = \langle f, v^{(k)} \rangle = 0$, the general solution of (4.4) can be written as

$$\tilde{u} = u + \sum_{i=1}^{k} c_i u^{(i)},$$

where $\tilde{u}$ is any particular solution of (4.4), and the $\{c_i\}$ are arbitrary constants.

3.   We have given no new recipe for discovering solutions! All we have done is to give criteria for determining whether or not solutions exist.

Example 1.   Consider the following set of three equations in three unknowns:

$$\left. \begin{array}{l} u_1 + u_2 + u_3 = 1, \\ 2\, u_1 - u_2 + u_3 = 3, \\ u_1 - 2\, u_2 \phantom{{}+ u_3} = 1, \end{array} \right\} \qquad \text{or} \qquad Au = f.$$

Then (4.6) becomes

$$\left. \begin{array}{l} v_1 + 2\, v_2 + v_3 = 0, \\ v_1 - v_2 - 2 v_3 = 0, \\ v_1 + v_2 \phantom{{}+ 2 v_3} = 0, \end{array} \right\} \qquad \text{or} \qquad A^* v = 0,$$

which has the one-dimensional null space $v_2 = -v_1, v_3 = v_1$. Thus $v^{(1)} = (1, -1, 1)$ is a solution of $A^*v = 0$, and there is no solution of this equation that is independent of $v^{(1)}$. Since $f = (1, 3, 1)$ we see that $\langle f, v^{(1)} \rangle = 1 - 3 + 1 = -1 \neq 0$, so that $Au = f$ has *no* solution.

Alternative Theorem for Integral Equations

Let $k(x, \xi)$ be a real-valued function defined on the square $a < x < b, a < \xi < b$. We consider the inhomogeneous integral equation

$$(4.7) \qquad \int_a^b k(x, \xi) u(\xi) \, d\xi - u(x) = f(x), \qquad a < x < b,$$

the related homogeneous equation

$$(4.8) \qquad \int_a^b k(x, \xi) u(\xi) \, d\xi - u(x) = 0, \qquad a < x < b,$$

and the adjoint homogeneous equation

$$(4.9) \qquad \int_a^b k(\xi, x) v(\xi) \, d\xi - v(x) = 0, \qquad a < x < b.$$

In Chapter 5 we shall prove the alternative theorem: (a) If (4.8) has only the trivial solution, so does (4.9), and then (4.7) has one and only one solution; (b) if (4.8) has nontrivial solutions, so does (4.9), and then (4.7) will have solutions if and only if

$$\int_a^b f(x) v(x) \, dx = 0$$

for every v which is a solution of (4.9).

Alternative Theorem for Boundary Value Problems

Consider first problem (4.1) with data $\{f; 0, 0, \ldots, 0\}$,

$$(4.10) \qquad Lu = f, \quad a < x < b; \qquad B_1 u = \cdots = B_p u = 0,$$

for which the following alternative theorem holds:

(a)   If the direct homogeneous problem (4.2) has only the trivial solution, so does the adjoint homogeneous problem (4.3), and (4.10) has one and only one solution.

(b)   If (4.2) has k independent solutions, (4.3) also has k independent solutions [although not necessarily the same as those of (4.3)]. Then (4.10) has solutions if and only if $\int_a^b fv^{(1)}\,dx = \cdots = \int_a^b fv^{(k)}\,dx = 0$, where $(v^{(1)},\ldots,v^{(k)})$ is a set of k independent solutions of (4.3). If the solvability conditions are satisifed, the general solution of (4.10) is of the form $\tilde{u} + \sum_{i=1}^k c_i u^{(i)}$, where $\tilde{u}$ is a particular solution of (4.10), the $\{c_i\}$ are arbitrary constants, and $u^{(1)},\ldots,u^{(k)}$ are k independent solutions of (4.2).

The necessity of the solvability conditions is easy to prove. Let u be a solution of (4.10), and v a solution of (4.3). Multiply (4.10) by v and (4.3) by u, subtract, and integrate from a to b to obtain

$$\int_a^b fv\,dx = \int_a^b (vLu - uL^*v)\,dx = J(u,v)\big]_a^b,$$

but $J(u,v)\big]_a^b = 0$, since u satisfies homogeneous boundary conditions and v satisfies the adjoint boundary conditions (which were designed exactly for the purpose of making $J\big]_a^b$ vanish). The sufficiency will be proved in Chapter 5.

Example 2. (a) Consider the problem

(4 11) $\quad -u'' = f, \quad 0 < x < 1; \qquad u(1) - u(0) = 0, \quad u'(1) - u'(0) = 0.$

The adjoint homogeneous problem is

$$-v'' = 0, \quad 0 < x < 1; \qquad v(1) - v(0) = 0, \quad v'(1) - v'(0) = 0,$$

which has the solutions $v = $ constant. Thus (4.11) has solutions if and only if

(4.12) $$\int_0^1 f\,dx = 0.$$

There is a simple physical explanation for this condition. We can regard (4.11) as governing the steady temperature $u(x)$ in an insulated thin ring of unit perimeter subject to sources of density $f(x)$. The periodic boundary conditions reflect the fact that the points $x=0$ and $x=1$ represent the same physical point in the ring, so that the temperature must have the same value at $x=0$ and $x=1$ and the gradient of the temperature must also have the same value at $x=0$ and $x=1$. Since heat cannot flow through the surface of the ring, conservation of heat shows that a steady state is possible only if the net heat input along the ring is 0, that is, if $\int_0^1 f(x)\,dx = 0$, which is just (4.12).

(b) For the problem

$$-u'' - \pi^2 u = f, \quad 0 < x < 1; \qquad u(1) + u(0) = 0, \qquad u'(1) + u'(0) = 0$$

there are two solvability conditions:

$$\int_0^1 f \cos \pi x \, dx = 0, \qquad \int_0^1 f \sin \pi x \, dx = 0.$$

Example 3. Consider the inhomogeneous problem

(4.13) $u' + u = f, \quad 0 < x < 1; \quad u(0) - eu(1) = 0.$

The adjoint homogeneous problem is (see Example 1, Section 3),

$$-v' + v = 0, \quad 0 < x < 1; \qquad -ev(0) + v(1) = 0,$$

with the nontrivial solution $v = Ae^x$. Therefore (4.13) has solutions if and only if $\int_0^1 fe^x \, dx = 0$. We can check this by elementary calculations. Solving the differential equation in (4.13), we find that $u = \int_0^x f(t)e^{t-x} \, dt + Ce^{-x}$; the boundary condition gives $C - \int_0^1 f(t)e^t \, dt - C = 0$, which is just the predicted solvability condition. If the condition is satisfied, then $u = \int_0^x f(t)e^{t-x} \, dt + Ce^{-x}$ is the general solution of (4.13).

Example 4. Consider next a problem in which there are *too many* boundary conditions:

(4.14) $Lu \doteq u' = f, \quad 0 < x < 1; \qquad u(0) = 0, \qquad u(1) = 0.$

To find the adjoint problem, we proceed as in Example 1, Section 3. Clearly $L^* = -D$, and Green's formula becomes

$$\int_0^1 (vLu - uL^*v) \, dx = uv \big]_0^1 = J(u,v) \big]_0^1 = u(1)v(1) - u(0)v(0).$$

Given the boundary conditions $u(0) = 0$, $u(1) = 0$, we look for homogeneous conditions in v which make $J(u,v)\big]_0^1$ vanish. But $J\big]_0^1$ is already 0! Thus, for any v, we have $J\big]_0^1 = 0$. Hence the adjoint problem is

$$L^*v = -\frac{dv}{dx} = 0, \qquad 0 < x < 1,$$

whose general solution is $v = \text{constant}$. Problem (4.14) has a solution if and only if f is orthogonal to a constant, that is, $\int_0^1 f(x) \, dx = 0$, a result that could also have been obtained by inspection from (4.14).

Next we turn to the completely inhomogeneous problem (4.1), repeated here for convenience:

(4.15) $Lu = f, \quad a < x < b; \qquad B_1 u = \gamma_1, \quad \ldots, \quad B_p u = \gamma_p.$

We can easily find a necessary condition for (4.15) to have solutions. Let v be a solution of the homogeneous adjoint problem (4.3). Multiply (4.15) by v and (4.3) by u, subtract, and integrate from a to b. This yields

$$\int_a^b (vLu - uL^*v)\, dx = J(u,v)]_a^b.$$

The left side reduces to $\int_a^b fv\, dx$, but the right side no longer vanishes because u satisfies inhomogeneous boundary conditions. Using (3.8), we can write

$$J(u,v)]_a^b = \gamma_1 B_{2p}^* v + \cdots + \gamma_p B_{p+1}^* v,$$

where the right side is completely known. Thus (4.15) can have solutions only if

(4.16) $$\int_a^b fv\, dx = \gamma_1 B_{2p}^* v + \cdots + \gamma_p B_{p+1}^* v,$$

a solvability condition which also turns out to be sufficient.

Example 5. Consider the problem

(4.17) $$u' + u = f, \quad 0 < x < 1; \qquad u(0) - eu(1) = \gamma_1.$$

Then $L^*v = -v' + v$, and

$$\begin{aligned}
J(u,v)]_0^1 &= u(1)v(1) - u(0)v(0) \\
&= [u(0) - eu(1)][-v(0)] + u(1)[v(1) - ev(0)] \\
&= (B_1 u)(B_2^* v) + (B_2 u)(B_1^* v).
\end{aligned}$$

Let v be the solution of the homogeneous adjoint problem

$$L^*v = 0, \qquad B_1^* v = 0,$$

whose solution (see Example 3) is $v = Ae^x$. With u the solution of (4.17) we find that

$$J(u,v)]_0^1 = \gamma_1 B_2^* v = -\gamma_1 v(0) = -\gamma_1 A.$$

Therefore (4.17) has solutions if and only if

$$A \int_0^1 fe^x\, dx = -\gamma_1 A, \qquad \text{that is,} \quad \int_0^1 fe^x\, dx = -\gamma_1.$$

Again we can check this easily by a straightforward calculation. The differential equation in (4.17) has the general solution

$$u = \int_0^x f(t)e^{t-x}\,dt + Ce^{-x},$$

so that, by imposing the boundary condition

$$C - e\int_0^1 f(t)e^{t-1}\,dt - eCe^{-1} = \gamma_1,$$

we obtain the previously found solvability condition

$$\int_0^1 f(t)e^t\,dt = -\gamma_1.$$

Exercises

4.1. Refer to Exercise 3.1. Find the solvability condition for the problem

$$D^4 u = f, \quad 0 < x < 1;$$
$$u(0) = u(1) = u''(0) = u''(1) = 0, \qquad u'(1) - u'(0) = \gamma.$$

Check your result by solving the differential equation with the four homogeneous conditions by using Green's function (4.10) of Chapter 1 and then trying to satisfy the last boundary condition.

4.2. Find the solvability conditions for

$$-u'' - u = f, \quad -\pi < x < \pi;$$
$$u(\pi) - u(-\pi) = \gamma_1, \qquad u'(\pi) - u'(-\pi) = \gamma_2.$$

4.3. Consider a system of m linear equations in n unknowns (if $m \neq n$ the system is *unbalanced*):

$$\left. \begin{array}{c} a_{11}u_1 + \cdots + a_{1n}u_n = f_1, \\ \vdots \\ a_{m1}u_1 + \cdots + a_{mn}u_n = f_m, \end{array} \right\} \quad \text{or} \quad Au = f,$$

where the matrix A has m rows and n columns. Thus A can be regarded as a transformation from n-tuples to m-tuples. The adjoint matrix A^* is obtained by transposing the rows and columns of A, so that A^* has n rows and m columns. The null spaces of A and A^* may now have different dimensions. Show that $Au=f$ has solutions if and only if f is orthogonal to all the solutions of $A^*v=0$ (note that f and v are both m-tuples). Give an example of a 2 by 1 matrix A such that the homogeneous equation $Au=0$ has only the trivial solution, but $Au=f$ has no solution for some f's and has a solution for other f's.

5. MODIFIED GREEN'S FUNCTIONS

Let us begin with a simple but representative example. The boundary value problem

$$(5.1) \qquad -u''=f, \quad 0<x<1; \qquad u'(0)=u'(1)=0$$

describes steady heat conduction in a rod whose lateral surface and *ends* are insulated. Unless the source term $f(x)$ satisfies the solvability condition $\int_0^1 f(x)\,dx=0$, there is no hope for a solution of (5.1). [If one considered the corresponding time-dependent heat conduction equation with a steady source term $f(x)$, the temperature would increase indefinitely with time unless $\int_0^1 f(x)\,dx=0$.] In particular, if $f=\delta(x-\xi)$, the solvability condition is *not* satisfied and there is no solution to $-g''=\delta(x-\xi)$, $0<x,\xi<1$, $g'(0,\xi)=g'(1,\xi)=0$. We want to construct the nearest thing to a Green's function to enable us to solve (5.1) when a solution exists (that is, when $\int_0^1 f\,dx=0$). This can be done by considering a modified accessory problem with the consistent data $\{\delta(x-\xi)-1,0,0\}$. We have compensated for the concentrated source at ξ by a uniform distribution of sinks along the rod, the net total heat input being 0. Therefore we examine the problem

$$(5.2) \qquad -g''=\delta(x-\xi)-1, \quad 0<x,\xi<1; \qquad g'(0,\xi)=g'(1,\xi)=0$$

which is consistent and has many solutions differing from one another by a constant. Since, for $x\neq\xi$, $-g'=-1$, we have

$$g(x,\xi)=\begin{cases} A+Bx+\dfrac{x^2}{2}, & 0<x<\xi, \\[2mm] C+Dx+\dfrac{x^2}{2}, & \xi<x<1. \end{cases}$$

The boundary conditions yield $B=0$, $D=-1$; continuity at $x=\xi$ gives $A=C-\xi$, so that

$$(5.3) \qquad g(x,\xi)=\begin{cases} C-\xi+\dfrac{x^2}{2}, & 0<x<\xi, \\[2mm] C-x+\dfrac{x^2}{2}, & \xi<x<1, \end{cases}$$

where C can of course depend on ξ. At this stage we would normally try to apply the jump condition: $g'(\xi+,\xi)-g'(\xi-,\xi)=-1$, but this is now *automatically* satisfied regardless of the value of C. Therefore (5.3) is the general solution of (5.2). Moreover, for each position ξ of the source one could choose a different C. In many calculations it is convenient to single out a particular g by requiring that

$$(5.4) \qquad \int_0^1 g(x,\xi)\,dx = 0 \qquad \text{for all } \xi, \quad 0<\xi<1.$$

Requirement (5.4) is *not* a solvability condition, but merely picks out a solution of (5.2) that turns out to be symmetric in x and ξ. This symmetric solution of (5.2) will be denoted by $g_M(x,\xi)$ and is known as the *modified Green's function*. By imposing (5.4) on (5.3), we find explicitly $C=(\xi^2/2)+\frac{1}{3}$ and

$$(5.5) \qquad g_M(x,\xi)=\begin{cases} \dfrac{1}{3}-\xi+\dfrac{x^2+\xi^2}{2}, & 0\leqslant x\leqslant\xi, \\[2mm] \dfrac{1}{3}-x+\dfrac{x^2+\xi^2}{2}, & \xi\leqslant x\leqslant 1. \end{cases}$$

Note that there are other choices of C in (5.3) which make g symmetric, so that (5.4) is sufficient but not necessary for symmetry (see Exercise 5.1).

One reason for the special importance of (5.5) is that this particular g can be obtained by a limiting process from an ordinary Green's function. Let λ be a real parameter, and consider Green's function $g(x,\xi;\lambda)$ satisfying

$$(5.6) \quad -g''-\lambda g=\delta(x-\xi), \quad 0<x,\xi<1; \quad g'(0,\xi)=g'(1,\xi)=0.$$

For $\lambda\neq n^2\pi^2$, $n=0,1,2,\ldots$, the corresponding homogeneous problem has only the trivial solution, so that the ordinary Green's function can be constructed; moreover, since the problem is self-adjoint, g will be symmetric in x and ξ. A straightforward calculation gives

$$(5.7) \quad g(x,\xi;\lambda)=-\frac{1}{\sqrt{\lambda}\,\sin\sqrt{\lambda}}\begin{cases} \cos\sqrt{\lambda}\,x\,\cos\sqrt{\lambda}\,(1-\xi), & 0\leqslant x\leqslant\xi, \\[2mm] \cos\sqrt{\lambda}\,\xi\,\cos\sqrt{\lambda}\,(1-x), & \xi\leqslant x\leqslant 1, \end{cases}$$

which is well defined for $\lambda\neq n^2\pi^2$ but becomes singular at every λ of the form $n^2\pi^2$. Near $\lambda=0$, $g(x,\xi;\lambda)$ behaves like $-1/\lambda$, so that

$$\tilde{g}\doteq g+\frac{1}{\lambda}$$

has a limiting value at $\lambda = 0$. Moreover, $\tilde{g}$ satisfies

(5.8) $-\tilde{g}'' - \lambda\tilde{g} = \delta(x-\xi) - 1, \quad 0 < x, \xi < 1; \qquad \tilde{g}'(0,\xi) = \tilde{g}'(1,\xi) = 0.$

For $\lambda \neq n^2\pi^2$, (5.8) has one and only one solution, and that solution obeys (5.4), as can be seen by integrating (5.8) from 0 to 1. As $\lambda \to 0$, the solution of (5.8) remains regular and tends to the solution of (5.2) which satisfies (5.4).

We pass now to the general self-adjoint problem of the second order. Assume that the homogeneous problem $Lu = 0$, $B_1 u = B_2 u = 0$ has nontrivial solutions of the form $Cu_1(x)$, where $u_1(x)$ is a normalized solution, that is, a solution for which

$$\int_a^b u_1^2(x)\,dx = 1.$$

The problem $Lg = \delta(x-\xi)$, $B_1 g = B_2 g = 0$ is inconsistent, so that we must introduce the *modified Green's function* g_M, chosen to satisfy

(5.9) $Lg_M = \delta(x-\xi) - u_1(x)u_1(\xi), \quad a < x, \xi < b; \qquad B_1 g_M = B_2 g_M = 0,$

(5.10) $\int_a^b g_M(x,\xi)u_1(x)\,dx = 0 \qquad \text{for every } \xi, \quad a < \xi < b.$

The BVP (5.9) is consistent, since

$$\int_a^b \left[\delta(x-\xi) - u_1(x)u_1(\xi) \right] u_1(x)\,dx = 0.$$

Condition (5.10) serves to single out a particular solution of (5.9), which turns out to be symmetric in x and ξ. The construction of g_M in any specific case proceeds as for (5.2).

Let us now show how the modified Green's function can be used to solve

(5.11) $Lu = f, \quad a < x < b; \qquad B_1 u = B_2 u = 0.$

Of course (5.11) is solvable only if $\int_a^b f u_1\,dx = 0$, which we assume holds. Multiply (5.9) by the solution u of (5.11), multiply (5.11) by g_M, subtract, and integrate from a to b. This gives

$$\int_a^b (uLg_M - g_M Lu)\,dx = u(\xi) - \int_a^b u_1(x)u_1(\xi)u(x)\,dx - \int_a^b g_M(x,\xi)f(x)\,dx.$$

The left side is equal to $J(u, g_M)]_a^b$, which vanishes by the boundary conditions. Therefore

$$u(\xi) = Cu_1(\xi) + \int_a^b g_M(x,\xi) f(x) \, dx,$$

and, using the symmetry of g_M, we find on relabeling x and ξ that

(5.12) $$u(x) = Cu_1(x) + \int_a^b g_M(x,\xi) f(\xi) \, d\xi.$$

This then is the general solution of (5.11). If we set $C=0$ in (5.2), we obtain a particular solution of (5.11) which is the *minimum norm* solution (see Exercise 5.7 and the pseudo inverse, Section 3, Chapter 9). For more information on these subjects, see the article by Loud, the book by Ben Israel and Greville, and the book by Nashed.

Exercises

5.1. Show that condition (5.10) is sufficient to ensure symmetry of the solution of (5.9). It is clear that the condition is *not* necessary; indeed, if C is independent of ξ, the addition of a term $Cu_1(x)u_1(\xi)$ to g_M does not spoil either symmetry or the fact that (5.9) is satisfied but (5.10) is no longer satisfied.

5.2. Let (L, B_1, B_2) be a self-adjoint problem of the second order, and suppose the completely homogeneous problem $Lu=0$, $a<x<b$, $B_1u = B_2u=0$ has two independent nontrivial solutions (this means that *every* solution of $Lu=0$ satisfies the boundary conditions). Develop the theory of the modified Green's function for this case. Find the modified Green's function for $L=(d^2/dx^2)+\pi^2$ with B_1, B_2 defined by $B_1u = u(0)+u(1)$, $B_2u = u'(0)+u'(1)$.

5.3. Find the nontrivial solutions of

$$D^4u=0, \quad 0<x<1; \quad u''(0)=u'''(0)=u''(1)=u'''(1)=0,$$

and give a physical interpretation in beam theory. Show that the problem is self-adjoint. Define and construct the modified Green's function. Solve $D^4u=f$ with the above homogeneous boundary conditions when f satisfies the solvability conditions.

5.4. Let L be a second-order differential operator, and suppose (L, B_1, B_2) is *not* self-adjoint. Assume that the homogeneous problem $Lu=0$, $B_1 u = B_2 u = 0$ has only one independent solution u_1 (so that the homogeneous adjoint problem has only one independent solution v_1). Develop the theory of the modified Green's function in this case. Construct g_M for $L = D^2$ with B_1, B_2 defined by $B_1 u = u(0) + u(1)$, $B_2 u = u'(0) - u'(1)$.

5.5. Show how to use g_M of (5.9)-(5.10) to solve the completely inhomogeneous problem $Lu=f$, $B_1 u = \gamma_1, B_2 u = \gamma_2$ when the appropriate solvability condition is satisfied.

5.6. Show that the problem

$$-u'' + q^2 u = 0, \quad -1 < x < 1; \quad u'(1) + hu(1) = 0; \quad u'(-1) - hu(-1) = 0$$

has only the trivial solution except when q^2 and h are related by a certain transcendental equation (here $q^2 \geqslant 0$, and h is real). Construct the modified Green's function for this case.

5.7. Given a fixed interval (a, b), the norm of a function $u(x)$ is defined as

$$\|u\| \doteq \left[\int_a^b u^2(x)\, dx \right]^{1/2}.$$

Two functions u, v are said to be orthogonal if $\int_a^b uv\, dx = 0$. If $u = v + w$, where v and w are orthogonal, the Pythagorean theorem holds: $\|u\|^2 = \|v\|^2 + \|w\|^2$.

Show that the two terms on the right side of (5.12) are orthogonal and hence that the choice $C = 0$ yields the solution of least norm.

5.8. The BVP

$$-u'' = f, \quad 0 < x < 1; \quad u(0) = u(1), \quad u'(0) = u'(1)$$

describes heat flow in a thin ring of unit circumference. Show that the modified Green's function for this problem is

$$g_M(x, \xi) = \tfrac{1}{12} + \frac{(x-\xi)^2}{2} - \tfrac{1}{2}|x-\xi|.$$

REFERENCES AND ADDITIONAL READING

Ben-Israel, A. and Greville, T. N. E., *Generalized inverses: theory and application*, Wiley-Interscience, New York, 1974.

Coddington, E. A. and Levinson, N., *Theory of ordinary differential equations*, McGraw-Hill, New York, 1955.

Loud, W. S., Some examples of generalized Green's functions and generalized Green's matrices, *SIAM Rev.* **12**, No. 2 (1970).

Nashed, M. Z., *Generalized inverses and applications*, Academic Press, New York, 1976.

4
Metric Spaces
and Hilbert Spaces

1. FUNCTIONS AND TRANSFORMATIONS

Since the terms "function," "transformation," "operator," and "mapping" are defined in precisely the same manner, a reasonable person might infer that they are used interchangeably, but in practice there are occasional distinctions that will be pointed out as they occur (see, for instance, Example 6 below). The idea of function pervades mathematics, beginning with the familiar real function of a real variable, which can be regarded as a rule f that assigns to each real number x an unambiguous real number $y = f(x)$.

We shall, however, need the concept of function in a more general setting.

Definition. Let D and E be arbitrary sets. A *function f* is said to be defined *on D* and to assume its values *in E* if to *each* element x belonging to D there is made to correspond exactly one element y in E. We then write $y = f(x)$ and say that f transforms or maps D *into E*; alternatively, we write $f: D \rightarrow E$.

Remarks

1. Of course y usually varies with x, but to a particular x in D there corresponds one and only one y (known as the *image* of x). We are therefore considering only single-valued functions. The element x is also called the *independent variable*, and y the *dependent variable*.

2. A function can be visualized as the black box of Figure 1.1. Every element x in D is an admissible input. The box transforms the input x into a well-defined output y belonging to E.

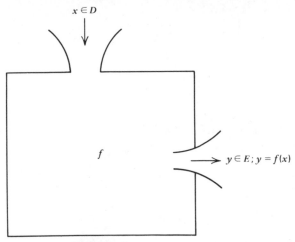

Figure 1.1

3.   The set D is known as the *domain* of f. The significance of the italicized word "on" in the definition of function is that every element x in D is admissible as an input, so that f is defined on all of D. On the other hand, not every element of E is necessarily an output. As x traverses D, the set of all outputs y forms a subset of E known as the *range* of f, denoted by R. Thus R consists of all the elements in E which are of the form $f(x)$ for some x in D. The significance of the italicized words "in" and "into" in the definition of function is that we are not committing ourselves as to whether or not the outputs cover all of E. If, in fact, it happens that $R = E$, we say that f is *onto* (or onto E). If we had had the foresight to choose the set E from the start to be just the set R, then f would automatically be onto. The reason why this choice is not usually made is that even for functions of the same type (such as ordinary real-valued functions of a real variable) the range varies from function to function; moreover, the range may be difficult to characterize precisely [see Example 1(d)], so that it is easier to regard R as a subset of a simple, known set E.

4.   Strictly speaking, one should distinguish between the function f— the box in Figure 1.1—and the output $f(x)$ corresponding to the input x. We shall, however, lapse into the traditional time-saving usage of calling both f and $f(x)$ by the name "function."

5.   There is nothing sacred about using the letters x, y, f in their particular roles in the definition. Any other letters will do, but in any specific

illustration we must make sure that the same letter is not given two different meanings.

6.   The function f is *one-to-one* if to each y in R there corresponds exactly one x in D such that $y=f(x)$. In other words, f is one-to-one if, whenever $f(x_1)=f(x_2)$, then $x_1=x_2$. If f is one-to-one, it can be "unraveled" by means of the inverse function f^{-1} (*not* $1/f$), whose domain D' is the range R of f and whose range R' is just the domain D of f. To each element y in D', the function f^{-1} assigns the unambiguous element x in R' for which $y=f(x)$.

Example 1.   If D is a subset of the real numbers and E is the set of real numbers, we are dealing with an ordinary real-valued function of a real variable. As specific examples consider the following:

(a) D is the real line, $f(x)=1/(x^2+1)$.

(b) D is the set $0 \leqslant x < \infty$, $f(x)=1/(x^2+1)$.

(c) D is the real line, $f(x)=x^3$.

(d) D is the real line, $f(x)=x^6+3x^5+2x^4+x$.

The functions in (a) and (b) are given by the same formula but are defined on different subsets of the real line and should be distinguished. In both cases R is the set $0<y \leqslant 1$, but only the function in (b) can be inverted. In (c) R is the real line and f is invertible. In (d) it is difficult to determine the range R precisely, but it is of the form $-a \leqslant y < \infty$ for some $a>0$ and f is not invertible. In (a), (c), (d) we say that f maps the real line *into* itself, but only in (c) are we allowed to use the word "onto" in place of "into."

Example 2.   If D is a subset of the real numbers and E is the set of complex numbers, f is known as a complex-valued function of a real variable [for instance, $f(x)=e^{ix}$].

Example 3.   If D and E are sets of complex numbers, f is known as a function of a complex variable.

Example 4.   If D is the set of ordered n-tuples of real numbers $x=(x_1,\ldots,x_n)$ and E is the set of real numbers, f is known as a function of n real variables. We write $y=f(x)$ or $y=f(x_1,\ldots,x_n)$.

Example 5.   If D is the set of ordered n-tuples of real numbers and E is the set of ordered m-tuples of real numbers, f is usually known as a *transformation, operator,* or *mapping* of D into E, and we tend to use capital letters such as A, F, and T instead of f. Of course, such a transformation can equally well be described by m functions $f_1, \ldots, f_m$ of n real variables.

Example 6. In many problems D and E are themselves sets of functions. This leads to an unavoidable complication in notation and terminology. The rule which assigns to each element in D an element of E is then called a *transformation* or operator. This sort of operator is usually denoted by a capital letter such as T. For instance, if D is the set of smooth, real-valued functions $u(t)$ defined on $-\infty < t < \infty$, the differentiation operator T assigns to each $u(t)$ the function $v(t) = du/dt$. In operator notation we would write $v = Tu$ or $v(t) = Tu(t)$. In this book the letters x and y have been avoided, for it is not clear whether x is better used as the input to the box T or as the independent variable within the input function itself. As we proceed, we will have occasion to use x in either capacity. For instance, we might let D be the set of all smooth functions $x(t)$ on $-\infty < t < \infty$, and E the set of all functions $y(t)$ on $-\infty < t < \infty$. The differentiation operator T would then assign to $x(t)$ the function $y(t) = dx/dt$, and we would write $y = Tx$ or $y(t) = Tx(t)$. Alternatively, we might let D be the set of all smooth functions $u(x)$ on $-\infty < x < \infty$, and E the set of all functions $v(x)$ on $-\infty < x < \infty$. The differentiation operator T then assigns to each $u(x)$ in D the element $v(x) = du/dx$ in E, for which we would write $v = Tu$ or $v(x) = Tu(x)$.

Example 7. If D is a set of functions and E a set of real or complex numbers, we use the name *functional* for a rule which assigns to each function in D a number in E. Either capital letters or lower case letters are used to denote functionals. For instance, if we let D be the set of continuous functions on a bounded interval $a \leqslant x \leqslant b$, then both

$$Tu \doteq \int_a^b u^3(x)\, dx \qquad \text{and} \qquad Fu \doteq u(b) - u(a)$$

are examples of functionals on D.

Our principal interest is in transformations T defined on function spaces. For further progress restrictions must be imposed on the nature of D, E, and T. The sets D and E, as they occur in most applications, are endowed with both algebraic and metric structures. The algebraic structure will usually be that of a linear space (also known as a vector space); even if D and E are not linear spaces, they will often be subsets of linear spaces. A linear space consists of elements, known as vectors, for which the operations of vector addition and scalar multiplication are defined. Any two vectors may be added to form a vector also in the space. Any vector may be multiplied by any scalar to yield a vector also in the space (the scalar field will always be either the field of real numbers or that of complex numbers). A linear space is therefore said to be closed under the operations of vector addition and scalar multiplication. Of course these operations must obey certain rules which are abstracted from the prototype of all linear spaces, namely, the usual three-dimensional space of directed line segments emanating from a common origin. The sets D and E will also be provided with a metric (or topological) structure. This gives us a way of measuring the distance separating any two elements, the distance satisfying certain axioms similar to the ones for the distance between points in ordinary Euclidean space. In the applications both structures will usually be present simultaneously. It is pedagogically preferable, however, to introduce the two concepts independently. In Chapter 9 and in Section 4 of the present chapter we discuss nonlinear transformations, but the rest deals mostly with *linear transformations*. A linear transformation T satisfies the condition $T(\alpha u_1 + \beta u_2) = \alpha T u_1 + \beta T u_2$ for any vectors u_1, u_2 in D and any scalars α, β.

2. LINEAR SPACES

The definition of a linear (or vector) space takes its cue from the vector structure of "ordinary" three-dimensional space V_3. In describing V_3, we shall admittedly use some undefined geometrical terms, but at this stage we must rely on intuition. The elements of V_3 are the directed line segments emanating from a common origin. Scalar multiplication by real numbers and vector addition (based on the parallelogram rule) are illustrated in Figure 2.1. The zero vector is the degenerate line segment coinciding with the origin O. Whether we think of the elements of V_3 as directed line segments or merely as the endpoints of these segments is to some extent a matter of taste, but obviously the terminology has been developed from the first point of view.

In Figure 2.1 the set of vectors αu as α ranges over the real numbers forms a line through the origin; the set of vectors $\alpha u + \beta v$ generates a plane through the origin. The property that characterizes V_3 as a three-dimensional space is the existence of three vectors u, v, w (for instance u, v in Figure 2.1 and a vector w not coplanar with them) such that each vector in the space can be expressed in one and only one way in the form $\alpha u + \beta v + \gamma w$.

Of course there are many other properties which we tend to associate automatically with V_3, such as the length of a vector and the angle between a pair of vectors, but we must now dismiss these metric properties from our mind to concentrate on the algebraic structure. The appropriate generalization is contained in the following definition.

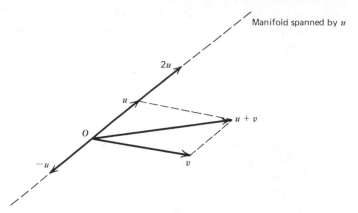

Figure 2.1

Definition. A *linear space* (or *vector space*) X is a collection of elements (also called *vectors*) u, v, w,..., in which each pair of vectors u, v can be combined by a process known as *vector addition* to yield a well-defined vector $u+v$ and in which each vector u can be combined with each number α by a process known as *scalar multiplication* to yield a well-defined vector αu. Vector addition and scalar multiplication satisfy the following axioms:

1. (a) $u+v=v+u$.
   (b) $u+(v+w)=(u+v)+w$.
   (c) There is a unique vector 0 such that $u+0=u$ for every u in X.
   (d) To each vector u, there corresponds a unique vector labeled $-u$ such that $u+(-u)=0$.
2. (a) $\alpha(\beta u)=(\alpha\beta)u$.
   (b) $1u=u$.
3. (a) $(\alpha+\beta)u=\alpha u+\beta u$.
   (b) $\alpha(u+v)=\alpha u+\alpha v$.

In these axioms u, v, w are arbitrary vectors in X, and α, β are arbitrary numbers.

Remarks

1. If the numbers α, β,... admitted as scalar multipliers are real numbers, we speak of a *real linear space* $X^{(r)}$; if complex numbers are permitted as multipliers, the space is a *complex linear space* $X^{(c)}$. When no superscript is used, either kind of space is envisaged.

2. No effort has been made to give a minimal set of axioms in the definition of a linear space.

3. The same symbol, 0, is used for both the zero vector and the number zero. The same symbol, $+$, is used for addition of vectors and addition of numbers. We hope no serious confusion will arise from this somewhat sloppy practice.

4. Axiom 1(b) shows that the triple sum is independent of order and may therefore be written as $u + v + w$ without ambiguity. The sum of any finite number of vectors $u_1, \ldots, u_k$ is also well defined without parentheses and regard to order. We have no way at this stage to define the sum of an infinite number of vectors. This requires the notion of convergence (or passage to the limit), which will come later.

5. The vector $u + (-v)$ is also written as $u - v$. It is easy to show that to each pair of vectors u, v there corresponds a unique vector w such that $u = v + w$ and that this vector w is just $u - v$.

6. One can also prove that $0u = 0$ (number zero multiplied by vector u is the zero vector) and that $(-1)u = -u$ [number minus one multiplied by vector u is equal to vector $-u$, whose existence is guaranteed by axiom 1(d)].

7. If X contains a nonzero vector u, it automatically contains an infinite number of vectors (those of the form αu as α ranges over all scalars).

Dependence and Independence of Vectors

Definition. Let $u_1, \ldots, u_k$ be vectors in the linear space X. A vector of the form $\alpha_1 u_1 + \cdots + \alpha_k u_k$ is said to be a *linear combination* of the vectors $u_1, \ldots, u_k$. The combination is said to be *nontrivial* if at least one of the α_i's is nonzero. A set of vectors $u_1, \ldots, u_k$ is *dependent* if there exists a nontrivial linear combination of them which is equal to the zero vector; otherwise the set is independent.

Remarks

1. We can restate the above as follows: the vectors $u_1, \ldots, u_k$ are dependent if there exist numbers $\alpha_1, \ldots, \alpha_k$ *not all* 0 such that

(2.1) $$\alpha_1 u_1 + \cdots + \alpha_k u_k = 0;$$

if the only solution of (2.1) is $\alpha_1 = \cdots = \alpha_k = 0$, the vectors are independent.

2. If $u_1, \ldots, u_k$ are dependent (or form a dependent set), at least one vector is a linear combination of the others.

3. If 0 belongs to a set of vectors, the set is dependent.

4. If the set $u_1, \ldots, u_k$ is dependent, so is any set which includes all the vectors $u_1, \ldots, u_k$.

5. For an infinite set of vectors there are two possible definitions of independence. The first is purely algebraic and states that an infinite set is algebraically independent if every finite subset is independent. Instead we could demand that the equation $\alpha_1 u_1 + \cdots + \alpha_k u_k + \cdots = 0$ be satisfied only for $\alpha_1 = \cdots = \alpha_k = \cdots = 0$. Obviously the latter definition requires the concept of an infinite series of vectors, which is not yet available (see Sections 5 and 6).

Dimension of a Linear Space

Definition. The linear space X is *n-dimensional* if it possesses a set of n independent vectors, but every set of $n + 1$ vectors is a dependent set. The space X is *∞-dimensional* if, for each positive integer k, one can find a set of k independent vectors in X.

At the risk of stating the obvious, we should point out that even a one-dimensional space contains an infinite number of elements. Next we turn to the concept of a basis in a finite-dimensional space. At this stage we introduce the notion in purely algebraic terms, which we shall not bother to extend to infinite-dimensional spaces (a definition of basis which brings the metric properties into play will be given in Section 6). The importance of a basis is that it enables us to express all the many vectors in X in terms of a relatively small, fixed set of vectors.

Definition. A finite set of vectors $h_1, \ldots, h_k$ is said to be a *basis* for the finite-dimensional space X if each vector in X can be represented in *one and only one way* as a linear combination of $h_1, \ldots, h_k$.

The following are nearly immediate consequences of the definition and the concept of independence. The proofs are left to the reader.

Theorem. The vectors in a basis are independent.

Theorem. In an n-dimensional space, any set of n independent vectors forms a basis.

Linear Manifolds

In ordinary, three-dimensional space V_3, we can find subsets, known as linear manifolds, which are themselves linear spaces under the same definition of vector addition and scalar multiplication used in V_3. These linear manifolds are just the lines and planes containing the origin (of course, the whole space and the set consisting only of the zero vector are also linear manifolds, though not very exciting ones).

Definition. A set M in a linear space X is a *linear manifold* (in X) if, whenever the vectors u and v belong to M, so does $\alpha u + \beta v$ for arbitrary scalars α and β.

A linear manifold must contain the zero vector and may be regarded as a linear space in its own right with vector addition and scalar multiplication inherited from X. It therefore makes sense to talk of the dimension of a linear manifold. One of the easiest ways to construct linear manifolds is by taking linear combinations of vectors in X.

Definition. Let U be a set of vectors (finite or infinite) in the linear space X. The *algebraic span* of U, denoted by $S(U)$, is the set of all finite linear combinations of vectors chosen from U.

Clearly $S(U)$ is always a linear manifold in X; other names for $S(U)$ are the *linear manifold generated by* U or the *linear hull of* U. If U consists of a single nonzero vector u in X, the algebraic span $S(U)$ is the set of all vectors of the form αu, that is, a one-dimensional linear manifold. If U consists of two independent vectors u and v, the algebraic span of U is the two-dimensional manifold of vectors having the form $\alpha u + \beta v$. If, on the other hand, U is a set of two dependent vectors u and v, not both 0, $S(U)$ is a one-dimensional manifold. Generally, if U is a set of k vectors, its algebraic span $S(U)$ is a linear manifold of dimension $\leqslant k$, with equality if and only if the k vectors are independent. If X is n-dimensional and U is a basis for X (that is, a set of n independent vectors), $S(U) = X$; of course it is also true that $S(X) = X$, but U has fewer elements than X.

Remark. If u and v are independent vectors in X, the set P of vectors of the form $u + \alpha v$ is *not* a linear manifold, since P does not contain the origin.

Examples of Linear Spaces

Example 1. In elementary algebra one frequently encounters the space $\mathcal{R}_n$ of all ordered n-tuples of real numbers. This space appears in a natural way in the solution of n linear equations in n unknowns. Let $u = (\xi_1, \ldots, \xi_n)$ and $v = (\eta_1, \ldots, \eta_n)$ be two arbitrary elements in $\mathcal{R}_n$, and let α be an arbitrary real number. The following definitions clearly make $\mathcal{R}_n$ a linear space:

$$u + v = (\xi_1 + \eta_1, \ldots, \xi_n + \eta_n),$$
$$\alpha u = (\alpha \xi_1, \ldots, \alpha \xi_n),$$
$$0 = (0, \ldots, 0),$$
$$-u = (-\xi_1, \ldots, -\xi_n).$$

The set of n vectors $(1, 0, \ldots, 0), (0, 1, 0, \ldots, 0), \ldots, (0, 0, \ldots, 0, 1)$ forms a basis in $\mathcal{R}_n$, which is therefore an n-dimensional real linear space. There are of course many other bases. Examples of linear manifolds are the set of all vectors of the form $(\xi_1, \xi_2, 0, \ldots, 0)$; the set of all vectors with $\xi_1 = \xi_2 = \cdots = \xi_n$; the set of all solutions of a fixed *homogeneous* system of n linear equations in n unknowns with real coefficients.

Example 2. The space $\mathcal{C}_n$ consists of ordered n-tuples of complex numbers with complex numbers α as scalar multipliers. Otherwise exactly the same definitions are used as in Example 1; the specific instances of a basis and of linear manifolds shown there carry over to $\mathcal{C}_n$. Thus $\mathcal{C}_n$ is a complex linear space of dimension n. It may at first seem paradoxical that $\mathcal{C}_n$ and $\mathcal{R}_n$ have the same dimension, since $\mathcal{C}_n$ appears much larger than $\mathcal{R}_n$. For instance, $\mathcal{C}_1$ is the set of all complex numbers, which is normally identified with the real plane. However, as a complex linear space, $\mathcal{C}_1$ has dimension 1.

Example 3. We are now going to create linear spaces whose elements u, $v, \ldots$ are real-valued functions $u(x), v(x), \ldots$ on $a \leqslant x \leqslant b$. All we do is look at ordinary operations on functions, but from a slightly more abstract point of view. The function or curve $u(x)$ is now regarded as a single entity —an element u in a space $X^{(r)}$. How do we add such entities? The element (or vector) $u + v$ is obtained from the elements u and v by adding their ordinates pointwise, as in Figure 2.2. Thus $u + v$ is just the curve or function usually written as $u(x) + v(x)$. The function which vanishes identically in $a \leqslant x \leqslant b$ plays the role of the 0 vector in the space. Scalar multiplication by real numbers is defined in the obvious way. With these

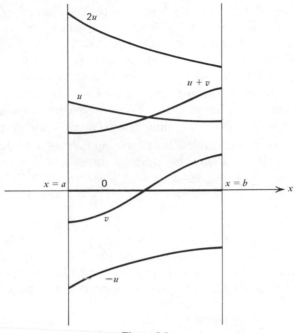

Figure 2.2

definitions we see that the space $X^{(r)}$ of all real-valued functions on $a \leqslant x \leqslant b$ is a real linear space. This space is much too general to be useful, but the following linear manifolds in $X^{(r)}$ will come up in applications [where they are usually regarded as linear spaces in their own right rather than as linear manifolds in $X^{(r)}$]:

(a) P_n, the set of all polynomials of degree *less* than n, is an n-dimensional space. The simplest basis in P_n is $h_1 = 1, h_2 = x, h_3 = x^2, \ldots, h_n = x^{n-1}$, where $h_1 = 1$ is the function equal to 1 on $a \leqslant x \leqslant b$.

(b) The set of all polynomials is an infinite-dimensional space. Any nontrivial polynomial can have only a finite number of zeros, so that the equation

$$\alpha_1 + \alpha_2 x + \alpha_3 x^2 + \cdots + \alpha_n x^{n-1} = 0, \qquad a \leqslant x \leqslant b,$$

can only have the solution $\alpha_1 = \alpha_2 = \cdots = \alpha_n = 0$. Thus, whatever n is, the set $1, x, x^2, \ldots, x^{n-1}$ is independent. The set of all polynomials therefore contains independent sets with arbitrarily large numbers of elements. From the definition of dimension, it follows that the space is infinite-dimensional.

(c) The set of all real-valued solutions of the equation $(d^2u/dx^2) - u = 0$, $a \leqslant x \leqslant b$, is a two-dimensional linear space, for which the pair $h_1 = e^x$ and $h_2 = e^{-x}$ is a basis. Another basis is the pair $\sinh x$ and $\cosh x$. Note that the set of all solutions of an inhomogeneous equation does *not* form a linear space, nor does the set of solutions of a nonlinear equation.

(d) The set of all real-valued continuous functions $u(x)$ on $a \leqslant x \leqslant b$ is an infinite-dimensional linear space. The set of functions which are even about $x = (a + b)/2$ is an infinite-dimensional linear manifold in this space.

(e) The set of all real-valued functions $u(x)$ on $a \leqslant x \leqslant b$ for which the Riemann integral $\int_a^b u^2(x)\,dx$ is finite is an infinite-dimensional linear space. All that has to be shown is that finite values for $\int_a^b u^2(x)\,dx$ and $\int_a^b v^2(x)\,dx$ imply a finite value for $\int_a^b (u+v)^2\,dx$. This follows from the Schwarz inequality (5.14).

Example 4. The set of complex-valued functions $u(x), v(x), \ldots$ on a real interval $a \leqslant x \leqslant b$ forms a complex linear space $X^{(c)}$ under the following definitions. The element $u + v$ is a complex-valued function whose value at x is obtained by adding the complex numbers $u(x)$ and $v(x)$. The element αu, where complex α's are then admitted, is the complex-valued function whose value at x is obtained by multiplying the complex numbers α and $u(x)$. Again the function 0 plays the role of the zero vector. The reader should have little difficulty in modifying the real spaces in Example 3 to complex spaces.

Example 5. Let V be a region in the Euclidean space of the k variables $x_1, \ldots, x_k$. The set of all real (or complex)-valued functions $u(x_1, \ldots, x_k)$ on V is an infinite-dimensional linear space.

Exercises

2.1. Prove the statements in Remarks 5 and 6 following the definition of a linear space.

2.2. Define a linear space as follows: elements are n-tuples of complex numbers, but only scalar multiplication by real numbers is allowed; otherwise the same definitions as in Example 1 of linear spaces hold. Verify that this is indeed a *real* linear space and that its dimension is $2n$.

2.3. Show from the definition of a linear space that $\alpha 0 = 0$; $u + v = u + w$ implies $v = w$; $\alpha u = \alpha v$ and $\alpha \neq 0$ imply $u = v$; $\alpha u = \beta u$ and $u \neq 0$ imply $\alpha = \beta$.

2.4. Let C^* be the set of all real-valued continuous functions on $-\infty < x < \infty$ which fail to have a first derivative at the fixed point $x = 0$. The operations are defined as usual for a function space. Is C^* a linear space?

2.5. Consider the set P_n^0 of all polynomials of degree less than n that have the value 0 at some fixed point, say $x = \xi$. With operations as in P_n, show that P_n^0 is a linear space. What is its dimension? Find a basis for P_n^0. Show that the set of polynomials that take on the value 1 at $x = \xi$ is not a linear space.

2.6. Consider the differential equation $d^2u/dx^2 = 0$ on $0 < x < 1$. What is the dimension of the linear space of solutions satisfying the boundary condition $u(0) = u(1)$; the boundary conditions $u(0) = u(1) = 0$; the boundary conditions $u'(0) = u'(1) = 0$?

2.7. Another way of showing that $1, x, \ldots, x^{n-1}$ are independent on $a \leqslant x \leqslant b$ is to consider the equation $\alpha_1 + \alpha_2 x + \cdots + \alpha_n x^{n-1} = 0$ on $a \leqslant x \leqslant b$, and to differentiate the equation $n-1$ times. We then have n homogeneous equations in the n unknowns $\alpha_1, \ldots, \alpha_n$. Since the determinant of the coefficients is easily seen to be different from 0, the required independence follows.

Let $c_1, \ldots, c_k$ be arbitrary *unequal* real numbers. Show that $x^{c_1}, \ldots, x^{c_k}$ are independent on any interval $0 < a \leqslant x \leqslant b$.

3. METRIC SPACES

The vectors encountered so far possess algebraic properties similar to those of ordinary three-dimensional vectors—they can be added and multiplied by scalars subject to the rules of a linear space—but the equally familiar notions of length and direction have not yet made their appearance. We wish to endow linear spaces of higher dimension with a structure as rich as the one of three-dimensional vector analysis. On the algebraic structure already introduced, we must superimpose a metric structure. We begin by investigating some properties of metric spaces whose elements or points are quite general and need not be vectors in a linear space, that is, we first study the metric structure independently of any algebraic structure.

Definition. A set X of *elements* (or *points*) $u, v, w, \ldots$ is said to be a *metric space* if to each pair of elements u, v there is associated a real number $d(u,v)$, the *distance* between u and v, satisfying

(3.1)
$$\begin{cases} d(u,v) > 0 & \text{for } u,v \text{ distinct,} \\ d(u,u) = 0, \\ d(u,v) = d(v,u), \\ d(u,w) \leqslant d(u,v) + d(v,w) \text{—triangle inequality.} \end{cases}$$

The function d is known as the *metric* or *distance function*.

Remarks

1.   The ordinary distance between points in the Euclidean plane and in Euclidean 3-space certainly satisfies these axioms, the last of which states that the length of one side of a triangle does not exceed the sum of the lengths of the other two sides.

2.   In the definition we could have limited ourselves to the two axioms: (a) $d(u,v) = 0$ if and only if u and v are not distinct; (b) $d(u,w) \leqslant d(v,u) + d(v,w)$. Indeed, placing $w = u$, we find that the distance between two points is nonnegative; letting $w = v$ and then reversing the roles of v and w, we find that distance is symmetric. It then also follows that the triangle inequality holds.

The distance function generates an automatic notion of convergence: a sequence of points $\{u_k\}$ converges to the point u if and only if the sequence of *real numbers* $\{d(u,u_k)\}$ converges to 0.

Definition.   We write $\lim_{k \to \infty} u_k = u$ or $u_k \to u$, and say that $\{u_k\}$ *converges* to u, or that $\{u_k\}$ has the *limit* u, if for each $\varepsilon > 0$ there exists an index N such that

$$d(u,u_k) \leqslant \varepsilon \qquad \text{whenever } k > N.$$

Often we deal with a sequence all of whose elements beyond a certain index get close to each other but for which it is not known a priori that the sequence converges to a particular element.

Definition.   A sequence $\{u_k\}$ is a *Cauchy* (or *fundamental*) sequence if for each $\varepsilon > 0$ there exists N such that

$$d(u_m, u_p) \leqslant \varepsilon \qquad \text{whenever } m,p > N.$$

Theorem. If a sequence $\{u_k\}$ converges, it is fundamental.

Proof. Let u be the limit of $\{u_k\}$. From the triangle inequality, $d(u_m, u_p) \leqslant d(u_m, u) + d(u, u_p)$. Since $u_k \to u$, there exists N such that $d(u_n, u) \leqslant \varepsilon/2$ whenever $n > N$. If p and m are chosen larger than N, then $d(u_m, u_p) \leqslant \varepsilon$ as required.

It might appear at first that the converse is also true. If all the points of a sequence beyond a certain index are close to each other, they should be close to some definite point u. This will be true if the point u has not been carelessly left out of the space.

Definition. A metric space X is *complete* if every Cauchy sequence of points from X converges to a limit in X.

Example of an Incomplete Space
The space *Rat* of all rational numbers $u, v, \ldots$ with the metric $d(u,v) = |u - v|$ is a metric space. Consider the following sequence in *Rat*:

$$u_1 = 1, \quad u_2 = 1 + \frac{1}{1!}, \ldots, \quad u_n = 1 + \frac{1}{1!} + \frac{1}{2!} + \cdots + \frac{1}{(n-1)!}, \ldots .$$

For $p > m \geqslant 1$ we have

$$|u_p - u_m| = \frac{1}{m!} + \frac{1}{(m+1)!} + \cdots + \frac{1}{(p-1)!}$$

$$= \frac{1}{m!}\left[1 + \frac{1}{(m+1)} + \cdots + \frac{1}{(m+1)(m+2)\cdots(p-1)}\right]$$

$$\leqslant \frac{1}{m!}\left(1 + \tfrac{1}{2} + \tfrac{1}{4} + \cdots\right) = \frac{2}{m!},$$

so that for all m, p sufficiently large we can make $|u_p - u_m|$ as small as we wish. The sequence is therefore a fundamental sequence in *Rat*, but we now show that it does not have a limit in *Rat*. In fact, suppose that $u_k \to p/q$, where p and q are integers and q is chosen larger than 2 (which can obviously always be done). From the definition of convergence there exists N such that

$$\left|\frac{p}{q} - \left(1 + \cdots + \frac{1}{k!}\right)\right| < \frac{1}{4q!} \qquad \text{whenever } k > N.$$

Since the sequence $\{u_k\}$ is strictly increasing, this means that

$$0 < \frac{p}{q} - \left(1 + \cdots + \frac{1}{k!}\right) < \frac{1}{4q!}, \qquad k > N.$$

Multiplying by $q!$ and rearranging, we obtain

$$0 < \text{integer} < \frac{1}{4} + \frac{q!}{(q+1)!} + \cdots \frac{q!}{k!},$$

and, since $q > 2$,

$$\frac{q!}{(q+1)!} + \cdots + \frac{q!}{k!} < \frac{1}{q+1} + \cdots + \frac{1}{(q+1)(q+2)\cdots(k)}$$

$$< \frac{1}{3} + \frac{1}{3^2} + \cdots = \frac{1}{2}.$$

The earlier inequality then yields $0 < \text{integer} < \frac{3}{4}$, which is a contradiction. Thus $\{u_k\}$ cannot converge in *Rat*. If, however, the sequence $\{u_k\}$ is viewed as a sequence in $\mathcal{R}_1$, the space of real numbers with the metric $|u - v|$, then $\{u_k\}$ converges to the real nmber e, which is not a rational number.

The difficulty in this example lies in the space of rational numbers, which, unfortunately, is full of gaps invisible to the naked eye. We fill in the gaps in the space *Rat* as follows. If a Cauchy sequence of rational numbers does not have a rational limit, we associate with this sequence an abstract element which plays the role of the limit. These abstract elements, known as irrational numbers, are used to enlarge the space of rational numbers, the larger space being called the space of real numbers. Is this space complete? The answer is yes, either by adopting this as an axiom of the real number system or by showing that it follows from some other "equivalent" axiom. In any event the completeness of the real number system is the rock on which analysis is based. Once the completeness of the real number system is accepted, it becomes relatively easy to show that every incomplete metric space can be completed (see Exercise 3.1).

Examples of Metric Spaces

Example 1. The real line becomes a metric space if the distance between two real numbers u and v is defined as $|u - v|$. We have already indicated in the preceding paragraphs that this space is complete.

Example 2. The set of all complex numbers $z = x + iy$ is a metric space under the definition

$$d(z_1, z_2) = |z_1 - z_2| = \left[(x_1 - x_2)^2 + (y_1 - y_2)^2 \right]^{1/2}.$$

Properties (3.1) are easy to verify, and the space can be shown to be complete.

Example 3. Consider the set of all ordered n-tuples, $u = (\xi_1, \ldots, \xi_n)$, $v = (\eta_1, \ldots, \eta_n), \ldots$, of real numbers. Different metrics can be introduced, yielding different metric spaces. Many of the interesting ones are special cases of the class

$$(3.2) \qquad d_p(u, v) = \left[|\xi_1 - \eta_1|^p + \cdots + |\xi_n - \eta_n|^p \right]^{1/p}, \qquad 1 \leqslant p < \infty.$$

Of particular interest are $p = 1$, $p = 2$, and $p = \infty$:

$$(3.3) \qquad d_1(u, v) = |\xi_1 - \eta_1| + \cdots + |\xi_n - \eta_n|,$$

$$(3.4) \qquad d_2(u, v) = \left[|\xi_1 - \eta_1|^2 + \cdots + |\xi_n - \eta_n|^2 \right]^{1/2},$$

$$(3.5) \qquad d_\infty(u, v) = \max_i |\xi_i - \eta_i|.$$

The first three properties in (3.1) are easily verified in all cases, and the triangle inequality is straightforward for d_1 and d_∞. For d_2 the triangle inequality follows from the Schwarz inequality (5.12). All of these spaces are also complete. Let us prove this statement under the metric d_p. Let $\{u_k = (\xi_1^{(k)}, \ldots, \xi_n^{(k)})\}$ be a fundamental sequence; then for each $\varepsilon > 0$ there exists N such that

$$d(u_k, u_m) = \left[|\xi_1^{(k)} - \xi_1^{(m)}|^p + \cdots + |\xi_n^{(k)} - \xi_n^{(m)}|^p \right]^{1/p} \leqslant \varepsilon$$

whenever $k, m > N$. This implies that

$$|\xi_1^{(k)} - \xi_1^{(m)}| \leqslant \varepsilon, \ldots, |\xi_n^{(k)} - \xi_n^{(m)}| \leqslant \varepsilon, \qquad k, m > N.$$

From the completeness of the real number system, we conclude that the sequences $\{\xi_1^{(k)}\}, \ldots, \{\xi_n^{(k)}\}$ all converge as $k \to \infty$. Let $\lim_{k \to \infty} \xi_i^{(k)} = \xi_i$; then $\lim_{k \to \infty} u_k = (\xi_1, \ldots, \xi_n)$, and the space is complete.

Example 4. For the set of ordered n-tuples of complex numbers, all of the special cases and proofs of Example 3 carry over without additional difficulty.

Example 5. Consider the set of all real-valued continuous functions $u(x)$ defined on $a \leqslant x \leqslant b$ with the distance function

$$(3.6) \qquad\qquad d_\infty(u,v) = \max_{a \leqslant x \leqslant b} |u(x) - v(x)|.$$

Properties (3.1) are easily verified, so that we have a metric space denoted by $C(a,b)$. Let $\{u_k(x)\}$ be a Cauchy sequence in $C(a,b)$; then for each $\varepsilon > 0$ there exists N such that

$$\max_{a \leqslant x \leqslant b} |u_k(x) - u_m(x)| < \varepsilon, \qquad m, k > N.$$

But this is just the Cauchy criterion for uniform convergence; hence there exists a function $u(x)$, necessarily continuous, to which the sequence $\{u_k(x)\}$ converges uniformly. The space $C(a,b)$ is therefore complete. The metric (3.6) is known as the *uniform* metric.

Example 6. Consider, as in Example 5, the set of all real-valued continuous functions $u(x)$ on the bounded interval $a \leqslant x \leqslant b$ but now with the metric:

$$(3.7) \qquad\qquad d_2(u,v) = \left[\int_a^b |u(x) - v(x)|^2 dx \right]^{1/2}.$$

This is easily seen to be a metric space, the triangle inequality again being a consequence of the Schwarz inequality, this time in the form (5.14). The space is, however, *not complete*. In fact, we have already seen in Chapter 1 that functions with jump discontinuities can be approximated in the mean-square sense (that is, in the metric d_2) by continuous functions. The sequence of continuous functions

$$u_k(x) = \frac{1}{2} + \frac{1}{\pi} \arctan kx, \qquad -1 \leqslant x \leqslant 1,$$

is a Cauchy sequence in the metric d_2 which also converges pointwise to the discontinuous function $u(x) = \frac{1}{2} + \frac{1}{2} \text{sgn} \, x$. Moreover, $\lim_{k \to \infty} d_2(u_k, u) = 0$, that is, u_k approaches u in the mean-square sense. There is no continuous function $v(x)$ for which $d_2(u,v) = 0$, so that there cannot exist a continuous function v for which $\lim_{k \to \infty} d_2(u_k, v) = 0$.

Example 7. The space $L_2^{(r)}(a,b)$, probably the most important in applications, is the *completion* of the set of continuous functions in the metric d_2 of Example 6. The difficulty with Example 6 is that the space of continuous functions is not extensive enough to accommodate the metric d_2, just as the space of rational numbers is not extensive enough for the metric $|u - v|$. We fill the "gaps" in the set of continuous functions by adjoining to the set abstract elements associated with nonconvergent Cauchy sequences. The procedure is described in Exercise 3.1. The resulting space is the space $L_2^{(r)}(a, b)$ of real-valued functions square integrable in the Lebesgue sense (see Chapter 0).

Starting instead with the set of complex-valued functions on $a \leqslant x \leqslant b$ and completing it with respect to the metric d_2, we obtain the space $L_2^{(c)}(a, b)$ of complex-valued functions square integrable in the Lebesgue sense.

Likewise we can consider the completion of the set of real- or complex-valued functions $u(x) = u(x_1, \ldots, x_k)$ defined on a k-dimensional domain D in the variables $x_1, \ldots, x_k$ under the metric

$$d_2(u, v) = \int_D |u(x) - v(x)|^2 \, dx_1 \ldots dx_k.$$

Example 8. The following metric space appears in coding theory. A message is an ordered n-tuple of binary digits such as $(0, 1, 1, 0, \ldots, 0)$, and so on. Let $x = (\xi_1, \ldots, \xi_n)$ and $y = (\eta_1, \ldots, \eta_n)$ be two messages. We define $d(x, y) = |\xi_1 - \eta_1| + \cdots + |\xi_n - \eta_n|$; in other words, d is just the total number of places in which the two messages differ. It is easy enough to verify that all the axioms for a metric space are satisfied and that the space is complete.

Additional Concepts in Metric Spaces

Let X be a metric space with metric d, X and d being fixed throughout. A subset S of X is then automatically a metric space with the metric inherited from X. Thus S has intrinsic properties when regarded as a metric space in its own right, but also has properties relative to X when viewed as a metric subset of X. This distinction should be kept in mind in the discussion below.

It is perhaps also useful to recall one of the distinctions between sets and sequences: in a set elements are listed only once, whereas in a sequence an element can be repeated.

The *closure* of S (in X) is the set $\bar{S}$ consisting of the limits of all sequences that can be constructed from S. $\bar{S}$ must contain all points u in S, since the sequence $u, u, u, \ldots$ clearly has limit u. $\bar{S}$ may in addition contain points of X not in S. If S is the set $0 < u < 1$ on the real line X with the usual distance between points, $\bar{S}$ is the set $0 \leqslant u \leqslant 1$. If S is the set $x^2 + y^2 + z^2 < 1$ in the usual 3-space, $\bar{S}$ is the set $x^2 + y^2 + z^2 \leqslant 1$.

A set S is *closed* (or closed in X if the underlying space is in doubt) if $\bar{S} = S$. Thus limiting processes do not take us out of a closed set. The set $\bar{S}$ is closed whether or not S is. A set containing only a finite number of elements is closed. The terms "closed" and "complete" are somewhat similar, the latter being an absolute notion whereas the former is relative. Regarded as a subset of itself, every set S is closed, but not necessarily complete. It is easy to see that *a subset S of a complete metric space X is itself a complete metric space if and only if it is closed in X*. A set S is *open* if, whenever u is in S, so is some neighborhood of u. Precisely: for each u in S there exists $a > 0$ such that all points v in X satisfying $d(u, v) < a$ are also in S. On the real line the set $0 < u < 1$ is open; in 3-space the set $x^2 + y^2 + z^2 < 1$ is open. A set does not have to be either open or closed.

Let $S \subset T$ be two subsets of X. S is *dense* in T if, for each element u in T and each $\varepsilon > 0$, there exists an element v of S such that $d(u, v) < \varepsilon$. Thus every element of T can be approximated to arbitrary precision by elements of S. A set is always dense in itself and in its closure. If S is dense in T and T is dense in U, then S is dense in U. The set of rational numbers is dense on the real line. The set of all polynomials is dense in the set of continuous functions under the metric of Example 5; this is just the content of the Weierstrass approximation theorem proved in Exercise 2.7, Chapter 2. The set of polynomials is also dense in the set of continuous functions under the metric of Example 6. In fact, if $u(x)$ is a continuous function and $p(x)$ a polynomial, we have

$$d_2^2(u, p) = \int_a^b |u - p|^2 \, dx \leqslant (b - a) \max_{a \leqslant x \leqslant b} |u - p|^2,$$

and since by the Weierstrass theorem we can find p such that $\max |u - p|$ is arbitrarily small, we can make d_2 arbitrarily small. Under the metric d_2 it can also be shown that continuous functions are dense in $L_2(a, b)$. Thus polynomials are dense in $L_2(a, b)$, that is, *any square-integrable function can be approximated in the mean-square sense by polynomials*. The statement remains valid for complex-valued functions.

A set S is *bounded* if there exists a number M such that $d(u, v) \leqslant M$ for all u, v in S.

A set is *compact* (in older terminology, *compact in itself*) if each sequence of points in S contains a subsequence which converges to a point in S. (Given a sequence $u_1, u_2, \ldots$ and a sequence of positive integers $k_1 < k_2 < \ldots$, the sequence $u_{k_1}, u_{k_2}, \ldots$ is said to be a subsequence of $u_1, u_2 \ldots$. Thus $1, 5, 9, 13, \ldots$ is a subsequence of $1, 3, 5, 7, \ldots$, but $1, 1, 1, \ldots$ is not.) The notions of compactness and boundedness are absolute ones depending only on the metric properties of S itself without reference to its relation to any underlying space X. Note that a compact subset of a metric space X is necessarily closed in X and bounded. In the n-dimensional spaces of Example 3, any bounded closed set is compact, but this will not be the case in some infinite-dimensional spaces. For instance, in the metric space $C(0,1)$ of Example 5 (continuous functions on $0 \leqslant x \leqslant 1$ with uniform metric), the set S of continuous functions satisfying $|u(x)| \leqslant 1$ is a bounded closed set that is *not* compact, because the sequence $1, x, x^2, \ldots$ chosen from S has no converging subsequence in the uniform metric. An infinite orthonormal set in $L_2(0,1)$ is another example of a bounded closed set which is not compact. A set S in a metric space X is *relatively compact* if $\bar{S}$ is compact. In other words, each sequence from S contains a subsequence which converges to an element of X (not necessarily in S). Thus on the real line any bounded set (closed, open, or neither) is relatively compact.

A *transformation* A from a metric space X into itself is a rule which assigns to each point u in X exactly one element $v = Au$ also in X. The transformation is *continuous at* u if, whenever $u_n \to u$, then $Au_n \to Au$. Equivalently: for each $\varepsilon > 0$ there exists $\delta > 0$ such that $d(u, v) < \delta$ implies $d(Au, Av) < \varepsilon$. Intuitively, this means that all inputs close to u yield outputs close to Au. The transformation is said to be *continuous* (or continuous on X) if it is continuous at every u in X. Similar definitions are formulated for functionals. Note that it makes no sense at this stage to talk about linear or nonlinear transformations, since we have no algebraic structure on the space. In the examples in the next section words like "linear" and "nonlinear" will occasionally sneak in even though we are dealing with metric spaces, the reason being that there is an implicit algebraic structure superimposed on the metric one.

The principal theorem about compact sets states that a continuous real-valued functional on a compact set X attains its maximum and minimum on the set. This is the natural extension of the familiar property that a continuous real-valued function on a bounded, closed interval on the real line must assume its extremal values on the interval.

Exercises

3.1. *Completion of a metric space X with metric d*

(a) Two Cauchy sequences $\{u_k\}$ and $\{u_k'\}$ in X are equivalent if and only if $d(u_k, u_k') \to 0$ as $k \to \infty$. We write $\{u_k\} \sim \{u_k'\}$. Show that, if $\{u_k\} \sim \{u_k'\}$ and $\{u_k'\} \sim \{u_k''\}$, then $\{u_k\} \sim \{u_k''\}$. The space of all Cauchy sequences in X is therefore partitioned into so-called equivalence classes. Each equivalence class consists of all Cauchy sequences equivalent to one another. These equivalence classes are denoted by capital letters such as U and V. By a representative of U we mean any Cauchy sequence $\{u_k\}$ in U.

(b) Let Y be the space of equivalence classes of Cauchy sequences defined in (a). We tentatively define a metric δ on Y by

$$\delta(U,V)= \lim_{k\to\infty} d(u_k,v_k),$$

where $\{u_k\}$ and $\{v_k\}$ are representatives of U and V, respectively. Show that this definition makes sense, that is, that the right side exists and is independent of the representatives used. Next, show that δ satisfies the axioms of a metric.

(c) Show that Y is complete in the metric δ.

(d) Each element u in X can be regarded as an element of Y by identifying u with the equivalence class of all Cauchy sequences that converge to u (one such sequence always exists: $u,u,u,\ldots$). If u,v are elements in X, and U,V are their respective equivalence classes, show that $d(u,v)=\delta(U,V)$. Thus X can be regarded as a subset of Y, and Y is therefore a completion of X. Show that X is dense in Y. In this sense the completion is unique.

3.2. (a) Is $d(u,v)=(u-v)^2$ an admissible metric on the real line?

(b) Is $d(u,v)=|\xi_1 - \eta_1|$ an admissible metric on the set of all ordered n-tuples of real numbers $(\xi_1,\ldots,\xi_n)$?

3.3. (a) Two metrics d,δ on the same set X are said to be equivalent if convergence to a limit in either metric entails convergence to the same limit in the other. Show that d and δ are equivalent if there exist positive constants α and β independent of u,v such that

$$\alpha d(u,v) \leqslant \delta(u,v) \leqslant \beta d(u,v).$$

(b) Let X be the set of continuously differentiable real functions $u(x)$ on $0\leqslant x \leqslant 1$, satisfying $u(0)=0$. For $u,v\in X$, define

$$d^2(u,v)= \int_0^1 (u-v)^2 dx, \qquad \delta^2(u,v)= \int_0^1 (u'-v')^2 dx,$$

$$D^2(u,v)=d^2+\delta^2,$$

and show that d,δ,D are all admissible metrics. Show that δ and D are equivalent. Of course, δ and d are *not* equivalent, as is seen by examining elements of the form $\sin \alpha x$. What about d and D?

3.4. (a) Show that

$$d(u,w) \leqslant d(u,v_1) + d(v_1,v_2) + d(v_2,v_3) + \cdots + d(v_n,w)$$

for any $u, w, v_1, \ldots, v_n$. The geometric interpretation is that the length of a side of a polygon does not exceed the sum of the lengths of the other sides.

(b) Show that the metric is a continuous function, that is,

$$\lim_{\substack{k \to \infty \\ j \to \infty}} d(u_k, v_j) = d(u, v) \qquad \text{whenever } u_n \to u \text{ and } v_n \to v.$$

3.5. Prove that the complement of an open set is closed and vice versa.

3.6. We have seen that polynomials and continuous functions are dense in $L_2(a,b)$. Thus the rather "wild" functions that are in L_2 can be approximated in the mean-square sense by "well-behaved" functions. Here are some similar results, for which the reader is invited to supply the proofs:

(a) The set of differentiable functions is dense in $L_2(a,b)$.

(b) The set of twice-differentiable functions which vanish at a and b is dense in $L_2(a,b)$.

(c) The set of test functions $C_0^\infty(a,b)$ is dense in $L_2(a,b)$ (see Exercise 1.1, Chapter 2).

Thus the completion in the metric d_2 of any of the sets (a), (b), (c) yields $L_2(a,b)$.

4. CONTRACTIONS

As we shall see further on, many functional, differential, and integral equations can be recast in the form

$$(4.1) \qquad\qquad u = Tu,$$

where T is a transformation of a metric space into itself. From this point of view, solutions of (4.1) are *fixed points* of the transformation, that is, elements u which are invariant under T. A popular—and often successful—method for solving (4.1) is the so-called *method of successive approximations* (or method of iteration). Starting with some "initial approximation" u_0, one defines successive approximations $u_1 = Tu_0, u_2 = Tu_1, \ldots$, in the hope that, for large n, u_n will be a good approximation to a fixed point u of T. This hope is justified if T is *continuous* and if the sequence $\{u_k\}$ tends to a limit. Indeed, let this limit be v; then $u_n = Tu_{n-1}$, $u_n \to v$, $u_{n-1} \to v$, and, by continuity, $Tu_{n-1} \to Tv$, so that $v = Tv$. Contractions (or contraction transformations) form an important class of transformations for which the method of successive approximations works.

Definition. A transformation T of a metric space into itself is *Lipschitz continuous* if there exists a constant ρ, independent of u and v, such that

(4.2) $d(Tu, Tv) \leqslant \rho d(u, v)$ for all u, v in X.

If (4.2) holds for some fixed $\rho < 1$, T is called a *contraction*.

Thus a contraction brings points uniformly closer together, shrinking the distance between them by at least the scale factor ρ. It is easy to see that a Lipschitz continuous transformation is continuous, and therefore a contraction is continuous. The example $\sqrt{|u|}$ on R_1 shows that a continuous transformation is not necessarily Lipschitz continuous.

Theorem. Let T be a contraction on a complete metric space X. Equation (4.1) then has one and only one solution, which can be obtained by iteration from any initial approximation whatever.

Proof. (a) *Uniqueness of fixed point.* Suppose $u = Tu$ and $v = Tv$, so that $d(u, v) = d(Tu, Tv)$. Since T is a contraction, $d(Tu, Tv) \leqslant \rho d(u, v)$ and therefore $d(u, v) \leqslant \rho d(u, v)$ for $\rho < 1$, which implies that $d(u, v) = 0$ and $u = v$.

(b) *Existence of fixed point.* Let u_0 be an initial approximation, and let the iterates be defined by $u_n = Tu_{n-1}$. We show that $\{u_k\}$ is a Cauchy sequence. First note that

$$d(u_m, u_{m+1}) = d(Tu_{m-1}, Tu_m) \leqslant \rho d(u_{m-1}, u_m) \leqslant \cdots \leqslant \rho^m d(u_0, u_1).$$

From Exercise 3.4 we have, for $p > m$,

$$d(u_m, u_p) \leqslant d(u_m, u_{m+1}) + \cdots + d(u_{p-1}, u_p)$$
$$\leqslant \rho^m d(u_0, u_1) + \rho^{m+1} d(u_0, u_1) + \cdots + \rho^{p-1} d(u_0, u_1)$$
$$= d(u_0, u_1) \rho^m (1 + \rho + \rho^2 + \cdots + \rho^{p-m-1}) \leqslant d(u_0, u_1) \frac{\rho^m}{1 - \rho},$$

where the condition $\rho < 1$ enabled us to replace the finite geometric sum by the sum of the infinite series. Thus $\{u_k\}$ is Cauchy; since X is complete, there exists u such that $u_k \to u$. The continuity of T then implies that $u = Tu$.

Example of Contraction Theorem. Suppose we have two maps of the United States, map A being larger than B. Map B is placed on top of map A, so that the United States in B fits entirely within the United States in A (see Figure 4.1). Our theorem states that there is one and only one place in

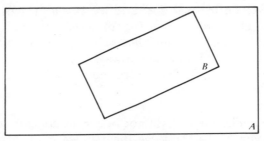

Figure 4.1

the United States whose position on the two maps will coincide (that is, there is one and only one place which will be found on map B directly above its position on map A). The transformation T is from A into itself: if u is a point in A, say Cincinnati, find Cincinnati on map B and look directly below on map A to find the point v which defined $v = Tu$. Observe also that the theorem tells us that we can determine the fixed point by iterating from any initial locality whatever.

Corollary 1. If a is a fixed point of an arbitrary transformation T, it is also a fixed point of T^k for every positive integer k. If, furthermore, T is a contraction, then so is T^k, and both T and T^k have the same fixed point a.

Corollary 2. Let T be a transformation, not assumed to be continuous, on a complete metric space X, and let T^k, for some positive integer k, be a contraction with fixed point a. Then T has the fixed point a and no other.

Proof of Corollary 2. Since $T^k a = a$, $T^{k+1} a = Ta$ and hence $T^k(Ta) = Ta$, so that Ta is also a fixed point of T^k. Since T^k is a contraction, it has only one fixed point; therefore $Ta = a$ and a is a fixed point of T. By Corollary 1 any other fixed point of T would also be a fixed point of T^k, but T^k, being a contraction, has a unique fixed point.

Remarks

1. Condition (4.2) with $\rho < 1$ is essential, as the following examples illustrate. The identity transformation satisfies (4.2) with $\rho = 1$ and has every point of X as a fixed point. In the plane a translation satisfies (4.2) with $\rho = 1$ and has no fixed point. Let X be a closed circular annulus in the plane, and let T be a rotation through an angle not a multiple of 2π; then T satisfies (4.2) with $\rho = 1$ but has no fixed point. Let X be the closed

interval $1 \leqslant u < \infty$ on the real line, and let $Tu = u + (1/u)$; then T transforms X into itself and satisfies $d(Tu, Tv) < d(u, v)$ for all $u \neq v$, but T has no fixed point; of course T does *not* satisfy (4.2) for some fixed $\rho < 1$; indeed for a given $\rho < 1$ one can always find a pair (u, v) such that $\rho d(u, v) < d(Tu, Tv) < d(u, v)$. Such a phenomenon cannot occur if X is compact (see Exercise 4.2).

2. In many applications T will not be a contradiction on the whole of its natural domain X. To apply the theorem we must then find a closed subset S of X (S is therefore a complete metric space in its own right) such that, for each $u \in S$, $Tu \in S$, and each pair $u, v \in S$, $d(Tu, Tv) \leqslant \rho d(u, v)$, with $\rho < 1$, independently of u, v.

Example 1. Let $f(u)$ be a real-valued continuous function of a real variable which transforms the closed interval $[a, b]$ into itself (thus f is defined on $a \leqslant u \leqslant b$, and its range is contained in $[a, b]$ as in Figure 4.2). The equation $u = f(u)$ then determines the points where the curve $f(u)$ crosses the diagonal. If $f(u)$ is a contraction [that is, if there exists $\rho < 1$ such that $|f(u) - f(v)| \leqslant \rho |u - v|$ for all u, v], there is exactly one such crossing point, and this point can be found by iteration from any initial try such as u_0 or $\tilde{u}_0$. The contraction condition will be satisfied if $|f'(u)| \leqslant \rho < 1$, $a \leqslant u \leqslant b$. If $f(u)$ is increasing, as in the figure, the iterations will approach the fixed point monotonically; but if $f(u)$ is decreasing, the iterations will alternate about the fixed point.

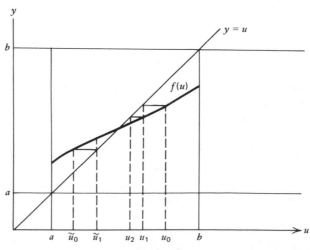

Figure 4.2

Root-finding problems of the form $g(u)=0$ occur frequently and can be translated into $u=f(u)$ with $f(u)=u-\lambda g(u)$, $\lambda\neq0$. Suppose $g(a)<0$, $g(b)>0$, and $0<\alpha\leqslant g'(u)\leqslant\beta$ on $a\leqslant u\leqslant b$. There is one and only one root of $g(u)$ in (a,b). Setting $\lambda=1/\beta$, and using the mean-value theorem, we see that $f(u)$ maps $[a,b]$ into itself and is a contraction. Starting with an initial approximation u_0, the iterates $u_n=f(u_{n-1})=u_{n-1}-\lambda g(u_{n-1})$ then converge to the root of $g(u)$. Of course numerical analysts employ a wider variety of iterative procedures; in particular, λ is often changed from step to step as in Newton's method. Many of these schemes can also be shown to have fixed points, but we are not going to pursue this subject any further.

Example 2. Contractions are also useful for the existence and uniqueness of solutions of simultaneous equations (Exercise 4.1).

Example 3. (a) Let $u\in C(a,b)$, the space of real-valued continuous functions on the bounded, closed interval $[a,b]$, and let $k(x,y,z)$ be a given real-valued continuous function on $a\leqslant x\leqslant b, a\leqslant y\leqslant b, -\infty<z<\infty$, satisfying the Lipschitz condition

$$(4.3) \qquad |k(x,y,z_2)-k(x,y,z_1)|\leqslant M|z_1 \quad z_2|,$$

where M is a positive constant independent of x,y,z_1,z_2. Substitution of $u(y)$ for z in $k(x,y,z)$ gives a continuous function of x and y for which $\int_a^b k(x,y,u(y))dy$ is in $C(a,b)$. The nonlinear Fredholm integral equation

$$(4.4) \qquad u(x)=\int_a^b k(x,y,u(y))\,dy+f(x), \qquad a\leqslant x\leqslant b,$$

where $f\in C(a,b)$ can therefore be written as $u=Tu$, where T is the transformation of $C(a,b)$ into itself defined by the right side of (4.4). Let us determine whether T is a contraction on the space $C(a,b)$ equipped with the uniform metric (3.6): $d_\infty(u,v)=\max_{a\leqslant x\leqslant b}|u(x)-v(x)|$. For any two continuous functions u and v we have

$$|(Tu)(x)-(Tv)(x)|\leqslant\int_a^b|k(x,y,u(y))-k(x,y,v(y))|dy,$$

so that, by (4.3),

$$|Tu-Tv|\leqslant M\int_a^b|u(y)-v(y)|dy\leqslant M(b-a)d_\infty(u,v).$$

Thus

$$d_\infty(Tu, Tv) \leqslant M(b-a)d_\infty(u,v),$$

and T is a contraction if

(4.5) $$M < \frac{1}{b-a},$$

as will happen if either M or $b-a$ is sufficiently small. If this condition is satisfied, any initial element $u_0(x)$ in $C(a,b)$ generates an iterative sequence through

(4.6) $$u_n(x) = f(x) + \int_a^b k(x,y,u_{n-1}(y))\,dy,$$

which tends uniformly as $n \to \infty$ to the one and only continuous solution of (4.4).

(b) A particular case of (4.4) results from setting $k(x,y,z) = \mu k(x,y)z$, where μ is a real number and $k(x,y)$ is continuous on the square $a \leqslant x,y \leqslant b$. Then (4.3) is satisfied for $M = |\mu|N$, where N is the maximum of $|k|$ on $a \leqslant x,y \leqslant b$.

Equation (4.4) becomes the *linear integral equation*

(4.7) $$u = \mu Ku + f, \qquad a \leqslant x \leqslant b,$$

where

(4.8) $$Ku = \int_a^b k(x,y)u(y)\,dy,$$

and K is known as an *integral operator* with *kernel* $k(x,y)$. From (4.5) we conclude that (4.7) has one and only one solution for

(4.9) $$|\mu| < \frac{1}{N(b-a)};$$

in particular, the equation $u = \mu Ku$ has only the trivial solution $u \equiv 0$ when (4.9) is satisfied. In the notation of operator theory, we write $\mu = 1/\lambda$, so that $\lambda u = Ku$ and we can then say that all eigenvalues of K must be less than $N(b-a)$ in absolute value.

The iterative scheme (4.6) applied to (4.7) becomes

$$u_n = f + \mu Ku_{n-1} = f + \mu K(f + \mu Ku_{n-2})$$
$$= \cdots = f + \mu Kf + \mu^2 K^2 f + \cdots + \mu^{n-1}K^{n-1}f + \mu^n K^n u_0.$$

By the contraction property u_n is known to converge uniformly to the solution of (4.7) when (4.9) is satisfied. We can also show easily that the term $\mu^n K^n u_0$ approaches 0 uniformly as $n \to \infty$, so that the solution of (4.7) is given by the uniformly convergent *Neumann series*

$$(4.10) \qquad u(x) = f + \sum_{j=1}^{\infty} u^j K^j f, \qquad |\mu| < \frac{1}{N(b-a)}.$$

If (4.7) is taken as an equation on $L_2(a,b)$, many results remain true. (See Chapter 6.)

Example 4. (a) Consider now the *Volterra* equation obtained by substituting x for the upper limit b in (4.4):

$$(4.11) \qquad u(x) = \int_a^x k(x,y,u(y))\,dy + f(x), \qquad a \leqslant x \leqslant b.$$

The function $k(x,y,z)$ needs to be defined only for $y \leqslant x$. We shall assume that k is continuous on G: $a \leqslant x \leqslant b, a \leqslant y \leqslant x, -\infty < z < \infty$, and that, on G, k satisfies the Lipschitz condition

$$(4.12) \qquad |k(x,y,z_2) - k(x,y,z_1)| \leqslant M|z_1 - z_2|.$$

Equation (4.11) is of the form $u = Tu$, where

$$Tu = f(x) + \int_a^x k(x,y,u(y))\,dy$$

defines a transformation of $C(a,b)$ into itself. We have

$$|(Tu)(x) - (Tv)(x)| \leqslant \int_a^x |k(x,y,u(y)) - k(x,y,v(y))|\,dy,$$

and, using (4.12),

$$|(Tu)(x) - (Tv)(x)| \leqslant M\int_a^x |u(y) - v(y)|\,dy$$

$$\leqslant M(x-a)\max_{a \leqslant y \leqslant x} |u(y) - v(y)| = M(x-a)d_\infty(u,v).$$

Therefore

$$|(T^2u)(x) - (T^2v)(x)| \leqslant M\int_a^x |(Tu)(y) - (Tv)(y)|\,dy \leqslant \frac{M^2(x-a)^2}{2}d_\infty(u,v),$$

and, proceeding step by step, we find that

$$(4.13) \qquad |(T^n u)(x) - (T^n v)(x)| \leqslant \frac{M^n (x-a)^n}{n!} d_\infty(u,v),$$

so that

$$(4.14) \qquad d_\infty(T^n u, T^n v) \leqslant \frac{M^n (b-a)^n}{n!} d_\infty(u,v).$$

By choosing n sufficiently large, we can ensure that $M^n(b-a)^n/n! < 1$ and therefore T^n is a contraction. Corollary 2 then tells us that T has one and only one fixed point. A consequence of Corollary 2 (see Exercise 4.8) is that the fixed point of T can be obtained by the usual iterative scheme, which in our case has the form (4.6) with x instead of b at the upper limit. Note that these results hold for arbitrarily large b as long as the Lipschitz condition (4.12) remains valid.

(b) If $k(x,y,z) = \mu k(x,y)z$ with k continuous on the triangular region $a \leqslant x \leqslant b, a \leqslant y \leqslant x$, then (4.11) becomes the linear Volterra equation

$$(4.15) \qquad u(x) = \mu \int_a^x k(x,y)u(y)\,dy + f(x), \qquad a \leqslant x \leqslant b.$$

The Lipschitz condition (4.12) is automatically satisfied, and we can therefore assert that (4.15) has one and only one solution and that, in particular, the homogeneous equation $(f=0)$ has only the trivial solution $u \equiv 0$, so that the Volterra equation has no eigenvalues.

Example 5. We shall consider the initial value problem for a (nonlinear) ordinary differential equation of the first order. Let $f(t,z)$ be a real-valued continuous function on the plane; we look for solutions $u(t)$, necessarily continuous, of

$$(4.16) \qquad \frac{du}{dt} = f(t, u(t)), \qquad u(t_0) = z_0.$$

With only the continuity assumption on f, it can be shown that (4.16) has at least one solution in some two-sided t interval about t_0. If, in addition, $f(t,z)$ satisfies a Lipschitz condition in z, there is *exactly* one solution in some two-sided t interval about t_0. We shall give a proof of this statement, but first let us see through simple examples that the conditions being imposed are not superfluous. Consider the IVP

$$(4.17) \qquad \frac{du}{dt} = 2\sqrt{|u|}, \qquad u(0) = 0.$$

Here $f(t,z)=2\sqrt{|z|}$ is independent of t (so-called *autonomous* equation). Clearly f is continuous in the t-z plane, but f does *not* satisfy a Lipschitz condition in any neighborhood of $z=0$. Problem (4.17) has the two solutions

$$u_1(t)\equiv 0, \qquad u_2(t)=\left\{\begin{array}{ll} t^2, & t\geqslant 0, \\ -t^2, & t\leqslant 0. \end{array}\right.$$

Note that $u_2(t)$ is continuous and differentiable everywhere (even at $t=0$); and that $u_2'(t)=2\sqrt{|u_2|}$ for all t. In fact, for each $a>0$,

$$u_a(t)=\left\{\begin{array}{ll} 0, & t\leqslant a, \\ (t-a)^2, & t\geqslant a, \end{array}\right.$$

is also a solution of (4.17).

Next consider the IVP

(4.18) $$\frac{du}{dt}=u^2, \qquad u(1)=1.$$

It is easily seen that this problem has one and only one solution, given by

$$u(t)=\frac{1}{2-t}, \qquad t<2.$$

Note, however, that the solution becomes infinite as $t\to 2$, so that the IVP (4.18) has one and only one solution, but only in the interval $-\infty<t<2$. Here $f(t,z)=z^2$ satisfies a Lipschitz condition for any *bounded* z interval but not for $-\infty<z<\infty$. This is enough to prevent the solution of (4.18) from being defined for all t; if the Lipschitz condition had held for $-\infty<z<\infty$, the IVP would have had one and only one solution for all t.

We now return to the general problem (4.16). We shall assume that $f(t,z)$ satisfies the Lipschitz condition

(4.19) $$|f(t,z_2)-f(t,z_1)|\leqslant M|z_2-z_1|$$

for all t,z_1,z_2 in the Lipschitz region G (large rectangle in Figure 4.3), defined by

(4.20) $$G:\ |t-t_0|\leqslant\tau, \qquad |z-z_0|\leqslant c.$$

Problem (4.16) is clearly equivalent to the Volterra integral equation

(4.21) $$u(t)=z_0+\int_{t_0}^{t}f(y,u(y))\,dy\doteq Tu,$$

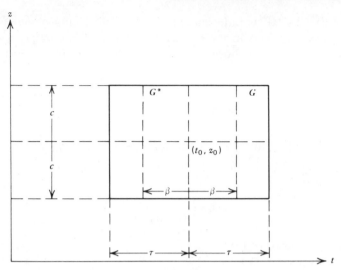

Figure 4.3

where t_0 is even allowed to be larger than t. This equation is a simple form of (4.11) with an obvious relabeling of some of the variables. As we saw in Example 4(a), (4.21) will have one and only one solution in $t_0 \leqslant t \leqslant t_0 + \tau$ if condition (4.19) holds for the region $t_0 \leqslant t \leqslant t_0 + \tau$ and *all* z_1, z_2. If (4.19) is also valid for $t_0 - \tau \leqslant t \leqslant \tau_0$ and all z_1, z_2, a trivial modification of the argument in Example 4(a) shows that we also have a unique solution for $t_0 - \tau \leqslant t \leqslant t_0$. Thus (4.16) *has one and only one solution in* $|t - t_0| < \tau$ *if* (4.19) *holds for* $|t - t_0| < \tau$ *and all* z_1, z_2. In particular, if the Lipschitz condition holds for all z_1, z_2 and for every finite τ, (4.16) has one and only one solution in $-\infty < t < \infty$; this will be true, for instance, if df/dz is bounded in each strip $|t - t_0| \leqslant \tau$, $-\infty < z < \infty$.

Unfortunately, however, a more frequent case is one where (4.19) holds only for the rectangle G defined by (4.20). Our proof of existence and uniqueness for (4.11) needs repair. Result (4.13) is a consequence of the fact, no longer true, that whenever $u(t)$ is in the Lipschitz region so is Tu. We now give the modified proof for (4.21).

Let $\max_{(t,z)\in G}|f(t,z)| = p$, and let $\beta = \min(\tau, c/p)$. The rectangle G^* defined by $|t - t_0| \leqslant \beta$, $|z - z_0| \leqslant c$ has the same height as G but is perhaps not as wide (see Figure 4.3). Consider the set C^* of continuous functions $u(t)$ defined on $|t - t_0| \leqslant \beta$ and such that $|u(t) - z_0| \leqslant c$. For short we say that C^* is the set of continuous functions lying in G^*. We claim that, if u is in C^*, so is Tu. Indeed it is clear that Tu is continuous and

$$|(Tu)(t) - z_0| = \left| \int_{t_0}^{t} f(y, u(y)) \, dy \right| \leqslant \left| \left[\int_{t_0}^{t} |f(y, u(y))| \, dy \right] \right| \leqslant p|t - t_0| \leqslant p\beta \leqslant c,$$

so that $Tu \in C^*$ (observe that the repeated absolute value signs in the first inequality are needed to cover the case where the upper limit t is less than t_0). Since G^* is part of the Lipschitz region G, it follows that, for $u, v \in C^*$, we have

$$|(Tu)(t) - (Tv)(t)| \leqslant \left| \left[\int_{t_0}^{t} |f(y, u(y)) - f(y, v(y))| \, dy \right] \right| \leqslant M \left| \left[\int_{t_0}^{t} |u - v| \, dy \right] \right|$$

$$\leqslant M|t - t_0| d_\infty(u, v) \leqslant M\beta d_\infty(u, v),$$

where $d_\infty(u, v)$ is the uniform metric on $|t - t_0| \leqslant \beta$.

Since Tu and Tv are in the Lipschitz region, we have

$$|(T^2 u)(t) - (T^2 v)(t)| \leqslant M \left| \left[\int_{t_0}^{t} |Tu - Tv| \, dy \right] \right| \leqslant M^2 d_\infty(u, v) \left| \left[\int_{t_0}^{t} |y - t_0| \, dy \right] \right|$$

$$\leqslant \frac{|t - t_0|^2}{2} M^2 d_\infty(u, v) \leqslant \frac{M^2}{2} \beta^2 d_\infty(u, v)$$

and

$$|T^n u - T^n v| \leqslant \frac{M^n}{n!} \beta^n d_\infty(u, v), \qquad d_\infty(T^n u, T^n v) \leqslant \frac{M^n \beta^n}{n!} d_\infty(u, v).$$

Thus T^n is a contraction of C^* for n sufficiently large. Since C^* is a closed subset of the complete metric space $C(t_0 - \beta, t_0 + \beta)$, it is a complete metric space in its own right. Therefore T^n has a unique fixed point, and by Corollary 2 so does T. We have therefore established that (4.16) has one and only one solution in some two-sided interval about $t = t_0$. The solution can be obtained as follows: start with any initial approximation $u_0(t)$ in C^* [it is not necessary but is nevertheless desirable to have $u_0(t_0) = z_0$; a commonly used initial choice is $u_0(t) \equiv z_0$], and define successively for $n = 1, 2, \ldots$

$$u_n(t) = z_0 + \int_{t_0}^{t} f(y, u_{n-1}(y)) \, dy.$$

Of course, for $n \geqslant 1$, $u_n(t_0) = z_0$; as $n \to \infty$, the sequence $\{u_n(t)\}$ converges uniformly to the solution of (4.16), at least on $|t - t_0| \leqslant \beta$.

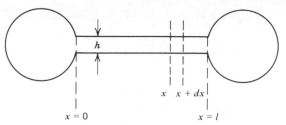

Figure 4.4

Example 6. A satellite web coupling can be idealized as a thin sheet connecting two cylindrical satellites. A cross section of the arrangement is shown in Figure 4.4. Both satellites remain at the same constant absolute temperature U while heat is radiated from the surface of the web into space at 0 absolute temperature. There is no temperature variation in the direction perpendicular to the paper. Since the web is relatively thin, we may also assume that the temperature is constant across the thickness of the web. Therefore, in the *steady state*, the temperature in the web will depend only on the x coordinate.

A heat balance for a web section one unit deep between x and $x + dx$ yields $-kh(d^2u/dx^2) + 2q = 0$, where k is the thermal conductivity, h the thickness, and q the heat radiated per unit area of the surface (the factor 2 being required because there is radiation from both the top and bottom surfaces). According to the Stefan-Boltzmann law, $q = \alpha u^4$, where α is a positive constant describing the radiation properties of the surface of the web. This leads to the nonlinear BVP

$$-khu'' = -2\alpha u^4, \quad 0 < x < l; \qquad u(0) = u(l) = U.$$

In terms of the dimensionless variables

$$x^* = \frac{x}{l}, \qquad u^* = \frac{u}{U},$$

the BVP becomes, after *dropping the asterisks* for convenience,

(4.22) $\qquad -u'' = -\lambda u^4, \quad 0 < x < 1, \qquad u(0) = u(1) = 1,$

where the *nondimensional positive* constant λ is given by

(4.23) $$\lambda = \frac{2\alpha l^2 U^3}{kh}.$$

We can translate (4.22) into an integral equation by using Green's function $g(x,\xi)$ of $-d^2/dx^2$ with *vanishing* boundary conditions at $x=0$ and $x=1$. The explicit expression $g(x,\xi)=x_<(1-x_>)$ was derived in Chapter 1. Writing $u=1-v$, we observe that v satisfies

$$-v''=\lambda u^4, \quad 0<x<1, \qquad v(0)=v(1)=0,$$

so that $v(x)=\lambda\int_0^1 g(x,y)u^4(y)\,dy$ and therefore

$$(4.24) \qquad u(x)=Tu \doteq 1-\lambda\int_0^1 g(x,y)u^4(y)\,dy, \qquad 0<x<1,$$

which is equivalent to the BVP (4.22).

Our goal is to solve (4.24), and at first an attempt might be made to view the problem as a special case of the nonlinear Fredholm equation of Example 3(a). In our case $k(x,y,z)=\lambda g(x,y)z^4$, and the Lipschitz condition (4.3) is not satisfied for all z. Fortunately it turns out that we will only need a Lipschitz condition to hold for $0\leqslant z\leqslant 1$, since the desired solution $u(x)$ of (4.22) or (4.24) is confined to that range; indeed an absolute temperature must be *nonnegative*, and the maximum principle applied to (4.22) shows that $u(x)\leqslant 1$ (as is even more obvious from the integral equation for u). This suggests that instead of dealing with the space $C(0,1)$ we should consider the subset S of $C(0,1)$ consisting of functions satisfying $0\leqslant u(x)\leqslant 1$. Clearly S is a complete metric space when equipped with the uniform metric. We want to show that the transformation T defined by the right side of (4.24) is a *contraction* on S. First we must show that T transforms elements of S into elements of S.

If $u\in S$, we have

$$(Tu)(x)=1-\lambda\int_0^1 g(x,y)u^4(y)\,dy \geqslant 1-\lambda\int_0^1 g(x,y)\,dy \geqslant 1-\frac{\lambda}{8},$$

so that $Tu\geqslant 0$ if $\lambda\leqslant 8$, and obviously $Tu\leqslant 1$. Thus, if $\lambda\leqslant 8$, T is a transformation on S. Is T a contraction?

If u and v are arbitrary elements in S, then

$$|v^4(y)-u^4(y)|\leqslant |v-u|(v^3+uv^2+vu^2+u^3)\leqslant 4d_\infty(u,v),$$

so that

$$|(Tv)(x)-(Tu)(x)|\leqslant\lambda\int_0^1 g(x,y)|v^4(y)-u^4(y)|\,dy\leqslant 4\lambda d_\infty(u,v)\int_0^1 g(x,y)\,dy.$$

Since the last integral is equal to $x(1-x)/2$, we find that

$$d_\infty(Tv, Tu) \leqslant \frac{\lambda}{2} d_\infty(u,v),$$

and T is a contraction if $\lambda < 2$. Hence, for $\lambda < 2$, (4.24) or, equivalently, (4.22) has one and only one solution in the range $0 \leqslant u(x) \leqslant 1$. The iteration scheme

$$-u_n'' = -\lambda u_{n-1}^4, \quad 0 < x < 1; \quad u_n(0) = u_n(1) = 0,$$

converges uniformly to the desired solution as long as the initial element u_0 satisfies $0 \leqslant u_0(x) \leqslant 1$. The sequence $\{u_n\}$ is not monotone, but in Chapter 9 we modify the procedure to generate a monotone sequence and we further prove that a unique solution in S exists for all $\lambda > 0$.

Exercises

4.1. Consider the set of n nonlinear equations

(4.25) $Su = u,$

where $u = (u_1, \ldots, u_n)$ is an n-tuple of real numbers and Su is the n-tuple $(S_1(u_1, \ldots, u_n), \ldots, S_n(u_1, \ldots, u_n))$, where each S_i is a real-valued function of the real variables $u_1, \ldots, u_n$.

We shall assume the Lipschitz condition, for $i = 1, \ldots, n$,

(4.26)

$$|S_i(u_1, \ldots, u_n) - S_i(v_1, \ldots, v_n)| \leqslant M_{i1}|u_1 - v_1| + \cdots + M_{in}|u_n - v_n|.$$

Using, as appropriate, the metrics (3.3), (3.4), (3.5), show that (4.25) has one and only one solution if *any* of the following conditions hold:

(4.27) $\displaystyle \max_j \sum_{i=1}^{n} M_{ij} < 1, \qquad \max_i \sum_{j=1}^{n} M_{ij} < 1, \qquad \sum_{i,j=1}^{n} M_{ij}^2 < 1.$

Specialize to the linear case

$$S_i(u_1, \ldots, u_n) = \sum_{j=1}^{n} a_{ij} u_j + f_i,$$

where the a_{ij} and f_i are constants. Conditions (4.27) can then be expressed in terms of the matrix (a_{ij}).

4.2. Let X be a compact metric space, and T a transformation on X satisfying $d(Tu, Tv) < d(u, v)$. Show that T is in fact a contraction, so that the phenomenon described in Remark 1 following the contraction theorem cannot occur.

4.3. On $0 \leqslant u < \infty$ consider the transformation $f(u) = u + e^{-u}$. Show that, for $u \neq v, d(f(u), f(v)) < d(u, v)$, yet f has no fixed point. Does this contradict the contraction theorem? If $f(u)$ is defined in the same way on $0 \leqslant u \leqslant b$, where b is finite, there is still no fixed point. Why?

4.4. Consider the Volterra equation (4.11) under the following conditions:
(a) $|k(x, y, z_2) - k(x, y, z_1)| \leqslant M |z_2 - z_1|$ for all x, y, z_1, z_2 in G, defined by $a \leqslant x \leqslant b, a \leqslant y \leqslant x, |z - \eta| \leqslant c$.
(b) $|f(x) - \eta| \leqslant c_1 < c$ for $a \leqslant x \leqslant b$.
(c) $\max_G |k(x, y, z)| \leqslant \dfrac{c - c_1}{b - a}$.
Show that (4.11) has one and only one continuous solution in $a \leqslant x \leqslant b$.

4.5. Consider the sets of all n-tuples of real-valued continuous functions $u(t) = (u_1(t), \ldots, u_n(t))$ on a fixed bounded interval $a \leqslant t \leqslant b$. Show that this set becomes a complete metric space under any of the following metrics:

$$d_1(u, v) = \sum_{i=1}^{n} \max_{a < t < b} |u_i(t) - v_i(t)|,$$

$$d_2(u, v) = \left\{ \sum_{i=1}^{n} \max_{a < t < b} |u_i(t) - v_i(t)|^2 \right\}^{1/2},$$

$$d_\infty(u, v) = \max_i \ \max_{a < t < b} |u_i(t) - v_i(t)|.$$

4.6. Let $f_1(t, z_1, \ldots, z_n), \ldots, f_n(t, z_1, \ldots, z_n)$ be n real-valued continuous functions of the real variables $t, z_1, \ldots, z_n$. Consider the first-order *system* of n (nonlinear) ordinary differential equations

$$\frac{du_1}{dt} = f_1(t, u_1(t), \ldots, u_n(t)), \ldots, \frac{du_n}{dt} = f_n(t, u_1(t), \ldots, u_n(t)),$$

with the initial conditions

$$u_1(t_0) = \eta_1, \ldots, u_n(t_0) = \eta_n.$$

In vector notation the IVP takes the form

(4.28) $$\frac{du}{dt} = f(t, u(t)), \qquad u(t_0) = \eta,$$

where $u(t) = (u_1(t), \ldots, u_n(t))$, $\eta = (\eta_1, \ldots, \eta_n)$, and

$$f(t, z) = (f_1(t, z_1, \ldots, z_n), \ldots, f_n(t, z_1, \ldots, z_n)).$$

Note that a single differential equation of order n can always be transformed to a first-order system of n equations.

Translate (4.28) into a vector Volterra integral equation. Assume that f satisfies the Lipschitz condition, for $i = 1, \ldots, n$,

(4.29) $$|f_i(t, z) - f_i(t, w)| \leqslant M_{i1}|z_1 - w_1| + \cdots + M_{in}|z_n - w_n|,$$

which holds for *all* z and w and for $|t - t_0| \leqslant \tau$.

Show that (4.28) has one and only one solution for all t on $|t - t_0| \leqslant \tau$. It is easiest to use the metric d_∞ of Exercise 4.5 and to note that (4.29) implies that there exists M such that

$$|f_i(t, z) - f_i(t, w)| \leqslant M \max_j |z_j - w_j|, \qquad i = 1, \ldots, n.$$

4.7.   (a) Consider the linear system of n equations

(4.30) $$\frac{du}{dt} = A(t)u(t) + q(t), \qquad u(t_0) = \eta,$$

which is of the form (4.28) with

$$f_i(t, u_1, \ldots, u_n) = \sum_{j=1}^{n} a_{ij}(t)u_j + q_i(t),$$

where the coefficients $a_{ij}(t)$ and the forcing terms $q_i(t)$ are assumed continuous. Show that the Lipschitz condition (4.29) is satisfied for any bounded interval whatever, so that (4.30) has one and only one solution for any bounded interval and hence for $-\infty < t < \infty$.

(b) Consider the IVP for a single linear differential equation of order n for an unknown scalar function $y(t)$. Set $u_1 = y, u_2 = y', \ldots, u_n = y^{(n-1)}$ to translate the differential equation into a first-order linear system for the vector $u = (u_1, \ldots, u_n)$, which will be of the form (4.30) if $a_n(t) \neq 0$, where $a_n(t)$ is the leading coefficient in the given ordinary differential equation. This proves the existence and uniqueness theorem for the IVP for a single scalar equation of any order.

4.8. Suppose T^n is a contraction on a complete metric space X with fixed point a. Then Corollary 2 guarantees that a is the one and only fixed point of T. Show that, if u_0 is any element in X, the sequence $u_0, u_1 = Tu_0, u_2 = Tu_1, \ldots$ tends to a.

4.9. Suppose that for each λ in $\lambda_1 \leqslant \lambda \leqslant \lambda_2$ the transformation T_λ (depending continuously on λ) is a contraction on a complete metric space X. Show that the fixed point of T_λ varies continuously with λ.

4.10. Consider the nonlinear BVP

$$(4.31) \qquad -u'' = \alpha \sin u, \quad 0 < x < 1, \quad u'(0) = u(1) = 0,$$

which governs the buckling of a suitably supported rod under a compressive thrust to which the parameter α is proportional. By introducing an appropriate Green's function, translate (4.31) into an integral equation

$$u(x) = Tu \doteq \alpha \int_0^1 g(x, \xi) \sin u(\xi) \, d\xi,$$

and show that T is a contraction in a suitably defined metric space when $\alpha < 2$. Thus for $\alpha < 2$ the only solution of (4.31) is $u = 0$, so that the critical load must exceed 2 (in fact, we show in Chapter 9 that the critical load is $\pi^2/4$).

5. NORM, INNER PRODUCT, BANACH SPACE, HILBERT SPACE

We now return to linear spaces. We shall define a notion of length or norm for the vectors of a linear space.

Definition. A *normed linear space* is a linear space (real or complex) in which a real-valued function $\|u\|$ (known as the *norm* of u) is defined, with the properties

$$(5.1) \qquad \begin{cases} (a) & \|u\| > 0, \quad u \neq 0, \\ (b) & \|0\| = 0, \\ (c) & \|\alpha u\| = |\alpha| \, \|u\|, \\ (d) & \|u + v\| \leqslant \|u\| + \|v\|. \end{cases}$$

Remarks

1.   Here $\|u\|$ plays the same role as the length of u in ordinary three-dimensional space. Property (c) states that the length of a scalar multiple of a vector u is just the appropriate multiple of the length of the vector u (note that the vector $-3u$ should have and has three times the length of u). Property (d) is a form of the triangle inequality. Clearly all the properties (5.1) hold for the usual three-dimensional vectors.

2.   A normed linear space is *automatically* a metric space with the metric defined by

$$(5.2) \qquad\qquad d(u,v) = \|u-v\|.$$

Let us see that (3.1) follows from (5.1) and (5.2). Clearly $d(u,v) > 0$ for u,v distinct and $d(u,u) = 0$. Since, by (5.1)(c), $\|u-v\| = \|v-u\|$, we have $d(u,v) = d(v,u)$. In (5.1)(d) we substitute $u-v$ for u and $v-w$ for v to obtain $\|u-w\| \leqslant \|u-v\| + \|v-w\|$ or $d(u,w) \leqslant d(u,v) + d(v,w)$. Thus (5.2) defines a suitable metric known as the *natural metric generated by the norm*. Observe that we can recover the norm from d by

$$(5.3) \qquad\qquad \|u\| = d(u,0).$$

It follows that $\|u\|$ is a continuous function of u, that is, if $u_k \to u$, then $\|u_k\| \to \|u\|$.

3.   A normed linear space may be viewed either as a linear space, as a metric space, or as both; hence its elements may be referred to interchangeably as *vectors* or *points*. This is a familiar situation in ordinary three-dimensional geometry, where the space consists of points with the usual Euclidean distance between them, but where each point is also the terminus of a well-defined vector emanating from a fixed origin. Points and vectors can therefore be identified. The distance $d(u,v)$ between two points is just the length of the difference of the two vectors, that is, $\|u-v\|$; in particular $d(u,0) = \|u\|$, so that the distance of a point to the origin is the length of the vector.

4.   Consider the real line with the metric $d(u,v)$, where $d(u,u) = 0$ and $d(u,v) = 1$ if $u \neq v$. It is easy to verify that properties (3.1) hold, yet there is no norm which generates this metric. Indeed, if such a norm existed, we would have from (5.3) that $\|u\| = 1$ for all $u \neq 0$, which clearly violates (5.1)(c). Suppose a metric $d(u,v)$ is given on a linear space. What additional conditions must d obey for the space to be a normed linear space?

The answer is simple; we must have

$$d(\alpha u, \alpha v) = |\alpha| d(u,v) \qquad \text{for all } u,v,\alpha.$$

The definition $\|u\| = d(u,0)$ then does the trick.

5. The metric and linear properties are compatible in the sense that, if $u_n \to u$ and $v_n \to v$, then $\alpha u_n + \beta v_n \to \alpha u + \beta v$; also, if $\alpha_n \to \alpha$, then $\alpha_n u \to \alpha u$.

Definition. A normed linear space which is complete in its natural metric (5.2) is called a *Banach space*.

Inner Product Spaces

In a normed linear space a vector has a length. We want to refine the structure further so that the angle between vectors is also defined. In particular, we want a criterion for determining whether or not two vectors are perpendicular. As in three-dimensional space, these notions are most easily derived from an inner product (or dot product or scalar product) between vectors.

Definition. An *inner product* $\langle u,v \rangle$ on a *real* linear space $X^{(r)}$ is a *real*-valued function of ordered pairs of vectors u,v with the properties

$$(5.4) \quad \begin{cases} \langle u,v \rangle = \langle v,u \rangle, \\ \langle \alpha u,v \rangle = \alpha \langle u,v \rangle, \\ \langle u+v,w \rangle = \langle u,w \rangle + \langle v,w \rangle, \\ \langle u,u \rangle > 0 \qquad \text{for } u \neq 0 \end{cases}$$

where u,v,w are arbitrary vectors and α is an arbitrary real number. Such a linear space is called a *real inner product space*.

Remark. It follows that $\langle u,0 \rangle = 0$ for every u. Although $\langle u,u \rangle$ is non-negative, $\langle u,v \rangle$ may be positive, negative, or 0. The requirement on $\langle u,u \rangle$ is made so that it will have a positive square root which can act as a norm.

The definition of an inner product on a complex vector space is somewhat different. Since the inner product will still be used to generate a norm, the property $\langle u,u \rangle > 0$ for $u \neq 0$ must be preserved. This forces some changes in (5.4); otherwise $\langle iv,iv \rangle = i\langle v,iv \rangle = i\langle iv,v \rangle = i^2 \langle v,v \rangle = -\langle v,v \rangle$, and it would be impossible to have $\langle u,u \rangle$ positive for all $u \neq 0$. The remedy is to change the first property in (5.4) to $\langle u,v \rangle = \overline{\langle v,u \rangle}$.

Definition. An *inner product* $\langle u,v \rangle$ on a *complex* linear space $X^{(c)}$ is a *complex-valued* function of ordered pairs u,v with the properties

(5.5)
$$\begin{cases} \langle u,v \rangle = \overline{\langle v,u \rangle} & \text{(this implies } \langle u,u \rangle \text{ real),} \\ \langle \alpha u,v \rangle = \alpha \langle u,v \rangle, \\ \langle u+v,w \rangle = \langle u,w \rangle + \langle v,w \rangle, \\ \langle u,u \rangle > 0 & \text{for } u \neq 0, \end{cases}$$

where u,v,w are arbitrary vectors in $X^{(c)}$ and α is an arbitrary complex number. The linear space $X^{(c)}$ provided with an inner product is called a *complex inner product space.*

An immediate consequence of the definition is that $\langle u,\alpha v \rangle = \bar{\alpha} \langle u,v \rangle$, where the appearance of the complex conjugate means that the inner product is not quite linear in the second term. The inner product is an instance of a bilinear form (see Exercise 5.7). We also note that $\langle u,0 \rangle = \langle 0,v \rangle = 0$. Observe also that if α is restricted to real values properties (5.5) reduce to (5.4). Unless otherwise stated, the results below are valid for both real and complex inner products; the derivations will, however, be given only for the more difficult complex case.

Theorem (Schwarz inequality). For any two vectors u,v in an inner product space,

(5.6)
$$|\langle u,v \rangle|^2 \leqslant \langle u,u \rangle \langle v,v \rangle,$$

with equality if and only if u and v are dependent.

Remark. Remember that $\langle u,v \rangle$ is a complex number, so that the absolute value sign in (5.6) is not wasted; without it we could not guarantee that the left side is real! On a real space the absolute value sign can be dropped.

Proof. For any two vectors u,w, the inequality $\langle u-w,u-w \rangle \geqslant 0$ yields

(5.6a) $$\langle u,u \rangle \geqslant \langle w,u \rangle + \langle u,w \rangle - \langle w,w \rangle = 2 \operatorname{Re} \langle w,u \rangle - \langle w,w \rangle,$$

with equality if and only if $w=u$.
   If $v=0$, then (5.6) is trivially true with the equality sign; if $v \neq 0$, set

$$w = \frac{\langle u,v \rangle}{\langle v,v \rangle} v$$

in (5.6a) to obtain the Schwarz inequality. Since $w=u$ if and only if u is a multiple of v, we have equality in (5.6) if and only if either $v=0$ or u is a multiple of v. Thus equality in (5.6) occurs if and only if u and v are dependent.

Corollary.

$$(5.7) \qquad \langle u+v,u+v\rangle^{1/2} \leqslant \langle u,u\rangle^{1/2} + \langle v,v\rangle^{1/2}.$$

Proof. Expand the real nonnegative quantity $\langle u+v,u+v\rangle$, and use (5.6).

It is therefore clear that $\langle u,u\rangle^{1/2}$ is an admissible norm on our linear space, since

$$\|u\| = \langle u,u\rangle^{1/2}$$

satisfies conditions (5.1). Thus an inner product space is automatically a normed linear space and hence a metric space with

$$d(u,v) = \|u-v\| = \langle u-v,u-v\rangle^{1/2},$$

and the Schwarz inequality (5.6) takes on the more attractive form

$$(5.8) \qquad |\langle u,v\rangle| \leqslant \|u\| \, \|v\|.$$

With a metric at our disposal we have the usual definition of convergence, which we repeat for convenience: a sequence $\{u_k\}$ is said to converge to u if for each $\varepsilon>0$ there exists $N>0$ such that $\|u_k - u\| < \varepsilon, k > N$. Moreover, since a linear structure exists together with the metric structure, we can now define convergence of series.

Definition. The series $u_1 + u_2 + \cdots + u_n + \cdots$ (also written as $\Sigma_{k=1}^{\infty} u_k$) is said to converge to u if the sequence of partial sums converges to u, that is, if for each $\varepsilon>0$ there exists $N>0$ such that $\|\Sigma_{k=1}^{n} u_k - u\| < \varepsilon, n > N$.

It follows from the Schwarz inequality that the *inner product is a continuous function of its arguments*. Thus $\lim_{n\to\infty}\langle u_n,v_n\rangle = \langle u,v\rangle$ whenever $u_n \to u$ and $v_n \to v$. In particular, $\|u_n\| \to \|u\|$ if $u_n \to u$. If $\Sigma_{k=1}^{\infty} u_k = u$, the series $\Sigma_{k=1}^{\infty}\langle u_k,v\rangle$ automatically converges to $\langle u,v\rangle$.

Definition. An inner product space which is complete in its natural metric is called a *Hilbert space*.

Definition. An n-dimensional inner product space is known as a *Euclidean space E_n*. A real n-dimensional Euclidean space is often denoted by R_n.

We shall see later that a Euclidean space is automatically complete; hence it is a Hilbert space. Of course many of the inner product spaces of interest are infinite-dimensional; such spaces are not automatically Hilbert spaces and may require the completion procedure described in Exercise 3.1 to convert them into Hilbert spaces. Before proceeding to a more systematic investigation of Hilbert spaces and operators defined on them, we try to substantiate an earlier claim that an inner product provides additional structure to a linear space.

First, we note the *parallelogram law*

$$(5.9) \qquad \|u+v\|^2 + \|u-v\|^2 = 2\|u\|^2 + 2\|v\|^2,$$

which states that the sum of the squares of the diagonals of the parallelogram built from u and v is equal to the sum of the squares of the sides. Relation (5.9) holds, not for arbitrary normed linear spaces, but only for those which can be provided with an inner product (see Exercise 5.4).

In an inner product space there is also a natural definition of perpendicularity.

Definition. Two vectors are *orthogonal* (or perpendicular) if $\langle u,v\rangle = 0$. A set of vectors, each pair of which is orthogonal, is called an *orthogonal set* (the set is *proper* if all vectors in it $\neq 0$); if in addition every vector in the set has unit norm, the set is *orthonormal*.

The terminology is justified by observing that $\langle u,v\rangle = 0$ implies that $\|u+v\|^2 = \langle u+v, u+v\rangle = \|u\|^2 + \|v\|^2$, which is just the Pythagorean theorem. It follows that, if the sum of two orthogonal vectors vanishes, each vector must be the zero vector.

Let H be a Hilbert space and M a linear manifold in H, that is, a set of elements in H such that, whenever $u,v \in M$, then $\alpha u + \beta v \in M$ for all scalars α, β. At first we consider only the simplest linear manifold, a line through the origin.

Definition. The set of vectors αv_0, where $v_0 \neq 0$ and α runs through the scalars (real or complex numbers, depending on the space), is called the *line* generated by v_0.

In the real case the line generated by v_0 contains only two unit vectors: $v_0/\|v_0\|$ and $-v_0/\|v_0\|$, but in the complex case the line contains all unit vectors of the form $e^{i\beta}v_0/\|v_0\|$ with β real.

Definition. The (orthogonal) *projection* of u on the line M generated by v_0 is the vector $v = \langle u, e \rangle e$, where e is a unit vector on the line (the various possible choices for e all yield the *same* vector v).

Without reference to any unit vector we can write the projection of u on the line generated by v_0 as

(5.10)
$$v = \frac{\langle u, v_0 \rangle}{\|v_0\|^2} v_0.$$

(In mathematical terminology a projection is always a vector, although in engineering it sometimes means the magnitude of v.) The definition agrees with intuition, for we can easily check that $u - v$ is orthogonal to all vectors in M. Thus u has been written as a sum of two vectors, one in M and the other perpendicular to M. Moreover, such a decomposition is unique, for suppose we had $u = v + w = v' + w'$, where both v and v' are in M and both w and w' are perpendicular to M; then by subtraction we find $0 = (v - v') + (w - w')$, where $(v - v')$ and $(w - w')$ are orthogonal. But if the sum of two orthogonal vectors vanishes, they must both vanish, so that $v = v'$ and $w = w'$. The vector v can also be unambiguously characterized as the vector in M which lies closest to u. These ideas are illustrated in Figure 5.1 for the real case, where we can also define the angle θ_{uv_0} between u and v_0 by

$$\cos \theta_{uv_0} = \frac{\langle u, v_0 \rangle}{\|u\| \, \|v_0\|}.$$

The Schwarz inequality shows that the right side is between -1 and 1, so that a real value of the angle between 0 and π is indeed defined.

These notions can be extended to closed linear manifolds in H. We recall that a closed linear manifold M has the property that, whenever

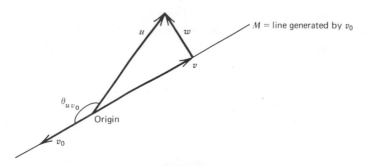

Figure 5.1

$u_1, u_2, \ldots$ is a sequence in M which converges in H, the limit u actually belongs to M. Example 3(b) below shows that linear manifolds in H are not necessarily closed. There are, however, classes of manifolds that are always closed: finite-dimensional manifolds and manifolds orthogonal to a given manifold.

Definition. Let M be a linear manifold (closed or not) in H. The set $M^\perp$ (*linear manifold orthogonal to* M) is defined as follows: u belongs to $M^\perp$ if and only if it is orthogonal to every vector in M.

The set $M^\perp$ is seen to be a linear manifold in fact as well as in name. Moreover, $M^\perp$ is closed. Indeed, let $u_1, u_2, \ldots$ be a sequence in $M^\perp$ with limit u in H; since $\langle u_k, f \rangle = 0$ for every f in M, continuity of the inner product shows that $\langle u, f \rangle = 0$, so that u is in $M^\perp$ and hence $M^\perp$ is closed.

We are now in a position to state the projection theorem, which plays a central role in both the theory and the applications of Hilbert spaces. Two proofs will be given. The one below is quite general, whereas the one in the next section is valid only for separable spaces (the definition of this concept will be given in due time).

Projection Theorem. Let M be a closed linear manifold in H. Each vector u in H can be written in one and only way as the sum of a vector in M and a vector in $M^\perp$. The vector in M is known as the (orthogonal) *projection* of u on M and can also be characterized as the unique vector in M which is closest to u.

Remark. Figure 5.1 illustrates the theorem when M is one-dimensional. It is also easy to visualize the meaning of the theorem when M is a two-dimensional linear manifold (that is, a plane through the origin) in a three-dimensional Hilbert space H. Let u be a fixed, arbitrary vector emanating from the origin, and drop a perpendicular from u to M. The theorem asserts the obvious geometric fact that u is the sum of its projection v lying in M and of the vector $w = u - v$, which is orthogonal to M. If we consider $\|u - z\|$ for all z lying in M with the same origin as u, then $z = v$ gives the smallest possible value of $\|u - z\|$; identifying vectors with their endpoints, we can say that v is the point (or vector) in M closest to u.

Proof. (a) We first show that there exists at least one vector in M which is closest to u. As z ranges through M, the set of real numbers $\|u - z\|$ is bounded below (since these numbers are nonnegative), so that we can define the nonnegative number

$$d = \inf_{z \in M} \|u - z\|,$$

and we wish to prove that the infimum is attained. By definition there exists a minimizing sequence $\{z_n\}$ in M such that $\|u - z_n\| \to d$. We apply the parallelogram law (5.9) with $u - z_m$ and $u - z_n$ playing the roles of u and v, respectively, to obtain

$$\|z_n - z_m\|^2 = 2\|u - z_m\|^2 + 2\|u - z_n\|^2 - 4\left\|u - \frac{z_n + z_m}{2}\right\|^2.$$

Since M is linear, the vector $(z_n + z_m)/2$ belongs to M and $\|u - (z_n + z_m)/2\|^2$ must exceed d^2. Therefore

$$\|z_n - z_m\|^2 \leqslant 2\|u - z_m\|^2 + 2\|u - z_n\|^2 - 4d^2,$$

and, as m, n, tend to infinity, the right side tends to 0 and $\{z_n\}$ is a Cauchy sequence. Since H is complete, $z_n \to v$ in H; moreover, $v \in M$ because M is closed.

(b) Let v be the limit of the minimizing sequence of part (a), and let w be the vector $u - v$. We want to show that w is orthogonal to M, or, equivalently, that $\langle w, z \rangle = 0$ for every z in M. This statement is obviously true for $z = 0$, so let us take z to be an arbitrary nonzero element of M. Then $v + \alpha z$ belongs to M for all complex α, and by the definition of v we have

$$\|w - \alpha z\|^2 = \|u - (v + \alpha z)\|^2 \geqslant \|u - v\|^2 = \|w\|^2,$$

so that $\|w - \alpha z\|^2$ has its minimum at $\alpha = 0$. Taking α real, we can write

$$\|w - \alpha z\|^2 = \|w\|^2 + \alpha^2 \|z\|^2 - \alpha \langle w, z \rangle - \alpha \langle z, w \rangle,$$

and setting the derivative with respect to α equal to 0 at $\alpha = 0$, we find that $\operatorname{Re}(w, z) = 0$ for every $z \in M$. Since iz also belongs to M, we have $\operatorname{Re}(w, iz) = 0$ or $\operatorname{Im}(w, z) = 0$. We can then conclude that $\langle w, z \rangle = 0$.

(c) We showed in part (b) that $u = v + w, v \in M, w \in M^{\perp}$. Such a decomposition is unique. Suppose we also had $u = v' + w'$, $v' \in M, w' \in M^{\perp}$; then $0 = (v - v') + (w - w')$ with $v - v'$ in M and $w - w'$ in $M^{\perp}$. If the sum of two orthogonal vectors vanishes, each of the vectors must vanish. It follows therefore that the vector v of part (a) is unique.

Remarks on the proof

1. If M is not closed, the infimum may not be attained. Take, for instance, $H = L_2(a, b)$ and $M =$ the set of all polynomials. If u is an element in L_2 but is not a polynomial, there exists for each $\varepsilon > 0$ an element v_ε in M for which $\|u - v_\varepsilon\| \leqslant \varepsilon$. Thus $\inf_{z \in M} \|u - z\| = 0$, but there is no element v in M for which $\|u - v\| = 0$.

2.   Exercise 5.5 shows that, if M is a closed linear manifold in a Banach space B, the infimum may not be attained because of the failure of the parallelogram law.

3.   To show that the minimizing sequence $\{z_n\}$ in part (a) was Cauchy, we only needed that $(z_m+z_n)/2$ be in M whenever z_m and z_n were. This is exactly the defining property of a *convex* set (if the midpoint of the line segment joining two points in the set is also in the set, the entire line segment is in the set). A linear manifold is a convex set, but the class of convex sets is much larger. One can prove (see Exercise 5.8) that, given $u \in H$ and a *closed* convex set C, there exists a unique element in C which is closest to u.

Corollary.   Let M be a closed linear manifold in H, with $M \neq H$. Then there exists a nonzero element in $M^\perp$.

Proof.   Let $u \neq 0$ be in H but not in M. Let the projection of u on M be v; then $w = u - v$ is a nonzero vector in $M^\perp$.

The Gram-Schmidt Orthogonalization Process

It is often more convenient to deal with orthonormal sets of vectors rather than just independent sets. The Gram-Schmidt process converts an independent set into an orthonormal set with the same span.

Let $h_1, h_2, \ldots$ be a finite or infinite sequence of vectors in H; for each $k \geq 1$ (but not exceeding the number of elements in the sequence) let $h_1, \ldots, h_k$ be *independent*. We shall construct an orthonormal set $e_1, e_2, \ldots$ such that, for each k, e_k is a linear combination of $h_1, \ldots, h_k$. Note that, if the original sequence has only a finite number of elements, the orthonormal sequence has the same number of elements; if the original sequence is infinite, so is the orthonormal sequence. The construction is as follows. The unit vector e_1 is defined as $h_1/\|h_1\|$. To obtain e_2 we proceed in two steps. First, we construct g_2 by subtracting from h_2 its projection on e_1, that is, $g_2 = h_2 - \langle h_2, e_1 \rangle e_1$; g_2 is clearly orthogonal to e_1 and is not the zero vector, since it is a linear combination of h_2 and h_1 with the coefficient of h_2 not 0; normalization then gives $e_2 = g_2/\|g_2\|$. Continuing in this way, we construct successively

$$(5.11) \qquad g_k = h_k - \sum_{j=1}^{k-1} \langle h_k, e_j \rangle e_j; \qquad e_k = \frac{g_k}{\|g_k\|}.$$

The construction goes on ad infinitum if the original sequence is infinite; otherwise it stops when the finite number of vectors in the original sequence has been exhausted. From (5.11) we see that e_k is a linear combination of $h_1, \ldots, h_k$, so that $e_1, \ldots, e_k$ all lie in the span M_k of $h_1, \ldots, h_k$. Now M_k is a k-dimensional manifold containing the k independent vectors $e_1, \ldots, e_k$, which must therefore span M_k.

An illustration of the process is given in Example 3(c) below.

Corollary. Every Euclidean space E_n has an orthonormal basis.

Proof. Use the Gram-Schmidt process on any basis for the space.

By introducing orthonormal bases in n-dimensional spaces, we can simplify proofs that might otherwise be awkward.

Theorem. Every Euclidean space is complete.

Proof. Let $\{u_k\}$ be a Cauchy sequence of elements in E_n, the index k ranging from 1 to ∞. With $e_1, \ldots, e_n$ an orthonormal basis for E_n, we can write

$$u_k = \sum_{j=1}^{n} \alpha_{k,j} e_j,$$

and, taking the inner product with e_i, we find that

$$\alpha_{k,i} = \langle u_k, e_i \rangle,$$

and, by Schwarz's inequality,

$$|\alpha_{k,i} - \alpha_{p,i}| = |\langle u_k - u_p, e_i \rangle| \leqslant \|u_k - u_p\|.$$

Thus, for i fixed, $\{\alpha_{k,i}\}$ is a Cauchy sequence of numbers (real or complex, depending on whether E_n is real or complex) and must therefore converge to a number, say α_i. It follows that

$$\lim_{k \to \infty} u_k = \sum_{i=1}^{n} \alpha_i e_i,$$

which is a vector in E_n, so that E_n is complete.

Corollary. Let M be a finite-dimensional manifold in a Hilbert space H (finite- or infinite-dimensional). Then M is closed in H.

Example 1. A concrete example of a real n-dimensional Euclidean space is the space of n-tuples of real numbers, which becomes a real inner product space with the definition

$$\langle u, v \rangle = \xi_1 \eta_1 + \cdots + \xi_n \eta_n,$$

generating the norm

$$\|u\| = \left(\xi_1^2 + \cdots + \xi_n^2 \right)^{1/2}$$

and the metric

$$d_2(u, v) = \|u - v\| = \left[(\xi_1 - \eta_1)^2 + \cdots + (\xi_n - \eta_n)^2 \right]^{1/2}.$$

The Schwarz inequality (5.6) becomes

(5.12)
$$\left[\sum_{k=1}^{n} \xi_k \eta_k \right]^2 \leqslant \left(\sum_{k=1}^{n} \xi_k^2 \right) \left(\sum_{k=1}^{n} \eta_k^2 \right).$$

This is just the space already encountered in Example 3, Section 3, with the metric (3.4). We showed there that the space is complete. Note that the other metrics considered in that example can be generated from a norm but *not* from an inner product. These spaces are complete normed spaces (that is, Banach spaces) but *not* Hilbert spaces (see Exercise 5.4).

Example 2. A concrete example of a complex Euclidean space is the space of n-tuples of complex numbers with the inner product defined as

$$\langle u, v \rangle = \sum_{k=1}^{n} \xi_k \bar{\eta}_k.$$

Conditions (5.5) are easily verified. The norm and metric are given by

$$\|u\| = \left[\sum_{k=1}^{n} |\xi_k|^2 \right]^{1/2}; \qquad d_2(u, v) = \|u - v\| = \left[\sum_{k=1}^{n} |\xi_k - \eta_k|^2 \right]^{1/2},$$

while the Schwarz inequality (5.6) becomes

(5.13)
$$\left| \sum_{k=1}^{n} \xi_k \bar{\eta}_k \right|^2 \leqslant \left(\sum_{k=1}^{n} |\xi_k|^2 \right) \left(\sum_{k=1}^{n} |\eta_k|^2 \right).$$

Example 3. (a) Consider the space $L_2^{(r)}(a, b)$ of real-valued functions $u(x)$ defined on $a \leqslant x \leqslant b$ for which the Lebesgue integral $\int_a^b u^2(x) \, dx$ exists and is finite. The inner product is defined by

$$\langle u,v \rangle = \int_a^b u(x)v(x)\,dx,$$

which meets all the conditions (5.4). The corresponding norm and metric are

$$\|u\| = \left[\int_a^b u^2\,dx \right]^{1/2}; \qquad d_2(u,v) = \|u-v\| = \left[\int_a^b (u-v)^2\,dx \right]^{1/2},$$

and the Schwarz inequality becomes

$$(5.14) \qquad \left| \int_a^b uv\,dx \right| \leqslant \left[\int_a^b u^2\,dx \right]^{1/2} \left[\int_a^b v^2\,dx \right]^{1/2}.$$

As has already been pointed out in Example 7, Section 3, $L_2^{(r)}(a,b)$ is complete with this metric. Therefore $L_2^{(r)}(a,b)$ is a Hilbert space.

(b) Let us look at some linear manifolds in $H = L_2^{(r)}(a,b)$. Any such manifold has an inner product structure inherited from H. The set M_1 of continuous functions on $a \leqslant x \leqslant b$ is a linear manifold in H; M_1 is *not* closed, but $\overline{M}_1 = H$. The set M_2 of polynomials on $a \leqslant x \leqslant b$ is a linear manifold which is not closed, but again $\overline{M}_2 = H$. The set M_3 of the elements in H which are even about $x = (a+b)/2$ is a closed linear manifold with $M_3 \neq H$. The set M_4 of polynomials even about $x = (a+b)/2$ is a linear manifold in H with $M_4 \neq \overline{M}_4 \neq H$. All these linear manifolds are infinite-dimensional. The set M_5 of all polynomials of degree < 17 is a finite-dimensional linear manifold in H and as such is automatically closed.

(c) Consider the set $(1, x, x^2, \dots)$ in $L_2^{(r)}(-1,1)$. We use a slight variant of the Gram-Schmidt process to convert the set into an orthogonal set. Instead of normalizing g_k in (5.11) it is customary to require that $g_k(1) = 1$. This leads to the orthogonal set of *Legendre polynomials*, whose first few elements are

$$\psi_1(x) = 1, \; \psi_2(x) = x, \; \psi_3(x) = \tfrac{1}{2}(3x^2 - 1), \; \psi_4(x) = \tfrac{1}{2}(5x^3 - 3x), \; \dots.$$

Example 4. (a) Let $L_2^{(c)}(a,b)$ be the set of complex-valued functions $u(x)$ defined on $a \leqslant x \leqslant b$ for which the Lebesgue integral $\int_a^b |u|^2\,dx$ exists and is finite. It is easily seen that

$$\langle u,v \rangle = \int_a^b u(x)\bar{v}(x)\,dx$$

is an admissible inner product, yielding the norm and metric

$$\|u\| = \left[\int_a^b |u|^2\right]^{1/2}; \qquad d_2(u,v) = \|u-v\| = \left[\int_a^b |u-v|^2\, dx\right]^{1/2}.$$

The Schwarz inequality takes the form

(5.15) $$\left|\int_a^b u\bar{v}\, dx\right| \leqslant \left[\int_a^b |u|^2\, dx\right]^{1/2}\left[\int_a^b |v|^2\, dx\right]^{1/2}.$$

It is a consequence of the properties of the Lebesgue integral that the space is complete. Thus $L_2^{(c)}(a,b)$ is a Hilbert space.

(b) Consider the set $L_2^{(c)}(D)$ of all complex-valued functions $u(x_1,\ldots,x_k)$ on a domain D in the k variables $x_1,\ldots,x_k$ for which the k-dimensional Lebesgue integral

$$\int_D |u(x_1,\ldots,x_k)|^2\, dx_1 \cdots dx_k$$

exists and is finite. This space is a Hilbert space with the inner product defined as

$$\langle u,v\rangle = \int_D u\bar{v}\, dx_1 \cdots dx_k.$$

Example 5. The space $C(a,b)$ of continuous functions (real- or complex-valued) on a bounded interval $a \leqslant x \leqslant b$ is a normed space under the definition

$$\|u\| = \max_{a \leqslant x \leqslant b} |u(x)|.$$

This norm generates the natural metric

$$d_\infty(u,v) = \|u-v\| = \max_{a \leqslant x \leqslant b} |u(x)-v(x)|,$$

under which $C(a,b)$ was shown to be complete in Example 5, Section 3. Thus $C(a,b)$ is a Banach space, but Exercise 5.5 shows that the norm *cannot* be derived from an inner product.

Example 6. Sobolev spaces. Let Ω be a bounded domain in R_n with boundary Γ. Consider the set S of real-valued functions which are continuous and have a continuous gradient on $\overline{\Omega}$. The bilinear form

$$\langle u, v \rangle \doteq \int_{\Omega} (\operatorname{grad} u \cdot \operatorname{grad} v + uv) \, dx$$

is clearly an admissible inner product on S. The completion of S in the norm generated by this inner product is known as the Sobolev space $H^1(\Omega)$. This Hilbert space and related ones play an important role in the theory of partial differential equations (see Section 4, Chapter 8).

Exercises

5.1. For the space of real n-tuples we defined in (3.2) a variety of metrics. The exponent $1/p$ is needed so that the metric will generate a norm from (5.3) satisfying condition (c) of (5.1). Show that all the conditions for a norm are satisfied by $d_p(u, 0)$, but that only d_2 yields a norm satisfying the parallelogram law (5.9). Therefore, by Exercise 5.4, only this last space is a Hilbert space.

5.2. Consider a linear space provided with a metric satisfying

$$d(\alpha u, \alpha v) = |\alpha| d(u, v)$$

for all vectors u, v and all scalars α. Show that $d(u, 0)$ is an admissible norm on the space.

5.3. Show that on any inner product space

$$\langle u, v \rangle + \langle v, u \rangle = \tfrac{1}{2} \big[\|u + v\|^2 - \|u - v\|^2 \big],$$

and that for a complex inner product space

$$\langle u, v \rangle - \langle v, u \rangle = \frac{i}{2} \big[\|u + iv\|^2 - \|u - iv\|^2 \big].$$

Consequently, in a complex inner product space

$$(5.16) \quad \langle u, v \rangle = \tfrac{1}{4} \big[\|u + v\|^2 - \|u - v\|^2 + i\|u + iv\|^2 - i\|u - iv\|^2 \big],$$

whereas in a real inner product space

$$(5.17) \qquad \langle u, v \rangle = \tfrac{1}{4} \big[\|u + v\|^2 - \|u - v\|^2 \big].$$

Thus in an inner product space knowledge of the norm of every element determines the inner product unambiguously. Therefore in any normed linear space *there can exist at most one inner product which generates the norm.*

5.4. *The Jordan-Von Neumann theorem.* Consider a *real* Banach space whose norm satisfies the parallelogram law (5.9). Define the real-valued function $f(u,v)$ from

$$f(u,v) = \tfrac{1}{4}\big[\,\|u+v\|^2 - \|u-v\|^2\,\big].$$

(a) Show that f is a continuous function of v.

(b) Show that, for any integer $k, f(ku,v)=kf(u,v)$, and hence that, for any real $\alpha, f(\alpha u, v) = \alpha f(u,v)$.

(c) Show that $f(u_1 + u_2, v) = f(u_1,v) + f(u_2,v)$.

In the light of these properties $f(u,v)$ is an admissible inner product with $f(u,u) = \|u\|^2$. By the final remark in Exercise 5.3, no other inner product can generate the given norm. Thus *the necessary and sufficient condition for a real Banach space to be a Hilbert space is that the norm satisfies the parallelogram law* (5.9). If the law is satisfied, the inner product can be calculated from the norm by (5.16).

5.5. Use the result of Exercise 5.4 to show that $C(a,b)$ with the uniform metric is not a Hilbert space even though it is a Banach space.

5.6. Consider the Hilbert space $H = L_2^{(r)}(-1,1)$.

(a) Let M be the linear manifold of even functions in H. What is $M^\perp$? What is the unique decomposition predicted by the projection theorem?

(b) Answer the same question if M is the linear manifold of all functions in H such that $\int_{-1}^{1} u\,dx = 0$.

(c) Is the set of functions such that $\int_{-1}^{1} u\,dx = 1$ a linear manifold?

5.7. A *bilinear form* $b(u,v)$ on a complex linear space X is a complex-valued function of pairs of elements u and v satisfying

$$b(\alpha u_1 + \beta u_2, v) = \alpha b(u_1,v) + \beta b(u_2,v),$$
$$b(u, \alpha v_1 + \beta v_2) = \bar{\alpha} b(u,v_1) + \bar{\beta} b(u,v_2).$$

It follows that $b(0,v) = b(u,0) = 0$. The bilinear form $b(u,v)$ generates a quadratic functional $b(u,u) = B(u)$, known as the *associated quadratic form*. The bilinear form is said to be symmetric if $b(u,v) = \overline{b(v,u)}$, in which case the associated quadratic form is *real*. If, moreover, $B(u) \geq 0$ for all u, we say that both the quadratic form and the bilinear form are *nonnegative*. If $B(u) > 0$ for all $u \neq 0$, the forms are *positive*. The simplest example of a positive, symmetric bilinear form is the inner product satisfying (5.5).

Show that any bilinear form satisfies the *polar identity*

$$(5.18) \quad b(u,v) = \tfrac{1}{4}\{ B(u+v) - B(u-v) + iB(u+iv) - iB(u-iv) \},$$

which should be compared with (5.16). If $b(u,v)$ is *nonnegative*, prove the Schwarz-like inequality

$$|b(u,v)|^2 \leq B(u)B(v).$$

Show that any positive bilinear form on X generates an admissible inner product $\langle u, v \rangle_b = b(u,v)$.

5.8. Let C be a closed convex set in a Hilbert space H, and let u be an arbitrary vector in H. Show that there exists one and only one element in C which is closest to u. (See Remark 3 following the projection theorem.)

5.9. In a Banach space show that $\|u\| - \|v\| \leq \|u-v\|$ and therefore, by interchanging u and v, that

$$(5.19) \qquad |\|u\| - \|v\|| \leq \|u-v\|.$$

6. SEPARABLE HILBERT SPACES AND ORTHONORMAL BASES

We recall that a Hilbert space H is a linear space on which an *inner product* is defined, the space being *complete* in the metric generated by the inner product. Our principal interest in this section is in infinite-dimensional spaces, particularly in $L_2(a,b)$, but the results we shall derive for the infinite-dimensional case will apply—with obvious simplifications—to Euclidean spaces also. This is hardly surprising, since much of the infinite-dimensional theory is motivated by analogy with the finite-dimensional case. We shall deal with separable Hilbert spaces (the condition of separability defined below ensures that our infinite-dimensional space is not too "large"), of which L_2 is a particular instance. The development will be presented for the general, abstract case because the notation is simpler and the geometrical analogy more vivid. In specific applications to L_2 we shall revert to functional notation with Examples 3 and 4 of Section 5 serving as a dictionary of translation.

Let U be a set of vectors in H. The algebraic span $S(U)$ has been defined as the set of all finite linear combinations of vectors drawn from U. In general the set $S(U)$ will not be closed in H; its closure $\bar{S}(U)$ is known as the *closed span* of U. If U happens to contain only a finite number of independent vectors, $S(U)$ is a finite-dimensional manifold and is therefore closed. In other cases $S(U)$ and $\bar{S}(U)$ may differ markedly; for instance, if U is the set $(1, x, x^2, \ldots)$ in $L_2(a, b)$, $S(U)$ is the set of all polynomials, whereas $\bar{S}(U)$ is the whole of $L_2(a, b)$ (see Section 3).

If the closed span of U coincides with H, we say that U is a *spanning set*; alternatively, we may say that the algebraic span of U is dense in H. By taking U to be the whole of H, it is trivially true that $S(U) = \bar{S}(U) = H$, but the goal is to find a relatively small set U which will be a spanning set for H.

Definition. The Hilbert space H is *separable* if it contains a *countable* spanning set U.

We recall that a set is countable if its elements can be placed into one-to-one correspondence with the positive integers or some subset of the positive integers (a set containing only a finite number of elements is countable; the rational numbers form a countable set, but the real numbers do not). Without the crucial word "countable" in the definition of separability, H itself could be used as the spanning set. The existence of a countable spanning set means that H is not too "large," although it will always contain an uncountable number of elements.

A finite-dimensional space is separable since any basis serves as a finite spanning set. Obviously an infinite-dimensional space cannot have a finite spanning set. Thus a separable, infinite-dimensional Hilbert space must contain a countably infinite set U: $\{u_1, u_2, \ldots\}$ such that each vector u in H can be approximated to any desired accuracy by a linear combination of a finite number of elements of U; in other words for each u in H and each $\varepsilon > 0$, there exist an index N and scalars $\alpha_1, \ldots, \alpha_N$ (*which usually depend on ε*) such that

$$(6.1) \qquad \left\| u - \sum_{k=1}^{N} \alpha_k u_k \right\| < \varepsilon.$$

A separable space H contains an algebraically independent spanning set (all that is necessary is to eliminate from the spanning set U the elements

u_k which are linear combinations of elements with lower indices). By the Gram-Schmidt procedure one can convert this spanning set into an orthonormal set with the same algebraic span and hence with the same closed span. *Thus a separable space H contains an orthonormal spanning set.*

It should be pointed out the condition (6.1) does not guarantee that u can be expanded in an infinite series $\sum_{k=1}^{\infty} \alpha_k u_k$. For (6.1) to be equivalent to this series expansion it must be possible to choose the coefficients $\{\alpha_k\}$ in (6.1) independently of ε, and this cannot always be done. The space $L_2(a,b)$ is separable since the countable set $\{1, x, x^2, \dots\}$ is a spanning set, but there are few elements of $L_2(a,b)$ that can be expanded in a power series $\sum_{k=1}^{\infty} \alpha_k x^{k-1}$; such a function would, at the very least, have to be infinitely differentiable, and most L_2 functions are not that smooth. This distinction, previously discussed in Exercise 2.7(d), Chapter 2, *fortunately disappears for orthonormal spanning sets.* (See Theorem 1.)

In addition to $L_2(a,b)$, the spaces $L_2(a, \infty)$, $L_2(-\infty, b)$, $L_2(-\infty, \infty)$ are all separable, but since polynomials do not belong to these spaces, the earlier argument must be modified (see Exercise 6.6).

From here on we confine ourselves to separable Hilbert spaces. We begin with the notion of a basis for such a space.

Definition. A set of vectors $\{h_1, \dots, h_n, \dots\}$ is said to be a *basis* (*Schauder basis*) for H if each vector u in H can be represented in one and *only one way* as

$$u = \sum_{k=1}^{\infty} \alpha_k h_k.$$

The numerical coefficients α_k are the *coordinates* of u in the basis.

Remarks

1. We have not yet shown that H has a basis.

2. The coordinates α_k are easily calculated if the basis is orthonormal. Suppose in fact that $\{e_1, \dots, e_n, \dots\}$ is an orthonormal basis, so that $u = \sum_{k=1}^{\infty} \alpha_k e_k$. Let us calculate α_i. Choose $n \geqslant i$, and let $s_n = \sum_{k=1}^{n} \alpha_k e_k$. Then $\langle s_n, e_i \rangle = \langle \sum_{k=1}^{n} \alpha_k e_k, e_i \rangle = \alpha_i$. Since $s_n \to u$, and the inner product is continuous, we have

$$(6.2) \qquad \langle u, e_i \rangle = \lim_{n \to \infty} \langle s_n, e_i \rangle = \alpha_i.$$

An orthonormal basis is also known as a *complete orthonormal set.*

Note that, if the orthonormal set is not a basis but u happens to be expressible as $\sum_{k=1}^{\infty} \alpha_k e_k$, it is still true that $\alpha_i = \langle u, e_i \rangle$. In the section on linear functionals we show how to calculate the coordinates in an oblique basis.

3. If $\{h_1, \ldots, h_n, \ldots\}$ is a Schauder basis, the equation $\alpha_1 h_1 + \cdots + \alpha_n h_n + \cdots = 0$ has only the solution $\alpha_1 = \cdots = \alpha_n = \cdots = 0$. Thus the set satisfies a stronger "independence" condition than the algebraic independence previously introduced. It is of course clear that any finite subset of the basis is independent, so that the basis is certainly algebraically independent.

4. In an infinite-dimensional space a basis necessarily consists of an infinite number of independent vectors, but not every infinite independent set is a basis. (For instance, removing a single element from a basis leaves an infinite independent set which is not a basis.)

In the next section we prove the existence of an orthonormal basis for H, but we stop here to prove the following preliminary theorem.

Riesz-Fischer Theorem.   Let $\{e_k\}$ be an orthonormal set (not necessarily a basis) in H, and let $\{\alpha_k\}$ be a sequence of complex numbers. Then $\sum_{k=1}^{\infty} \alpha_k e_k$ and $\sum_{k=1}^{\infty} |\alpha_k|^2$ converge or diverge together.

Proof.   Let $\{s_n\}$ be the sequence of partial sums for the series $\sum_{k=1}^{\infty} \alpha_k e_k$, and let $\{t_n\}$ be the sequence of partial sums for the series $\sum_{k=1}^{\infty} |\alpha_k|^2$. Note that $\{s_n\}$ is a sequence of elements in H, whereas $\{t_n\}$ is a sequence of nonnegative real numbers. These sequences are related by

$$(6.3) \qquad \|s_p - s_m\|^2 = \left\| \sum_{k=m+1}^{p} \alpha_k e_k \right\|^2 = \sum_{k=m+1}^{p} |\alpha_k|^2 = (t_p - t_m), \qquad p > m.$$

If $\sum \alpha_k e_k$ converges, $\{s_n\}$ is a Cauchy sequence in H and (6.3) shows that $\{t_n\}$ is a Cauchy sequence of numbers and must therefore converge. Conversely, if $\sum |\alpha_k|^2$ converges, $\{t_n\}$ is a Cauchy sequence of numbers and (6.3) now shows that $\{s_n\}$ is a Cauchy sequence in H. Since H is complete, $\{s_n\}$ must converge.

Problem of Best Approximation

We wish to approximate an arbitrary vector u in H by a linear combination of the given independent set $\{u_1, \ldots, u_k\}$, that is, we wish to find the element $\Sigma_{i=1}^{k}\alpha_i u_i$ closest to u in the sense of the metric of H. Since the set of linear combinations of $\{u_1,\ldots,u_k\}$ is a k-dimensional linear manifold M_k (and hence necessarily closed), the problem is a special case of that considered in the projection theorem. Our purpose here is to study the question more deeply and concretely, when H is separable. By the Gram-Schmidt procedure we can construct from $\{u_1,\ldots,u_k\}$ an orthonormal basis $\{e_1,\ldots,e_k\}$ for M_k. We are then interested in finding the element $\Sigma_{i=1}^{k}\beta_i e_i$ which is closest to u. The advantage of using an orthonormal basis rather than an oblique one is that the calculation of the coefficients is much simpler. For any $\beta_1,\ldots,\beta_k$ we have

$$(6.4) \quad \left\| u - \sum_{i=1}^{k} \beta_i e_i \right\|^2 = \|u\|^2 + \sum_{i=1}^{k} \beta_i \bar{\beta}_i - \sum_{i=1}^{k} \beta_i \langle e_i, u \rangle - \sum_{i=1}^{k} \bar{\beta}_i \langle u, e_i \rangle$$

$$= \|u\|^2 + \sum_{i=1}^{k} |\langle u, e_i \rangle - \beta_i|^2 - \sum_{i=1}^{k} |\langle u, e_i \rangle|^2.$$

It is clear that the minimum is attained for $\beta_i = \langle u, e_i \rangle$. These optimal values of the coefficients are known as the *Fourier coefficients* of u with respect to the set $\{e_1,\ldots,e_k\}$ and will be permanently denoted by γ_i; thus $\gamma_i = \langle u, e_i \rangle$, $i = 1,\ldots,k$. The uniquely determined best approximation to u in M_k is the *Fourier sum* $v_k = \Sigma_{i=1}^{k} \gamma_i e_i = \Sigma_{i=1}^{k} \langle u, e_i \rangle e_i$. Geometrically, the single term $\langle u, e_i \rangle e_i$ is the orthogonal projection of u on the line generated by e_i. The Fourier sum is the orthogonal projection of u on the linear manifold M_k. We easily confirm by explicit calculation that $w_k = u - v_k$ is orthogonal to M_k, as predicted by the projection theorem.

Our best approximation to u will usually be far from perfect. In fact, the square of the distance between u and the Fourier sum v_k is

$$\left\| u - \sum_{i=1}^{k} \langle u, e_i \rangle e_i \right\|^2 = \|u\|^2 - \sum_{i=1}^{k} |\langle u, e_i \rangle|^2,$$

from which we deduce that

$$(6.5) \quad \sum_{i=1}^{k} |\langle u, e_i \rangle|^2 \leqslant \|u\|^2.$$

Naturally the best approximation to u improves as the orthonormal set is taken larger. Suppose we include another vector e_{k+1} in our orthonormal set of vectors; the new best approximation is the Fourier sum $\Sigma_{i=1}^{k+1} \langle u, e_i \rangle e_i$, which is just the previous one with the added term

$\langle u, e_{k+1}\rangle e_{k+1}$ corresponding to the extra axis e_{k+1}. *What is crucial is that previously calculated coefficients do not have to be changed.* This suggests the possibility of using infinite orthonormal sets, with the best approximation to u being an infinite series in $\{e_k\}$. If we had used independent, but not orthonormal, vectors $u_1, \ldots, u_k$ to approximate u, we would find that the inclusion of an additional vector u_{k+1} in the approximating pool usually leads to entirely new coefficients in the best approximation.

Before proceeding to consider the problem of best approximation when the set $\{e_k\}$ is infinite, we pause to show that an orthonormal spanning set for H is necessarily a basis.

Theorem 1.   Let $\{e_1, \ldots, e_n, \ldots\}$ be an orthonormal spanning set for H. Then the set is a basis for H.

Proof.   We must show that, for each u in H, $u = \sum_{k=1}^{\infty} \langle u, e_k \rangle e_k$, that is, we must show that for each u and each $\varepsilon > 0$ there exists N such that $\|u - \sum_{k=1}^{n} \langle u, e_k \rangle e_k\| < \varepsilon$ whenever $n > N$. To prove this, choose N and $\alpha_1, \ldots, \alpha_N$ such that $\|u - \sum_{k=1}^{N} \alpha_k e_k\| < \varepsilon$, which can be done since $\{e_i\}$ is a spanning set. Now $\sum_{k=1}^{N} \alpha_k e_k$ lies in M_N, the linear manifold generated by $e_1, \ldots, e_N$. The best approximation to u in M_N is given by $\sum_{k=1}^{N} \langle u, e_k \rangle e_k$, so that

$$\left\| u - \sum_{k=1}^{N} \langle u, e_k \rangle e_k \right\| \leqslant \left\| u - \sum_{k=1}^{N} \alpha_k e_k \right\| < \varepsilon.$$

If $n > N$, we have

$$\left\| u - \sum_{k=1}^{n} \langle u, e_k \rangle e_k \right\| \leqslant \left\| u - \sum_{k=1}^{N} \langle u, e_k \rangle e_k \right\|,$$

and the theorem has therefore been proved.

Corollary.   Let $\{e_1, \ldots, e_n, \ldots\}$ be an orthonormal set which is not necessarily a spanning set for H. Let the closed span of this set be denoted by M. Then the set is a basis for M.

We now return to the problem of best approximation when the approximating set is a countably infinite orthonormal set $\{e_1, \ldots, e_n, \ldots\}$. Then, according to the preceding corollary, each element in the closed span M of the set can be expressed as an infinite series $\sum \beta_k e_k$. Let u be an element of H. We want to find the vector in M which lies closest to u. We suspect that this best approximation is the *Fourier series*

$$(6.6) \qquad v = \sum_{i=1}^{\infty} \langle u, e_i \rangle e_i = \sum_{i=1}^{\infty} \gamma_i e_i.$$

To show that this series converges we first note that the sequence of partial sums for the series $\sum_{i=1}^{\infty} |\gamma_i|^2$ is monotonically increasing and, by (6.5), is bounded above by the fixed number $\|u\|^2$. Thus the series $\sum_{i=1}^{\infty} |\gamma_i|^2$ converges, and so does $\sum_{i=1}^{\infty} \gamma_i e_i$ (by the Riesz-Fischer theorem). Moreover, we note *Bessel's inequality*

$$(6.7) \qquad \sum_{i=1}^{\infty} |\gamma_i|^2 = \sum_{i=1}^{\infty} |\langle u, e_i \rangle|^2 \leqslant \|u\|^2$$

and its trivial consequence

$$(6.8) \qquad \lim_{i \to \infty} \langle u, e_i \rangle = 0.$$

That v in (6.6) gives the best approximation to u among vectors in M now follows by letting k tend to infinity in (6.4). It is also easy to see that the vector $w = u - v$ is orthogonal to M.

What if the closed manifold M in H is given to us without the a priori knowledge that it is spanned by a countable orthonormal set? Since M is a subset of a separable space, it can be shown that it is itself separable and is therefore spanned by a countable set which can be made orthonormal by the Gram-Schmidt process. Thus M is the span of an orthonormal set which, by the corollary, must be a basis for M. We are therefore in a position to state a refined version of the projection theorem.

Projection Theorem for Separable Spaces. Let H be a separable space and M a closed linear manifold in H. Then M has an orthonormal basis $\{e_k\}$, and each vector u in H can be decomposed in one and only one way as the sum of a vector v in M and a vector w in $M^\perp$. The vector v is the one and only vector in M lying closest to u; moreover, v has the explicit form (6.6).

Characterization of an Orthonormal Basis

Let H be a separable space. By definition an orthonormal set $\{e_k\}$ is a basis for H if each vector u in H has a unique representation $u = \sum_{k=1}^{\infty} \alpha_k e_k$. From this it follows that $\alpha_k = \langle u, e_k \rangle$, so that $\{e_k\}$ is an orthonormal basis if and only if each u in H can be written as $u = \sum_{k=1}^{\infty} \langle u, e_k \rangle e_k$.

We now give a number of equivalent criteria for an orthonormal set to be a basis for H.

Theorem 2.   If *any* of the following criteria is met, the orthonormal set $\{e_k\}$ is a basis for H.

   1.   For each u in H, $u = \sum_{k=1}^{\infty} \langle u, e_k \rangle e_k$. This is just a direct consequence of the definition of an orthonormal basis. It says that each u is equal to its Fourier series or, equivalently, that the best approximation to u in terms of the $\{e_k\}$ is perfect.
   2.   The set $\{e_k\}$ is maximal. (An orthonormal set $\{e_1, \ldots, e_n, \ldots\}$ is *maximal* if there exists no element e in H such that $\{e, e_1, \ldots, e_n, \ldots\}$ is an orthonormal set.)
   3.   The only element u in H for which *all* the Fourier coefficients $\langle u, e_1 \rangle, \ldots$ are 0 is the element $u = 0$.
   4.   *Parseval's identity.* For each u in H,

$$(6.9) \qquad \|u\|^2 = \sum_{k=1}^{\infty} |\langle u, e_k \rangle|^2.$$

Remark.   If H is finite-dimensional of dimension n, there is a very simple way of characterizing an orthonormal basis: it must have exactly n elements. Note, however, that any one of criteria 1 through 4 also characterizes an orthonormal basis in n-space; each can be regarded as an indirect way of saying that there are n elements in the orthonormal set. Unfortunately counting the elements is not good enough in an infinite-dimensional space H; admittedly an orthonormal set must have an infinite number of elements to be a basis, but that is not sufficient. On the other hand, criteria 1 through 4 carry over to the infinite-dimensional case.

Proof.
   (a) Criterion 2 implies 3. Assuming criterion 3 false, there would exist $\psi \neq 0$ such that $\langle \psi, e_i \rangle = 0$ for all i. Then $\psi / \|\psi\|$ would be a unit element orthogonal to $e_1, \ldots, e_n, \ldots$ and could therefore be used to enlarge the orthonormal set, thereby violating criterion 2.
   (b) Criterion 3 implies 1. We have previously shown that $\sum_{k=1}^{\infty} \langle u, e_k \rangle e_k$ converges to some element v in H; we must prove that $v = u$ if criterion 3 holds. Now it is clear that v and u have the same Fourier series, so that $\langle u - v, e_k \rangle = 0$ for all k; therefore criterion 3 guarantees $u - v = 0$ or $u = v$.
   (c) Criterion 1 implies 4. This is automatic by passing to the limit as $k \to \infty$ in the identity $\|u - \sum_{i=1}^{k} \langle u, e_i \rangle e_i\|^2 = \|u\|^2 - \sum_{i=1}^{k} |\langle u, e_i \rangle|^2$.

(d) Criterion 4 implies 2. Assuming criterion 2 false, there exists e with $\|e\|=1$ and $\langle e,e_k\rangle=0$ for all k. But then criterion 4 would tell us that $\|e\|^2=\Sigma_{k=1}^{\infty}|\langle e,e_k\rangle|^2$, an evident contradiction.

Remark. In view of Theorem 1, it is clear that criterion 1 is equivalent to: for each u in H and each $\varepsilon>0$ there exist $N>0$ and $c_1,\ldots,c_N$ such that $\|u-\Sigma_{k=1}^{N}c_ke_k\|<\varepsilon$.

Orthonormal Bases on Concrete Functional Spaces

In $L_2^{(c)}(a,b)$ the elements u are complex-valued functions $u(x)$ that are Lebesgue square integrable on the interval $a\leqslant x\leqslant b$. We revert to functional notation in this section. Let $\{e_1(x),\ldots,e_n(x),\ldots\}$ be an orthonormal basis for $L_2^{(c)}(a,b)$. The criterion for orthonormality takes the form

$$(6.10) \qquad \int_a^b e_i(x)\bar{e}_j(x)\,dx=\begin{cases} 0, & i\neq j, \\ 1, & i=j. \end{cases}$$

Since $\{e_i\}$ is an orthonormal basis, each $u(x)$ in $L_2^{(c)}(a,b)$ can be expanded as a Fourier series

$$(6.11) \qquad u(x)=\sum_{k=1}^{\infty}\gamma_ke_k(x), \qquad \gamma_k=\int_a^b u(x)\bar{e}_k(x)\,dx,$$

where convergence in (6.11) stands for mean-square convergence

$$(6.12) \qquad \lim_{n\to\infty}\int_a^b\left|u(x)-\sum_{k=1}^{n}\gamma_ke_k(x)\right|^2 dx=0.$$

If we want (6.11) *to hold in some other mode such as pointwise convergence or uniform convergence, it is not enough to know that* $\{e_i(x)\}$ *is a basis* (see Section 6, Chapter 0, for relations among various modes of convergence). Such questions require delicate analysis, and the desired type of convergence may be achieved only if $u(x)$ belongs to some class of functions such as differentiable functions or functions of bounded variation. Nevertheless the methods used to show that $\{e_i(x)\}$ is a basis can often be adapted—at the cost of additional effort—to prove that (6.11) holds in some stronger sense. Since mean-square convergence is by far the most important type of convergence in applications, we will usually be satisfied to show that an orthonormal set $\{e_i(x)\}$ is a basis, which will often not be easy to do. We remark that it is enough to show that (6.11) holds in the sense of (6.12) for any subset M of functions $u(x)$ that is dense in L_2. In fact, it is enough to show that each u in M can be approximated (in the L_2 norm) to arbitrarily prescribed accuracy by finite linear combinations of the set $\{e_i(x)\}$. The following three methods are particularly important for showing that an orthonormal set is a basis:

1. Adapt the Weierstrass approximation theorem (see Exercises 6.3 and 6.4, where trigonometric orthonormal sets are treated in this way).

2. Use the general theorem on eigenfunctions of compact self-adjoint operators (Theorem 5, Section 3, Chapter 6).

3. Refer to the theory of delta sequences in Section 2, Chapter 2.

Let us explain what is meant by the third method. Consider the sequence $\{t_n(x)\}$ of partial sums of (6.11). We have

$$t_n(x)= \sum_{k=1}^n \gamma_k e_k(x)= \sum_{k=1}^n \left(\int_a^b u(y)\bar{e}_k(y)\,dy \right) e_k(x),$$

which, since we are dealing with finite sums, can be rewritten as

(6.13) $$t_n(x)= \int_a^b S_n(x,y)u(y)\,dy,$$

where

$$S_n(x,y) \doteq \sum_{k=1}^n e_k(x)\bar{e}_k(y)$$

is known as the kernel of (6.13). We must then show that as $n\to\infty$ (6.13) approaches $u(x)$. Admittedly we only want to prove this in the L_2 sense, but it is possible to try from (6.13) to analyze the possible convergence of t_n in some other mode, although we shall not do so. What is needed roughly is to show that, as $n\to\infty$, $S_n(x,y)\to\delta(x-y)$, where δ is the Dirac delta function of Chapter 2. We shall use this approach to show later that the set (6.15) is in fact an orthonormal basis.

Remarks

1. In some problems [see, for instance, (6.17)] it is more convenient to index the orthonormal set starting from $k=0$. On occasion [see (6.15)] the index is even allowed to run from $-\infty$ to ∞, the infinite series $\sum_{k=-\infty}^\infty \gamma_k e_k(x)$ being interpreted as $\lim_{n\to\infty}\sum_{k=-n}^n \gamma_k e_k(x)$.

2. If $\{e_k(x)\}$ is an orthonormal basis, so is $\{\bar{e}_k(x)\}$.

3. If $\{e_k(x)\}$ is an orthonormal basis and $\{\alpha_k\}$ is a sequence of nonzero complex numbers, $\{\alpha_k e_k(x)\}$ is an orthogonal basis. Note that, when we expand an element u in this new basis, the coordinates of u may not tend to 0 as the index tends to infinity.

4. If $\{e_k(x)\}$ is an orthonormal basis for $L_2(-1,1)$, then

$$\left\{\left(\frac{2}{b-a}\right)^{1/2} e_k\left[\frac{2}{b-a}\left(x-\frac{a+b}{2}\right)\right]\right\}$$

is an orthonormal basis for $L_2(a,b)$.

5. In $L_2^{(r)}(a,b)$ all functions are real-valued, and a basis consists of real functions. Formulas (6.10) through (6.12) are still valid, but the complex conjugates may be omitted. A basis for $L_2^{(r)}(a,b)$ is also a basis for $L_2^{(c)}(a,b)$, and any real basis for $L_2^{(c)}(a,b)$ is automatically a basis for $L_2^{(r)}(a,b)$.

6. In applications to differential equations (see Chapter 7), one frequently encounters sets of functions that are *orthogonal with weight* $s(x)$, that is, such that

$$(6.14) \qquad \int_a^b s(x)e_k(x)\bar{e}_j(x)\,dx = \begin{cases} 0, & k\neq j, \\ 1, & k=j, \end{cases}$$

where $s(x)>0$ in $a<x<b$. Although the set $\{s^{1/2}e_k\}$ is orthonormal in the sense of (6.10), it is often preferable to introduce a new inner product $\langle u,v\rangle_s = \int_a^b s(x)u(x)\bar{v}(x)\,dx$ and its corresponding norm, $\|u\|_s = [\int_a^b s(x)|u(x)|^2\,dx]^{1/2}$. With these definitions the space of functions for which $\|u\|_s$ exists and is finite is a Hilbert space, and our general analysis can be applied. If $\{e_k(x)\}$ is an orthonormal basis in the sense of (6.14), then every u with $\|u\|_s$ finite can be expanded in the form

$$u(x) = \sum_{k=1}^{\infty} \gamma_k e_k(x), \qquad \gamma_k = \langle u,e_k\rangle_s = \int_a^b s(x)u(x)\bar{e}_k(x)\,dx,$$

where the series converges in the sense of

$$\lim_{n\to\infty} \int_a^b s(x)\left|u(x)-\sum_{k=1}^{n}\gamma_k e_k(x)\right|^2 dx = 0.$$

Example 1. We have already encountered the Legendre polynomials (Example 3, Section 5) as the outcome of the Gram-Schmidt procedure of orthogonalization on the monomials $1, x, x^2, \ldots$ on $-1 \leqslant x \leqslant 1$. Since the monomials form a spanning set, the Legendre polynomials form an orthogonal basis.

Example 2. The following sets of trigonometric functions are orthonormal bases on the respective intervals shown:

(a)   On any interval of length 2π:

(6.15)

$$\left(\frac{1}{2\pi}\right)^{1/2}, \ \left(\frac{1}{2\pi}\right)^{1/2} e^{ix}, \ \left(\frac{1}{2\pi}\right)^{1/2} e^{-ix}, \ \left(\frac{1}{2\pi}\right)^{1/2} e^{2ix}, \ \left(\frac{1}{2\pi}\right)^{1/2} e^{-2ix}, \ \ldots.$$

(b)   On any inverval of length 2π:

(6.16)

$$\left(\frac{1}{2\pi}\right)^{1/2}, \ \left(\frac{1}{\pi}\right)^{1/2} \cos x, \ \left(\frac{1}{\pi}\right)^{1/2} \sin x, \ \left(\frac{1}{\pi}\right)^{1/2} \cos 2x, \ \left(\frac{1}{\pi}\right)^{1/2} \sin 2x, \ \ldots.$$

(c)   On $0 \leqslant x \leqslant \pi$:

(6.17) $\left(\frac{1}{\pi}\right)^{1/2}, \quad \left(\frac{2}{\pi}\right)^{1/2} \cos x, \quad \left(\frac{2}{\pi}\right)^{1/2} \cos 2x, \qquad \ldots$

(d)   On $0 \leqslant x \leqslant \pi$:

(6.18) $\left(\frac{2}{\pi}\right)^{1/2} \sin x, \quad \left(\frac{2}{\pi}\right)^{1/2} \sin 2x, \qquad \ldots$

(a$_1$)   On any interval of length $2L$:

(6.19) $\left\{\left(\frac{1}{2L}\right)^{1/2} e^{ik\pi x/L}\right\}, \qquad k = 0, \pm 1, \pm 2, \ldots.$

(b$_1$)   On any interval of length $2L$:

(6.20)

$$\left(\frac{1}{2L}\right)^{1/2}, \quad \left\{\left(\frac{1}{L}\right)^{1/2} \cos \frac{k\pi x}{L}\right\}, \quad \left\{\left(\frac{1}{L}\right)^{1/2} \sin \frac{k\pi x}{L}\right\}, \qquad k = 1, 2, \ldots.$$

(c₁)   On $0 \leqslant x \leqslant L$:

(6.21) $\left(\dfrac{1}{L}\right)^{1/2},\quad \left\{\left(\dfrac{2}{L}\right)^{1/2}\cos\dfrac{k\pi x}{L}\right\},\qquad k=1,2,\ldots.$

(d₁)   On $0 \leqslant x \leqslant L$:

(6.22) $\left\{\left(\dfrac{2}{L}\right)^{1/2}\sin\dfrac{k\pi x}{L}\right\},\qquad k=1,2,\ldots.$

Expansions using the functions in (a₁) or (b₁) are known as *full-range* expansions, whereas those using (c₁) or (d₁) are *half-range* expansions. It is a simple exercise in integration to prove that each set is orthonormal on the stated interval. We shall prove that (6.15) is a basis by showing that the kernel in (6.13) is a delta sequence. It then follows from Exercise 6.5 that all other listed sets are also bases.

Since set (6.15) is indexed from $n = -\infty$ to $n = \infty$, we shall want to show that

$$\lim_{n\to\infty}\left\|u(x)-\sum_{k=-n}^{n}\gamma_k\frac{e^{ikx}}{(2\pi)^{1/2}}\right\|=0,$$

where

$$\gamma_k=\frac{1}{(2\pi)^{1/2}}\int_{-\pi}^{\pi}u(y)e^{-iky}\,dy.$$

The sequence of partial sums $\{t_n(y)\}$ is defined from

$$t_n(x)=\sum_{k=-n}^{n}\gamma_k\frac{e^{ikx}}{(2\pi)^{1/2}}=\frac{1}{2\pi}\int_{-\pi}^{\pi}s_n(x-y)u(y)\,dy,$$

where

$$s_n(x)=\sum_{k=-n}^{n}e^{ikx}.$$

We showed in Theorem 3, Section 3, Chapter 2 that $\{s_n(x)\}$ is a delta sequence in $-\pi \leqslant x < \pi$. In fact, if the 2π-periodic extension of u is continuously differentiable, $t_n(x)\to u(x)$ uniformly on $(-\infty,\infty)$ and hence certainly on $-\pi \leqslant x < \pi$. Therefore not only is set (6.15) a basis but also we have results on uniform convergence of trigonometric Fourier series.

Example 3.   By combining orthonormal bases in one dimension, we can construct orthonormal bases in higher dimensions. Suppose $\{e_k(x)\}$ is an orthonormal basis for $L_2^{(c)}(a,b)$ and $\{f_k(y)\}$ is an orthonormal basis for $L_2^{(c)}(c,d)$; then the set $\{e_1(x)f_1(y), e_1(x)f_2(y), e_2(x)f_1(y), e_2(x)f_2(y),\ldots\}$, consisting of *all products* of pairs from the original bases, is an orthonormal basis for $L_2^{(c)}(D)$, where D is the rectangle $a \leqslant x \leqslant b, c \leqslant y \leqslant d$. The proof is omitted. Note that the diagonal set $\{e_1(x)f_1(y), e_2(x)f_2(y),\ldots\}$ is orthonormal but is not a basis for $L_2^{(c)}(D)$.

As an example of this construction, we deduce from (6.22) that $\{(2/ab)\sin(j\pi x/a)\sin(k\pi y/b)\}$ is an orthonormal basis for the rectangle $0 \leqslant x \leqslant a$, $0 \leqslant y \leqslant b$. Repeated application of the procedure enables us to construct bases in higher dimensions.

Example 4.   Other examples of orthonormal bases such as Hermite polynomials and Bessel functions are given in Chapter 7.

Exercises

6.1.   Let D be a dense subset of H, and suppose that every element of D is in the closed span of an orthonormal set. Show the orthonormal set is a basis for H.

As an illustration of this theorem, the following procedure is often used to prove that an orthonormal set $\{e_k(x)\}$ is a basis for $H \doteq L_2(a,b)$. The set D of continuous functions is known to be dense in H. Suppose that elements of D can be uniformly approximated by linear combinations of elements of $\{e_k\}$. This means that for each g in D and each $\varepsilon > 0$, we can find a linear combination h of a finite number of elements of $\{e_k\}$ such that $d_\infty(g,h) < \varepsilon$, where d_∞ is the uniform metric of Example 5, Section 3. It then follows that $\|g - h\| < \varepsilon(b-a)^{1/2}$ so that every continuous function is in the closed span of $\{e_k\}$. Hence $\{e_k\}$ is a basis for H. Thus the only thing that has to be proved for the set $\{e_k\}$ is that continuous functions can be uniformly approximated by finite linear combinations of $\{e_k\}$.

6.2.   Show that a necessary and sufficient condition for the orthonormal set $\{e_i\}$ to be a basis for H is that

$$\langle u,v \rangle = \sum_{i=1}^{\infty} \langle u,e_i \rangle \langle e_i,v \rangle \qquad \text{for all } u,v \text{ in } H.$$

6.3.   (a) Show that $\cos^k x$ is a linear combination of $1, \cos x,\ldots,\cos kx$.

(b) Let $f(x)$ be a complex-valued continuous function on $0 \leqslant x \leqslant \pi$. Make the admissible change of variables $y = \cos x$, and use the

Weierstrass approximation theorem on $-1 \leqslant y \leqslant 1$ to show that $f(x)$ can be uniformly approximated by linear combinations of 1, $\cos x, \ldots, \cos kx$. It then follows from Exercise 6.1 that set (6.17) is an orthonormal spanning set for $L_2(0, \pi)$ and therefore a basis.

6.4. (a) Show that $\sin^k x$ is a linear combination of 1, $\sin x, \cos x, \ldots$, $\sin kx, \cos kx$.

   (b) Let $f(x,y)$ be a complex-valued continuous function defined on the disk $x^2 + y^2 \leqslant 1$. An extension of the Weierstrass theorem shows that $f(x,y)$ can be uniformly approximated by linear combinations of $x^m y^k$. Introducing polar coordinates, we see that $f(\cos\theta, \sin\theta)$, which is the value of $f(x,y)$ on the circumference of the disk, can be uniformly approximated by linear combinations of $\cos^m \theta \sin^k \theta$ on $0 \leqslant \theta \leqslant 2\pi$. Use Exercises 6.3(a) and 6.4(a) to show that (6.16) is an orthonormal basis over the stated interval.

6.5. Show that sets (6.15) and (6.16) lead to the same Fourier series by giving explicit relations between the coefficients of the two series

$$u(x) = \sum_{k=-\infty}^{\infty} \gamma_k \frac{e^{ikx}}{(2\pi)^{1/2}} \quad \text{and}$$

$$u(x) = a_0 \left(\frac{1}{2\pi}\right)^{1/2} + \sum_{k=1}^{\infty} \left[a_k \frac{\cos kx}{(\pi)^{1/2}} + b_k \frac{\sin kx}{(\pi)^{1/2}} \right].$$

These relations are

$$a_0 = \gamma_0, \qquad a_k = \frac{1}{\sqrt{2}}(\gamma_k + \gamma_{-k}), \qquad b_k = \frac{i}{\sqrt{2}}(\gamma_k - \gamma_{-k})$$

and, therefore,

$$\gamma_0 = a_0, \qquad \gamma_k = \frac{1}{\sqrt{2}}(a_k - ib_k) \quad \text{for } k > 0,$$

$$\gamma_k = \frac{1}{\sqrt{2}}(a_{-k} + ib_{-k}) \quad \text{for } k < 0.$$

Show that these results are in agreement with (3.4), Chapter 2.

Since set (6.15) has been shown to be a basis, it follows that the orthonormal set (6.16) is also a basis. Since the sets have period 2π, they are bases on any 2π interval. If $u(x)$ is an odd function, that is,

$u(-x) = -u(x)$, then all the $\{a_k\}$ vanish. Since u is arbitrary in $(0,\pi)$, we have shown that set (6.18) is a basis on $(0,\pi)$. Now let $u(x)$ be even, so that $u(x) = u(-x)$. This leads to $b_k = 0$ for all k. Hence set (6.17) is a basis on $(0,\pi)$. A simple change of variables lets us extend the results to the remaining sets, (6.19) through (6.22).

6.6. We wish to show that $L_2^{(c)}(0,\infty)$ is separable. A function $u(x)$ belongs to $L_2^{(c)}(0,\infty)$ if $\int_0^\infty |u(x)|^2\,dx < \infty$. [Note that this does not imply that $\lim_{x\to\infty} u(x) = 0$.] Show that this space is separable by proving that the countable set $H(k-x)x^m$, where $k = 1, 2, \ldots, m = 0, 1, 2, \ldots$, and H is the Heaviside function, is dense in $L_2^{(c)}(0,\infty)$.

6.7. Let $\{e_1, \ldots, e_n, \ldots\}$ be an orthonormal basis for H. The following scheme constructs a related spanning set for H which is *not* a basis. The spanning set in question is $\{h_1, \ldots, h_n, \ldots\}$, where $h_n = e_1 + (e_{n+1}/n+1)$.

(a) Show that the set $\{h_i\}$ is algebraically independent.

(b) Show that the sets $\{h_i\}$ and $\{e_i\}$ have the same closed span (observe that e_1 is in the span of $\{h_i\}$ since $\lim_{n\to\infty} h_n = e_1$).

(c) Show that e_1 cannot be expanded in a series in the set $\{h_i\}$. Assume $e_1 = \alpha_1 h_1 + \cdots$, and take the inner product with respect to each of the e_k's to arrive at a contradiction.

7. LINEAR FUNCTIONALS

Let M be a linear manifold in the separable complex Hilbert space H. If to each u in M there corresponds a complex number, denoted by $l(u)$, satisfying the condition

(7.1)

$$l(\alpha u + \beta v) = \alpha l(u) + \beta l(v) \quad \text{for all } u, v \text{ in } M \text{ and all complex numbers } \alpha, \beta,$$

we say that l is a *linear functional* on M. The domain M will often be the entire Hilbert space. Note that a linear functional satisfies $l(0) = 0$ and $l(\sum_{i=1}^k \alpha_i u_i) = \sum_{i=1}^k \alpha_i l(u_i)$.

Remark. On a real Hilbert space a linear functional is real-valued and satisfies (7.1) for real α, β.

Example 1. The functional whose value at u is $\|u\|$ is not linear.

Example 2.   Let f be a fixed vector in H, and associate with each u in H the complex number $l(u) = \langle u, f \rangle$. Then l is a linear functional—in fact, the prototype of linear functionals, as we shall see below.

Example 3.   The functionals defined on H by $l(u) = \langle u, f \rangle + \alpha$ and $l(u) = \langle f, u \rangle$ are not linear.

A linear functional is *bounded* on its domain M if there exists a constant c such that, *for all u in M, $|l(u)| \leqslant c \|u\|$.* The smallest constant c for which the inequality holds for all u in M is known as the *norm* of l and is denoted by $\|l\|$.

To say that l is continuous at a point u in its domain means roughly that $l(v)$ is near $l(u)$ whenever v is near u. Precisely: for each $\varepsilon > 0$ there exists $\delta > 0$ such that $|l(v) - l(u)| < \varepsilon$ whenever v is in M and $\|u - v\| < \delta$. An equivalent definition is the following: the functional l is *continuous* at u if, whenever $\{u_n\}$ is a sequence in M with limit u, then $l(u_n) \to l(u)$. A functional continuous at every point of M is said to be continuous on M.

Theorem.   A linear functional continuous at the origin is continuous on its entire domain of definition M.

Proof.   Let $\{u_n\}$ be a sequence in M with limit u in M. Then $u_n - u \to 0$ and hence, by hypothesis, $l(u_n - u) \to l(0) = 0$. By the linearity of l, $l(u_n - u) = l(u_n) - l(u)$, and therefore $l(u_n) \to l(u)$ as desired.

Theorem.   Boundedness and continuity are equivalent for linear functionals.

Proof.   (a) Let l be bounded; then $|l(u_n) - l(u)| = |l(u_n - u)| \leqslant \|l\| \, \|u_n - u\|$. As $u_n \to u$, we have therefore $l(u_n) \to l(u)$ and l is continuous.

(b) Let l be continuous. If l is unbounded, there exists a sequence $\{u_n\}$ of nonzero elements in M such that $|l(u_n)| \geqslant n \|u_n\|$; hence the sequence $v_n = u_n / n \|u_n\|$, which tends to 0, has the property $|l(v_n)| \geqslant 1$, which violates continuity at the origin.

We now give an example of an unbounded linear functional on a dense subset of an infinite-dimensional Hilbert space. Let $H = L_2(-1, 1)$, and let M be the subset of continuous functions. The functional $l(u) = u(0)$ which assigns to each continuous function its value at the origin is clearly linear. Now let $f(x)$ be a real continuous function such that $f(-1) = f(1) = 0$, and $f > 0$ in $-1 < x < 1$. Define the sequence $u_k(x) = f(kx)$ when $|x| \leqslant k^{-1}$ and $u_k(x) = 0$ elsewhere. Since $\int_{-1}^{1} u_k^2(x)\, dx = (1/k) \int_{-1}^{1} f^2(x)\, dx$, we see that

$u_k(x) \to 0$ in the L_2 norm. We have, however, $u_k(0) = f(0)$, so that the numerical sequence $u_k(0)$ does not tend to 0 as $k \to \infty$. Thus $l(u)$ is not continuous at 0 and is therefore unbounded. On the other hand, every linear functional defined on an n-dimensional manifold in H is bounded. This is an immediate consequence of the following result.

Theorem.   Every linear functional on E_n is bounded.

Proof.   It suffices to show that $l(u_k) \to 0$ whenever $u_k \to 0$. In terms of the orthonormal basis $\{e_1, \ldots, e_n\}$, $u_k = \sum_{i=1}^{n} \alpha_{k,i} e_i$, where $\alpha_{k,i} = \langle u_k, e_i \rangle$. The continuity of the inner product shows that, if $u_k \to 0$, then $\lim_{k \to \infty} \alpha_{k,i} = 0$ for $i = 1, \ldots, n$. Hence $\lim_{k \to \infty} l(u_k) = \lim_{k \to \infty} \sum_{i=1}^{n} \alpha_{k,i} l(e_i) = 0$.

One of the fundamental theorems of Hilbert space (akin in importance to the projection theorem, Section 5, to which it is closely related) is the Riesz representation theorem, which tells us that every bounded linear functional on H is the inner product with respect to some fixed vector in H.

Riesz Representation Theorem.   To each continuous linear functional l defined on the whole of H corresponds an unambiguously defined vector f such that

$$(7.2) \qquad\qquad l(u) = \langle u, f \rangle \qquad \text{for every } u \text{ in } H.$$

Proof.   The set of vectors N for which $l(u) = 0$ is easily seen to be a linear manifold. If N coincides with H, there is no problem since $f = 0$ clearly satisfies (7.2). Suppose then that N is a proper subset of H. According to the corollary following the projection theorem, there exists a nonzero vector in $N^\perp$; by multiplication by an appropriate scalar one can choose a vector f_0 in $N^\perp$ such that $\|f_0\| = 1$ and $l(f_0)$ is *real*. Indeed, for any u the vector $l(u)f_0 - l(f_0)u$ is clearly in N and therefore orthogonal to f_0. Hence $l(u)\langle f_0, f_0 \rangle - l(f_0)\langle u, f_0 \rangle = 0$, so that, since $l(f_0)$ is real, $l(u) = \langle u, l(f_0)f_0 \rangle$. The required f in (7.2) is then $l(f_0)f_0$. To prove uniqueness suppose f and g satisfy (7.2) for all u. Then $\langle u, f - g \rangle = 0$ for all u, and by choosing $u = f - g$, we find that $\|f - g\| = 0$ or $f = g$.

Remark.   Two other proofs of this theorem are outlined in Exercises 7.2 and 7.3. Observe that every bounded linear functional on $L_2^{(c)}(a, b)$ is of the form $\int_a^b u(x) \bar{f}(x) \, dx$ for some f.

Representation of Bounded Linear Functionals: Dual Bases

A linear functional l on E_n can be characterized by its values on a basis. If $\{h_1,\ldots,h_n\}$ is such a basis, the n complex numbers $l(h_1),\ldots,l(h_n)$ enable us to calculate $l(u)$ for each l in E_n by the formula

$$(7.3) \qquad l(u) = l\left(\sum \alpha_i h_i\right) = \sum l(\alpha_i h_i) = \sum \alpha_i l(h_i).$$

Conversely, if we are given n complex numbers $l_1,\ldots,l_n$, there exists one and only one linear functional l on E_n such that $l(h_1)=l_1,\ldots,l(h_n)=l_n$. To see this all that is necessary is to read (7.3) from right to left: express u in the basis $\{h_i\}$ as $\sum \alpha_i h_i$, and then set $l(u)=\sum \alpha_i l_i$ to define the linear functional l.

The situation is more delicate in an infinite-dimensional space since now continuity of the linear functional is required for the second equality in (7.3). Thus a *continuous* linear functional is characterized completely by its values on a basis. Given a sequence of complex numbers $\{l_i\}$, when is there a bounded linear functional such that $l(h_i)=l_i$? A sufficient condition is $\sum |l_i|^2 < \infty$, for we can then define $l(\sum \alpha_i h_i)$ as $\sum \alpha_i l_i$, this series being convergent by the Schwarz inequality for series; moreover, l is uniquely determined.

We have already frequently noted that the coordinates of a vector u expanded in an orthonormal basis $\{e_i\}$ are easily calculated by $\alpha_i = \langle u, e_i \rangle$. If the basis is an oblique one, say $\{h_i\}$, the calculation of the coordinates of u is more difficult unless we introduce the so-called *dual* or *reciprocal* basis $\{h_i^*\}$ with the properties

$$(7.4) \qquad \langle h_i, h_j^* \rangle = \begin{cases} 0, & i \neq j, \\ 1, & i = j. \end{cases}$$

Let us now prove the existence of such a basis. We can treat the h_j^* one at a time. For j fixed, the preceding discussion shows that there exists a unique bounded linear functional l_j with the property $l_j(h_i)=0$, $i \neq j$, and $l_j(h_j)=1$. By the Riesz representation theorem there must exist a unique vector h_j^* such that $l_j(u)=\langle u, h_j^* \rangle$. Clearly h_j^* has the properties $\langle h_i, h_j^* \rangle = 0$, $i \neq j$, $\langle h_j, h_j^* \rangle = 1$. Obviously we can perform this construction for every index j, so that (7.4) is indeed satisfied. Now let $\{h_i\}$ be a basis in H, and let $\{\alpha_i\}$ be the coordinates of a vector u in this basis, that is, $u=\sum \alpha_i h_i$. Taking the inner product with respect to h_j^* and using the continuity of the inner product, we find this simple expression for the coordinates:

$$(7.5) \qquad \alpha_j = \langle u, h_j^* \rangle.$$

Exercises

7.1. Let l be a bounded linear functional on H; does there exist an element f^* such that $l(u) = \langle f^*, u \rangle$? *Hint*: Look at $l(\alpha u)$.

7.2. Prove the Riesz representation theorem by the following alternative method. The underlying idea is that, if $l(u)$ were really of the form $\langle u, f \rangle$, then $\max |l(u)|/\|u\|$ would occur for $u = \alpha f$.

(a) Since l is bounded, let $\|l\| =$ supremum over nonzero u of $|l(u)|/\|u\|$. Everything is trivial if $\|l\| = 0$, so assume $\|l\| > 0$. Show, by using the parallelogram law, that there exists an element u_0 with $\|u_0\| = 1$ for which $l(u_0) = \|l\|$.

(b) Now show that the vector f in Riesz's theorem is $\|l\| u_0$. This means proving $l(u) - \langle u, \|l\| u_0 \rangle = 0$ for all u or $\langle z, \|l\| u_0 \rangle = 0$, where $z = l(u) u_0 / \|l\| - u$. Since $l(z) = 0$, it suffices to show that $l(z) = 0$ implies $\langle z, \|l\| u_0 \rangle = 0$. Prove this in a manner similar to the way in which orthogonality is proved in the projection theorem.

The uniqueness proof is the same as that in the text.

7.3. Another proof of the Riesz representation theorem consists of exhibiting the element f explicitly in terms of an orthonormal basis $\{e_i\}$. For each u in H we have $u = \Sigma \langle u, e_i \rangle e_i$, and $l(u) = \Sigma \langle u, e_i \rangle l(e_i)$. On the other hand, for any pair of vectors u, f, we have

$$\langle u, f \rangle = \left\langle \sum \langle u, e_i \rangle e_i, \sum \langle f, e_i \rangle e_i \right\rangle = \sum \langle u, e_i \rangle \langle \overline{f, e_i} \rangle.$$

Thus the desired f has the property $\bar{l}(e_i) = \langle f, e_i \rangle$, and therefore $f = \Sigma_i \bar{l}(e_i) e_i$. It remains only to show that this definition makes sense [that is, that $\|f\|$ is finite or, equivalently, that $\Sigma |l(e_i)|^2$ converges; assume the contrary and arrive at a contradiction by considering elements of the form $\Sigma_{k=1}^{N} \bar{l}(e_k) e_k$].

REFERENCES AND ADDITIONAL READING

Halmos, P. R., *A Hilbert space problem book*, Van Nostrand, Princeton, N. J., 1967.

Lorch, E. R., *Spectral theory*, Oxford University Press, New York, 1962.

Kolmogorov, A. N. and Fomin, S. V., *Functional analysis*, Vols. I and II, Graylock, Albany, N. Y., 1957, 1961.

Naylor, A. W. and Sell, G. R., *Linear operator theory in engineering and science*, Holt, Rinehart, and Winston, New York, 1971.

Riesz, F. and Sz.-Nagy, B., *Functional Analysis*, Ungar, New York, 1955.

5
Operator Theory

1. BASIC IDEAS AND EXAMPLES

Whatever linear problem we are trying to solve, whether a set of algebraic equations, a differential equation, or an integral equation, there are advantages of clarity, generality, and geometric visualization in setting the problem in a suitable Hilbert space framework. To do so does, however, require some introductory definitions, and the impatient reader may find it refreshing to refer from time to time to the set of examples that follow in short order. Throughout we shall be dealing with a separable Hilbert space.

A (linear) *transformation* or *operator* A from the Hilbert space H into itself is a correspondence which assigns to each element u in H a well-defined element v, written Au, also belonging to H with the property

$$(1.1) \quad A(\alpha u + \beta v) = \alpha A u + \beta A v \qquad \text{for all } u, v \text{ in } H \text{ and all scalars } \alpha, \beta.$$

Note that a transformation takes vectors into vectors, whereas a functional maps vectors into scalars. A slightly more general definition of a linear transformation is obtained if we allow A to be defined only on a linear manifold D_A in H. Here D_A is the *domain* of A; often $D_A = H$, but it is not always possible to define an operator on the whole of H. (For instance, if H is L_2, a differential operator can be defined only on part of H.) A linear operator always satisfies $A0 = 0$, and the linearity property (1.1) can be extended to *finite* sums. The set of all images (that is, the set of all vectors of the form Au for some u in D_A) is labeled R_A and is known as the *range* of A. The set N_A of all vectors for which $Au = 0$ is called the *null space* of A (the null space is the set of all solutions of the homogeneous equation associated with A). The sets N_A and R_A are always linear manifolds.

A transformation A is *bounded* on its domain if, for all u in D_A, there exists a constant c such that $\|Au\| \leqslant c\|u\|$. Thus the ratio of the "output" norm to the "input" norm is bounded above. The smallest number c which satisfies the inequality for all u in D_A is the *norm* of A, written as $\|A\|$. We see with no difficulty that

$$(1.2) \qquad \|A\| = \sup_{\|u\| \neq 0} \frac{\|Au\|}{\|u\|} = \sup_{\|u\| = 1} \|Au\|.$$

Even if A is bounded, the supremum may not be attained for any element u (see Example 4). If $\|A\| = 0$, then A is the zero operator. To prove that a number m is the norm of an operator A, one must show that $\|Au\| \leqslant m\|u\|$ for all $u \in D_A$ *and* either that there exists $u_0 \neq 0$ in D_A such that $\|Au_0\| = m\|u_0\|$ or that there exists a sequence $\{u_n\} \in D_A$ with $\|u_n\| \neq 0$ and $\|Au_n\| - m\|u_n\| \to 0$. To prove that A is unbounded one must exhibit a sequence $\{u_n\}$ with $\|u_n\| \neq 0$ such that $\|Au_n\|/\|u_n\| \to \infty$.

A transformation A is *continuous at the point* u in D_A if, whenever $\{u_n\}$ is a sequence in D_A with limit u, then $Au_n \to Au$. A transformation is continuous (on its domain) if it is continuous at every point in D_A. The following theorems have proofs similar to those of corresponding theorems for functionals.

Theorem.   If A is continuous at the origin, it is continuous on all of D_A.

Theorem. A is continuous if and only if it is bounded.

Theorem.   In E_n all linear transformations are bounded.

Representation of Bounded Linear Operators; Matrices

A linear transformation A from $E_n^{(c)}$ into itself is completely characterized by its (vector) values on a fixed basis $h_1, \ldots, h_n$. Indeed, if u is a vector in $E_n^{(c)}$, we can write $u = \sum_{j=1}^n \alpha_j h_j$ and $v \doteq Au = \sum_{j=1}^n \alpha_j Ah_j$, so that knowledge of the n vectors $Ah_1, \ldots, Ah_n$ enables us to calculate Au. We can go further: each vector Ah_j is itself in $E_n^{(c)}$ and can in turn be written as

$$(1.3) \qquad Ah_j = \sum_{i=1}^n a_{ij} h_i \qquad \text{(summation on \textit{first index} in } a_{ij}\text{)},$$

where the complex numbers a_{ij} are defined unambiguously from (1.3) and can be calculated from

$$(1.4) \qquad a_{ij} = \langle Ah_j, h_i^* \rangle \qquad (= \langle Ae_j, e_i \rangle \text{ in an orthonormal basis}),$$

in which $\{h_i^*\}$ is the dual basis introduced in (7.4), Chapter 4. Thus

$$v = Au = \sum_{j=1}^{n} \alpha_j Ah_j = \sum_{i=1}^{n} \left(\sum_{j=1}^{n} a_{ij}\alpha_j \right) h_i,$$

and, if we denote the ith coordinate of v in the $\{h_i\}$ basis by β_i, we have

(1.5) $\qquad \beta_i = \sum_{j=1}^{n} a_{ij}\alpha_j \qquad$ (summation on *second index* in a_{ij}).

The set of n^2 complex numbers a_{ij} completely characterizes the transformation A. These numbers are usually displayed as a square array:

$$\begin{bmatrix} a_{11} & \cdots & a_{1n} \\ \vdots & & \vdots \\ a_{n1} & \cdots & a_{nn} \end{bmatrix},$$

known as the *matrix* of A in the basis $\{h_i\}$. The shorter notation $[a_{ij}]$ will often be used. By tradition the first index in a_{ij} gives the *row* location; the second, the *column* location. Note that the matrix is used in different ways in (1.3) and (1.5). To obtain the vector Ah_j, we "multiply" the jth *column* of A by $(h_1, \ldots, h_n)$; whereas to find the ith coordinate of the image vector v in terms of the coordinates of u, we "multiply" the ith *row* of A by $(\alpha_1, \ldots, \alpha_n)$.

The relation between a transformation and a matrix is as follows.

Relative to a given basis, each linear transformation A determines a unique matrix $[a_{ij}]$ whose entries can be calculated from (1.3), and, conversely, each matrix (that is, each square array of n^2 numbers) generates a unique linear transformation via (1.3).

Everything said holds also for transformations on $E_n^{(r)}$, but then the entries in the matrix are real.

When dealing with an infinite-dimensional separable Hilbert space H, we find, just as we did for linear functionals, that the analogy with E_n can be carried out successfully only for bounded operators. Let A be a *bounded* linear operator defined on the whole of H, and, for simplicity, let $\{e_i\}$ be an *orthonormal* basis. We can then write $u = \sum_j \alpha_j e_j$ and $v \doteq Au = A(\sum_j \alpha_j e_j) = \sum_j \alpha_j Ae_j$, where the continuity of A has been used in the last equality. Each vector Ae_j can be written as $\sum_i a_{ij}e_i$, where $a_{ij} = \langle Ae_j, e_i \rangle$. It follows that $\sum_i |a_{ij}|^2 < \infty$ (and it can also be shown that $\sum_j |a_{ij}|^2 < \infty$). Also $\beta_i = \langle v, e_i \rangle = \langle Au, e_i \rangle = \langle A\sum_j \alpha_j e_j, e_i \rangle = \sum_j \alpha_j \langle Ae_j, e_i \rangle = \sum_j a_{ij}\alpha_j$. Thus, relative to a fixed basis, each bounded linear transformation generates a unique infinite matrix $[a_{ij}]$ each of whose rows and columns is square summable. These conditions are, however, not sufficient for a set of numbers $[a_{ij}]$ to be the matrix of a bounded linear transformation (see Exercise 1.1).

Examples of Linear Operators

The following examples deserve careful examination; they will be referred to frequently as we take up new material.

Example 1. (a) The zero transformation 0 is defined by $v \doteq 0u \doteq 0$ for every u in H. The transformation is clearly linear and bounded and has norm 0. The range consists of the single element 0, and the null space is H.

   (b) The identity I is a transformation defined by $v \doteq Iu \doteq u$ for each u in H. The transformation is linear, $\|I\| = 1$, $R_I = H$, $N_I = \{0\}$.

Example 2. Let M be a closed linear manifold in H, and for each u in H define $v = Pu$ as the (orthogonal) projection of u on M. P is linear, and $\|v\| \le \|u\|$ with equality if and only if u is in M. We have $\|P\| = 1$, $R_P = M$, $N_P = M^\perp$.

Example 3. (a) Let A be a linear transformation from $E_n^{(c)}$ into itself. We have already shown that A is necessarily bounded, but it is not always easy to calculate $\|A\|$. Suppose A has the matrix $[a_{ij}]$ in the *orthonormal* basis $\{e_1, \ldots, e_n\}$. Then by (1.5) and (5.13), Chapter 4, we have

$$\|v\|^2 = \sum_{i=1}^{n} |\beta_i|^2 = \sum_{i=1}^{n} \left| \sum_{j=1}^{n} a_{ij} \alpha_j \right|^2 \le \sum_{i,j=1}^{n} |a_{ij}|^2 \|u\|^2.$$

It follows that

$$(1.6) \qquad \|A\| \le \left[\sum_{i,j=1}^{n} |a_{ij}|^2 \right]^{1/2},$$

but the upper bound can be grossly conservative. If A happens to have *diagonal* form in the orthonormal basis, life becomes much simpler (which is a strong incentive for studying the possibility of choosing a basis in which A is diagonal). Then $a_{ii} = m_i$ and $a_{ij} = 0$, $i \ne j$, so that

$$\|v\|^2 = \sum_{i=1}^{n} |m_i \alpha_i|^2 \le m^2 \|u\|^2, \qquad m = \max |m_i|.$$

This shows that $\|A\| \leqslant m$, but since $\|Ae_i\| = |m_i|$, we see that there is a unit vector for which $\|Ae\| = m$. Thus $\|A\| = m$. Although we have carefully refrained from using the word "eigenvalue," it is clear that $\|A\|$ is closely related to the eigenvalues of A. This question and the characterization of N_A and R_A will be taken up later. We now look at two simple cases of linear transformations on E_2.

(b) On the real plane $E_2^{(r)}$, consider the transformation A which rotates each vector counterclockwise through the angle θ. To find the matrix of A relative to a right-handed orthonormal basis (e_1, e_2), we have to draw a sketch from which we conclude that

$$Ae_1 = (\cos\theta)e_1 + (\sin\theta)e_2, \qquad Ae_2 = (-\sin\theta)e_1 + (\cos\theta)e_2,$$

and therefore

$$A = \begin{bmatrix} \cos\theta & -\sin\theta \\ \sin\theta & \cos\theta \end{bmatrix}.$$

The coordinates (β_1, β_2) of the rotated vector are related to the coordinates (α_1, α_2) of the original vector by

$$\beta_1 = \alpha_1 \cos\theta - \alpha_2 \sin\theta, \qquad \beta_2 = \alpha_1 \sin\theta + \alpha_2 \cos\theta.$$

We could have considered a linear transformation on the complex space $E_2^{(c)}$ with the same matrix as above relative to some orthonormal basis. The formulas would remain the same, but the geometric interpretation would be lost. In either case $N_A = \{0\}$, $R_A = E_2$, $\|A\| = 1$.

(c) On $E_2^{(c)}$ or $E_2^{(r)}$ consider the transformation A whose matrix, relative to an orthonormal basis (e_1, e_2), is

$$\begin{bmatrix} 0 & 1 \\ 0 & 0 \end{bmatrix}.$$

The coordinates of the image vector are related to those of the original by

$$\beta_1 = \alpha_2, \qquad \beta_2 = 0.$$

Both N_A and R_A consist of all vectors proportional to e_1; also $\|A\| = 1$.

Example 4. On $L_2^{(c)}(0, 1)$ consider the transformation corresponding to multiplication by the independent variable x. Thus $v(x) \doteq Au \doteq xu(x)$. The transformation is defined for all elements of $L_2^{(c)}(0, 1)$ and is linear. From

$$\|Au\|^2 = \int_0^1 x^2 |u|^2 \, dx \leqslant \int_0^1 |u|^2 \, dx = \|u\|^2$$

we infer that $\|A\| \leqslant 1$ and that A is therefore bounded. Next we show that $\|A\| = 1$ by exhibiting a family $u_\varepsilon(x)$ with the property $\|Au_\varepsilon\|/\|u_\varepsilon\| \to 1$ as $\varepsilon \to 0$. Let $u_\varepsilon(x)$ be 0 except on the small interval $1 - \varepsilon \leqslant x \leqslant 1$, where $u_\varepsilon(x) = 1$. Then $\|u_\varepsilon\|^2 = \varepsilon$ and $\|Au_\varepsilon\|^2 = \varepsilon[1 - \varepsilon + (\varepsilon^2/3)]$, which shows that u_ε has the desired property. Although $\|A\| = 1$, there is no nonzero element in $L_2^{(c)}(0,1)$ for which $\|Au\| = \|u\|$. Note that $N_A = \{0\}$ but that R_A is *not* all of $L_2^{(c)}(0,1)$.

Example 5. In the interval $0 < x < 1$ we want to consider the transformation defined by $v(x) \doteq Au \doteq (1/x)u(x)$. Since $1/x$ becomes infinite at $x = 0$, the transformation is not defined for all u in $L_2^{(c)}(0,1)$. On the other hand, if u vanishes sufficiently fast at $x = 0$, then u/x will be in $L_2^{(c)}(0,1)$. Thus the natural domain of definition for this operator is the linear manifold D_A of functions $u(x)$ in $L_2^{(c)}(0,1)$ for which u/x is also in $L_2^{(c)}(0,1)$. D_A clearly includes the functions u in $L_2^{(c)}(0,1)$ which vanish identically in a neighborhood of $x = 0$. Since such functions are dense in $L_2^{(c)}(0,1)$, D_A is certainly dense in $L_2^{(c)}(0,1)$. We now show that A is *unbounded* on D_A. Consider the function $u_\varepsilon(x)$ in D_A defined by $u_\varepsilon = 0$, $0 < x < \varepsilon$; $u_\varepsilon = 1$, $\varepsilon < x < 1$. Then $\|u_\varepsilon\|^2 = 1 - \varepsilon$ and $\|Au_\varepsilon\|^2 = (1 - \varepsilon)/\varepsilon$. As $\varepsilon \to 0$, the ratio $\|Au_\varepsilon\|/\|u_\varepsilon\|$ becomes arbitrarily large, so that A is unbounded. We have $N_A = \{0\}$, $R_A = L_2^{(c)}(0,1)$.

Example 6. On $L_2^{(c)}(a,b)$ consider the *integral operator* defined by

$$(1.7) \qquad\qquad v(x) \doteq Ku \doteq \int_a^b k(x,y)u(y)\,dy.$$

Here $k(x,y)$ is a given function defined over the square $a \leqslant x,\ y \leqslant b$. The function $k(x,y)$ is known as the *kernel* of the operator. We have already considered such operators on the space of continuous functions in Section 4, Chapter 4. Although a function of two variables is involved in an intermediate stage, the operator K takes a function of one variable into another function of one variable. If $u(x)$ is in $L_2^{(c)}(a,b)$, so will be $v(x)$ as long as $k(x,y)$ satisfies some very mild conditions. If, for instance,

$$(1.8) \qquad\qquad \int_a^b \int_a^b |k^2(x,y)|\,dx\,dy < \infty,$$

then $k(x,y)$ is said to be a *Hilbert-Schmidt kernel*, and the operator K maps $L_2^{(c)}(a,b)$ into itself and is bounded. In fact, by the Schwarz inequality,

$$|v(x)|^2 = \left| \int_a^b k(x,y)u(y)\,dy \right|^2$$

$$\leqslant \int_a^b |k(x,y)|^2\,dy \int_a^b |u(y)|^2\,dy = \|u\|^2 \int_a^b |k(x,y)|^2\,dy.$$

Therefore

$$\|v\|^2 \leqslant \|u\|^2 \left[\int_a^b \int_a^b |k^2(x,y)|\,dx\,dy \right],$$

so that K is bounded:

(1.9)
$$\|K\| \leqslant \left[\int_a^b \int_a^b |k^2(x,y)|\,dx\,dy \right]^{1/2}.$$

Example 7. On $L_2^{(c)}(0,1)$ consider the operator defined by

(1.10)
$$v(x) \doteq Au \doteq \int_0^x u(y)\,dy.$$

The operator is defined for all u in $L_2^{(c)}(0,1)$ since square integrability implies that u is integrable over any subinterval of $(0,1)$. This operator is linear and can be regarded as a special case of Example 6 with $a=0$, $b=1$ and $k(x,y)=H(x-y)$, where $H(x)$ is the usual Heaviside function. From (1.9) we find that $\|A\| \leqslant [\int_0^1 \int_0^1 H^2(x-y)\,dx\,dy]^{1/2} = (\frac{1}{2})^{1/2}$. The exact expression for $\|A\|$ is $2/\pi$. (See Exercise 1.4.) We observe that $N_A = \{0\}$ and that R_A consists of differentiable functions vanishing at $x=0$ with the derivative in $L_2^{(c)}(0,1)$.

Example 8. We make a preliminary attempt to define a differentiation operator A for functions on $0 \leqslant x \leqslant 1$. Since functions in $L_2(0,1)$ may not even be continuous, we cannot define the operator for the whole of L_2. As a first try, let D_A be the set of functions in L_2 with a continuous derivative on $0 \leqslant x \leqslant 1$, and define

$$v(x) \doteq Au \doteq \frac{du}{dx}.$$

Both D_A and R_A are subsets of L_2, and D_A is a dense linear manifold in $L_2(0,1)$. The operator A is *unbounded* on D_A: let $u_n = \sin n\pi x$; then $\|u_n\|^2 = \frac{1}{2}$ and $\|Au_n\|^2 = n^2\pi^2/2$, so that $\|Au_n\|/\|u_n\|$ is unbounded as $n \to \infty$. Note that N_A is the set of constant functions and that R_A (by the way we have defined A) is the set of continuous functions on $0 \leqslant x \leqslant 1$.

When faced with solving the inhomogeneous equation $du/dx = f$, $u \in D_A$, we note two unpleasant features. First, the solution, when it exists, is not unique since we can add a constant function and still have a solution. This difficulty is easily overcome by restricting the domain of the operator to functions satisfying an appropriate "boundary condition" such as $u(0) = 0$. The second, and more serious, trouble is that we are able to solve the inhomogeneous equation only for continuous f rather than for every f in L_2. Admittedly this is, in part, a self-imposed restriction since we have defined D_A so that the range contains only continuous functions. We shall see in Example 11 (after the topic of closed operators) how to reformulate the problem.

Example 9. The shift operator. Let H be an infinite-dimensional separable space, and let $\{e_1, e_2, \dots, e_n, \dots\}$ be an orthonormal basis in H. Define the operator A by giving its effect on the basis: A transforms each unit vector in the next one on the list (*right shift*), that is, $Ae_i = e_{i+1}$. The image of $u = \sum_{i=1}^{\infty} \alpha_i e_i$ is therefore $Au = \sum_{i=1}^{\infty} \alpha_i e_{i+1}$. It is clear that A is linear and is defined on the whole of H, and that $\|Au\| = \|u\|$ for every u; hence $\|A\| = 1$. The range of A is the closed manifold orthogonal to e_1, that is, R_A consists of all vectors whose first coordinate is 0. R_A is closed but not dense in H. The null space of A consists only of the element 0. The matrix of A consists of 0's except for the diagonal directly below the main diagonal, which consists of 1's.

Example 10. A modified shift. Again $\{e_i\}$ is an orthonormal basis, but now define $Be_k = (1/k^2)e_{k+1}$. Then $\|B\| = 1$, $N_B = \{0\}$, and R_B consists of all vectors $\sum_{i=1}^{\infty} \beta_i e_i$ with $\beta_1 = 0$ and $\sum_{k=1}^{\infty} k^4 |\beta_{k+1}|^2 < \infty$. We observe that R_B is not closed, but that $\overline{R_B} = M_1^{\perp}$, where M_1 is the linear manifold generated by e_1. *Proof.* If $u \in \overline{R_B}$, there exists $\{u_n\}$ in R_B with $u_n \to u$. Since $\langle u_n, e_1 \rangle = 0$, it follows by continuity of the inner product that $\langle u, e_1 \rangle = 0$, so that $\overline{R_B} \subset M_1^{\perp}$. For $u \in M_1^{\perp}$, consider the truncated vector $[u]_n$ obtained from u by setting equal to 0 all coordinates with indices larger than n. Then $[u]_n$ is in R_B, and since $[u]_n \to u$, u must lie in $\overline{R_B}$. Thus $\overline{R_B} \supset M_1^{\perp}$, which, together with the reverse inclusion, shows that $\overline{R_B} = M_1^{\perp}$.

Exercises

1.1. Let A be a bounded linear transformation on H, and let $\{e_k\}$ be an orthonormal basis. If $[a_{ij}]$ is the matrix of A in this basis, we have shown that $\sum_i |a_{ij}|^2 < \infty$ for each j, that is, each column is square summable. By considering the adjoint (Section 4), it can also be shown that each row is square summable. Show that these two conditions, even taken together, are not sufficient for A to be bounded. (*Hint:* Take a diagonal matrix whose elements increase indefinitely along the diagonal.) Show that a sufficient (but not necessary) condition for A to be bounded is that $\sum_{i,j} |a_{ij}|^2 < \infty$.

1.2. On $L_2^{(c)}(0,1)$ consider the transformation A defined by $Au = f(x)u(x)$, where $f(x)$ is a fixed function. Show that, if f is continuous on $0 \leqslant x \leqslant 1$, then $\|A\| = \max_{0 \leqslant x \leqslant 1} |f(x)|$.

1.3. *Holmgren kernels.* Consider the integral operator of Example 6, but, instead of the Hilbert-Schmidt condition (1.8), assume that

(1.11) $$\max_{a \leqslant z \leqslant b} \int_a^b \int_a^b |k(x,y)||k(x,z)| \, dx \, dy \doteq M < \infty.$$

Show that K is a bounded operator and that $\|K\| \leqslant M^{1/2}$. On $L_2(-\infty, \infty)$ the example $k(x,y) = h(x-y)$, where $\int_{-\infty}^{\infty} |h(x)| < \infty$, shows that there are Holmgren kernels that are not Hilbert-Schmidt.

1.4. Consider the operator of Example 7.
  (a) Show that the set

$$\left\{ e_n(x) = \sqrt{2} \, \cos \frac{2n-1}{2} \pi x \right\}, \qquad n = 1, 2, \ldots,$$

  is an orthonormal basis in $L_2^{(c)}(0,1)$. *Hint:* The set $\cos(k\pi x/2)$, $k = 0, 1, 2, \ldots$, is a basis for $L_2(0,2)$; any function in $(0,1)$ can be extended to $(0,2)$ with the property $f(x+1) = -f(1-x)$.
  (b) Expand u in the set $\{e_n(x)\}$ of part (a) to show that $\|Au\| \leqslant (2/\pi)\|u\|$.
  (c) Show $\|A\| = 2/\pi$ by exhibiting a function u such that $\|Au\| = (2/\pi)\|u\|$.

1.5. Obtain the results corresponding to those of Example 9 for the left-shift operator defined by $Ae_i = e_{i-1}, i = 2, \ldots; Ae_1 = 0$.

1.6. Consider the two-sided basis

$$\left\{ e_n(x) = \frac{1}{\sqrt{2}} e^{2n\pi ix} \right\}, \qquad n = 0, \pm 1, \pm 2, \ldots, \qquad \text{on } L_2^{(c)}(-1,1).$$

Define a linear operator by $Ae_n = e_{n+1}$. Show that this operator can be represented as a multiplication by a function. Find $\|A\|, N_A, R_A$.

CLOSED OPERATORS

Definition. The linear operator B is said to be an extension of the linear operator A if $D_B \supset D_A$ and $Bu = Au$ for each u in D_A.

Bounded linear operators defined on a linear manifold (closed or not) in H can always be extended to the whole of H without affecting continuity or changing the norm.

(a) If D_A is closed and $D_A \neq H$, we construct the extension by defining the new operator B to be 0 on $(D_A)^\perp$ and to coincide with A on D_A, letting linearity take care of everything else. Precisely: If $u \in H$, write $u = v + w$, $v \in D_A$, $w \in D_A^\perp$; then Bu is defined as Av. The operator B is easily seen to be linear on H and $\|B\| = \|A\|$.

(b) If D_A is not closed, we first extend A to $\overline{D}_A$ by "continuity." Let $\{u_n\}$ be a sequence in D_A with limit u in $\overline{D}_A$. Since A is bounded, the sequence $\{Au_n\}$ is Cauchy and has a limit, say f. It is natural to set Au equal to f. To see that this definition depends only on u and not on $\{u_n\}$, let $\{v_n\}$ be another sequence in D_A with limit u. Then $u_n - v_n \to 0$, and by continuity $A(u_n - v_n) \to 0$, so that $\{Au_n\}$ and $\{Av_n\}$ have the same limit, f. In this way we extend A to $\overline{D}_A$ and then by (a) to H.

Thus, if in a problem we are permitted to change the domain of a given bounded operator A, we may as well take the operator as a bounded operator *defined on the whole of H*.

Suppose now that A is an *unbounded* linear operator on a linear manifold D_A in H. Let $\{u_n\}$ be a sequence of elements in D_A for which $\lim_{n \to \infty} u_n = 0$; such a sequence will be called a *null sequence in D_A*. We cannot have $Au_n \to 0$ for *all* null sequences in D_A, for otherwise A would be continuous at 0 and hence bounded. (On the other hand, there will always be some null sequences for which $Au_n \to 0$. For instance, let $u \neq 0$ be in D_A and $\{\alpha_n\}$ be a sequence of scalars such that $\alpha_n \to 0$; then $\{u_n \doteq \alpha_n u\}$ is a null sequence in D_A for which $Au_n \to 0$.) Thus there must exist some null sequences in D_A for which $Au_n \to f \neq 0$ or for which Au_n has no limit. The first possibility leads to very pathological operators which are *never encountered* in the study of differential or integral equations. We may therefore confine ourselves to the following class of operators.

Definition. A linear operator is *closable* if for every null sequence $\{u_n\}$ in D_A either $Au_n \to 0$ or Au_n has no limit.

Note that the definition includes all bounded operators and presumably unbounded operators such as those of Examples 5 and 8, Section 1. Let us examine the differentiation operator of Example 8. For any null sequence $\{u_n\}$ in D_A we are to show that, if $Au_n \to f$ (that is, if $u_n' \to f$), then $f = 0$. If $u_n' \to f$, then $\int_0^1 u_n' \bar{z} \, dx \to \langle f, z \rangle$ for every z in L_2, so certainly for every z in the class M of continuously differentiable functions with $z(0) = z(1) = 0$; hence for z in M we obtain, through integration by parts,

$$- \int_0^1 u_n \bar{z}' \, dx \rightarrow \langle f, z \rangle,$$

which, by the continuity of the inner product, gives $\langle f, z \rangle = 0$ for every z in M. Since M is dense in H, we have $f = 0$, so that A on D_A is *closable*. This means that, if $\{u_n(x)\}$ is a sequence of functions in D_A with $u_n \rightarrow 0$, then either $u_n' \rightarrow 0$ or u_n' has no limit [the latter possibility occurs if $u_n = (\sin nx)/n$, for instance].

The advantage of a closable operator is that it can be usefully extended by the following simple procedure. Let $\{u_n\} \in D_A$ with $u_n \rightarrow u$ (which may or may not be in D_A), and suppose $Au_n \rightarrow f$. Let $\{v_n\}$ be another sequence in D_A which approaches the same limit u; then $u_n - v_n$ is a null sequence in D_A, so that either $A(u_n - v_n) \rightarrow 0$ or this sequence has no limit. Thus either $Av_n \rightarrow f$ or Av_n has no limit. We therefore yield gracefully to the temptation to include u in the domain of the operator and to let f be the image of u. If we do this for all convergent sequences $\{u_n\}$ in D_A for which $\{Au_n\}$ has a limit, we obtain a new operator known as the *closure* of A and denoted by $\tilde{A}$. It is clear that $\tilde{A}$ is a linear operator defined on a domain $D_{\tilde{A}}$ which includes D_A and that $Au = Au$ whenever $u \subset D_A$; thus $\tilde{A}$ is an extension of A. The operator $\tilde{A}$ belongs to the important class of closed operators.

Definition. Let A be a linear operator on the linear manifold D_A. We say that A is *closed* if it has the following property: Whenever $\{u_n\}$ is in D_A, and $u_n \rightarrow u$, and $Au_n \rightarrow f$, then u is in D_A and $Au = f$.

Remarks

1.   If A is closable, $\tilde{A}$ is closed.

2.   A bounded operator defined on the whole of H (or even defined on a closed set) is closed.

3.   *A closed operator on a closed domain is bounded.* We shall not prove this, but we analyze some of its consequences. It means first that a closed, *unbounded* operator can never be defined on the whole of H. Differential operators are the most important among closed, unbounded operators, and they are usually defined on domains *dense in H*. The only unbounded operators that can be defined on the whole of H are "highly discontinuous" and never occur in the study of differential or integral equations.

4.   Since a closed operator does not necessarily have a closed domain or a closed range, the reader may wonder how it got its name. The graph of an operator A is the set of ordered pairs (u, Au) with u in D_A. The graph can be viewed as a subset of a new normed space consisting of pairs (u, v) with $u, v \in H$ and $\|(u, v)\| \doteq \|u\| + \|v\|$. It then turns out that the original operator A is closed if and only if its graph is a closed set in the new normed space.

5.   The null space of a closed operator is a closed set.

Now let us examine again the differentiation operator A of Example 8, Section 1. This operator was defined on the domain D_A of continuously differentiable functions on $0 \leqslant x \leqslant 1$. We have shown that the operator is closable, but it is easy to see that it is not closed. In Exercise 2.1 we exhibit a sequence of very smooth functions $\{u_n(x)\}$ such that u_n tends in the L_2 sense to a function u that is only piecewise differentiable, while at the same time u_n' tends to a limit v. Thus u belongs to $D_{\tilde{A}}$ and $\tilde{A}u = v$, but u does not belong to D_A. It is negligent of us to leave out piecewise smooth functions from the original domain, but even then the differentiation operator would not be closed. To see exactly what the significance of $\tilde{A}$ is, let $\{u_n\}$ be a sequence in D_A such that, simultaneously,

$$(2.1) \qquad \|u_n - u\| \to 0 \qquad \text{and} \qquad \|u_n' - v\| \to 0.$$

Then by definition $u(x)$ belongs to the domain of $\tilde{A}$ and $\tilde{A}u = v$. Of course such a definition is appealing only if $\tilde{A}$ can be characterized concretely as a generalization of differentiation with v the derivative of u.

Since $u_n'(x)$ is continuous, we have

$$(2.2) \qquad u_n(x) = \int_0^x u_n'(t)\, dt + u_n(0).$$

The L_2 convergence of u_n' to v implies that $\int_0^x u_n'\, dt$ converges in L_2 to $\int_0^x v\, dt$. Equation (2.2) then shows that the sequence of constants $\{u_n(0)\}$ must converge in L_2; obviously the limit is equivalent to a constant, say α, so that (2.2) yields

$$(2.3) \qquad u(x) = \int_0^x v(t)\, dt + \alpha \qquad \text{almost everywhere.}$$

Since the right side of (2.3) is continuous, $u(x)$ can be redefined, if necessary, on a set of measure 0 to make the equation valid for all x in $0 \leqslant x \leqslant 1$. As an immediate consequence we see that $\alpha = u(0)$ and that $u(x)$

and $v(x)$ are related by

(2.4) $$u(x) = \int_0^x v(t)\,dt + u(0), \qquad 0 \le x \le 1.$$

If v is continuous, u is differentiable everywhere and $u' = v$, but even if v is only integrable, (2.4) provides a useful extension of the notion of differentiability. It follows from (2.4) that u' exists almost everywhere and that $u' = v$ almost everywhere; in addition u is the integral of its derivative. [Everyone should know Cantor's example of a nonconstant continuous function whose derivative is equal to 0 almost everywhere. Obviously such a function does not satisfy (2.4).] A function satisfying (2.4) is said to be *absolutely continuous* (for which we often use the abbreviation a.c.).

Therefore the closure $\tilde{A}$ of the differentiation operator has a domain $D_{\tilde{A}}$ consisting of all a.c. functions with first derivative in L_2. For a differential operator A of order p the domain of the closure $\tilde{A}$ consists of functions $u(x)$ whose first $(p-1)$ derivatives are continuous and such that $u^{(p-1)}$ is a.c. with $u^{(p)}$ in L_2. This exact delineation of the smoothness of the functions in the domain is not of great importance to us; usually it suffices to know that the domain is the largest class of functions for which the differential operator makes sense and the image is in L_2.

We shall see in the next section how boundary conditions behave under closure.

Exercises

2.1. On $L_2^{(c)}(0,1)$, let $u(x) = |x - \tfrac{1}{2}|$, and let $\{u_n(x)\}$ be the sequence of partial sums of the Fourier cosine series of $u(x)$. Show that $\|u_n - u\|$ $\to 0$ and that $\|u_n' - v\| \to 0$, where $v = \mathrm{sgn}(x - \tfrac{1}{2})$. Since u is not continuously differentiable, it does not belong to the domain of the differentiation operator A as originally defined. Of course u belongs to $D_{\tilde{A}}$ and $\tilde{A}u = v$. This exercise shows that piecewise smooth functions belong to $D_{\tilde{A}}$.

2.2. In (5.34), Chapter 2, we gave another generalization of the notion of differentiation. We said that a function u satisfies $u' = f$ on $(0,1)$ if, for each test function $\phi(x)$ with support in $(0,1)$,

$$\int_0^1 f\phi\,dx = -\int_0^1 u\phi'\,dx.$$

Show that this definition is equivalent to saying that u is absolutely continuous on $0 \le x \le 1$.

3. INVERTIBILITY—THE STATE OF AN OPERATOR

Suppose A is a given linear operator (not necessarily bounded) on a linear manifold D_A in H. The central problem of operator theory is the solution ("inversion") of the inhomogeneous equation

$$(3.1) \qquad\qquad\qquad Au = f,$$

where f is an arbitrary, given element in H and we are looking for solution(s) u lying in D_A. In a perfect world (3.1) would have one and only one solution for each f in H, and the solution u would depend continuously on the "data" f. In the language of Chapter 1, such a problem would be well posed. If this were always true, there might result a sharp drop in the employment of mathematicians, so that perhaps we should accept as a partial blessing the difficulties we are about to encounter in the analysis of (3.1).

We can divide the problem into two overlapping parts: the question of invertibility, and the question of characterizing the range of A. We take up invertibility first. If the mapping A is one-to-one from its domain to its range, then for each $f \in R_A$ there exists one and only one solution u (in D_A) of (3.1). This correspondence enables us to define the inverse operator A^{-1} by $u = A^{-1}f$, and A^{-1} is clearly linear. We are also interested in whether or not A^{-1} is bounded (if A^{-1} is bounded, it is continuous and therefore the solution u depends continuously on the "data" f). On E_n the inverse, if it exists, must be bounded, but on an infinite-dimensional space *the inverse of a bounded operator is not necessarily bounded*. The two very simple theorems that follow characterize, respectively, the existence of an inverse and the existence of a bounded inverse. Throughout, the linear operator A is defined on a fixed domain D_A, a linear manifold in H.

Theorem 1. A^{-1} exists (A has an inverse, A is one-to-one) if and only if the homogeneous equation $Au = 0$ has only the zero solution (that is, the null space N_A contains only the zero element).

Proof. If A is one-to-one, then $Au = 0$ can have at most one solution, and since $u = 0$ is obviously a solution, it is the only one. If $Au = 0$ has only the zero solution, we must show that $Av = f$ can have at most one solution; if $Av_1 = f$ and $Av_2 = f$, the linearity of A and of its domain implies that $A(v_1 - v_2) = 0$, so that $v_1 - v_2 = 0$ or $v_1 = v_2$.

Definition. The operator A (on D_A) is *bounded away from 0* if there exists $c > 0$ such that $\|Au\| \geqslant c\|u\|$ for all $u \in D_A$.

Remark. Obviously, if A is bounded away from 0, then $Au=0$ has only the zero solution and therefore A^{-1} exists. Moreover, A^{-1} must be bounded, for, if we set $v=Au$, then $\|v\|=\|Au\|\geqslant c\|u\|$, so that $\|u\|\leqslant c^{-1}\|v\|$, and since $u=A^{-1}v$, we find that $\|A^{-1}\|\leqslant c^{-1}$. This leads easily to the following theorem.

Theorem 2. The operator A (on D_A) has a bounded inverse if and only if it is bounded away from 0.

Remark. Thus, if A fails to have a bounded inverse, there exists a sequence $\{u_n\}$ of unit elements such that $Au_n\to 0$.

We now turn to the characterization of R_A. By definition of the range, (3.1) can be solved, although perhaps not uniquely, if and only if $f\in R_A$. It would be nice if R_A were all of H, but that is not the case in many of the examples we have just given. It turns out that it is often difficult to describe the range precisely (see, for instance, Examples 4,6,7,10, Section 1), but easier to characterize its closure $\bar{R}_A$. Either this closure is all of H, or it is a proper subset of H (in which case there exist nonzero elements in H orthogonal to $\bar{R}_A$ and hence to R_A). Often R_A and H are both infinite-dimensional, so that it is easier to talk about the dimension of $R_A^{\perp}$, the so-called *codimension* of R_A.

We are now ready to present at least a coarse-grained classification of operators. The *state* of an operator will be represented by a Roman numeral (I, II, or III), followed by an Arabic numeral (1 or 2). The Roman numeral describes the invertibility properties of A, whereas the Arabic numeral tells us whether or not $\bar{R}_A=H$.

$$(3.2)\quad\begin{cases} \text{I.} & \text{Bounded inverse.} \\ \text{II.} & A^{-1}\text{ exists but is unbounded.} \\ \text{III.} & A^{-1}\text{ does not exist } (Au=0\text{ has a nontrivial solution).} \\ \text{1.} & \bar{R}_A=H. \\ \text{2.} & \bar{R}_A\neq H. \end{cases}$$

A further refinement in the classification will sometimes be made. The subscript c or n on the Arabic numeral tells us that the range is closed or not closed, respectively. Only state $(I,1_c)$ represents the ideal operator with a bounded inverse and $R_A=H$. Such an operator is said to be *regular*. The other states characterize operators with various assortments of ills. Often one tries to adjust the definition of A and D_A (say, by closure) so that state $(I,1_c)$ is achieved (see Example 1, for instance).

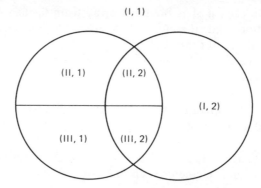

Figure 3.1 The diagram shows the possible states of operators as given in (3.2). The diagram can also be interpreted as showing how, for a fixed A, the state of $A - \lambda I$ may vary with λ (without suggesting that every region is necessarily traversed as λ varies): II is the continuous spectrum, III the point spectrum, 2 the compression spectrum, $II \cup III$ the approximate spectrum, and $(I, 1)$ the resolvent set.

The possible states of an operator are illustrated in Figure 3.1 without regard to subscripts. In the interior of the left circle something is wrong with the inverse, whereas in the interior of the right circle $\bar{R}_A \neq H$. The table below shows the states for the examples of Section 1.

	Example	State	Example	State
	1(a)	$(III, 2_c)$	5	$(I, 1_c)$
	1(b)	$(I, 1_c)$	7	$(II, 1_n)$
(3.3)	2	$(III, 2_c)$	8	$(III, 1_n)$
	3(b)	$(I, 1_c)$	9	$(I, 2_c)$
	4	$(II, 1_n)$	10	$(II, 2_n)$

Taking subscripts into account, there are 12 possible states in symbols if not in fact. Four of these can be eliminated for closed operators. We need the following theorems, the first of which (unproved) has already been discussed in Remark 3 of Section 2.

Theorem 3. A closed operator on a closed domain is bounded.

Theorem 4. If A is closed and A^{-1} exists, then A^{-1} is closed.

Proof. Let $\{f_n\}$ be a sequence in R_A with $f_n \to f$ and $u_n \doteq A^{-1}f_n \to u$; we must show that f is in R_A and $A^{-1}f = u$. Since $Au_n = f_n$, $f_n \to f$, and $u_n \to u$, it follows from the fact that A is closed that u must be in the domain of A and that $Au = f$; therefore f is in R_A and $u = A^{-1}f$.

Theorem 5. Let A be closed, and let A^{-1} exist; then R_A is closed if and only if A^{-1} is bounded.

Proof. (a) Let A^{-1} be bounded, and let $\{f_n\}$ be a sequence in R_A with $f_n \to f$. The boundedness of A^{-1} shows that $u_n \doteq A^{-1}f_n$ is a Cauchy sequence which must therefore have a limit, say u, in H. Thus $\{u_n\} \in D_A$, $u_n \to u$, and $Au_n \to f$; hence, since A is closed, $u \in D_A$ and $Au = f$. This shows that f is in R_A and therefore R_A is closed.

(b) By Theorem 4, A^{-1} is closed; if also R_A is closed, then, by Theorem 3, A^{-1} is bounded.

This means that states $(I, 1_n)$, $(I, 2_n)$, $(II, 1_c)$, and $(II, 2_c)$ cannot occur for a closed operator, leaving us with eight states: $(I, 1_c)$, $(I, 2_c)$, $(II, 1_n)$, $(II, 2_n)$, and all four states associated with III. States $(III, 1_c)$ and $(III, 2_n)$ are possible for closed operators even though they are not included in (3.3). State $(III, 1_c)$ is obtained when we close the operator in Example 8, Section 1, and state $(III, 2_n)$ occurs in Exercises 3.1 and 3.3.

There are important classes of operators on infinite-dimensional spaces for which much more can be said. We must postpone that discussion until we have studied adjoints in the next section. For an operator on $E_n^{(c)}$, however, the situation is exceedingly simple. First of all, R_A is always a closed set. If the homogeneous equation has only the trivial solution, then $R_A = H$ and $Au = f$ has the solution $u = A^{-1}f$, where A^{-1} is itself a linear operator on $E_n^{(c)}$ and hence bounded. If the homogeneous equation has a nontrivial solution, we know that $R_A \neq H$ (in fact, the codimension of R_A is equal to the dimension of N_A). Thus an operator on $E_n^{(c)}$ can only be in state $(I, 1_c)$ or in state $(III, 2_c)$; see Theorem 3, Section 5.

Let us now look at the differentiation operator d/dx on various domains. We shall be particularly interested in the closed operators associated with various boundary conditions.

Example 1. (a) Let D_1 be the subset of $L_2(0, 1)$ consisting of continuously differentiable functions $u(x)$ on $0 \leqslant x \leqslant 1$ with $u(0) = 0$. Define $A_1 u = du/dx$ for $u \in D_1$. This operator is closable, and its *closure* T_1 is an extension of A_1 with domain $D_{T_1} \supset D_{A_1}$. We claim that the boundary condition $u(0) = 0$ survives in D_{T_1}. In fact, suppose $\{u_n\}$ is in D_{A_1}, $u_n \to u$, and $u_n' \to f$; then by definition of the closure we have $u \in D_{T_1}$ and $T_1 u = f$. We now show that $u(0) = 0$; indeed, since $u_n(0) = 0$, we have $u_n(x) = \int_0^x u_n'(y)\,dy$, so that by taking limits $u(x) = \int_0^x f(y)\,dy$ and $u(0) = 0$. It is also clear that R_{T_1} consists of all elements in L_2. In fact, T_1^{-1} is closed because T_1 is; furthermore, it follows from the inversion formula $u(x) = \int_0^x f(y)\,dy$ and Example 7, Section 1, that T_1^{-1} is bounded, so that Theorem 5 guarantees that R_{T_1} is closed. Since R_{A_1} is already dense in L_2, R_{T_1} must also be dense in L_2 and $R_{T_1} = L_2$. Thus T_1 is in state $(I, 1_c)$, that is, T_1 is regular.

(b) If instead of the boundary condition in (a) we impose $u(0) = au(1)$, where a is a fixed real number, we have a new operator A_2 which is also closable and whose closure is denoted by T_2. The boundary condition survives in the process of closure. If $a \neq 1$, the one and only solution of $T_2 u = f$ is $u = \int_0^1 g(x, y) f(y)\,dy$, where $g = a/(1 - a)$ for $x < y$, and $g = 1/(1 - a)$ for $x > y$. Obviously g is just Green's function for the problem. We see that $R_{T_2} = L_2$ and that T_2 is regular. If $a = 1$, then $u = C$ is a solution of the homogeneous equation and T_2 is in state $(III, 2_c)$.

(c) If instead of the boundary condition in (a) we impose the two boundary conditions $u(0) = 0$ *and* $u(1) = 0$, we obtain an operator A_3 which is closable, with closure T_3. Again, boundary conditions survive the process of closure, but now $R_{T_3} \neq L_2$. The unique solution of $T_3 u = f$ is as in (a), but a solution is now possible only if f satisfies the solvability condition $\int_0^1 f(x)\,dx = 0$. Thus T_3 is in state $(I, 2_c)$.

(d) If we impose the boundary condition $u'(0) = 0$ instead of $u(0) = 0$, we discover a new phenomenon. The operator A_4 is closable, but the boundary condition $u'(0) = 0$ *disappears* in the process of closure, so that the closure T_4 of A_4 is just the same as the closure of the differentiation operator without boundary conditions. Thus T_4 is in state $(III, 1_c)$.

We have the following general principle (stated somewhat loosely) for linear differential operators of order n: boundary conditions involving derivatives of order less than n survive closure, but those involving derivatives of order greater than or equal to n disappear in closure.

Example 2. Let Ω be an open set in R_n, and let Γ be its boundary. We are interested in the following boundary value problem:

$$-\Delta u = -\left(\frac{\partial^2 u}{\partial x_1^2} + \cdots + \frac{\partial^2 u}{\partial x_n^2} \right) = f(x), \qquad x = (x_1, \ldots, x_n) \in \Omega,$$

$$u|_{x \text{ on } \Gamma} = 0.$$

This is just the Dirichlet problem for Laplace's equation (actually we have, for convenience, written the equation with the negative Laplacian on the left side; obviously the change $f \to -f$ restores the equation to the form $\Delta u = f$). To treat this problem in a suitable functional framework, we first define an operator $A = -\Delta$ on the domain D_A of functions having continuous partial derivatives of order 2 on $\overline{\Omega}$, and vanishing on Γ. The closure T of this operator is defined on a domain $D_T \supset D_A$. Although it is possible to develop the theory in this manner, the modern theory of partial differential equations uses a different approach (see Section 4, Chapter 8).

Exercises

3.1. Let $\{e_i\}$ be an orthonormal basis for $L_2(a,b)$, and let the operator A be defined by

$$Ae_1 = 0; \qquad Ae_k = \frac{e_k}{k}, \quad k \geqslant 2.$$

Show that A is in state $(\text{III}, 2_n)$.

3.2. Let A be a linear transformation from E_n to E_m. Describe all possible states of A.

3.3. On $L_2(0,\pi)$ consider the integral operator K with kernel

$$k(x,y) = \sum_{n=1}^{\infty} \frac{\cos nx \cos ny}{n^2}.$$

Show that K is in state $(\text{III}, 2_n)$.

4. ADJOINT OPERATORS

Consider first a bounded operator A defined on the whole Hilbert space H. With v a fixed element in H, we can regard $\langle Au, v \rangle$ as a complex number which varies with u; it is clear that $\langle Au, v \rangle$ is a bounded linear functional in u, and so, by the Riesz representation theorem of Chapter 4, there must exist a well-defined element g in H such that

$$(4.1) \qquad \langle Au, v \rangle = \langle u, g \rangle \qquad \text{for all } u \text{ in } H.$$

The element g depends on v, and we write $g = A^*v$, where A^* is seen to be a linear operator defined on the whole of H. The operator A^* is known as the *adjoint* of A; A^* is bounded and $\|A^*\| = \|A\|$ (see Theorem 1, Section 8). If u, v are any two elements in H, we have

$$(4.2) \qquad \langle Au, v \rangle = \langle u, A^*v \rangle.$$

The matrix representation $[a_{ij}^*]$ of A^* in an orthonormal basis $\{e_i\}$ is easily calculated in terms of the matrix $[a_{ij}]$ of A in the same basis. From (1.4) we have

$$a_{ij} = \langle Ae_j, e_i \rangle = \langle e_j, A^*e_i \rangle = \overline{\langle A^*e_i, e_j \rangle},$$

so that

(4.3) $a_{ij} = \overline{a_{ji}^*}$ (or $a_{ij}^* = \bar{a}_{ji}$).

If we are dealing with a real Hilbert space, the matrix entries are real and the complex conjugate may be omitted in (4.3). The matrix representation is a satisfactory way to describe adjoints in finite-dimensional spaces and in some infinite-dimensional problems like Example 9, Section 1, where an easy calculation shows that A^* is the *left* shift defined by $A^*e_i = e_{i-1}, i > 1$; $A^*e_1 = 0$. In other problems a more striking form for the adjoint is found by using (4.2) directly. Turning to Example 6, Section 1, we have, for any u, z,

$$\langle Ku, z \rangle = \int_a^b dx \, \bar{z}(x) \int_a^b k(x,y) u(y) \, dy = \int_a^b u(y) \, dy \int_a^b k(x,y) \bar{z}(x) \, dx$$

$$= \int_a^b u(x) \, dx \int_a^b k(y,x) \bar{z}(y) \, dy,$$

where the second equality results from a change in the order of integration, whereas the last equality is merely a convenient relabeling of variables enabling us to identify the last term as $\int_a^b u(x) \overline{K^*z} \, dx$, where

(4.4) $K^*z = \int_a^b \bar{k}(y,x) z(y) \, dy.$

Thus K^* is also an integral operator on $L_2(a,b)$ with kernel

(4.5) $k^*(x,y) = \bar{k}(y,x),$

a formula which bears comparison with (4.3) for matrices.

As a specific illustration of these formulas, consider the operator A of Example 7, Section 1, which can be regarded as an integral operator on $L_2(0,1)$ with kernel $k(x,y) = H(x-y)$. Thus $k^*(x,y) = H(y-x)$, and (4.4) becomes

$$A^*z = \int_0^1 H(y-x) z(y) \, dy = \int_x^1 z(y) \, dy.$$

We now turn to unbounded operators and confine ourselves to closed operators defined on domains dense in H. Let A on D_A be such an operator, and, as before, let v be a fixed element in H. As u varies over D_A, $\langle Au, v \rangle$ takes on various numerical values, so that $\langle Au, v \rangle$ is a functional (possibly unbounded) in u. Although no longer guaranteed by the Riesz theorem, it may happen that for some elements v we can write

$$(4.6) \qquad \langle Au, v \rangle = \langle u, g \rangle \qquad \text{for all } u \text{ in } D_A.$$

Any pair (v, g) for which (4.6) holds for all u in D_A will be called an *admissible pair*. The pair $(0, 0)$ is admissible, but it is not clear that there are other admissible pairs. It is true, however, that g is unambiguously determined from v: if (4.6) were true for the same v and for different g_1 and g_2, we would have $\langle u, g_1 - g_2 \rangle = 0$ for all u in D_A, and since D_A is dense in H, $g_1 - g_2 = 0$. Thus, for a given v, there is at most one g for which (4.6) holds. Considering *all* admissible pairs, we see that g has a functional relation to v, and therefore we can write $g = A^* v$. The operator A^* is the *adjoint* of A; it has a well-defined domain D_{A^*} which includes at least the element 0, and $A^* 0 = 0$. The operator A^* is linear, and we have

$$(4.7) \qquad \langle Au, v \rangle = \langle u, A^* v \rangle \qquad \text{for all } u \text{ in } D_A, \quad v \text{ in } D_{A^*}.$$

It can also be shown that A^* is closed and that its domain D_{A^*} is dense in H.

We say that A on D_A is *self-adjoint* if $D_{A^*} = D_A$ and $A^* = A$ on their common domain of definition. If A is bounded, $D_{A^*} = D_A = H$, and therefore A is self-adjoint if and only if

$$(4.8) \qquad \langle Au, v \rangle = \langle u, Av \rangle \qquad \text{for all } u, v \in H.$$

An operator is *symmetric* on its domain of definition if

$$(4.9) \qquad \langle Au, v \rangle = \langle u, Av \rangle \qquad \text{for all } u, v \in D_A.$$

Every self-adjoint operator is symmetric and every *bounded* symmetric operator is self-adjoint, but there are unbounded symmetric operators which are *not* self-adjoint. (See Example 3 below.) A bounded operator will be self-adjoint if its matrix representation in an orthonormal basis satisfies $a_{ij} = \bar{a}_{ji}$. For a bounded integral operator K the condition of self-adjointness is that the kernel satisfies $k(x, y) = \bar{k}(y, x)$.

If α is a scalar, we define αA and $A + \alpha I$ in the obvious way: $(\alpha A)(u) = \alpha Au$, $(A + \alpha I)(u) = Au + \alpha u$; both operators are defined on the same domain D_A as A. We then have

(4.10) $$(\alpha A)^* = \bar{\alpha} A^*, \qquad (A + \alpha I)^* = A^* + \bar{\alpha} I,$$

with $(\alpha A)^*$ and $(A + \alpha I)^*$ both having the same domain D_{A^*} as A^*.

Properties of the Adjoint

1. If A is closed and D_A dense in H, A^* is closed and D_{A^*} is dense in H. Moreover, $A^{**} = A$.

2. If A is closable, both A and its closure have the same adjoint (therefore, if A is closable on a dense domain D_A, we can find the closure of A by calculating A^{**} instead).

Example 1. We look at the closed operator T_1 of Example 1(a), Section 3, where $T_1 u = du/dx$ and D_{T_1} is the linear manifold in $L_2^{(c)}(0, 1)$ consisting of absolutely continuous functions $u(x)$ with $u'(x)$ in $L_2^{(c)}(0, 1)$ and $u(0) = 0$. The important thing is that D_{T_1} contains all sufficiently smooth functions with $u(0) = 0$. We want to calculate T_1^*. We must find *all* admissible pairs (v, g) satisfying

$$\langle T_1 u, v \rangle = \langle u, g \rangle \qquad \text{for all } u \text{ in } D_{T_1},$$

that is,

(4.11) $$\int_0^1 \bar{v} u' \, dx = \int_0^1 u \bar{g} \, dx \qquad \text{for all } u \text{ in } D_{T_1}.$$

We give two methods (one heuristic, the other precise) for finding the adjoint. The types of argument are similar to those used in variational approaches (see Section 4, Chapter 8).

Heuristic Method

We integrate by parts on the left side of (4.11) (this assumes v absolutely continuous with v' in L_2) to obtain, using the fact that $u(0) = 0$,

(4.12) $$\int_0^1 (\bar{v}' + \bar{g}) u \, dx - \bar{v}(1) u(1) = 0 \qquad \text{for all } u \text{ in } D_{T_1}.$$

Let us examine the consequences of this equation as u varies in D_{T_1}. If we consider first the subset M of functions on D_{T_1} that also vanish at $x = 1$, we see that the integral in (4.12) vanishes for $u \in M$. This implies that

$\bar{v}' + \bar{g} = 0$, so that the integral term must vanish identically; if we then turn to the subset of functions u in D_{T_1} for which $u(1) \neq 0$, we find that $v(1) = 0$. Thus (v, g) is admissible if $g = -v'$ *and* $v(1) = 0$ and v is absolutely continuous with v' in L_2. We therefore suspect that

(4.13) $\qquad T_1^* v = -\dfrac{dv}{dx}, \qquad D_{T_1^*}: v(1) = 0, \quad v \text{ a.c., } v' \in L_2.$

Of course it is not clear that we have found *all* admissible pairs in this way.

Precise Method

We start from (4.11), but do not assume anything about v. We integrate by parts on the *right* side of (4.11); with $G(x) = -\int_x^1 g(y)\,dy$, we see that G' exists, $G' = g$, and $G(1) = 0$, so that, since $u(0) = 0$,

(4.14) $\qquad \displaystyle\int_0^1 u' [\bar{v} + \bar{G}]\, dx = 0 \qquad \text{for all } u \text{ in } D_{T_1}.$

It is perhaps obvious that this implies $\bar{v} + \bar{G} = 0$, but we can prove it as follows. Let $u = \int_0^x [v(y) + G(y)]\,dy$; then $u \in D_{T_1}$ and $u' = v + G$; substituting this particular u in (4.14), we see that $\int_0^1 |v + G|^2\,dx = 0$ and therefore

$$v + G = 0 \qquad \text{or} \qquad v = \int_x^1 g(y)\,dy.$$

Thus v is necessarily differentiable with $v' = -g$ and $v(1) = 0$, showing that (4.13) does in fact define the adjoint T_1^*. In other examples we shall be satisfied with the heuristic method for determining the adjoint. Note that T_1 is not self-adjoint for two reasons: first, the formal differentiation in T_1^* is $-(d/dx)$ rather than d/dx, and, second, the boundary condition on D_{T_1} is transferred to the other end of the interval in $D_{T_1^*}$. For the operator $S_1 = -i(d/dx) = -iT_1$, defined on the same domain as T_1, we obtain, using (4.10),

$$S_1^* v = -i\dfrac{dv}{dx}, \qquad D_{S_1^*}: v(1) = 0, \quad v \text{ a.c., } v' \in L_2.$$

Since the boundary condition is not the same as for S_1, we still find that S_1 is not self-adjoint, although at least S_1^* consists of the same formal differentiation as S_1.

Example 2. In the calculation of the adjoint for the closed operator T_2 of Example 1(b), Section 3, with the boundary condition $u(0) = au(1)$, we see that (4.11) is to hold for all u in D_{T_2}. From this we conclude that

$$\int_0^1 (\bar{v}' + \bar{g})u\,dx - u(1)[\bar{v}(1) - a\bar{v}(0)] = 0 \qquad \text{for all } u \text{ in } D_{T_2}.$$

Therefore admissible pairs (v, g) satisfy

$$g = -v', \qquad v(1) = av(0).$$

Thus

$$T_2^* v = -\frac{dv}{dx}, \qquad D_{T_2^*} \colon v(1) = av(0), \quad v \text{ a.c., } v' \in L_2.$$

If we let $S_2 = -iT_2$, then

$$S_2^* v = -i\frac{dv}{dx}, \qquad D_{S_2^*} \colon v(1) = av(0), \quad v \text{ a.c., } v' \in L_2.$$

Therefore S_2 is *self-adjoint* if and only if $a = 1$ or $a = -1$.

Example 3.   Consider the adjoint for the operator T_3 of Example 1(c), Section 3, where "too many" boundary conditions have been imposed. We still have (4.11), but now for $u \in D_{T_3}$. Integration by parts yields

$$\int_0^1 (\bar{v}' + \bar{g})u\,dx = 0 \qquad \text{for all } u \in D_{T_3},$$

from which we conclude that $v' = -g$, but that no boundary conditions are needed on v. Thus

$$T_3^* v = -\frac{dv}{dx}, \qquad D_{T_3^*} \colon v \text{ a.c., } v' \in L_2.$$

For the operator $S_3 = -iT_3$ we have

$$S_3 u = -i\frac{du}{dx}, \qquad D_{S_3} \colon u(0) = u(1) = 0, \quad u \text{ a.c., } u' \in L_2,$$

$$S_3^* v = -i\frac{dv}{dx}, \qquad D_{S_3^*} \colon v \text{ a.c., } v' \in L_2.$$

The operator S_3 is an example of a symmetric operator which is not self-adjoint. It is clear that, if $u, v \in D_{S_3}$, then $\langle S_3 u, v \rangle = \langle u, S_3 v \rangle$, but $S_3^* \neq S_3$. In fact, S_3^* is an extension of S_3. For any symmetric operator A it is always true that A^* is an extension of A.

Exercises

4.1. If A, B are bounded operators on H, show that

$$(AB)^* = B^*A^*.$$

Even if A, B are both self-adjoint, the product AB may not be.

4.2. Let A by a symmetric operator on $E_n^{(c)}$, and let $\{h_k\}$ be a *nonorthogonal* basis in which A has the matrix $[a_{ij}]$. What is the matrix of A^* in the same basis?

4.3. Let $H = L_2^{(c)}(0, 1)$, and let D be the linear manifold of elements $u(x)$ with u' absolutely continuous and u'' in L_2. Consider the operator $A = -(d^2/dx^2)$ on the domain D_A of functions in D satisfying the boundary conditions $u(0) = u(1) = u'(0) = 0$. Is A symmetric? What is the adjoint of A?

4.4. Let $a(x)$ be continuous but *not* differentiable on $0 \leqslant x \leqslant 1$. Let $A = a(x)(d/dx)$ be an operator defined on the domain D_A consisting of elements $u(x)$ in $L_2^{(c)}(0, 1)$ that are absolutely continuous with u' in L_2. Find the adjoint operator A^*.

5. SOLVABILITY CONDITIONS

Let A be a linear operator on a domain D_A dense in H. We want to characterize the range of A: for what forcing terms f can we solve the inhomogeneous equation

$$(5.1) \qquad Au = f, \quad u \in D_A?$$

The statement $u \in D_A$ is solely for emphasis; it is already incorporated in the equation $Au = f$ since A is defined only on D_A. It is convenient to study at the same time the adjoint homogeneous equation

$$(5.2) \qquad A^*v = 0, \quad v \in D_{A^*},$$

where again the statement $v \in D_{A^*}$ is for emphasis only. Observe that (5.2) always has the solution $v = 0$ and perhaps some others.

If u is a solution of (5.1) and v a solution of (5.2), we have, by taking inner products,

$$\langle Au, v \rangle = \langle f, v \rangle.$$

From (4.7) and (5.2) it follows that $\langle Au, v \rangle = \langle u, A^*v \rangle = 0$, so that a *necessary* condition for (5.1) to have solution(s) is that

$$(5.3) \qquad\qquad \langle f, v \rangle = 0 \qquad \text{for all } v \text{ satisfying (5.2).}$$

Of course the *solvability condition* (5.3) has content only if (5.2) has nontrivial solutions. Some questions immediately come to mind. Is (5.3) sufficient for solvability of (5.1)? How many nontrivial solutions does (5.2) have? If (5.1) has solutions, how many are there?

We shall establish the sufficiency of (5.3) for a particular class of operators, but first we prove a general result which characterizes $(R_A)^\perp$ instead of R_A.

Theorem. The orthogonal complement of the range of A is the null space of the adjoint:

$$(5.4) \qquad\qquad (R_A)^\perp = N_{A^*}.$$

Proof. (a) Let $z \in N_{A^*}$, so that $A^*z = 0$. Then $\langle u, A^*z \rangle = 0$ for every u in H, and hence surely for every u in D_A; hence $\langle Au, z \rangle = 0$ for every $u \in D_A$, so that z is in $(R_A)^\perp$.

(b) Let $z \in R_A^\perp$; then $\langle Au, z \rangle = 0 = \langle u, 0 \rangle$ for every u in D_A. Thus $(z, 0)$ is an admissible pair, which means that $z \in D_{A^*}$ and $A^*z = 0$. Therefore $z \in N_{A^*}$.

Combining (a) and (b), we see that the theorem has been proved.

By taking orthogonal complements we find that $(R_A)^{\perp\perp} = (N_{A^*})^\perp$. But $(R_A)^{\perp\perp} = \bar{R}_A$, so that

$$(5.5) \qquad\qquad \bar{R}_A = (N_{A^*})^\perp.$$

This says that an element f is in the closure of R_A if and only if it is orthogonal to all solutions of (5.2). It does not tell us quite what we want, but it is the best we can say in general. If, however, R_A is a closed set, $R_A = \bar{R}_A$ and we have the following necessary and sufficient solvability condition.

Theorem 1 (Solvability for operators with closed range). If A has a closed range, $Au = f$ has solution(s) if and only if f is orthogonal to every solution of the adjoint homogeneous equation, that is,

$$(5.6) \qquad\qquad R_A = (N_{A^*})^\perp.$$

When does A have a closed range? Clearly, if A is an operator on E_n or, more generally, an operator with a finite-dimensional range, A obviously has a closed range. Theorem 5, Section 3, gives an important criterion for an operator to have a closed range. This theorem can be generalized as follows.

Theorem 2. Let A be closed, and let A be bounded away from 0 on $N_A^\perp \cap D_A$; then R_A is closed.

Remark. The hypothesis tells us there exists $c > 0$ such that $\|Au\| \geqslant c\|u\|$ for all $u \in N_A^\perp \cap D_A$. This means that the operator A *restricted to* $N_A^\perp \cap D_A$ has a bounded inverse; in particular, if N_A is $\{0\}$, the criterion is just that A on D_A has a bounded inverse.

Proof. Let $f_n \in R_A$ with $f_n \to f$. Let $Av_n = f_n$, and define $u_n = v_n - Pv_n$, where Pv_n is the projection of v_n on N_A; then u_n is in $D_A \cap N_A^\perp$ and $Au_n = f_n$. Now $\|u_n - u_m\| \leqslant (1/c)\|Au_n - Au_m\| = (1/c)\|f_n - f_m\|$, so that $\{u_n\}$ is a Cauchy sequence in D_A and therefore $u_n \to u$, u in H. Therefore we have $Au_n \to f$, $u_n \to u$; by the definition of a closed operator this means that $u \in D_A$, $Au = f$, so that R_A is closed.

We are now in a position to prove the complete Fredholm alternative for various classes of operators. We shall do so for operators A on E_n; the proofs given are not the simplest possible but are the ones most easily extended to wider classes of operators (see Section 7). The theorem deals with the relationship among the three equations

$$Au = 0, \qquad Au = f, \qquad A^*v = 0.$$

Theorem 3 (Fredholm alternative). Let A be an operator on $E_n = H$. Then *either* of the following holds:

(a) $Au = 0$ has only the zero solution, in which case $A^*v = 0$ has only the zero solution and $Au = f$ has precisely one solution for each f in H.

(b) N_A has dimension k, in which case N_{A^*} also has dimension k and $Au = f$ has solutions if and only if f is orthogonal to N_{A^*}.

Remark. We can elaborate a little on part (b). Let $\{u_1, \ldots, u_k\}$ be a basis for N_A, and let $\{v_1, \ldots, v_k\}$ be a basis for N_{A^*}. Then the solvability conditions for the equation $Au = f$ consist of the k equations $\langle f, v_1 \rangle = \cdots = \langle f, v_k \rangle = 0$, which guarantee that f is orthogonal to N_{A^*}. *If these conditions are satisfied, $Au = f$ has many solutions, all of which are of the* form

$$u = \tilde{u} + c_1 u_1 + \cdots + c_k u_k,$$

where $c_1, \ldots, c_k$ are arbitrary constants and $\tilde{u}$ is any particular solution of $Au = f$.

Proof of Theorem 3. The proof consists of three steps.

1. $Au = f$ has solution(s) if and only if $\langle f, v \rangle = 0$ for every solution of $A^*v = 0$, that is, $R_A = (N_{A^*})^\perp$. This is just Theorem 1, which requires only that the range be closed, a trivial fact for operators on E_n.

2. If $R_A = H$, then $N_A = \{0\}$ and vice versa. (We already know from step 1 that $N_{A^*} = \{0\}$ is necessary and sufficient for $R_A = H$. Thus, if we manage to prove step 2, we will also have shown that

$$N_A = \{0\} \Leftrightarrow N_{A^*} = \{0\}.$$

We now give the proof.

(a) Suppose $R_A = H$. We must show that $N_A = \{0\}$. Assuming the contrary, there exists $u_1 \neq 0$ with $Au_1 = 0$. Consider successively the equations $Au_2 = u_1, Au_3 = u_2, \ldots, Au_p = u_{p-1}, \ldots$. By hypothesis each equation is solvable, and $u_p \neq 0$, $A^{p-1} u_p = u_1$, $A^p u_p = 0$. Thus u_p belongs to the null space N_p of A^p. Obviously $N_p \supset N_{p-1}$, and the inclusion is strict since u_p belongs to N_p but not to N_{p-1}. We are thus led to an infinite sequence of spaces $N_1, N_2, \ldots$ of increasing dimension, a circumstance which violates the fact that H is finite-dimensional. Therefore we must have $N_A = \{0\}$.

(b) Assume $N_A = \{0\}$. We must show that $R_A = H$. The hypothesis implies that $N_A^\perp = H$. Reversing the roles of A and A^* in step 1 (which is permissible since $A^{**} = A$), we have $N_A^\perp = R_{A^*} = H$. Applying part (a) above to A^*, we find that $N_{A^*} = \{0\}$.

3. N_A and N_{A^*} have the same dimension. By step 2 we know that, if one of these has dimension 0, so does the other. Thus we need to consider only the case where both dimensions are positive. We shall proceed by contradiction. Let v, v^* be the respective dimensions, and assume, without loss of generality, that $v^* > v$. Choose orthonormal bases $\{u_1, \ldots, u_v\}$ and $\{v_1, \ldots, v_{v^*}\}$ for N_A and N_{A^*}, respectively. Consider the operator B defined by

$$Bu = Au - \sum_{j=1}^{v} \langle u_j, u \rangle v_j.$$

Then $\langle Bu, v_k \rangle = -\langle u_k, u \rangle$ for $k \leqslant \nu$, and $\langle Bu, v_k \rangle = 0$ for $k > \nu$. If $Bu = 0$, then $\langle u_k, u \rangle = 0$ for $k \leqslant \nu$ and therefore $Au = 0$ for $u \in N_A$; since u is also orthogonal to a basis for N_A, it follows that $u = 0$. Thus $Bu = 0$ implies that $u = 0$, but by step 2 we can then solve $Bu = v_{\nu+1}$, which yields $\|v_{\nu+1}\|^2 = \langle Bu, v_{\nu+1} \rangle = 0$, a contradiction.

These three steps together prove the Fredholm alternative, already stated without proof in Chapter 3.

Remarks

1. Theorem 3 can be phrased in the language of systems of algebraic equations. We consider the relationship among these three systems:

$$(5.7) \qquad \sum_{j=1}^{n} a_{ij} u_j = 0, \qquad i = 1, \ldots, n,$$

$$(5.8) \qquad \sum_{j=1}^{n} a_{ij} u_j = f_i, \qquad i = 1, \ldots, n,$$

$$(5.9) \qquad \sum_{j=1}^{n} \bar{a}_{ji} v_j = 0, \qquad i = 1, \ldots, n.$$

The theorem states that *either* of the following holds:

(a) The homogeneous system (5.7) has only the trivial solution $u_1 = \cdots = u_n = 0$, in which case (5.9) has only the trivial solution and (5.8) has precisely one solution.

(b) The homogeneous system (5.7) has k independent nontrivial solutions, say $(h_1^{(1)}, \ldots, h_n^{(1)}), \ldots, (h_1^{(k)}, \ldots, h_n^{(k)})$, in which case (5.9) also has k independent nontrivial solutions, say $(p_1^{(1)}, \ldots, p_n^{(1)}), \ldots, (p_1^{(k)}, \ldots, p_n^{(k)})$, and (5.8) has solutions if and only if $\sum_{i=1}^{n} f_i \bar{p}_i^{(j)} = 0, j = 1, \ldots, k$. If these solvability conditions are satisfied, the general solution of (5.8) is of the form $u = u_p + \sum_{i=1}^{k} c_i h^{(i)}$, where u_p is any particular solution of (5.8).

As we know from elementary considerations, alternative (a) occurs if the determinant of $[a_{ij}]$ does *not* vanish. Alternative (b) occurs if the determinant vanishes. The number k of independent solutions of (5.9) is then $n - r$, where r is the *rank* of $[a_{ij}]$.

2. In Section 7 the theorem will be extended to operators A of the form $K - \lambda I$, where K is compact and $\lambda \neq 0$.

6. THE SPECTRUM OF AN OPERATOR

The principal purpose in studying the spectrum is to make precise and to generalize, so far as possible, the method of eigenfunction expansion. The program can be successfully completed for self-adjoint operators, culminating in the spectral theorem for such operators. In applications it is not enough, however, to stop at the spectral theorem; the important thing is to construct explicitly the eigenfunctions (or the generalization referred to earlier). Although we give some restricted forms of the spectral theorem, we shall devote greater energy in Chapter 7 to methods for constructing the spectrum and the appropriate series or integral expansions needed in specific problems.

Let A be a closed operator whose domain is a linear manifold D_A dense in H. Our attention will be focused on the operator $A - \lambda I$, where λ is an arbitrary *complex* number and I is the identity operator. For any λ we may and shall take the domain of $A - \lambda I$ to be the same as that of A. This domain will be denoted simply by D. The range of $A - \lambda I$ will, however, usually vary with λ, and we shall therefore denote the range by $R(\lambda)$. Since $A - \lambda I$ is closed and defined on a dense set, its adjoint $A^* - \bar{\lambda} I$ is defined on a domain D^* which is dense in H. We now classify all points λ in the complex plane according to whether or not $A - \lambda I$ is regular.

Definition. The values of λ for which $A - \lambda I$ is regular, that is, $A - \lambda I$ has a bounded inverse and $R(\lambda) = H$, form the *resolvent set* of A. All other values of λ comprise the *spectrum* of A. It can be shown that the resolvent set is *open* and therefore the spectrum is *closed*.

Perhaps it is time to revise our earlier attitude, which implied that being in the resolvent set was marvelous and being in the spectrum was bad. If there were no spectrum at all, we could not develop anything resembling an eigenfunction expansion!

There are many competing classification schemes for further division of the spectrum, all of them based on the states of closed operators as given in Section 3. The one used here is a slight modification of the scheme proposed by Halmos and seems intuitively the simplest. Its disadvantage is that it divides the spectrum into sets which are *not* necessarily disjoint.

Definition. Values of λ in the spectrum are classified as follows:

1. λ belongs to the *point spectrum* if $A - \lambda I$ is in state III [that is, $(A - \lambda I)u = 0$ has nontrivial solutions; in other words, λ is an *eigenvalue* of A and any corresponding nontrivial solution u is an *eigenvector* of A

belonging to λ; the (geometric) multiplicity of λ is the dimension of the null space $N(\lambda)$ of $A - \lambda I$].

2. λ belongs to the *continuous spectrum* if $A - \lambda I$ is in state II [that is, $(A - \lambda I)^{-1}$ exists but is unbounded].

3. λ belongs to the *compression spectrum* if $A - \lambda I$ is in state 2 [that is, $\overline{R}(\lambda) \neq H$; the range has been compressed; thus $R(\lambda)$ has a nonzero dimension, the *deficiency* of λ, which is just the codimension of $R(\lambda)$].

The point spectrum and the continuous spectrum are disjoint sets in the λ plane. Their union is the *approximate point spectrum*. For λ to be in the approximate spectrum, $A - \lambda I$ must fail to have a bounded inverse or, equivalently, $A - \lambda I$ must not be bounded away from 0. Thus λ is in the approximate spectrum if and only if there is a sequence of unit elements $\{u_n\}$ such that $Au_n - \lambda u_n \to 0$. (This criterion is automatically satisfied if there exists $u \neq 0$ such that $Au - \lambda u = 0$; just choose $u_n = u / \|u\|$.) The approximate point spectrum often overlaps the compression spectrum (see Figure 3.1). In fact, for transformations on $E_n^{(c)}$ we have shown (Theorem 3, Section 5) that, if λ is an eigenvalue of multiplicity m, it also has deficiency m. Thus for a transformation on $E_n^{(c)}$ either λ is in the resolvent set or it belongs *simultaneously* to the point and compression spectrum. On an arbitrary Hilbert space, we can prove the following theorem.

Theorem 1. If λ has deficiency m in the compression spectrum of A, then $\bar{\lambda}$ is an eigenvalue of A^* with multiplicity m, and conversely.

Proof. We have shown [see (5.4)] that the orthogonal complement of the range is equal to the null space of the adjoint. Apply this theorem to the operator $A - \lambda I$.

Theorem 2. If A is symmetric, the following hold:

(a) $\langle Au, u \rangle$ is real for all $u \in D$.

(b) The approximate spectrum (which consists of the eigenvalues and of the continuous spectrum) is real.

(c) Eigenvectors corresponding to different eigenvalues are orthogonal.

Proof. (a) For u in D, $\langle Au, u \rangle = \langle u, Au \rangle = \overline{\langle Au, u \rangle}$, the first equality from symmetry and the second from the definition of inner product. Thus $\langle Au, u \rangle$ is real.

(b) If $Au = \lambda u$, then $\langle Au, u \rangle = \lambda \|u\|^2$, and for $\|u\| \neq 0$ we have λ real. With $\lambda = \xi + i\eta$, ξ and η real, we find that

$$\|(A - \lambda I)u\|^2 = \langle (Au - \xi u) - i\eta u, (Au - \xi u) - i\eta u \rangle = \|Au - \xi u\|^2 + \eta^2 \|u\|^2,$$

where the symmetry of $A - \xi I$ has been used to eliminate cross terms. Thus $\|(A - \lambda I)u\|^2 \geqslant \eta^2 \|u\|^2$, and $A - \lambda I$ is bounded away from 0 for $\eta \neq 0$. By Theorem 2, Section 3, $A - \lambda I$ has a bounded inverse and the approximate spectrum is therefore confined to the real axis.

(c) Let $\lambda \neq \mu$ and $Au = \lambda u$, $Av = \mu v$. Then $\langle Au, v \rangle - \langle u, Av \rangle = (\lambda - \bar{\mu})\langle u, v \rangle$. The left side vanishes by symmetry. Since μ is real, $\lambda \neq \bar{\mu}$ and therefore $\langle u, v \rangle = 0$.

For a self-adjoint operator we can say more.

Theorem 3. If A is self-adjoint, every λ with Im $\lambda \neq 0$ is in the resolvent set. The point spectrum (necessarily real) coincides with the compression spectrum, and the multiplicity of any eigenvalue is equal to its deficiency.

Proof. Since $A = A^*$, A is certainly symmetric. By Theorem 2, λ with Im$\lambda \neq 0$ is either in the resolvent set or the compression spectrum. The latter would imply that A^* (hence A) has $\bar{\lambda}$ as an eigenvalue, which is a contradiction. The whole spectrum is therefore real. By Theorem 1 the point and compression spectra coincide, and the multiplicity of an eigenvalue is equal to its deficiency. We may therefore omit reference to the compression spectrum.

In what follows, we refer to our standard examples, Examples 1 through 10 of Section 1.

Example 1: See Example 1, Section 1. (a) A is the zero operator. The only point in the spectrum is $\lambda = 0$, which is an eigenvalue of infinite multiplicity.

(b) A is the identity. The only point in the spectrum is $\lambda = 1$, an eigenvalue of infinite multiplicity.

Example 2: See Example 2, Section 1. Let P be the operator defined by $v = Pu$, where Pu is the (orthogonal) projection on a closed manifold M in H. The operator P is easily seen to be self-adjoint. The spectrum of P consists of the two eigenvalues $\lambda = 0$ and $\lambda = 1$. The null space of $\lambda = 1$ consists of all vectors in M, and the null space of $\lambda = 0$ consists of all vectors in $M^{\perp}$. If λ is not an eigenvalue, the solution of $(P - \lambda I)u = f$ is $Pf/(1 - \lambda) - (f - Pf)/\lambda$, so that $P - \lambda I$ has a bounded inverse and λ is in the resolvent set.

Example 3: See Example 3(b), Section 1. With $\theta \neq k\pi$, the eigenvalue problem leads to the equations

$$(\cos\theta - \lambda)u_1 - (\sin\theta)u_2 = 0,$$
$$(\sin\theta)u_1 + (\cos\theta - \lambda)u_2 = 0,$$

so that nontrivial solutions are possible if and only if

$$(\cos\theta-\lambda)^2+\sin^2\theta=0 \qquad \text{or} \qquad \lambda=e^{i\theta}, \quad e^{-i\theta}.$$

When this operator is viewed in $E_2^{(r)}$, its spectrum is empty; on $E_2^{(c)}$, however, there are two eigenvalues, $\lambda_1=e^{i\theta}$ and $\lambda_2=e^{-i\theta}$, with respective eigenvectors e_1-ie_2 and e_1+ie_2.

Example 4: See Example 3(c), Section 1. This example has many striking features. The eigenvalue problem reduces to the pair of equations

$$u_2=\lambda u_1, \qquad 0=\lambda u_2,$$

from which we conclude that $\lambda=0$ is the only eigenvalue; the corresponding null space has dimension 1 and is spanned by the eigenvector whose coordinates are $(1,0)$. The range $R(0)$ consists of all vectors with vanishing second coordinates, so that $R(0)=N(0)$, a surprising result [on E_n we always have $R_A=(N_{A*})^{\perp}$, and if A is self-adjoint, $R_A=N_A^{\perp}$; the case $R_A=N_A$ therefore cannot occur for self-adjoint transformations].

Example 5: See Example 4, Section 1. On $L_2^{(c)}(0,1)$, let $Au \doteq xu(x)$. This operator is clearly self-adjoint. The homogeneous equation $xu(x)=\lambda u(x)$ has only the trivial solution, so that A has *no eigenvalues*. For f in L_2 the inhomogeneous equation

$$xu-\lambda u=f$$

has the formal solution $u=f/(x-\lambda)$, which is a genuine L_2 solution whenever λ is *not* in the real interval $0\leqslant\lambda\leqslant1$. If λ is in that interval, the solution belongs to L_2 only if f vanishes to sufficiently high order at $x=\lambda$; the inverse of $A-\lambda I$ is seen to be unbounded, and $R(\lambda)$ is dense in L_2. Thus every value of λ in $0\leqslant\lambda\leqslant1$ is in the continuous spectrum—state $(II,1_n)$—and every other value of λ is in the resolvent set.

Remark. For some purposes it is useful to view $xu=\lambda u$ as a distributional equation. For λ in $[0,1]$ the equation then admits the solution $c\delta(x-\lambda)$, which can be regarded as a "singular" eigenfunction.

Example 6. On $L_2^{(c)}(0,1)$ consider the bounded self-adjoint operator defined by

$$(6.1) \qquad Au \doteq xu(x)+\int_0^1 u(x)\,dx,$$

which is similar to the one that arises for neutron transport in a slab. Since

A is self-adjoint, its spectrum is real. Let us search first for eigenvalues and eigenfunctions by examining the equation

$$(6.2) \qquad xu + \int_0^1 u(x)\,dx = \lambda u, \qquad 0 < x < 1.$$

If $\int_0^1 u(x)\,dx = 0$, we find that $u \equiv 0$; we therefore assume that $\int_0^1 u\,dx \neq 0$. In view of the fact that we are dealing with a linear homogeneous equation we may as well normalize u, so that

$$(6.3) \qquad \int_0^1 u\,dx = 1,$$

and then (6.2) gives

$$(6.4) \qquad u = \frac{1}{\lambda - x},$$

which belongs to L_2 if λ is outside $0 \leqslant \lambda \leqslant 1$. To satisfy (6.3) we need $\lambda > 1$, and, performing the required integration, we obtain

$$(6.5) \qquad \frac{\lambda}{\lambda - 1} = e,$$

which has the one and only solution

$$(6.6) \qquad \lambda_0 = \frac{e}{e - 1},$$

the corresponding eigenfunction being

$$(6.7) \qquad u(x) = \frac{1}{\lambda_0 - x}.$$

Next we show that the segment $0 \leqslant \lambda \leqslant 1$ is in the continuous spectrum. It suffices to show that $A - \lambda I$ is not bounded away from 0. We must exhibit a sequence $\{u_n\}$ such that $\|Au_n - \lambda u_n\| / \|u_n\|$ tends to 0 as $n \to \infty$. Let λ be fixed, $0 < \lambda < 1$, and for n sufficiently large take

$$u_n(x) = \begin{cases} 1, & \lambda < x < \lambda + \dfrac{1}{n}, \\[2mm] -1, & \lambda - \dfrac{1}{n} < x < \lambda, \\[2mm] 0, & |x - \lambda| \geqslant \dfrac{1}{n}. \end{cases}$$

Then $Au_n - \lambda u_n = (x - \lambda)u_n$, and

$$\|Au_n - \lambda u_n\|^2 = 2 \int_\lambda^{\lambda + 1/n} (x - \lambda)^2 dx = \frac{2}{3n^3},$$

$$\|u_n\|^2 = \frac{2}{n},$$

so that

$$\lim_{n \to \infty} \left(\frac{\|Au_n - u_n\|}{\|u_n\|} \right) = 0.$$

A slight modification is needed for $\lambda = 0$ and $\lambda = 1$.

As in Example 5, it is sometimes useful to interpret (6.2) distributionally. Setting $\int_0^1 u(x)\, dx = 1$, we have

$$(x - \lambda)u = -1, \qquad 0 \leqslant x \leqslant 1,$$

an equation which has already been discussed [(4.39), Chapter 2], and whose solution is (if $0 \leqslant \lambda \leqslant 1$)

$$u = C\delta(x - \lambda) - \text{pf} \frac{1}{x - \lambda},$$

where $C(\lambda)$ can be calculated from the normalization (6.3).

Example 7: See Example 4, Section 1. Let $Au = \int_0^x u(y)\, dy$ on $L_2^{(c)}(0, 1)$. The equation $Au = \lambda u$ implies, for $\lambda \neq 0$, that $\lambda u' = u$, $u(0) = 0$, and therefore $u \equiv 0$; for $\lambda = 0$, $Au = 0$ implies that $u = 0$. Thus A has *no* eigenvalues. Consider the equation $Au - \lambda u = f$; setting $z = \lambda u + f$, we see that z is differentiable and $z(0) = 0$. Differentiation gives $z' = u$, or, for $\lambda \neq 0$, $z' = (z - f)/\lambda$ and

$$z = -\frac{e^{x/\lambda}}{\lambda} \int_0^x f(y)e^{-y/\lambda}\, dy,$$

(6.8) $\quad u = \frac{z - f}{\lambda} = -\frac{f(x)}{\lambda} - \frac{e^{x/\lambda}}{\lambda^2} \int_0^x f(y)e^{-y/\lambda}\, dy = (A - \lambda I)^{-1}f.$

One can now verify that this function u satisfies the original equation $Au - \lambda u = f$, and it is the only solution. Furthermore, (6.8) shows clearly that $(A - \lambda I)^{-1}$ is bounded. For $\lambda = 0$ the equation $Au = f$ has solutions only if f is differentiable and $f(0) = 0$; this set is dense in $L_2(0, 1)$, and the inversion formula is $u = f'$. Thus A^{-1} is unbounded. Recapitulating: Every value of $\lambda \neq 0$ is in the resolvent set, and $\lambda = 0$ is in the continuous spectrum [state $(\text{II}, 1_n)$].

Example 8: See Example 9, Section 1. For the right-shift operator, if $u = \sum_{k=1}^{\infty} \alpha_k e_k$, then $Au = \sum_{k=1}^{\infty} \alpha_k e_{k+1}$. The eigenvalue problem is therefore equivalent to the system $0 = \lambda \alpha_1, \alpha_1 = \lambda \alpha_2, \alpha_2 = \lambda \alpha_3, \ldots$. If $\lambda \neq 0$, then $\alpha_1 = 0$ and therefore all $\alpha_k = 0$. If $\lambda = 0$, all α_k are again 0. Therefore A has no eigenvalues. Let us now look at the inhomogeneous equation $Au - \lambda u = f$ or $-\lambda \alpha_1 e_1 + \sum_{k=2}^{\infty} (\alpha_{k-1} - \lambda \alpha_k) e_k = \sum_{k=1}^{\infty} f_k e_k$. If $\lambda = 0$ *and* $f_1 = 0$, there is one and only one solution, given by $\alpha_1 = f_2, \alpha_2 = f_3, \ldots$. Thus $\lambda = 0$ is in the compression spectrum and has deficiency 1, and A^{-1} is bounded. If $\lambda \neq 0$, we obtain

$$\alpha_1 = -\frac{f_1}{\lambda}$$

and the recurrence relations $\alpha_k = (\alpha_{k-1} - f_k)/\lambda$.

If the series $\sum \alpha_k e_k$ obtained in this way converges, that is, if $\sum |\alpha_k|^2 < \infty$, it provides a solution to the inhomogeneous equation. Rather than ascertaining the convergence directly, we obtain useful information by studying the adjoint A^*, which is the left shift $A^* e_1 = 0, A^* e_{k+1} = e_k, k = 1, 2, \ldots$. Since A has no eigenvalues, the compression spectrum of A^* is empty. The eigenvalue problem for A^* is $\sum_{k=1}^{\infty} \alpha_{k+1} e_k = \lambda \sum_{k=1}^{\infty} \alpha_k e_k$, or $\lambda \alpha_k = \alpha_{k+1}, k = 1, 2, \ldots$. In terms of α_1, these equations have the solution $\alpha_k = \lambda^{k-1} \alpha_1$, and these will form the coordinates of a vector in H if and only if $\sum |\alpha_k|^2 < \infty$, that is, if and only if $|\lambda| < 1$, which is then the point spectrum of A^* and the compression spectrum of A. The circle $|\lambda| = 1$ must be in the spectrum (since the spectrum is a closed set) for both A and A^* but cannot be in the compression spectrum and so must be in the continuous spectrum. For $|\lambda| < 1$ we have $\|Af - \lambda f\| \geq |\,\|Af\| - \|\lambda f\|\,| = (1 - |\lambda|)\|f\|$, so that $A - \lambda I$ is bounded away from 0 and λ cannot be in the continuous spectrum.

Example 9: See Example 1, Section 3 (a) Let

$$T_1 u = \frac{du}{dx}, \qquad D_{T_1}: u \text{ a.c.}, \quad u' \in L_2(0, 1), u(0) = 0.$$

Then

$$T_1^* v = -\frac{dv}{dx}, \qquad D_{T_1^*}: v \text{ a.c.}, \quad v' \in L_2(0, 1), v(1) = 0.$$

The only solution of $T_1u=\lambda u, u\in D_{T_1}$, is $u\equiv 0$, so that the point spectrum of T_1 is empty. The inhomogeneous equation $T_1u-\lambda u=f, u\in D_{T_1}$, has the unique solution

$$u(x)=e^{\lambda x}\int_0^x e^{-\lambda y}f(y)\,dy,$$

so that $T_1-\lambda I$ has a bounded inverse for each λ, and $R(\lambda)=H$. *Therefore the spectrum of T_1 is empty.*

(b) Let

$$S_2u=-i\frac{du}{dx}, \qquad D_{S_2}:u \text{ a.c.,} \quad u'\in L_2(0,1), u(0)=au(1),$$

where a is a given real constant $\neq 0$. Then

$$S_2^*v=-i\frac{dv}{dx}, \qquad D_{S_2^*}:v \text{ a.c.,} \quad v'\in L_2(0,1), v(1)=av(0).$$

Thus S_2 is self-adjoint if $a=1$ or $a=-1$. Consider the case $a=1$; then the eigenvalue problem is $-iu'=\lambda u, u(0)=u(1)$, from which we conclude that $\lambda=2n\pi$, where n is an integer $-\infty<n<\infty$. Each eigenvalue has multiplicity 1, the null space being spanned by the eigenfunction $e^{2n\pi i x}$. Note that the *eigenfunctions form an orthonormal basis for $L_2(0,1)$.* Consider the inhomogeneous equation $S_2u-\lambda u=f$, which has the explicit form $-iu'-\lambda u=f$, $u(0)=u(1)$. If $\lambda\neq 2n\pi$, the one and only solution can be written in terms of a Green's function, and the corresponding integral representation of the solution shows that $(S_2-\lambda I)^{-1}$ is bounded. Thus $\lambda\neq 2n\pi$ is in the resolvent set. The case $a=-1$ proceeds along similar lines.

(c) Let

$$S_3u=-i\frac{du}{dx}, \qquad D_{S_3}:u \text{ a.c.,} \quad u'\in L_2(0,1), u(0)=u(1)=0.$$

Then

$$S_3^*v=-i\frac{dv}{dx}, \qquad D_{S_3^*}:u \text{ a.c.,} \quad u'\in L_2(0,1).$$

Note that S_3 is symmetric and has no point spectrum. Every value of λ is in the compression spectrum of S_3 and is an eigenvalue of S_3^*.

Spectral Theory on $E_n^{(c)}$

For an operator on $E_n^{(c)}$ there are only two possibilities: either λ is in the resolvent set, or λ simultaneously is an eigenvalue and is in the compression spectrum. We may therefore omit further reference to the compression spectrum.

We can easily show that every operator A on $E_n^{(c)}$ must have at least one eigenvalue. Let $\{h_k\}$ be a basis in $E_n^{(c)}$. If the coordinates of u in this basis are $\{\alpha_i\}$, then, according to (1.5), those of Au are $\{\beta_i = \sum_j a_{ij}\alpha_j\}$, where $[a_{ij}]$ is the matrix of A in the basis $\{h_i\}$. Thus the eigenvalue problem $Au - \lambda u = 0$ is equivalent to the homogeneous system

(6.9)
$$(a_{11}-\lambda)\alpha_1 + a_{12}\alpha_2 + \cdots + a_{1n}\alpha_n = 0,$$
$$\vdots$$
$$a_{n1}\alpha_1 + a_{n2}\alpha_2 + \cdots + (a_{nn}-\lambda)\alpha_n = 0,$$

which has a nontrivial solution if and only if

(6.10)
$$\det(A-\lambda I) = \begin{vmatrix} a_{11}-\lambda & a_{12} & \cdots & a_{1n} \\ \vdots & & & \\ a_{n1} & a_{n2} & \cdots & a_{nn}-\lambda \end{vmatrix} = 0.$$

Clearly $\det(A-\lambda I)$ is a polynomial of degree n in λ which must therefore have n zeros (not necessarily distinct), which we label $\lambda_1,\ldots,\lambda_n$. Each distinct λ_k, when substituted in (6.9), gives rise to a nontrivial vector solution $(\alpha_1^{(k)},\ldots,\alpha_n^{(k)})$. We are therefore entitled to call this λ_k an eigenvalue, and we have shown that *every operator on $E_n^{(c)}$ has at least one eigenvalue.* [An operator on $E_n^{(r)}$ does not necessarily have an eigenvalue, as illustrated by the rotation operator of Example 3 with $\theta \neq k\pi$.] To a k-fold zero of (6.10) may correspond as few as one or as many as k independent eigenvectors. For instance, the identity operator has $\lambda = 1$ as an n-fold zero, and since every vector in $E_n^{(c)}$ is a corresponding eigenvector, we can obviously choose n independent ones. On the other hand, Example 4 has only the eigenvalue $\lambda = 0$, which is a double zero of (6.10) but has geometric multiplicity equal to 1.

It is important to know when A has enough eigenvectors to form a basis (that is, when A has n independent eigenvectors). Exercise 6.7 asks the reader to show that, if A has n *distinct* eigenvalues, the eigenvectors form a basis. More important for us is the fact that the eigenvectors of any *symmetric* operator can be chosen to form an orthonormal basis even if not all the eigenvalues are distinct. We now prove this assertion.

Let A be a symmetric operator on $E_n^{(c)}$. We have already established that A has at least one eigenvalue, say λ_1. Since A is symmetric, λ_1 is real. Let e_1 be a unit eigenvector corresponding to λ_1, and let M_1 be the one-dimensional linear manifold generated by e_1. Every vector u in $E_n^{(c)}$ can be decomposed in one and only one way as a sum $v + w$, where $v \in M_1$ and $w \in M_1^\perp$. Clearly $Av = \lambda_1 v$, so that every vector in M_1 remains in M_1 after the transformation A. We say that M_1 is *invariant* under A. $M_1^\perp$ is also invariant under A since $\langle Aw, v \rangle = \langle w, Av \rangle = \lambda_1 \langle w, e_1 \rangle = 0$. We can therefore consider A as an operator on the $(n-1)$-dimensional space $M_1^\perp$. This operator is clearly symmetric and has therefore at least one eigenvalue λ_2 (not necessarily different from λ_1). Let e_2 be a unit eigenvector corresponding to λ_2; since e_2 is in $M_1^\perp$, e_2 is orthogonal to e_1. Let M_2 be the two-dimensional linear manifold generated by e_1 and e_2. We see that M_2 and $M_2^\perp$ are invariant under A and that A can therefore be regarded as a symmetric operator on M_2 with at least one real eigenvalue λ_3 and a unit eigenvector e_3. We continue this procedure for n steps until $M_n^\perp$ is empty.

We have therefore generated in this way n eigenvalues $\lambda_1, \ldots, \lambda_n$, which are *not necessarily distinct*, but to which correspond n orthonormal eigenvectors $e_1, \ldots, e_n$. Such a set of vectors is of course a basis for $E_n^{(c)}$. We have

(6.11)
$$Ae_\lambda = \lambda_k e_k, \qquad k = 1, \ldots, n,$$
$$\langle e_i, e_j \rangle = 0, \quad i \neq j; \qquad \langle e_i, e_i \rangle = 1.$$

The matrix of A in this basis has diagonal form: the eigenvalues appear along the principal diagonal, and all off-diagonal elements vanish.

The set of eigenvectors of (6.11) provides a convenient way for solving the inhomogeneous equation $Au = f$ or, even more generally, the equation

(6.12)
$$Au - \lambda u = f,$$

where A is a symmetric operator on $E_n^{(c)}$, f is a given element in $E_n^{(c)}$, and λ is a given complex number. Since the eigenvectors (6.11) of A form a basis, we can write

$$f = \sum_{k=1}^{n} \beta_k e_k, \qquad \text{where } \beta_k = \langle f, e_k \rangle.$$

If (6.12) has a solution u, we can express u as

$$u = \sum_{k=1}^{n} \alpha_k e_k,$$

and, on substitution in (6.12), we find that

$$\sum_{k=1}^{n} [(\lambda_k - \lambda)\alpha_k - \beta_k] e_k = 0,$$

from which we conclude that

$$(6.13) \qquad\qquad (\lambda_k - \lambda)\alpha_k = \beta_k, \qquad k = 1, 2, \ldots, n.$$

If λ is *different* from $\lambda_1, \ldots, \lambda_n$, (6.13) can be solved for each α_k, and we obtain the prospective solution

$$(6.14) \qquad\qquad u = \sum_{k=1}^{n} \frac{\beta_k}{\lambda_k - \lambda} e_k,$$

which is easily checked to be a solution of (6.12) in fact as well as in form. Of course, if λ is equal to some eigenvalue λ_m (say, nondegenerate), then there is no solution of (6.13) for α_m unless $\beta_m = 0$. If $\beta_m = 0$, then α_m is arbitrary. For $k \neq m$ we still have $\alpha_k = \beta_k / (\lambda_k - \lambda)$. Thus a necessary and sufficient condition for (6.12) to have solutions is that $\langle f, e_m \rangle = 0$; and, assuming that this solvability condition is satisfied, all solutions of (6.12) are of the form

$$(6.15) \qquad\qquad u = \sum_{k \neq m} \frac{\beta_k}{\lambda_k - \lambda} e_k + c e_m,$$

where c is arbitrary. A slight but obvious modification is needed if λ is equal to a degenerate eigenvalue.

Exercises

6.1. Let P be the orthogonal projection on a closed manifold M in the Hilbert space H. We saw in Example 2 that the only points in the spectrum are $\lambda = 0$ and $\lambda = 1$, which are eigenvalues. What are the multiplicities of these eigenvalues? What are the solvability conditions for the equations $Pu = f$ and $Pu - u = f$? Find the corresponding solutions if the solvability conditions are satisfied.

6.2. Consider the problem

$$u' = f, \quad 0 < x < 1; \qquad u(0) = 0, \quad u'(0) = 0.$$

Under what conditions on f is the problem solvable? What is the adjoint homogeneous problem? Why does the Fredholm alternative not apply?

6.3. Carry out the analysis in Example 9(b) for the case $a = -1$. Construct Green's function when λ is not in the point spectrum.

6.4. On $L_2^{(c)}(0,1)$ let $Au = a(x)u(x)$, where $a(x)$ is a fixed continuous function. Describe the spectrum.

6.5. On $L_2^{(c)}(-1,1)$ consider the operator

$$Au \doteq xu(x) + \theta \int_{-1}^{1} u(x) \, dx,$$

where θ is a given real number. Describe completely the spectrum of A.

6.6. On $E_n^{(c)}$ consider the nearest thing to a shift operator. Let A be defined by its effect on an orthonormal basis $\{e_1, \ldots, e_n\}$ as

$$Ae_i = e_{i+1}, \quad i = 1, \ldots, n-1; \qquad Ae_n = e_1.$$

Find the spectrum and discuss the limit as $n \to \infty$ (compare with Example 8).

6.7. Prove that an operator with n distinct eigenvalues on $E_n^{(c)}$ has n independent eigenvectors.

7. COMPACT OPERATORS

Definition. An operator K on H is said to be *compact* (or *completely continuous*) if K transforms bounded sets into relatively compact sets (see Section 3, Chapter 4), that is, whenever $\{u_n\}$ is a sequence in H with $\|u_n\| < M$, then $\{Ku_n\}$ contains a subsequence which converges to some point in H.

A compact operator is necessarily bounded, for otherwise there would exist a sequence with $\|u_n\| = 1$ and $\|Ku_n\| \to \infty$; by eliminating superfluous elements if necessary, we can suppose that $\|Ku_{n+1}\| > \|Ku_n\|$, and clearly $\{Ku_n\}$ does not contain a convergent subsequence.

Every bounded set in a finite-dimensional space is relatively compact (the Bolzano-Weierstrass theorem), so that every bounded operator on E_n is compact and every bounded operator on H having finite-dimensional range is compact. Arbitrary bounded operators on infinite-dimensional spaces are not necessarily compact. In fact, the identity operator is clearly bounded but is not compact since it transforms the infinite orthonormal set $\{e_i\}$ into itself and $\{e_1, e_2, \dots\}$ is a bounded sequence containing no convergent subsequence. The most important compact operators on infinite-dimensional spaces are integral operators of the Hilbert-Schmidt type (see Example 6, Section 1).

Compact operators seem to be very obliging since they transform sequences that are merely bounded into "nicer" sequences having convergent subsequences. This is a two-edged sword, however, since unavoidably there will be trouble with the inverse. An operator which strikes a better balance between itself and its inverse is $I + K$, where I is the identity. Here one can think of the operator K as being a perturbation of I. Operators of the form $I + K$ (or, more generally, of the form $K - \lambda I$, where λ is a complex number $\neq 0$) will have invertibility properties similar to those of operators on finite-dimensional spaces.

Theorem 1. If K is compact and $\{e_n\}$ is an infinite orthonormal sequence in H, then $\lim_{n \to \infty} K e_n = 0$.

Proof. If the contrary were true, there would exist a subsequence $\{f_n\}$ of $\{e_n\}$ such that $\|Kf_n\| > \varepsilon$ for all sufficiently large n. Since K is compact, a subsequence $\{g_n\}$ can be extracted from $\{f_n\}$ such that Kg_n converges, say to u. This element u is not 0 since $\|Kg_n\| > \varepsilon$ for n large. The continuity of the inner product shows that $\langle Kg_n, u \rangle \to \|u\|^2 \neq 0$; also $\langle Kg_n, u \rangle = \langle g_n, K^*u \rangle$, which tends to 0 as $n \to \infty$ by (6.8), Chapter 4. This contradiction proves the theorem.

Corollary. If K is a compact, invertible operator on an infinite-dimensional space, its inverse is *unbounded*.

Proof. If $\{e_n\}$ is an infinite orthonormal set, we have $Ke_n \to 0$ by Theorem 1, and, since $\|e_n\| = 1$, K is not bounded away from 0. Therefore if K is invertible its inverse is unbounded.

How can we tell whether an operator is compact? It is not always easy to apply the definition directly. The following theorem tells us that K is compact if it can be approximated in norm by compact operators (which, in applications, will often be operators with finite-dimensional ranges).

Theorem 2. The operator A is compact if there exists a sequence of compact operators $\{K_n\}$ such that $\|A - K_n\| \to 0$ as $n \to \infty$.

Proof. It suffices to show that each bounded sequence $\{u_k\}$ contains a subsequence $\{v_k\}$ for which Av_k converges. Since K_1 is compact, there exists a subsequence of $\{u_k\}$, say $\{u_k^{(1)}\}$, for which $K_1 u_k^{(1)}$ converges. In the bounded sequence $\{u_k^{(1)}\}$ we can find a subsequence $\{u_k^{(2)}\}$ such that $K_2 u_k^{(2)}$ converges; proceeding in this manner, we find a subsequence $\{u_k^{(n)}\}$ of $\{u_k^{(n-1)}\}$ for which $K_n u_k^{(n)}$ converges as $k \to \infty$. Consider now the diagonal sequence

$$u_1^{(1)}, \qquad u_2^{(2)}, \qquad u_3^{(3)}, \qquad \ldots, \qquad u_n^{(n)}, \ldots,$$

which is a subsequence of $\{u_k\}$ transformed into a convergent sequence by each of the operators $K_1, K_2, \ldots$. We claim that A also transforms this diagonal sequence into a convergent sequence. Indeed, we have

$$\|Au_n^{(n)} - Au_m^{(m)}\| \leqslant \|Au_n^{(n)} - K_k u_n^{(n)}\| + \|K_k u_n^{(n)} - K_k u_m^{(m)}\| + \|K_k u_m^{(m)} - Au_m^{(m)}\|$$

$$\leqslant \|A - K_k\|(\|u_n^{(n)}\| + \|u_m^{(m)}\|) + \|K_k(u_n^{(n)} - u_m^{(m)})\|.$$

Since $\{u_k\}$ is bounded, so is $\{u_n^{(n)}\}$, and there exists c such that $\|u_n^{(n)}\| + \|u_m^{(m)}\| < c$ for all m, n. By hypothesis we can choose k so large that $\|A - K_k\| \leqslant \varepsilon/2c$. With k fixed in this way we choose m and n so large that $\|K_k u_n^{(n)} - K_k u_m^{(m)}\| \leqslant \varepsilon/2$; this can be done since the sequence $K_k u_n^{(n)}$ converges as $n \to \infty$. Therefore for m and n sufficiently large we have

$$\|Au_n^{(n)} - Au_m^{(m)}\| \leqslant \varepsilon,$$

so that $Au_n^{(n)}$ is a Cauchy sequence and hence converges, completing the proof of the theorem.

Theorem 2 will be applied when we take up the theory of integral equations. We now turn to the Fredholm alternative for operators of the form $A = K - \lambda I$, K compact, $\lambda \neq 0$.

Theorem (Fredholm alternative). Let H be a separable Hilbert space, K a compact operator on H, and λ a complex number $\neq 0$; let $A = K - \lambda I$ ($A^* = K^* - \bar{\lambda} I$). Then *either* the following alternatives holds (compare with Theorem 3, Section 5, for E_n):

(a) $Au = 0$ has only the zero solution (λ is not an eigenvalue of K), in which case $A^* v$ has only the zero solution ($\bar{\lambda}$ is not an eigenvalue of K^*) and $Au = f$ has precisely one solution for each f in H.

(b) N_A has finite dimension K (λ is an eigenvalue of K with multiplicity k), in which case N_{A^*} also has dimension k ($\bar{\lambda}$ is an eigenvalue of K^* with multiplicity k) and $Au = f$ has solutions if and only if f is orthogonal to N_{A^*}. If $\{u_1, \ldots, u_k\}$ is a basis for N_A (that is, a set of independent eigenvectors of K corresponding to the eigenvalue λ), and if $\{v_1, \ldots, v_k\}$ is a basis for N_{A^*} (that is, a set of independent eigenvectors of K^* corresponding to the eigenvalue $\bar{\lambda}$), the solvability conditions are $\langle f, v_1 \rangle = \cdots = \langle f, v_k \rangle = 0$; and, if these conditions are satisfied, the general solution of $Au = f$ is $u = \tilde{u} + c_1 u_1 + \cdots + c_k u_k$, where $c_1, \ldots, c_k$ are arbitrary constants and $\tilde{u}$ is any particular solution of $Au = f$.

We indicate how the steps of the proof of Theorem 3, Section 5, have to be carried out.

1.   We must show that R_A is a closed set, so that we will have $R_A = N_{A^*}^\perp$. Since A is closed and, moreover $D_A = H$, it suffices, by Theorem 2, Section 5, to show that $\|Au\| \geq c\|u\|$ for some fixed $c > 0$, and all $u \in N_A^\perp$. If this were not true, there would exist $\{u_i\} \in N_A^\perp$, $\|u_i\| = 1$, $\|Au_i\| \to 0$, that is, $Ku_i - \lambda u_i \to 0$. But since K is compact, we can restrict ourselves to a sequence $\{u_i\}$ such that Ku_i converges; then $Ku_i - \lambda u_i \to 0$ implies that λu_i converges, and since $\lambda \neq 0$, u_i converges, say to u. It then follows that Ku_i converges to Ku and therefore $Ku - \lambda u = 0$ and $u \in N_A$; u is also the limit of a sequence of elements of the closed set $N_A^\perp$, so that $u \in N_A^\perp$; hence $u = 0$. However, since u is the limit of a sequence of unit elements, $\|u\| = 1$. Therefore A must be bounded away from 0 on $N_A^\perp$, and R_A is closed and $R_A = N_{A^*}^\perp$.

2.   If $R_A = H$, then $N_A = \{0\}$ and vice versa. We already know from step 1 that $N_{A^*} = \{0\}$ is necessary and sufficient for $R_A = H$.

(a) Suppose $R_A = H$. We proceed as in Section 5, but now it is not obvious that the strict inclusion $N_p \supset N_{p-1}$ for all p leads to a contradiction. The proof is left for Exercise 7.1 and shows that $N_A = \{0\}$.

(b) This is exactly the same as in Section 5.

3. N_A and N_{A^*} have the same *finite* dimension. First we show that N_A has a finite dimension. Since N_A is closed, let $\{e_i\}$ be an orthonormal basis for N_A; then, if the dimension of N_A were infinite, we would have $Ke_i \to 0$ by Theorem 1. Now $Ke_i = \lambda e_i$, so that $\lambda e_i \to 0$, which is impossible for $\lambda \neq 0$. The rest of the proof is exactly as in Section 5.

Remark.   In the alternative theorem for E_n, the common dimension of N_A and N_{A^*} is an integer which obviously cannot exceed n. In the theorem just proved k can be any finite integer.

Exercises

7.1.   This exercise fills in the missing step in 2(a) of the proof of the Fredholm alternative. Let $A = K - \lambda I$, where $\lambda \neq 0$ and K is a compact operator on H. We want to show that, if $R_A = H$, then N_A

contains only the zero element. As in Theorem 3, Section 5, we assume that there exists $u_1 \neq 0$ with $Au_1 = 0$ and construct successively $u_2, u_3, \ldots$, where $Au_p = u_{p-1}$. Each $u_p \neq 0$ and $A^p u_p = 0$, so that u_p is a nonzero element in the null space N_p of A^p. It is easy to see that $N_p \supset N_{p-1}$ and that the inclusion is strict; it is therefore possible to choose $u_p \in N_p$ with $\|u_p\| = 1$ and $u_p \perp N_{p-1}$. For $n > m$ consider $\|Ku_n - Ku_m\| = \|\lambda u_n + g\|$, where $g = Au_n - Au_m - \lambda u_m$. Show that $g \in N_{n-1}$ and $\|Ku_n - Ku_m\|^2 \geqslant |\lambda| > 0$, which contradicts compactness.

7.2. Let K be a compact operator. Then we know from the Fredholm alternative that the multiplicity of any nonzero eigenvalue is finite. More can be said: Let M_ε be the linear manifold consisting of the eigenvectors corresponding to *all* eigenvalues λ with $|\lambda| \geqslant \varepsilon$, where ε is any positive number; then M_ε is finite-dimensional. Prove the theorem for the case when K is self-adjoint.

7.3. We say that u_n *converges weakly* to u if $\langle u_n, h \rangle \to \langle u, h \rangle$ for every h in H. If $\{e_n\}$ is an infinite orthonormal set in H, then $e_n \to 0$ *weakly* (but of course e_n does not converge). It is easy to see that convergence entails weak convergence.

Prove that, if K is compact, it maps every weakly convergent sequence into a convergent sequence.

8. EXTREMAL PROPERTIES OF OPERATORS

Let A be an operator, possibly unbounded, defined on a domain D_A dense in the separable Hilbert space H. We are interested in finding bounds for the spectrum of A, particularly its point spectrum, if any. If (λ, u) is an eigenpair, that is, if λ is an eigenvalue with corresponding nontrivial eigenvector u, then

$$(8.1) \qquad\qquad Au = \lambda u,$$

and hence

$$(8.2) \qquad\qquad |\lambda| = \frac{\|Au\|}{\|u\|}, \qquad \lambda = \frac{\langle Au, u \rangle}{\|u\|^2}.$$

For any nonzero element v in D_A we define the *Rayleigh quotient*

$$(8.3) \qquad\qquad R(v) \doteq \frac{\langle Av, v \rangle}{\|v\|^2}.$$

In general $R(v)$ may take on complex values, but if A is symmetric, $R(v)$ is real for all v in D_A. If (λ, u) is an eigenpair, we have $R(u) = \lambda$.

Bounded Operators

We may take the domain of a bounded operator A to be the entire Hilbert space H. We recall definition (1.2):

$$(8.4) \qquad \|A\| = \sup_{\|v\| = 1} \|Av\| = \sup_{v \neq 0} \frac{\|Av\|}{\|v\|}$$

and introduce the new definition

$$(8.5) \qquad M_A \doteq \sup_{\|v\| = 1} |\langle Av, v \rangle| = \sup_{v \neq 0} |R(v)|,$$

where in both definitions we must use "sup" instead of "max" since the supremum is not necessarily attained. We shall see in Theorem 3 that M_A is necessarily finite. It follows from (8.2) that every eigenvalue λ satisfies

$$|\lambda| \leqslant \|A\|, \qquad |\lambda| \leqslant M_A.$$

One of our goals is to find out whether these inequalities can be transformed into equalities for the eigenvalues of largest modulus. We shall need a sequence of theorems.

Theorem 1. $\|A^*\| = \|A\|$.

Proof. For each v in H, the Schwarz inequality gives

$$0 \leqslant \|Av\|^2 = \langle Av, Av \rangle = \langle v, A^*Av \rangle \leqslant \|v\| \, \|A^*Av\| \leqslant \|v\| \, \|A^*\| \, \|Av\|.$$

Thus $\|Av\| \leqslant \|A^*\| \, \|v\|$, and by reversing the roles of A and A^*, we also find that $\|A^*v\| \leqslant \|A\| \, \|v\|$. Together, these inequalities prove the theorem.

Theorem 2. If λ is in the spectrum of A, then $|\lambda| \leqslant \|A\|$.

Proof. The result has already been proved for eigenvalues. If λ is in the compression spectrum of A, $\bar{\lambda}$ is an eigenvalue of A^*; hence $|\lambda| = |\bar{\lambda}| \leqslant \|A^*\| = \|A\|$. If λ is in the continuous spectrum of A, there exists a sequence $\{v_n\}$ with $\|v_n\| = 1$ such that $w_n \doteq Av_n - \lambda v_n \to 0$. Hence

$$|\lambda| = \|Av_n - w_n\| \leqslant \|Av_n\| + \|w_n\| \leqslant \|A\| + \|w_n\|,$$

and, since $\|w_n\| \to 0$, we have $|\lambda| \leqslant \|A\|$.

Theorem 3. $M_A \leqslant \|A\|$.

Proof. By the Schwarz inequality

$$|\langle Av, v \rangle| \leqslant \|Av\| \|v\| \leqslant \|A\| \|v\|^2.$$

For an arbitrary bounded operator the numbers M_A and $\|A\|$ are quite different. For instance, the operator A describing a rotation through an angle $\pi/2$ [see Example 3(b), Section 1] has $\langle Av, v \rangle = 0$ for every v and $\|A\| = 1$. However, we have the following theorem.

Theorem 4. If A is *symmetric*, $M_A = \|A\|$.

Proof. It suffices to show that $M_A \geqslant \|A\|$. The symmetry of A guarantees that $R(v)$ is real. Together with the definition of M_A, this implies that for all v, w

$$\langle A(v+w), v+w \rangle \leqslant M_A \|v+w\|^2, \qquad \langle A(v-w), v-w \rangle \geqslant -M_A \|v-w\|^2,$$

from which we obtain

$$\langle A(v+w), v+w \rangle - \langle A(v-w), v-w \rangle \leqslant M_A(\|v+w\|^2 + \|v-w\|^2).$$

The left side is equal to $2\langle Av, w \rangle + 2\langle Aw, v \rangle$. Using the parallelogram law on the right side, we find that

$$\langle Av, w \rangle + \langle Aw, v \rangle \leqslant M_A(\|v\|^2 + \|w\|^2).$$

If $Av \neq 0$, we may substitute $w = Av(\|v\|/\|Av\|)$ to obtain

$$\langle Av, Av \rangle + \langle AAv, v \rangle \leqslant 2M_A \|v\| \|Av\|.$$

Since A is symmetric, the left side is $2\|Av\|^2$, yielding $\|Av\| \leqslant M_A \|v\|$, an equality which remains valid if $Av = 0$. We have thus proved Theorem 4.

We now confine ourselves to symmetric operators. It turns out that parts of the theory are just as easily developed for unbounded operators that are either bounded below or bounded above.

Symmetric Operators

Let A be a symmetric operator defined on a domain D_A dense in H. Since $R(v)$ is now real for all v in D_A, it makes sense to look at $\sup R(v)$ and $\inf R(v)$ rather than $\sup |R(v)|$. Of course it is possible that $\sup R(v) = +\infty$

or $\inf R(v) = -\infty$. If both are finite, the operator is bounded; if only one is finite, the operator is said to be *bounded from one side*.

Definition.   The symmetric operator A is *bounded below* if there exists a constant c such that

(8.6) $$\langle Av, v \rangle \geqslant c\|v\|^2 \qquad \text{for all } v \in D_A.$$

We say that A is *nonnegative* if c can be chosen $\geqslant 0$ in (8.6) and *strictly positive* (or *coercive*) if c can be chosen > 0. An intermediate concept is that of a *positive* operator: $\langle Av, v \rangle > 0$ for all $v \neq 0$ in D_A. The operator A is *bounded above* if there exists a constant C such that

$$\langle Av, v \rangle \leqslant C\|v\|^2 \qquad \text{for all } v \in D_A,$$

with similar definitions for nonpositive, negative, and strictly negative operators.

We note that an operator is bounded if and only if it is bounded both above and below.

Example.   Let A be the operator $-d^2/dx^2$ defined on the domain D_A of functions $u(x)$ with two continuous derivatives on $0 \leqslant x \leqslant 1$ and satisfying $u(0) = u(1) = 0$. Then A is a strictly positive, unbounded operator.
Indeed, we have

$$\langle Au, u \rangle = -\int_0^1 u'' \bar{u}\, dx = \int_0^1 |u'|^2\, dx \geqslant 0,$$

which shows that A is nonegative. Moreover, since $u(0) = 0$,

$$|u(x)|^2 = \left|\int_0^x u'(t)\, dt\right|^2 \leqslant \int_0^x u^2\, dt \int_0^x |u'(t)|^2\, dt \leqslant x\int_0^1 |u'(t)|^2\, dt,$$

and therefore

$$\int_0^1 |u'(t)|^2\, dt \geqslant 2\|u\|^2,$$

which proves that A is strictly positive. The eigenvalues of A are $\lambda_n = n^2\pi^2$, $n = 1, 2, \ldots$. Since $\lambda_n \to \infty$ as $n \to \infty$, (8.2) shows that the operator is unbounded.

For operators that are bounded from one side, we can only hope to find extremal principles for the eigenvalues at the bounded end of the spectrum.

Definition

(8.7)
$$L_A \doteq \inf_{\substack{v \in D_A \\ v \neq 0}} R(v) = \inf_{\substack{v \in D_A \\ \|v\|=1}} \langle Av, v \rangle,$$

(8.8)
$$U_A \doteq \sup_{\substack{v \in D_A \\ v \neq 0}} R(v) = \sup_{\substack{v \in D_A \\ \|v\|=1}} \langle Av, v \rangle.$$

Remark. If A is bounded below, L_A is finite; if A is bounded above, U_A is finite. For a bounded operator both L_A and U_A are finite, and

(8.9)
$$M_A = \|A\| = \max\left(|U_A|, |L_A|\right).$$

Thus the larger of U_A and $-L_A$ is equal to $\|A\|$.

Theorem 5. Let A be symmetric and bounded below. If there is an element u in D_A for which the infimum in (8.7) is attained, (L_A, u) is an eigenpair and L_A is the lowest eigenvalue of A.

Proof. By assumption $R(v)$ is a minimum for $v = u$, and therefore, for any $\eta \in D_A$ and any real number ε, we have

$$R(u + \varepsilon\eta) \geqslant R(u) = L_A.$$

It follows that

$$\left[\frac{d}{d\varepsilon} R(u + \varepsilon\eta)\right]_{\varepsilon=0} = 0.$$

Performing the required calculation, we find that

$$\langle \eta, Au \rangle + \langle Au, \eta \rangle - L_A[\langle \eta, u \rangle + \langle u, \eta \rangle] = 0, \qquad \eta \in D_A.$$

Since $i\eta$ also belongs to D_A, we may substitute $i\eta$ for η to obtain

$$\langle \eta, Au \rangle - \langle Au, \eta \rangle - L_A[\langle \eta, u \rangle - \langle u, \eta \rangle] = 0.$$

Adding this equation to the preceding one, we have

$$\langle \eta, Au \rangle - L_A \langle \eta, u \rangle = 0,$$

that is,

$$\langle \eta, Au - L_A u \rangle = 0 \qquad \text{for every } \eta \in D_A.$$

Since D_A is dense in H, it follows that $Au - L_A u = 0$, so that (L_A, u) is an eigenpair. It remains only to prove that A has no eigenvalue smaller than L_A. Let (λ, ϕ) be any eigenpair. Then, from (8.2), $\lambda = R(\phi)$. Since the minimum of R is L_A, we have $\lambda \geqslant L_A$, thereby completing the proof.

Theorem 6. Let A be symmetric and bounded above. If there is an element u in D_A for which the supremum in (8.8) is attained, (U_A, u) is an eigenpair and U_A is the largest eigenvalue of A.

Proof. The operator $-A$ is symmetric and bounded below. By applying Theorem 5, we find that $-U_A$ is the lowest eigenvalue of $-A$ and hence U_A is the largest eigenvalue of A.

In applications it is relatively easy to determine whether A is bounded above or below or both. The difficulty arises in trying to establish whether or not the supremum or infimum is actually attained. Take, for instance, the multiplication operator on $L_2^{(c)}(0,1)$ defined by $Av = \int_0^1 xv(x)dx$. Then

$$R(v) = \int_0^1 x|v|^2 dx \Big/ \int_0^1 |v|^2 dx,$$

and clearly $0 \leqslant R(v) \leqslant 1$. In this case it can be seen that $U_A = 1$ and $L_A = 0$ by choosing sequences with support near $x = 1$ and $x = 0$, respectively. It can also be shown that neither the supremum nor the infimum of R is achieved for a nonzero element in D_A, so that the fact that A has no eigenvalues does not contradict Theorems 5 and 6. We note that in this last example L_A and U_A belong to the continuous spectrum. More generally, we have the following theorem.

Theorem 7. If A is bounded and symmetric, L_A and U_A are both in the approximate spectrum.

Proof. To show that U_A belongs to the approximate spectrum, we must exhibit a sequence $\{v_k\}$ of unit elements such that $Av_k - U_A v_k \to 0$. From the definition of U_A there exists a sequence of unit elements $\{v_k\}$ such that

$\langle Av_k, v_k \rangle \to U_A$. Consider first the case where A is a positive operator. We then have $U_A = \|A\|$ and

$$\|Av_k - U_A v_k\|^2 = \|Av_k\|^2 - 2U_A\langle Av_k, v_k \rangle + U_A^2$$
$$\leqslant \|A\|^2 - 2U_A\langle Av_k, v_k \rangle + U_A^2$$
$$= 2U_A^2 - 2U_A\langle Av_k, v_k \rangle,$$

which yields the desired result,

$$(8.10) \qquad\qquad Av_k - U_A v_k \to 0.$$

If A is not positive, we choose a positive number α so that $B \doteq A + \alpha I$ is positive. Clearly

$$\sup_{\|v\|=1} \langle Bv, v \rangle = U_A + \alpha,$$

and the maximizing sequence $\{v_k\}$ for A is also a maximizing sequence for B: $\langle Bv_k, v_k \rangle \to U_A + \alpha$. Hence we can apply (8.10) to B to obtain

$$Bv_k - (U_A + \alpha)v_k \to 0$$

or

$$Av_k - U_A v_k \to 0.$$

We have shown that U_A belongs to the approximate spectrum. The proof for L_A is similar.

If the symmetric operator A is not only bounded but also *compact*, the supremum (8.7), if nonzero, is actually attained. The same is true for the infimum (8.8).

Theorem 8. If A is a symmetric compact operator, then, if $U_A \neq 0$, it is an eigenvalue, and if $L_A \neq 0$, it is also an eigenvalue.

Proof. Since A is bounded and symmetric, U_A is in the approximate spectrum by Theorem 7. There exists a sequence $\{v_k\}$ of unit elements such that $Av_k - U_A v_k \to 0$. Because A is compact, we can redefine $\{v_k\}$ by eliminating superfluous elements of the original sequence, so that $\{Av_k\}$ converges. Thus $U_A v_k$ converges, and, if $U_A \neq 0$, v_k converges to a unit element v which, by the continuity of A, satisfies

$$Av - U_A v = 0.$$

A similar proof holds for L_A if $L_A \neq 0$.

Corollary. If A is a nontrivial symmetric, compact operator, at least one of the number $\|A\|$, $-\|A\|$ is an eigenvalue.

More detailed information on the spectrum of compact, symmetric operators will be presented in the next chapter.

REFERENCES AND ADDITIONAL READING

Halmos, P. R., *A Hilbert space problem book*, Van Nostrand, Princeton, N. J., 1967.

Lorch, E. R., *Spectral theory*, Oxford University Press, New York, 1962.

Kolmogorov, A. N. and Fomin, S. V., *Functional analysis*, Vols. I and II, Graylock, Albany, N. Y., 1957, 1961.

Naylor, A. W. and Sell, G. R., *Linear operator theory in engineering and science*, Holt, Rinehart, and Winston, New York, 1971.

Riesz, F. and Sz.-Nagy, B. *Functional analysis*, Ungar, New York, 1955.

6
Integral Equations

1. INTRODUCTION

Integral equations arise directly in various mathematical and applied settings, but for our purposes their principal function is to provide alternative formulations of boundary value problems. Some of the advantages of the integral equation approach are as follows:

1. The integral operator appearing in the equation is a bounded operator and often compact, whereas the differential operator is unbounded.
2. The boundary conditions are incorporated in the integral equation through its kernel, which is a Green's function.
3. Associated with the integral equation are variational principles and approximation schemes that complement those arising from the formulation as a differential equation.
4. Some BVPs for partial differential equations can be translated into integral equations of lower dimensionality.

The following examples of integral equations give some idea of the scope of the present chapter; in each case the unknown function appears under the integral sign.

Example 1. Abel's integral equation

$$(1.1) \qquad \int_0^y \frac{u(\eta)}{\sqrt{y-\eta}}\, d\eta = f(y), \qquad y>0,$$

where $f(y)$ is a *prescribed* function for $y>0$, and $u(y)$ is to be found for $y>0$. The problem arises in finding a curve C lying in the first quadrant, terminating at the origin, and having the following property: a particle starting from rest at the elevation y slides down C under the influence of

347

gravity in the prescribed time $f(y)$. By using conservation of energy, we can then derive (1.1) with $u(y)=(2g)^{-1/2}\,ds/dy$, where g is the acceleration of gravity and $s(y)$ is the arclength along C measured from $y=0$. If (1.1) is viewed purely mathematically, $f(y)$ may have to satisfy some conditions for a solution to exist. The physical interpretation imposes the further restriction that the solution u should generate a feasible arclength. This means that $u \geqslant (2g)^{-1/2}$ and therefore, from (1.1), that $f(y) \geqslant (2y/g)^{1/2}$, which states the obvious fact that the prescribed time of descent along a curve must exceed the time of free fall from the same elevation.

Example 2. The Fourier transform of $f(x)$ is defined as

$$(1.2) \qquad f^{\hat{}}(\omega)=\int_{-\infty}^{\infty} e^{i\omega x} f(x)\,dx, \quad -\infty < \omega < \infty.$$

To recover $f(x)$ from $f^{\hat{}}(\omega)$ we must solve the integral equation (1.2). The answer is given by the inversion formula (4.3), Chapter 2:

$$(1.3) \qquad f(x)=\frac{1}{2\pi}\int_{-\infty}^{\infty} e^{-i\omega x} f^{\hat{}}(\omega)\,d\omega.$$

Example 3. Let $H=L_2^{(c)}(-\infty,\infty)$, and let S_Ω be the linear manifold in H consisting of Ω-band-limited functions; thus $f(t)\in S_\Omega$ if $\int_{-\infty}^{\infty}|f(t)|^2\,dt < \infty$ and if its Fourier transform $f^{\hat{}}(\omega)$ vanishes for $|\omega| \geqslant \Omega$. (See Section 4, Chapter 2). For functions of *unit energy* in S_Ω, Parseval's relation gives $(1/2\pi)\int_{-\Omega}^{\Omega}|f^{\hat{}}(\omega)|^2\,d\omega=1=\int_{-\infty}^{\infty}|f(t)|^2\,dt$. We then pose the problem of finding the function of unit energy in S_Ω that maximizes the energy $\int_{-T}^{T}|f(t)|^2\,dt$ in a fixed, preassigned time interval $|t| \leqslant T$. The optimal function $f(t)$ is the eigenfunction corresponding to the largest eigenvalue of

$$(1.4) \qquad \lambda f(t)=\int_{-T}^{T} \frac{\sin\Omega(t-s)}{\pi(t-s)} f(s)\,ds, \quad -\infty < t < \infty.$$

It suffices to regard (1.4) as an equation on $-T < t < T$ to determine all eigenpairs $(\lambda_n, f_n(t))$ on $-T < t < T$. With f_n known for $|t| < T$, the right side of (1.4) has unambiguous meaning for all t, and we take this to be $\lambda_n f_n(t)$ for $|t| \geqslant T$.

Example 4. The normal modes $u(x)$ and normal frequencies λ of a vibrating membrane whose edge is fixed satisfy

$$(1.5) \qquad -\Delta u=\lambda u, \quad x\in\Omega; \quad u|_\Gamma=0,$$

where Ω is the plane domain covered by the membrane and Γ is its boundary. The equivalent integral equation is

(1.6) $$u(x) = \lambda \int_{\Omega} g(x,\xi) u(\xi)\, d\xi, \qquad x \in \Omega,$$

where $g(x,\xi)$ is Green's function satisfying (4.25), Chapter 1. Note that x and ξ are position vectors in the plane, so that (1.6) is an integral equation involving functions of two independent variables. If Ω is a unit disk and polar coordinates r, ϕ are introduced, (1.6) becomes

$$u(r,\phi) = \int_0^1 r'\, dr' \int_0^{2\pi} g(r,\phi; r',\phi') u(r',\phi')\, d\phi',$$

where $g(r,\phi; r',\phi')$ is given explicitly by (3.13), Chapter 8.

Example 5. The concentration $u(x)$ of a substance diffusing in an absorbing medium between parallel walls satisfies

(1.7) $$-u'' + q(x)u = f(x), \quad 0 < x < 1; \qquad u(0) = \gamma_1, \quad u(1) = \gamma_2,$$

where γ_1, γ_2 are the prescribed concentrations at the walls, $f(x)$ is the given source density, and $q(x)$ is the known absorption coefficient. If we knew Green's function for the differential operator in (1.7) we could write the solution u explicitly in the form (2.16), Chapter 3. Usually this Green's function is not known, however, so that we try to formulate the problem in terms of the simpler Green's function $g(x,\xi)$ satisfying

(1.8) $$-g'' = \delta(x-\xi), \quad 0 < x, \xi < 1; \qquad g(0,\xi) = g(1,\xi) = 0.$$

This function $g(x,\xi)$ is given explicitly by (1.3), Chapter 1. Regarding $q(x)u$ as an additional inhomogeneous term, we can translate (1.7) into the integral equation

(1.9) $$u(x) = \gamma_1 + (\gamma_2 - \gamma_1)x + \int_0^1 g(x,\xi) [f(\xi) - q(\xi) u(\xi)]\, d\xi, \qquad 0 < x < 1.$$

If $q(x)$ is small, (1.9) can be used as the basis of a regular perturbation scheme (see Chapter 9). If, on the other hand, q is very large, which is equivalent to a small diffusion constant, the techniques of singular perturbation will be needed.

Example 6. It is shown in Exercise 1.1 that, under the assumption of isotropic scattering, the neutron transport equation in a homogeneous slab

$a < x < b$ can be transformed into the integral equation

$$(1.10) \qquad u(x) = \lambda \int_a^b E_1(|x - \xi|) u(\xi) \, d\xi + f(x), \qquad a < x < b,$$

where λ and $f(x)$ are given, and $E_1(x)$ is the exponential integral (1.31). Equation (1.10) with $a = 0$, $b = \infty$ was one of the first applications of the Wiener-Hopf method (see Noble; Carrier, Krook, and Pearson; and Stakgold).

Example 7. The steady temperature $u(x)$ in a homogeneous medium occupying the domain $\Omega \subset R_3$ with prescribed temperature $f(x)$ on the boundary Γ satisfies

$$(1.11) \qquad\qquad -\Delta u = 0, \quad x \in \Omega; \qquad u|_\Gamma = f(x).$$

This BVP is equivalent to the integral equation

$$(1.12) \qquad\qquad \int_\Gamma \frac{1}{4\pi |x - \xi|} I(\xi) \, dS_\xi = f(x), \qquad x \in \Gamma,$$

where, once $I(x)$ has been found on Γ, we can calculate $u(x)$ in Ω from

$$(1.13) \qquad\qquad u(x) = \int_\Gamma \frac{1}{4\pi |x - \xi|} I(\xi) \, dS_\xi.$$

Note that (1.12) is an integral equation on the two-dimensional surface Γ, whereas the original problem (1.11) is three-dimensional.

Example 8. A nonlinear integral equation. As we saw in Example 6, Section 4, Chapter 4, the problem of radiation from the web coupling between two satellites leads to the nonlinear differential equation for the temperature $u(x)$:

$$(1.14) \qquad -u'' - \lambda u^4 = 0, \quad 0 < x < 1; \qquad u(0) = u(1) = 1.$$

By means of Green's function (1.8), the problem can be translated into the nonlinear integral equation

$$(1.15) \qquad\qquad u(x) = 1 - \lambda \int_0^1 g(x, \xi) u^4(\xi) \, d\xi,$$

which was solved by successive approximations for a range of values of λ.

Integral Operators

The principal ingredient in a linear integral equation is a linear integral operator K defined by

$$(1.16) \qquad z(x) = Ku = \int_{\Omega} k(x,\xi)u(\xi)\,d\xi, \qquad x \in \Omega.$$

Here Ω is a region, possibly unbounded, in R_n; $x = (x_1,\ldots,x_n)$ and $\xi = (\xi_1,\ldots,\xi_n)$ are points in Ω; and $d\xi = d\xi_1 \ldots d\xi_n$ is an n-dimensional element volume. The complex-valued functions $u(x)$ and $z(x)$, defined on Ω, are regarded as elements of a linear (function) space X to be specified later. The transformation K maps X into itself and is clearly linear:

$$K(\alpha u_1 + \beta u_2) = \alpha K u_1 + \beta K u_2$$

for all scalars α, β and all elements u_1 and u_2 in X. The transformation takes place through the intermediary of a *kernel* $k(x,\xi)$, a complex-valued function defined on the $2n$-dimensional Cartesian product $\Omega \times \Omega$. In many problems $k(x,\xi)$ has a natural definition for x outside Ω, so that K can be viewed as mapping functions u defined on Ω into functions z defined on some other region. The drawback to this point of view comes when we try to invert K. If z is assigned on a region Ω' larger than Ω, (1.16) will usually not have any solution, whereas if z is specified on a region smaller than Ω, the data is insufficient to determine u (see, however, Example 3).

For simplicity, but with no real loss of generality, we confine ourselves to problems where Ω is an interval, possibly unbounded, on the real axis. Thus our integral operator K transforms functions defined on $a \leqslant x \leqslant b$ into functions defined on the same interval through the formula

$$(1.17) \qquad z(x) = Ku = \int_{a}^{b} k(x,\xi)u(\xi)\,d\xi, \qquad a \leqslant x \leqslant b,$$

where $k(x,\xi)$ is a given function on the square $a \leqslant x, \xi \leqslant b$. We shall try to develop an L_2 theory of integral equations, so that we can take advantage of all the structure of a Hilbert space. Thus the domain of K will be $L_2^{(c)}(a,b)$, the set of all complex-valued functions $u(x)$ with finite L_2 norm, that is,

$$\|u\|^2 = \int_{a}^{b} |u(x)|^2 dx < \infty.$$

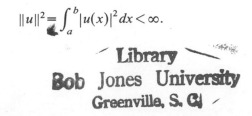

This norm is generated by the inner product

$$\langle u,v \rangle = \int_a^b u(x)\bar{v}(x)\,dx.$$

If K operates on an arbitrary element u of $L_2^{(c)}(a,b)$, how can we guarantee that $z = Ku$ is again in $L_2^{(c)}(a,b)$? This places only very mild restrictions on the kernel $k(x,\xi)$. We have shown that Hilbert-Schmidt kernels (Example 6, Section 1, Chapter 5) generate integral operators that are not only bounded but even compact; the Holmgren kernels of Exercise 1.3, Chapter 5, also generate bounded operators. We recapitulate some of these definitions below and also extend the theory.

Recall that an operator K on $L_2^{(c)}(a,b)$ is bounded if there exists a constant M such that $\|Ku\| \leqslant M\|u\|$ for every u in $L_2^{(c)}(a,b)$. The norm of K is then defined as

$$\|K\| = \sup_{u \neq 0} \frac{\|Ku\|}{\|u\|}.$$

The operator K is compact if it transforms bounded sets into relatively compact sets.

Definition. A kernel of the form $\sum_{i=1}^n p_i(x)\bar{q}_i(\xi)$ with n finite and $\int_a^b |p_i(x)|^2\,dx < \infty$, $\int_a^b |q_i(x)|^2\,dx < \infty$ is said to be *separable* (or *degenerate*), and so is the corresponding integral operator defined by (1.17).

Definition. A kernel $k(x,\xi)$ for which

$$(1.18) \qquad \int_a^b \int_a^b |k(x,\xi)|^2\,dx\,d\xi < \infty$$

is a *Hilbert-Schmidt* (or *H.-S.*) *kernel*, and the corresponding integral operator is known as an *H.-S. operator*.

Definition. A kernel $k(x,\xi)$ for which

$$(1.19) \qquad \sup_{a < \eta \leqslant b} \int_a^b \int_a^b |k(x,\xi)||k(x,\eta)|\,dx\,d\xi < \infty$$

is said to be a *Holmgren kernel*.

It is clear that a separable kernel is Hilbert-Schmidt. We showed in (1.9), Chapter 5, that for an H.-S. operator the following estimate holds:

$$(1.20) \qquad \|K\| \leqslant \left[\int_a^b \int_a^b |k(x,\xi)|^2\,dx\,d\xi \right]^{1/2}.$$

A Holmgren kernel also generates a bounded operator (see Exercise 1.3, Chapter 5). We now come to the principal feature of H.-S. operators.

Theorem. An H.-S. operator is compact.

Proof. (a) We first show that a separable operator S is compact. For each f in $L_2^{(c)}(a,b)$ we have

$$Sf = \int_a^b \sum_{i=1}^n p_i(x)\bar{q}_i(\xi) f(\xi)\, d\xi = \sum_{i=1}^n c_i p_i(x).$$

The range of S is finite-dimensional and S is clearly bounded, so that it is compact (see Section 7, Chapter 5).

(b) It can be shown that, if $\{e_i(x)\}$ is an orthonormal basis for $L_2^{(c)}(a,b)$, then $\{e_i(x)\bar{e}_j(\xi)\}$ is an orthonormal basis for $L_2^{(c)}(D)$, where D is the square $a \le x, \xi \le b$.

(c) Let $k(x,\xi)$ be an H-S. kernel; we can write $k(x,\xi) = \sum_{i,j=1}^\infty k_{ij}e_i(x)\bar{e}_j(\xi)$, where $k_{ij} = \int_a^b \int_a^b k(x,\xi)\bar{e}_i(x)e_j(\xi)\,dx\,d\xi$. By Parseval's identity

$$\sum_{i,j=1}^\infty |k_{ij}|^2 = \int_a^b \int_a^b |k(x,\xi)|^2 dx\, d\xi,$$

and therefore

$$\sum_{i,j=n+1}^\infty |k_{ij}|^2 = \int_a^b \int_a^b \left| k(x,\xi) - \sum_{i,j=1}^n k_{ij}e_i(x)\bar{e}_j(\xi) \right|^2 dx\, d\xi.$$

By choosing n sufficiently large the last double integral can be made smaller than ε. With n so chosen let S_ε be the separable operator generated by the kernel $s_\varepsilon(x,\xi) = \sum_{i,j=1}^n k_{ij}e_i(x)\bar{e}_j(\xi)$. Clearly $K - S_\varepsilon$ is Hilbert-Schmidt and is generated by $k(x,\xi) - s_\varepsilon(x,\xi)$. Therefore, by (1.20), we have

$$\|K - S_\varepsilon\|^2 \le \int_a^b \int_a^b |k(x,\xi) - s_\varepsilon(x,\xi)|^2 dx\, d\xi < \varepsilon.$$

Thus K can be approximated in norm by compact operators and must itself be compact by Theorem 2, Section 7, Chapter 5.

An H.-S. kernel $k(x,\xi)$ always generates a *compact operator*, even though $k(x,\xi)$ may be unbounded as a function of x and ξ. For instance, let $a = 0, b = 1, k(x,\xi) = 1/|x-\xi|^\alpha$ with $0 < \alpha < \frac{1}{2}$; then (1.18) is satisfied, so that k generates an H.-S. operator; if, however, $\frac{1}{2} \le \alpha < 1$, the operator is no longer Hilbert-Schmidt but is nevertheless compact (see Exercise 1.2). The kernel $e^{-|x-\xi|}$ on $-\infty < x, \xi < \infty$ is not Hilbert-Schmidt and generates an operator that is *not* compact; the operator is bounded, however, because it is a Holmgren operator (see Exercise 1.3, Chapter 5). The kernel $e^{-x^2 - \xi^2}$ on $-\infty < x, \xi < \infty$ is Hilbert-Schmidt.

If K is a bounded operator, its adjoint K^* is defined from

$$(1.21) \qquad \langle Ku, v \rangle = \langle u, K^* v \rangle \qquad \text{for all } u, v \in L_2^{(c)}(a, b).$$

By Theorem 1, Section 8, Chapter 5, we see that K^* is also bounded and that

$$(1.22) \qquad \|K^*\| = \|K\|.$$

If K is an integral operator with kernel $k(x, \xi)$, then K^* is also an integral operator whose kernel $k^*(x, \xi)$ satisfies

$$(1.23) \qquad k^*(x, \xi) = \bar{k}(\xi, x).$$

The operator K is *self-adjoint* (or *symmetric*, the words being synonymous for bounded operators) if $K = K^*$. The kernel of a self-adjoint operator satisfies

$$(1.24) \qquad k(x, \xi) = \bar{k}(\xi, x).$$

The operator $K^2 = KK$ is defined from

$$K^2 u = \int_a^b k(x, \xi_1) \, d\xi_1 \int_a^b k(\xi_1, \xi) u(\xi) \, d\xi$$
$$= \int_a^b k_2(x, \xi) u(\xi) \, d\xi,$$

where

$$(1.25) \qquad k_2(x, \xi) = \int_a^b k(x, \xi_1) k(\xi_1, \xi) \, d\xi_1.$$

Thus K^2 is itself an integral operator whose kernel $k_2(x, \xi)$ can be obtained from k by (1.25). Similarly K^m is an integral operator with kernel

$$(1.26) \qquad k_m(x, \xi) = \int_a^b \cdots \int_a^b k(x, \xi_1) k(\xi_1, \xi_2) \cdots k(\xi_{m-1}, \xi) \, d\xi_1 \cdots d\xi_{m-1},$$

requiring $m - 1$ integrations. The kernels (1.25) and (1.26) are known as *iterated kernels.* We note that $k_m(x,\xi)$ can also be written as

$$(1.27) \qquad k_m(x,\xi) = \int_a^b k(x,\xi_1)k_{m-1}(\xi_1,\xi)\,d\xi_1.$$

The iterated kernels will be symmetric if $k(x,\xi)$ is symmetric. In any event we find that

$$\|K^m\| \leqslant \|K\|^m.$$

Exercises

1.1. Consider neutron transport in the slab $a < Z < b$ when neutrons are allowed to leave the reactor on reaching the boundary but no neutrons are allowed to enter the slab through the boundary. With an isotropic source, (5.20), Chapter 0, reduces to

$$(1.28) \qquad \mu\frac{\partial\psi}{\partial Z} + \psi = \frac{c}{2}\int_{-1}^{1}\psi(Z,\mu)\,d\mu + Q^*(Z), \qquad a < Z < b,$$

$$(1.29) \qquad \psi(a,\mu) = 0, \quad \mu \geqslant 0; \qquad \psi(b,\mu) = 0, \quad \mu \leqslant 0.$$

Show that the integrodifferential equation (1.28) subject to the boundary conditions (1.29) can be transformed into an integral equation by the following steps.

For $\mu \neq 0$ multiply (1.28) by $e^{Z/\mu}$. Integrate from $Z = z$ to $Z = b$ when $\mu < 0$, and from $Z = a$ to $Z = z$ when $\mu > 0$, to obtain

$$(1.30) \qquad \psi(z,\mu) = \begin{cases} \dfrac{1}{\mu}\displaystyle\int_a^z e^{(Z-z)/\mu}w(Z)\,dZ, & \mu > 0, \\[2ex] -\dfrac{1}{\mu}\displaystyle\int_z^b e^{(Z-z)/\mu}w(Z)\,dZ, & \mu < 0, \end{cases}$$

where $w(Z)$ is the right side of (1.28). Integrate (1.30) over μ, and change the order of integration to obtain

$$u(z) \doteq \int_{-1}^{1}\psi(z,\mu)\,d\mu = \int_a^b E_1(|z - Z|)w(Z)\,dZ,$$

where

$$(1.31) \qquad E_1(x) \doteq \int_1^\infty \frac{e^{-xt}}{t}\,dt = \int_x^\infty \frac{e^{-y}}{y}\,dy, \qquad x > 0.$$

In view of the definition of w, it follows that $u(z)$ satisfies (1.10).

1.2. Consider the kernel $k(x,y)=1/|x-y|^{\alpha}$, $a \leqslant x,y \leqslant b$, where a and b are finite and $\alpha < 1$.

(a) The corresponding integral operator K is Hilbert-Schmidt for $\alpha < \frac{1}{2}$. Show that K is compact for $\alpha < 1$. *Hint:* Use Theorem 2, Section 7, Chapter 5, which states that K is compact if for each $\varepsilon > 0$ one can write K as the sum of a compact operator H and a bounded operator L with $\|L\| < \varepsilon$. The splitting required is of the type

$$k(x,y)=h(x,y)+l(x,y),$$

where l coincides with k on $|x-y| < \eta$ and *vanishes* for $|x-y| > \eta$.

(b) Show that the iterated kernel $k_2(x,y)$ is bounded for $\alpha < \frac{1}{2}$ and satisfies $|k_2| \leqslant c/|x-y|^{2\alpha-1}$ for $\frac{1}{2} < \alpha < 1$.

1.3. Let D be a bounded domain in R_n, $n \geqslant 3$, and let

$$k(x,\xi)=\frac{a(x,\xi)}{|x-\xi|^{n-2}}, \qquad x,\xi \in D,$$

where $a(x,\xi)$ is continuous for x and ξ in $\overline{D}$.

(a) Show that for $n=3$ the operator K is Hilbert-Schmidt.

(b) Show that for $n \geqslant 3$ the operator K is compact (use the hint in Exercise 1.2).

2. FREDHOLM INTEGRAL EQUATIONS

We shall consider linear integral equations of the form

$$(2.1) \qquad Ku-\lambda u=f, \quad \text{that is,} \quad \int_a^b k(x,\xi)u(\xi)\,d\xi-\lambda u(x)=f(x), \quad a \leqslant x \leqslant b.$$

Here the unknown function $u(x)$ always appears under the integral sign and, unless $\lambda = 0$, also appears outside the integral. The kernel $k(x,\xi)$, the inhomogeneous term $f(x)$, and the complex parameter λ are regarded as given, although we shall want to study the dependence of the solution(s) on f and λ.

For $\lambda \neq 0$, (2.1) is known as a *Fredholm equation of the second kind* and, for $\lambda = 0$, as a *Fredholm equation of the first kind*. If $f(x) \equiv 0$, we have the *eigenvalue problem*

$$(2.2) \qquad Ku-\lambda u=0 \qquad \text{or} \qquad \int_a^b k(x,\xi)u(\xi)\,d\xi=\lambda u(x), \quad a \leqslant x \leqslant b,$$

which, for most values of λ, can be expected to have the unique solution $u(x) \equiv 0$; the interest is in finding the exceptional values of λ (*eigenvalues*) for which (2.2) has nontrivial solutions $u(x)$ (*eigenfunctions*). Since K is a linear operator, the set of all eigenfunctions corresponding to the *same* eigenvalue forms a linear manifold. The dimension of this manifold is the *multiplicity* of the eigenvalue.

Our theory will be developed principally for the case where K is a compact integral operator (say, generated by an H.-S. kernel), but we will occasionally consider problems where K is only a bounded operator.

Before trying to solve (2.1) in a more or less explicit fashion, let us recall the alternative theorem in Section 7, Chapter 5, which tells us about existence, uniqueness, and solvability conditions. As we know, an important role is played by the adjoint homogeneous problem

$$(2.3) \quad K^*v - \bar{\lambda}v = 0 \quad \text{or} \quad \int_a^b \bar{k}(\xi, x)v(\xi)\,d\xi = \bar{\lambda}v(x), \quad a \leqslant x \leqslant b;$$

this features the operator $K^* - \bar{\lambda}I$, which is the adjoint of $K - \lambda I$ appearing in (2.1) and (2.2).

Theorem 1. Let λ be fixed, $\lambda \neq 0$, and let K be *compact*. Then *either* of the following alternatives holds:

(a) The number λ is not an eigenvalue of K and the number $\bar{\lambda}$ is not an eigenvalue of K^*. For each f, (2.1) has one and only one solution.

(b) The number λ is an eigenvalue of K having finite multiplicity k, say. The number $\bar{\lambda}$ is an eigenvalue of K^* having the same multiplicity. Equation (2.1) has solution(s) if and only if f is orthogonal to all k independent solutions of (2.2).

Remarks

1. The case $\lambda = 0$ must be treated separately. The equation $Ku = f$ will have solutions only if f is orthogonal to the null space of K^*, but this condition will usually not be sufficient because the range of K may not be closed.

2. $\lambda = 0$ can be an eigenvalue of infinite multiplicity (see Example 1 below).

3. If K is bounded but not compact, even a nonzero eigenvalue can have infinite multiplicity.

Example 1. Consider the separable kernel

$$(2.4) \qquad k(x,\xi) = \sum_{j=1}^{n} p_j(x)\bar{q}_j(\xi), \qquad 0 \leqslant x, \xi \leqslant 1,$$

where each of the two sets $\{p_j\}$ and $\{q_j\}$ is assumed to be an independent set. (This is no restriction, for otherwise the kernel could be rewritten in this form with a smaller n.) We denote by M_P and M_Q the n-dimensional linear manifolds spanned by $\{p_j\}$ and $\{q_j\}$, respectively. If u is any vector in H, we have

$$(2.5) \qquad Ku = \int_0^1 k(x,\xi)u(\xi)\,d\xi = \sum_{j=1}^{n} \langle u, q_j \rangle p_j(x).$$

(a) *The eigenvalue problem.* We see that (2.2) becomes

$$(2.6) \qquad \sum_{j=1}^{n} \langle u, q_j \rangle p_j(x) = \lambda u(x), \qquad 0 \leqslant x \leqslant 1.$$

Obviously $\lambda = 0$ is an eigenvalue. In view of the independence of the set $\{p_j\}$ the corresponding eigenfunctions must satisfy

$$\langle u, q_1 \rangle = \cdots = \langle u, q_n \rangle = 0, \qquad \text{that is, } u \in M_Q^{\perp}.$$

For λ to be a nonzero eigenvalue of (2.6) we must have $u(x) = \sum_{i=1}^{n} c_i p_i(x)$, which, on substitution in (2.6), leads to

$$(2.7) \qquad \sum_{j=1}^{n} \langle p_j, q_i \rangle c_j = \lambda c_i, \qquad i = 1, \ldots, n,$$

which is the eigenvalue problem for the n by n matrix $\alpha_{ij} = \langle p_j, q_i \rangle$. Thus any nonzero eigenvalue of (2.6) must be a nonzero eigenvalue of (2.7). Conversely, we can check that any nonzero eigenvalue λ of (2.7) is an eigenvalue of (2.6).

Let us look at some special cases with $n = 2$. If all four matrix elements vanish, (2.7) has no nonzero eigenvalues. This is the case for the kernel $\sin \pi x \sin 2\pi \xi + \sin 3\pi x \sin 4\pi \xi$, which therefore has only the zero eigenvalue with eigenfunctions u such that

$$\int_0^1 u(x)\sin 2\pi x\,dx = \int_0^1 u(x)\sin 4\pi x\,dx = 0.$$

If $k(x,\xi) = \sin \pi x \sin \pi \xi + \theta \cos \pi x \cos \pi \xi$, θ given $\neq 0$, then (2.7) becomes the pair of equations

$$\tfrac{1}{2}c_1 = \lambda c_1, \qquad \tfrac{1}{2}\theta c_2 = \lambda c_2,$$

from which we find (if $\theta \neq 1$) that there are two eigenvalues, $\lambda_1 = \tfrac{1}{2}$, $\lambda_2 = \theta/2$, with corresponding eigenfunctions $u_1 = A_1 \sin \pi x$, $u_2 = A_2 \cos \pi x$, where A_1 and A_2 are arbitrary. If, however, $\theta = 1$, then $\lambda = \tfrac{1}{2}$ is a double eigenvalue with the eigenfunctions $A_1 \sin \pi x + A_2 \cos \pi x$. Regardless of the value of $\theta(\neq 0)$, we have $\lambda = 0$ as an eigenvalue with eigenfunctions satisfying

$$\int_0^1 u(x) \sin \pi x \, dx = \int_0^1 u(x) \cos \pi x \, dx = 0.$$

(b) *The inhomogeneous equation.* If $\lambda = 0$, (2.1) becomes

$$(2.8) \qquad \sum_{j=1}^{n} \langle u, q_j \rangle p_j(x) = f(x), \qquad 0 \leqslant x \leqslant 1.$$

A solution is possible *only if f is in M_P.* If this is the case, write $u = v + w$, with $v \in M_Q$, $w \in M_Q^{\perp}$. Equation (2.8) then states that

$$(2.9) \qquad \sum_{j=1}^{n} \langle v, q_j \rangle p_j = f, \qquad f \in M_P.$$

For each f in M_P there is therefore a unique v in M_Q satisfying (2.9). In fact, if $f = \sum_{j=1}^{n} \xi_j p_j$, we need only find $v \in M_Q$ such that $\langle v, q_j \rangle = \xi_j$, $j = 1, \ldots, n$, which has the unique solution $v = \sum_{j=1}^{n} \xi_j q_j^*$, where $\{q_j^*\}$ is the dual basis (for M_Q) of $\{q_j\}$. Thus the general solution of (2.8) is $u = v + w$ with an arbitrary $w \in M_Q^{\perp}$.

If $\lambda \neq 0$, (2.1) becomes

$$(2.10) \qquad \sum_{j=1}^{n} \langle u, q_j \rangle p_j(x) - f(x) = \lambda u(x), \qquad 0 \leqslant x \leqslant 1,$$

so that u must be of the form $\sum_{i=1}^{n} c_i p_i - f/\lambda$. Substituting in (2.10), we obtain the inhomogeneous algebraic system

$$(2.11) \qquad \sum_{j=1}^{n} \langle p_j, q_i \rangle c_j - \lambda c_i = \frac{1}{\lambda} \langle f, q_i \rangle, \qquad i = 1, \ldots, n.$$

The corresponding homogeneous system is (2.7), and the usual theory for algebraic systems holds. In particular, if λ is not an eigenvalue of (2.7), (2.11) has a unique solution for each f. If this solution is $(c_1, \ldots, c_n)$, we can easily verify that (2.10) has the unique solution $u = \sum_{i=1}^{n} c_i p_i(x) - f(x)/\lambda$.

Example 2. Consider the H.-S. kernel

$$(2.12) \qquad k(x, \xi) = \begin{cases} 1 - \xi, & 0 \leqslant x \leqslant \xi, \\ 1 - x, & \xi \leqslant x \leqslant 1, \end{cases}$$

which is a continuous function on the square $0 \leqslant x, \xi \leqslant 1$ but has a discontinuous gradient along the diagonal $x = \xi$. This behavior suggests that k might be Green's function for a differential operator of the second order. If ξ is regarded as a fixed parameter, a simple calculation shows that

(2.13)

$$-\frac{d^2 k}{dx^2} = \delta(x - \xi), \quad 0 < x, \xi < 1; \qquad \frac{dk}{dx}(0, \xi) = 0, \quad k(1, \xi) = 0;$$

and k is indeed a Green's function. Note that (2.12) is the one and only solution of (2.13). From our previous work on differential equations, we can therefore conclude that the one and only solution $v(x)$ of

$$(2.14) \qquad -v'' = f, \quad 0 < x < 1; \qquad v'(0) = 0, \quad v(1) = 0$$

is

$$(2.15) \qquad v(x) = \int_0^1 k(x, \xi) f(\xi) \, d\xi \doteq Kf,$$

where K is the integral operator generated by the kernel $k(x, \xi)$. Thus K is the *inverse* of the differential operator $-(d^2/dx^2)$ defined on a domain of suitably smooth (precisely: v' absolutely continuous, v'' square integrable) functions $v(x)$ satisfying $v'(0) = v(1) = 0$. Note that the null space N_K consists only of the zero element. Indeed, if $Kf = 0$, then $v = 0$ and $-v'' = 0 = f$. The operator K in (2.15) is defined for all f in L_2, but the range R_K consists only of functions v which are sufficiently smooth and satisfy $v'(0) = v(1) = 0$. It is true however, that $\overline{R}_K = L_2^{(c)}(0, 1)$.

We shall now look at integral equations involving K. In the analysis we often transform the problem to an equivalent differential equation which we then solve. This may seem to deflate our earlier claim that integral equations are supposed to shed light on differential equations, rather than the other way around. For certain specific problems it may be simpler to solve the differential equation, but it is often easier to prove general theorems within the theory of integral equations.

(a) Consider first the eigenvalue problem

(2.16) $$Ku - \lambda u = 0.$$

For $\lambda \neq 0$, u is in R_K and is therefore twice differentiable. From (2.14) we have $(-Ku)'' = u$, so that

$$-u'' = \frac{1}{\lambda} u, \quad 0 < x < 1; \quad u'(0) = u(1) = 0,$$

which we write (with $\theta = 1/\lambda$) in the standard eigenvalue form

(2.17) $$-u'' = \theta u, \quad 0 < x < 1; \quad u'(0) = u(1) = 0.$$

Thus the nonzero eigenvalues of (2.16) and (2.17) are *reciprocals* of each other, while the corresponding eigenfunctions are the same. It is straightforward to solve (2.17) explicitly. We find that

$$\theta_n = \frac{(2n-1)^2 \pi^2}{4}, \quad u_n(x) = c \cos(2n-1)\frac{\pi x}{2}.$$

Therefore the nonzero eigenvalues and normalized eigenfunctions of (2.16) are

(2.18) $$\lambda_n = \frac{4}{(2n-1)^2 \pi^2}, \quad e_n(x) = \sqrt{2} \cos(2n-1)\frac{\pi x}{2}, \quad n = 1, 2, \ldots.$$

We have already seen that N_K consists only of the zero element, so that $\lambda = 0$ is *not* an eigenvalue of (2.16). It is worth noting, however, that 0 is a limit-point of the eigenvalues λ_n and that the set $\{e_n(x)\}$ is an orthonormal basis for $L_2^{(c)}(0,1)$. These properties are shared by all symmetric compact operators for which 0 is not an eigenvalue.

(b) Let us now investigate the inhomogeneous equation of the second kind

(2.19) $$Ku - \lambda u = f, \quad \lambda \neq 0.$$

Set $z = \lambda u + f$, and note that $z \in R_K$. Thus $-z'' = u$, and

(2.20) $$-z'' - \frac{z}{\lambda} = -\frac{f}{\lambda}, \quad 0 < x < 1; \quad z'(0) = z(1) = 0.$$

Through the relation $z = \lambda u + f$, any solution u of (2.19) generates a solution z of (2.20) and vice versa. If $\lambda \neq \lambda_n$ [see (2.18)], then (2.20) has a unique solution which can be obtained by using Green's function $g(x, \xi; \lambda)$

satisfying

(2.21) $-g'' - \dfrac{g}{\lambda} = \delta(x-\xi), \quad 0 < x, \xi < 1; \qquad g'|_{x=0} = g|_{x=1} = 0.$

If necessary, g can be calculated explicitly from (2.10), Chapter 3. In any event we have

(2.22) $u = \lambda^{-1}(z - f) = -\lambda^{-1}\left[\lambda^{-1}\displaystyle\int_0^1 g(x,\xi;\lambda) f(\xi)\, d\xi + f(x)\right].$

The solution can also be expressed in another form by expanding (2.19) or (2.20) in terms of the eigenfunctions $\{e_n\}$, a procedure which will be carried out systematically in Section 4; see (4.8) and (4.10).

If λ in (2.19) coincides with an eigenvalue, say λ_k, then (2.20) and (2.19) can be solved only if $\langle f, e_k \rangle = 0$. The solution is then not unique. Problem (2.21) is inconsistent, and a modified Green's function must be used (see Section 5, Chapter 3).

If $\lambda = 0$ in (2.19), we have an equation of the first kind. We have already seen that R_K consists only of twice-differentiable functions $v(x)$ satisfying $v'(0) = v(1) = 0$. If $f(x)$ is such a function, $Ku = f$ has the one and only solution $u = -f''$.

Example 3.   Consider the kernel

$$k(x,\xi) = \frac{e^{-|x-\xi|}}{2} \qquad \text{on the plane } -\infty < x, \xi < \infty.$$

Since

$$\int\!\!\!\int_{-\infty}^{\infty} |k(x,\xi)|^2\, dx\, d\xi = \int_{-\infty}^{\infty} dx \int_{-\infty}^{\infty} \frac{e^{-2|y|}}{4}\, dy = \infty,$$

we do *not* have an H.-S. kernel. However, the criterion of Exercise 1.3, Chapter 5, for Holmgren kernels is satisfied because

$$\max_{-\infty < z < \infty} \frac{1}{4}\int\!\!\!\int_{-\infty}^{\infty} e^{-|x-y|}e^{-|x-z|}\, dx\, dy = 1.$$

Thus the integral operator K generated by the kernel k is bounded, and $\|K\| \leqslant 1$. Exercise 2.2 shows that $\|K\| = 1$. Since K is also symmetric, it is self-adjoint and its spectrum is confined to the real axis.

(a) The eigenvalue problem $Ku - \lambda u = 0$ becomes

(2.23) $\displaystyle\int_{-\infty}^{\infty} \frac{e^{-|x-\xi|}}{2} u(\xi)\, d\xi - \lambda u(x) = 0, \qquad -\infty < x < \infty.$

The integral term is the convolution of the functions $e^{-|x|}/2$ and $u(x)$, so that Fourier transforms are indicated. Using the results of Chapter 2 (Example 1, Section 4, and Exercise 4.5), we find, with ω the real transform variable, that

$$(2.24) \qquad u^{\hat{}}(\omega)\left[\frac{1}{1+\omega^2}-\lambda\right]=0.$$

If λ is not in $0<\lambda\leqslant 1$, the bracketed term does not vanish. If $\lambda=1$, the bracketed term vanishes at $\omega=0$, and, if $0<\lambda<1$, it vanishes at $\omega=\omega_1$ and $\omega=\omega_2$, where

$$(2.25) \qquad \omega_{1,2}=\pm\left(\frac{1}{\lambda}-1\right)^{1/2}.$$

In any event the only L_2 solution of (2.24) is $u^{\hat{}}=0$, so that (2.23) has *no eigenvalues*. If for the moment we permit distributional solutions, then, for $0<\lambda<1$, we can have

$$u^{\hat{}}(\omega)=c_1\delta(\omega-\omega_1)+c_2\delta(\omega-\omega_2),$$

which, on inversion, gives

$$(2.26) \qquad u(x)=c_1e^{-i\omega_1 x}+c_2e^{-i\omega_2 x}.$$

This function is not in L_2 but nevertheless satisfies (2.23) in the classical sense. Thus one can view the functions $e^{-i\omega_1 x}$ and $e^{-i\omega_2 x}$ as "pseudo eigenfunctions" corresponding to the value $\lambda(0<\lambda<1)$. A more satisfactory description is that the interval $0\leqslant\lambda\leqslant 1$ is in the *continuous spectrum* of K. Since eigenvalues have already been ruled out, it suffices to show that there exists a sequence $\{u_n\}$ such that $\|Ku_n-\lambda u_n\|/\|u_n\|$ tends to 0 as $n\to\infty$. Now

$$(Ku_n)^{\hat{}}=\frac{1}{1+\omega^2}u_n^{\hat{}},$$

and by Parseval's theorem

$$\frac{\|Ku_n-\lambda u_n\|}{\|u_n\|}=\frac{\|(Ku_n)^{\hat{}}-\lambda u_n^{\hat{}}\|}{\|u_n^{\hat{}}\|}=\frac{\|[1/(1+\omega^2)-\lambda]u_n^{\hat{}}(\omega)\|}{\|u_n^{\hat{}}(\omega)\|}.$$

It remains only to show that the interval $0\leqslant\lambda\leqslant 1$ is in the continuous spectrum for the multiplication operator $1/(1+\omega^2)$ on $L_2^{(c)}(-\infty,\infty)$.

This is essentially the content of Example 5, Section 6, Chapter 5, and of Exercise 6.4, Chapter 5. All one has to do is to take for $u_n^\wedge(\omega)$ the function which has the value 1 in a small interval about a point where $1/(1+\omega^2)-\lambda=0$, and has the value 0 elsewhere [for the case $\lambda=0$ this means that $u_n^\wedge(\omega)$ is taken to be 1 outside a large interval on the ω axis].

An alternative approach is to translate (2.23) into a differential equation. It is easy to see that $k(x,\xi)$ is a Green's function:

$$(2.27) \qquad -\frac{d^2k}{dx^2}+k=\delta(x-\xi), \quad -\infty<x,\xi<\infty; \quad k(\pm\infty,\xi)=0.$$

Thus (2.23) is equivalent to $u-\lambda(-u''+u)=0$. If $\lambda=0$, $u=0$, and for $\lambda\neq0$ we have

$$(2.28) \qquad -u''+u=\theta u, \qquad \theta=\frac{1}{\lambda}, \quad -\infty<x<\infty.$$

For $\theta\neq1$ the general solution of (2.28) is $A\exp(\sqrt{1-\theta}\,x)$ $+B\exp(-\sqrt{1-\theta}\,x)$, which is never in $L_2^{(c)}(-\infty,\infty)$ for any value of θ in the complex plane (unless $A=B=0$). For $\theta=1$ the solution is $A+Bx$, which is not in L_2 unless $A=B=0$. Thus we see that (2.28) has no eigenvalues, and neither does (2.23).

(b) The inhomogeneous equation $Ku-\lambda u=f$ becomes

$$(2.29) \qquad \int_{-\infty}^{\infty}\frac{e^{-|x-\xi|}}{2}u(\xi)\,d\xi-\lambda u(x)=f(x), \qquad -\infty<x<\infty.$$

Given the element f in L_2 and the complex number λ, we want to find the L_2 solution(s) u of (2.29). In the absence of eigenvalues the solution, if any, is necessarily unique. Taking Fourier transforms, we obtain

$$u^\wedge\left[\frac{1}{1+\omega^2}-\lambda\right]=f^\wedge \quad \text{or} \quad u^\wedge=-\frac{f^\wedge}{\lambda}\left[\frac{\omega^2+1}{\omega^2+1-(1/\lambda)}\right], \quad \lambda\neq0.$$

In the second equality the term in brackets is further decomposed as

$$1+\frac{1}{\lambda}\frac{1}{\omega^2+1-(1/\lambda)},$$

where the second term now vanishes at $\omega=\pm\infty$. Thus we can write

$$(2.30) \qquad u^\wedge=-\frac{f^\wedge}{\lambda}-\frac{f^\wedge}{\lambda^2}\left[\frac{1}{\omega^2+1-(1/\lambda)}\right].$$

If λ is real but *not* in $0 \leqslant \lambda \leqslant 1$, then $\omega^2 + 1 - (1/\lambda)$ does not vanish on the real axis. In fact, we have

$$\omega^2 + 1 - \frac{1}{\lambda} = (\omega - \beta_1)(\omega - \beta_2), \qquad \beta_{1,2} \doteq \pm i\left(1 - \frac{1}{\lambda}\right)^{1/2}.$$

We are therefore entitled to set

$$\frac{1}{\omega^2 + 1 - (1/\lambda)} \doteq b\hat{\ }(\omega),$$

where $b\hat{\ }$ is the Fourier transform of the L_2 function

$$b(x) = \frac{1}{2\pi} \int_{-\infty}^{\infty} e^{-i\omega x} b\hat{\ }(\omega) \, d\omega,$$

which can be evaluated as in Section 4, Chapter 2, by using a contour in the lower half-plane when $x > 0$ and one in the upper half-plane when $x < 0$. A straightforward calculation gives

$$b(x) = \frac{1}{2[1 - (1/\lambda)]^{1/2}} e^{-|x|[1 - (1/\lambda)]^{1/2}},$$

and after use of the convolution theorem (2.30) yields

$$(2.31) \qquad u(x) = -\frac{f(x)}{\lambda} - \frac{1}{\lambda^2} \int_{-\infty}^{\infty} b(x - \xi) f(\xi) \, d\xi,$$

a result which can also be obtained by using a Green's function technique on the BVP equivalent to (2.29). Even if λ is complex, $\lambda \notin [0,1]$, (2.31) provides the one and only solution of (2.29) as long as $[1 - (1/\lambda)]^{1/2}$ in the definition of $b(x)$ is chosen to have a positive real part. If λ is in the continuous spectrum, that is, if $\lambda \in [0,1]$, then (2.29) will not have an L_2 solution for every f in L_2. In fact, the bracketed term in (2.30) then has singularities on the real axis at $\omega_{1,2}$ given by (2.25) and is no longer an L_2 function of ω; however, if its product with f is in L_2 (as happens if f vanishes rapidly enough at $\omega_{1,2}$), then (2.31) can still be used. Such questions will be examined systematically in Chapter 7.

Example 4. Let us take the same kernel as in Example 3 but over the finite square $-1 \leqslant x, \xi \leqslant 1$. The kernel is then Hilbert-Schmidt and again satisfies the differential equation (2.27) but with boundary conditions at the endpoints. For $x > \xi$, $k = e^{-(x - \xi)}$ and $k' = -e^{-(x - \xi)}$; therefore $k(1, \xi) = e^{-(1-\xi)}$, $k'(1, \xi) = -e^{-(1-\xi)}$. For $x < \xi$, $k = e^{x - \xi}$ and $k' = e^{x - \xi}$, so that

$k(-1,\xi) = e^{-1-\xi}, k'(-1,\xi) = e^{-1-\xi}$. The boundary conditions can then be taken as

$$k'(1,\xi) + k(1,\xi) = 0, \qquad k'(-1,\xi) - k(-1,\xi) = 0.$$

The integral equation

(2.32) $$\int_{-1}^{1} e^{-|x-\xi|} u(\xi)\, d\xi = \lambda u(x), \qquad -1 < x < 1,$$

is therefore equivalent to

(2.33) $-u'' + u = \theta u, \quad -1 < x < 1; \qquad u'(1) + u(1) = u'(-1) - u(-1) = 0,$

where $\theta = 1/\lambda$. Problem (2.33) yields an infinite set of eigenvalues with eigenfunctions forming an orthonormal basis.

Example 5. Consider the Fourier transform as an integral operator on $L_2^{(c)}(-\infty, \infty)$. Instead of denoting this operator by a capital letter in front of the function on which it acts, we use the caret (^) as a superscript following the function. Thus

(2.34) $$f^{\wedge}(x) = \int_{-\infty}^{\infty} e^{ixy} f(y)\, dy$$

is regarded as defining an integral operator ^ from $L_2^{(c)}(-\infty, \infty)$ into itself. It follows from the inversion formula for Fourier transforms that

(2.35) $\qquad f^{\wedge\wedge}(x) = 2\pi f(-x) \qquad$ and $\qquad f^{\wedge\wedge\wedge\wedge}(x) = 4\pi^2 f(x).$

The eigenvalue problem for the ^ operator is

(2.36) $$u^{\wedge}(x) = \lambda u(x),$$

which implies that $u^{\wedge\wedge\wedge\wedge}(x) = \lambda^4 u(x)$. In view of (2.35) the only possible eigenvalues of (2.36) are

(2.37) $$\lambda = \pm\sqrt{2\pi}, \qquad \pm i\sqrt{2\pi}.$$

To find the corresponding eigenfunctions let $f(x)$ be an even function in $L_2^{(c)}(-\infty, \infty)$. Then $f^{\wedge}(x)$ is also an even function of x. Let

$$u = f + \alpha f^{\wedge},$$

so that

$$u^\wedge(x)=f^\wedge(x)+\alpha f^{\wedge\wedge}(x)=f^\wedge(x)+2\pi\alpha f(-x)=f^\wedge(x)+2\pi\alpha f(x)$$
$$=2\pi\alpha\left[f(x)+\frac{1}{2\pi\alpha}f^\wedge(x)\right].$$

Thus $u^\wedge$ will satisfy (2.36) if $1/2\pi\alpha=\alpha$, that is, $\alpha=\pm(1/\sqrt{2\pi})$. The eigenvalues corresponding to these choices of α are $\pm\sqrt{2\pi}$. Clearly each of these eigenvalues has infinite multiplicity. Start with any even function, add $1/\sqrt{2\pi}$ times its transform, and the resulting function will satisfy (2.36) with $\lambda=\sqrt{2\pi}$. For example,

$$u=e^{-a|x|}+\frac{1}{\sqrt{2\pi}}\frac{2a}{a^2+x^2}$$

satisfies (2.36) with $\lambda=\sqrt{2\pi}$ for *any* value of $a>0$. These functions are independent, so that $\sqrt{2\pi}$ has infinite multiplicity. Similarly

$$u=e^{-ax^2}-\frac{1}{\sqrt{2a}}e^{-x^2/4a}$$

satisfies (2.36) with $\lambda=-\sqrt{2\pi}$. It can be shown that the eigenfunctions can be expressed in a natural way in terms of an orthonormal basis of Hermite functions (see Dym and McKean).

Exercises

2.1. Solve (2.19) by expanding in the orthonormal basis (2.18). Make sure to treat the special cases $\lambda=\lambda_n$ and $\lambda=0$.

2.2. Consider the integral operator K generated by a difference kernel:

$$Ku=\int_{-\infty}^{\infty}k(x-\xi)u(\xi)\,d\xi.$$

Observe that the kernel depends only on $x-\xi$ and stems from a function $k(x)$ of a single variable. Let $v=Ku$, and let $k^\wedge,u^\wedge,v^\wedge$ denote the Fourier transforms of $k(x),u(x),v(x)$, respectively. In the transform variable we have

$$v^\wedge(\omega)=k^\wedge(\omega)u^\wedge(\omega),$$

so that the integral operator is reduced to multiplication by the function $k^\wedge(\omega)$.

Show that

$$\|K\| = \sup_{-\infty < \omega < \infty} |k^{\wedge}(\omega)|,$$

which applied to Example 3 gives $\|K\| = 1$. Another way of obtaining $\|K\| = 1$ for Example 3 is to first show that $\|K\| \leqslant 1$ (as was done in the text) and then to exhibit an L_2 sequence $\{u_n(x)\}$ such that

$$\lim_{n \to \infty} \frac{\|Ku_n\|}{\|u_n\|} = 1.$$

Show that the sequence

$$u_n(x) = \begin{cases} 1, & |x| < n, \\ 0, & |x| > n, \end{cases}$$

has the required property. This sequence was suggested by the fact that $u(x) = 1$ is a pseudo eigenfunction associated with $\lambda = 1$, the largest value in the spectrum.

2.3. Find all the eigenvalues and eigenfunctions in Example 4, and discuss the inhomogeneous equation.

2.4. Consider the eigenvalue problem $Ku - \lambda u = 0$ on $L_2^{(c)}(-1, 1)$, where K is generated by the kernel $k(x, \xi) = 1 - |x - \xi|$. Translate into a differential equation with boundary conditions to find the eigenvalues and eigenfunctions.

2.5. In Example 5 exhibit eigenfunctions corresponding to the eigenvalues $i\sqrt{2\pi}$ and $-i\sqrt{2\pi}$.

2.6. Consider the kernel

$$k(x, \xi) = \sum_{n=1}^{\infty} \frac{\sin(n+1)x \sin n\xi}{n^2}, \qquad 0 \leqslant x, \xi \leqslant \pi.$$

Show that this is an unsymmetric H.-S. kernel and that the corresponding integral operator K has *no* eigenvalue. Compute $\|K\|$. Show that every value of λ except $\lambda = 0$ is in the *resolvent* set [that is, $Ku - \lambda u = f$ has a unique solution for each f, and $(K - \lambda I)^{-1}$ is bounded]. Show that $\lambda = 0$ is in the compression spectrum (compare with the modified shift of Example 10, Section 1, Chapter 5) and in the continuous spectrum.

2.7. Give an example of a bounded self-adjoint integral operator with a nonzero eigenvalue of infinite multiplicity. *Hint*: modify Example 5 by using a cosine or sine transform instead of a complex-exponential transform.

3. THE SPECTRUM OF A SELF-ADJOINT COMPACT OPERATOR

On a finite-dimensional space a nontrivial operator has at least one eigenvalue; if the operator is also self-adjoint, the set of all eigenvectors forms an orthonormal basis. Neither of these results holds for infinite-dimensional spaces such as $L_2^{(c)}(a,b)$ even if we confine ourselves to bounded integral operators. Indeed, the Volterra operator of Example 7, Section 6, Chapter 5, and the modified shift operator of Exercise 2.6 are not only bounded but also compact; neither of these unsymmetric operators has eigenvalues. The integral operator of Example 3, Section 2, is bounded (but not compact), is self-adjoint, and has no eigenvalues. However, if we restrict ourselves—as we shall—to compact self-adjoint integral operators, much of the theory for finite-dimensional spaces can be extended.

Let $k(x,\xi)$ be a symmetric kernel on $a \leqslant x, \xi \leqslant b$, that is, $k(x,\xi) = \bar{k}(\xi,x)$; then k generates a self-adjoint operator K on $L_2^{(c)}(a,b)$. If k satisfies the Hilbert Schmidt condition (1.18), K will be a compact operator. The eigenvalues are real, and eigenfunctions corresponding to different eigenvalues are orthogonal. To exclude the trivial operator we shall assume that $\|K\| \neq 0$. The existence of a nonzero eigenvalue then follows from the extremal principle of Theorem 8, Section 8, Chapter 5, and from its corollary. The eigenvalue λ_1 so characterized is the one of largest modulus, and $|\lambda_1| = \|K\|$. We restate the principle as follows.

Theorem 1. Let K be a compact self-adjoint operator. Then

$$(3.1) \qquad \qquad \sup_{\|v\|=1} |\langle Kv, v \rangle|$$

is attained for a normalized eigenfunction $e_1(x)$ of K, corresponding to an eigenvalue λ_1 whose absolute value is just the maximum in question. No eigenvalue of K can have a larger modulus. We have $|\lambda_1| = \|K\|$ and therefore $\|Kv\| \leqslant |\lambda_1| \|v\|$ for every v in H.

Eigenvectors corresponding to other eigenvalues are orthogonal to e_1. This suggests maximizing $|\langle Kv, v \rangle|$ over unit functions orthogonal to e_1 and thus obtaining another eigenpair (λ_2, e_2) with $|\lambda_2| \leqslant |\lambda_1|$. Let M_1 be the manifold (line) generated by e_1, and let $M_1^{\perp}$ be the manifold (necessarily closed) of elements orthogonal to e_1. We then have the following theorem.

Theorem 2.

(3.2)
$$\sup_{\substack{\|v\|=1 \\ v \in M_1^{\perp}}} |\langle Kv, v \rangle|$$

is attained for a normalized eigenfunction $e_2(x)$ of K corresponding to an eigenvalue λ_2 whose modulus is the maximum in question. Moreover, $|\lambda_2| \leqslant |\lambda_1|$ and

(3.3)
$$\|Kv\| \leqslant |\lambda_2| \|v\| \qquad \text{for every } v \text{ in } M_1^{\perp}.$$

Remark. It is clear that the supremum (3.1) cannot be less than the supremum (3.2), since the functionals are the same but the latter supremum is over a subspace of the former. It is possible, however, for the suprema to be equal if the eigenvalue of largest modulus is degenerate or if both $\|K\|$ and $-\|K\|$ are eigenvalues.

Proof. $M_1^{\perp}$ is itself a Hilbert space. We claim that K can be regarded as an operator on $M_1^{\perp}$, that is, K transforms elements of $M_1^{\perp}$ into elements of $M_1^{\perp}$. Indeed, if $v \in M_1^{\perp}$, we have $\langle v, e_1 \rangle = 0$ and

$$\langle Kv, e_1 \rangle = \langle v, Ke_1 \rangle = \lambda_1 \langle v, e_1 \rangle = 0.$$

Moreover, it is clear that K is compact and self-adjoint on $M_1^{\perp}$. Therefore we can apply Theorem 1 to the operator K on the Hilbert space $M_1^{\perp}$ to obtain Theorem 2.

We can now continue in this way to characterize successive eigenpairs $(\lambda_1, e_1), \ldots, (\lambda_n, e_n)$, where $|\lambda_1| \geqslant |\lambda_2| \geqslant \cdots \geqslant |\lambda_n|$. The set $(e_1, \ldots, e_n)$ is an orthonormal set. Let M_n be the manifold generated by $(e_1, \ldots, e_n)$, and let $M_n^{\perp}$ be the orthogonal manifold, that is, the closed subspace of H consisting of functions orthogonal to all the elements $e_1(x), \ldots, e_n(x)$. We can then characterize the next eigenpair as follows.

Theorem 3.

(3.4)
$$\sup_{\substack{\|v\|=1 \\ v \in M_n^{\perp}}} |\langle Kv, v \rangle|$$

is attained for a normalized eigenfunction $e_{n+1}(x)$ of K corresponding to an eigenvalue λ_{n+1} whose modulus is the maximum in question. Moreover, $|\lambda_{n+1}| \leqslant |\lambda_n|$, and

(3.5)
$$\|Kv\| \leqslant |\lambda_{n+1}| \|v\| \qquad \text{for every } v \in M_n^{\perp}.$$

Proof. The proof of Theorem 3 is similar to that of Theorem 2.

There are two possible outcomes of the succession of maximum principles.

1. At some finite stage (say, $n+1$) we have for the first time that the supremum is 0. Thus $|\lambda_n| > 0$, but the supremum (3.4) is 0. All successive suprema obviously also vanish. Thus we have n nonzero eigenvalues (counted with multiplicity); clearly $\lambda = 0$ is an eigenvalue of infinite multiplicity since every element in $M_n^\perp$ is a corresponding eigenfunction.

2. At no finite stage do we get a vanishing supremum. We therefore generate an infinite sequence of eigenvalues of positive modulus

$$(3.6) \qquad |\lambda_1| \geqslant |\lambda_2| \geqslant \cdots \geqslant |\lambda_n| \geqslant \cdots > 0$$

with a corresponding orthonormal set of eigenfunctions

$$(3.7) \qquad e_1(x), \qquad e_2(x), \qquad \ldots, \qquad e_n(x), \qquad \ldots .$$

In case 1, the operator K is *separable*. Indeed, let z be any element in H, and decompose z by the projection theorem as the sum of an element in M_n and one in $M_n^\perp$:

$$z = \langle z, e_1 \rangle e_1 + \cdots + \langle z, e_n \rangle e_n + v,$$

where $v \in M_n^\perp$. Since the supremum (3.4) is 0, (3.5) shows that $Kv = 0$. Thus

$$(3.8) \qquad Kz = \sum_{j=1}^{n} \lambda_j \langle z, e_j \rangle e_j = \sum_{j=1}^{n} \lambda_j e_j(x) \int_a^b z(\xi) \bar{e}_j(\xi) \, d\xi$$

$$= \int_a^b k(x, \xi) z(\xi) \, d\xi,$$

where

$$(3.9) \qquad k(x, \xi) = \sum_{j=1}^{n} \lambda_j e_j(x) \bar{e}_j(\xi).$$

In case 2 we have an infinite sequence of nonzero eigenvalues of decreasing modulus. We show first that $\lambda_n \to 0$. If not, the sequence $\{e_n/\lambda_n\}$ would be bounded and $K(e_n/\lambda_n) = e_n$; by the compactness of K, $\{e_n\}$ would have to contain a convergent subsequence, which is clearly false. Thus $\lambda_n \to 0$.

The question immediately arises whether the orthonormal set $\{e_j\}$ of eigenfunctions corresponding to nonzero eigenvalues forms a basis. Obviously it does not in case 1; even in case 2, where the set of eigenfunctions is infinite, it need not form a basis. We shall show, however, that the eigenfunctions corresponding to nonzero eigenvalues always form a basis for the *range* of K. This is an immediate conclusion from (3.8) for case 1. We need therefore consider only case 2. Let f be an arbitrary element of H; we want to show that

$$(3.10) \qquad\qquad \lim_{n\to\infty} \left\| Kf - \sum_{k=1}^{n} \langle Kf, e_k \rangle e_k \right\| = 0.$$

An easy calculation gives

$$Kf - \sum_{k=1}^{n} \langle Kf, e_k \rangle e_k = K\left(f - \sum_{k=1}^{n} \langle f, e_k \rangle e_k \right).$$

The element $v \doteq f - \sum_{k=1}^{n} \langle f, e_k \rangle e_k$ belongs to $M_n^{\perp}$, so that (3.5) yields

$$\|Kv\| \leqslant |\lambda_{n+1}| \left\| f - \sum_{k=1}^{n} \langle f, e_k \rangle e_k \right\| \leqslant |\lambda_{n+1}| \|f\|.$$

Since $\lambda_n \to 0$ as $n \to \infty$, we have proved (3.10). The result is contained in the following theorem.

Theorem 4   (Expansion theorem).   Any function in the range of K can be expanded in a Fourier series in the eigenfunctions of K corresponding to nonzero eigenvalues. These eigenfunctions $\{e_k\}$ must therefore form an orthonormal basis for R_K, but not necessarily for H. Thus, for every f in H, we have

$$(3.11) \qquad Kf = \Sigma \langle Kf, e_k \rangle e_k = \Sigma \langle f, Ke_k \rangle e_k = \Sigma \lambda_k \langle f, e_k \rangle e_k,$$

even though the series $\Sigma \langle f, e_k \rangle e_k$ may not represent f. All equalities are of course understood in the L_2 sense.

We can now show that the extremal principles yield all the nonzero eigenvalues and their corresponding eigenfunctions. Suppose there existed an eigenfunction v with eigenvalue $\lambda \neq 0$ not listed among the λ_k's. Then $Kv = \lambda v$, and by Theorem 4

$$Kv = \sum_{k=1}^{\infty} \langle Kv, e_k \rangle e_k = \lambda \sum_{k=1}^{\infty} \langle v, e_k \rangle e_k.$$

But v is orthogonal to the e_k's so that $Kv=0$, contradicting the assumption that v is an eigenfunction for a nonzero eigenvalue. Thus the sequence (3.6) contains all nonzero eigenvalues. Since $\lambda_n \to 0$, the multiplicity of every nonzero eigenvalue is finite (see also Section 7, Chapter 5).

Suppose next that f is an arbitrary function not necessarily in R_K. The Riesz-Fischer theorem (Section 6, Chapter 4) guarantees that $\Sigma\langle f,e_k\rangle e_k$ converges to some element, say g, in H. The elements Kf and Kg have the same expansion $\Sigma\lambda_k\langle f,e_k\rangle e_k$, so that the element $h=g-f$ is in the null space of K. Thus an arbitrary function f can be decomposed in one and only one way as

(3.12) $$f=h+\Sigma\langle f,e_k\rangle e_k, \qquad \text{where } Kh=0.$$

Of course h is just the projection of f on the null space of K. Thus h is an eigenfunction of K corresponding to the zero eigenvalue and is orthogonal to all the e_k's. We can restate the content of (3.12) as follows.

Theorem 5. The set of all eigenfunctions of K (including those corresponding to the zero eigenvalue) forms a basis for H. The set $\{e_n\}$ of eigenfunctions corresponding to nonzero eigenvalues forms a basis for H if and only if $\lambda=0$ is *not* an eigenvalue of K.

This theorem has an important application to self-adjoint boundary value problems. Let L be a formally self-adjoint operator of order p with real coefficients (p has to be even), and let $B_1,\ldots,B_p$ be p boundary functionals of the form (3.2), Section 3, Chapter 3, so that $(L,B_1,\ldots,B_p)$ is self-adjoint. Consider the eigenvalue problem

(3.13) $$Lu=\theta u, \quad a<x<b; \qquad B_1u=\cdots=B_pu=0.$$

We may assume without loss of generality that $\theta=0$ is not an eigenvalue (otherwise we merely add a suitable term $\theta'u$ to both sides of $Lu=\theta u$ and relabel the operator and the eigenvalue parameter.) We can then construct Green's function $g(x,\xi)$ satisfying

$$Lg=\delta(x-\xi), \quad a<x,\xi<b; \qquad B_1g=\cdots=B_pg=0.$$

The function $g(x,\xi)$ is real and symmetric, and (3.13) can be translated into the integral equation

$$u(x)=\theta\int_a^b g(x,\xi)u(\xi)\,d\xi, \qquad a<x<b,$$

which is of the form $Gu=\lambda u, \lambda=1/\theta$. Since G is a symmetric, compact integral operator, Theorem 5 applies. Now the eigenfunctions of (3.13) are just the eigenfunctions of G corresponding to $\lambda \neq 0$. We claim that $\lambda = 0$ is not an eigenvalue of G. Indeed, the BVP

$$Lz = u, \quad a < x < b; \qquad B_1 z = \cdots = B_p z = 0$$

has the one and only solution

$$z = Gu,$$

and, if $z = 0$, then $Lz = 0$ and $u = 0$.

Thus the eigenfunctions of G corresponding to $\lambda \neq 0$ and hence the eigenfunctions of (3.13) form an orthonormal basis.

Theorem 6. The eigenfunctions of a self-adjoint boundary value problem form an orthonormal basis.

Definite and Indefinite Operators

Let K be a *compact symmetric operator* defined on the whole of H. We recall from Section 8, Chapter 5, that K is *nonnegative* if $\langle Kv, v \rangle \geqslant 0$ for all v and *positive* if $\langle Kv, v \rangle > 0$ for all $v \neq 0$. A compact operator cannot be coercive in the sense of Section 8, Chapter 5. The fact that an operator is positive does *not* mean that its kernel is a positive function of x and ξ (see, however, Exercise 4.5). All eigenvalues of a positive operator are positive, and all eigenvalues of a nonnegative operator are nonnegative. A compact symmetric operator K, all of whose eigenvalues are positive, is a positive operator (the statement remains true if "nonnegative" is substituted for "positive" throughout). The reader can supply the definitions of nonpositive and negative operators and the obvious consequences. If, however, $\langle Kv, v \rangle$ is positive for some elements v and negative for others, K is said to be *indefinite*.

If K is a positive (or nonnegative) compact operator, we can reformulate the extremal principles as follows:

$$(3.14) \qquad \lambda_1 = \sup_{\|v\|=1} \langle Kv, v \rangle, \ldots, \lambda_{n+1} = \sup_{\substack{\|v\|=1 \\ v \in M_n^\perp}} \langle Kv, v \rangle, \ldots,$$

where M_n is the manifold spanned by the eigenfunctions $e_1(x), \ldots, e_n(x)$. Similarly, for a nonpositive operator K, we have

$$\lambda_1 = \inf_{\|v\|=1} \langle Kv, v \rangle, \ldots, \lambda_{n+1} = \inf_{\substack{\|v\|=1 \\ v \in M_n^\perp}} \langle Kv, v \rangle, \ldots.$$

Even if K is an indefinite compact operator, we can develop separate principles for its positive and negative eigenvalues. Instead of listing all nonzero eigenvalues by decreasing modulus as in (3.6), we define two sequences of eigenvalues: $\lambda_1^+, \lambda_2^+, \ldots$ are the positive eigenvalues in decreasing order, and $\lambda_1^-, \lambda_2^-, \ldots$ are the negative eigenvalues in increasing order. The corresponding eigenfunctions are denoted by $\{e_n^+\}$ and $\{e_n^-\}$, respectively. By the expansion theorem (Theorem 4) we have, for any v in H,

$$\langle Kv, v \rangle = \Sigma \lambda_k |\langle v, e_k \rangle|^2 = \Sigma \lambda_k^+ |\langle v, e_k^+ \rangle|^2 + \Sigma \lambda_k^- |\langle v, e_k^- \rangle|^2.$$

It therefore follows that

$$\lambda_1^- \|v\|^2 \leqslant \langle Kv, v \rangle \leqslant \lambda_1^+ \|v\|^2.$$

One easily establishes the principles

$$(3.15) \qquad \sup_{\substack{\|v\|=1 \\ v \in (M_n^+)^\perp}} \langle Kv, v \rangle = \lambda_{n+1}^+, \qquad \inf_{\substack{\|v\|=1 \\ v \in (M_n^-)^\perp}} \langle Kv, v \rangle = \lambda_{n+1}^-,$$

where M_n^+ is the manifold spanned by $(e_1^+, \ldots, e_n^+)$, and M_n^- the one spanned by $(e_1^-, \ldots, e_n^-)$.

In the general case of an indefinite, symmetric, compact operator both $\{\lambda_n^-\}$ and $\{\lambda_n^+\}$ will be infinite sequences; the number 0 may or may not be an eigenvalue. In special cases either or both of the sequences $\{\lambda_n^-\}, \{\lambda_n^+\}$ may contain only a finite number of terms or be empty. Of course, if both sequences are finite, then 0 must be an eigenvalue of infinite multiplicity. Figure 3.1 illustrates the two different methods of indexing the eigenvalues of a compact self-adjoint operator.

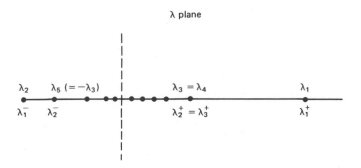

Figure 3.1 If there are only finitely many nonzero eigenvalues, $\lambda = 0$ is an eigenvalue (of infinite multiplicity). If there are an infinite number of nonzero eigenvalues, $\lambda = 0$ either is an eigenvalue (of finite or infinite multiplicity) or is in the continuous spectrum.

Bilinear Expansions

Let K be the self-adjoint integral operator on $L_2^{(c)}(a,b)$, generated by the symmetric H.-S. kernel $k(x,\xi)$. The nonzero eigenvalues and corresponding orthonormal eigenfunctions are indexed as in (3.6) and (3.7). Thus we have

$$(3.16) \qquad\qquad Ke_n = \lambda_n e_n \qquad \text{and} \qquad K^m e_n = \lambda_n^m e_n.$$

The self-adjoint integral operator K^m is generated by the symmetric iterated kernel $k_m(x,\xi)$ given by (1.26) or (1.27). Let us consider for fixed ξ the Fourier series of $k_m(x,\xi)$ in the orthonormal set $\{e_n(x)\}$:

$$k_m(x,\xi) \sim \sum_n \left(\int_a^b k_m(x,\xi)\bar{e}_n(x)\,dx \right) e_n(x).$$

Since $k_m(x,\xi) = \bar{k}_m(\xi,x)$ and $\int_a^b k_m(\xi,x)e_n(x)\,dx = \lambda_n^m e_n(\xi)$, we have

$$k_m(x,\xi) \sim \sum_n \lambda_n^m \bar{e}_n(\xi) e_n(x).$$

For $m \geq 2$, (1.27) shows that $k_m(x,\xi)$ is in R_K for each fixed ξ, so that the series converges to $k_m(x,\xi)$ in the mean square in the variable x for each fixed ξ. Obviously the same is true in the variable ξ for each fixed x. Moreover, one can also show that, if $k_2(x,\xi)$ is continuous, the following stronger theorem holds.

Theorem 7. Let $k_2(x,\xi)$ be continuous on $a \leq x,\xi \leq b$; then the mth iterated kernel has the *bilinear expansion*

$$(3.17) \qquad\qquad k_m(x,\xi) = \sum_n \lambda_n^m e_n(x)\bar{e}_n(\xi),$$

and the convergence is uniform on the square $a \leq x,\xi \leq b$.

Remark. Even if $k(x,\xi)$ is unbounded, it may happen that $k_2(x,\xi)$ is continuous as, for instance in Exercise 1.2. For $m=1$, (3.17) does not necessarily hold in the sense of uniform convergence even if $k(x,\xi)$ is continuous. We do, however, have the following theorem.

Theorem 8   (Mercer's theorem). If $k(x,\xi)$ is continuous and if all but a finite number of eigenvalues are of one sign, then

$$(3.18) \qquad k(x,\xi) = \Sigma \lambda_n e_n(x)\bar{e}_n(\xi) \qquad \text{uniformly on } a \leq x,\xi \leq b.$$

From (3.17) we conclude that

$$\sum_n \lambda_n^m = \int_a^b k_m(x,x)\,dx,$$

and, if K is a *positive* operator, we obtain the *trace inequality*

$$(3.19) \qquad \lambda_1 \leqslant \left[\int_a^b k_m(x,x)\,dx \right]^{1/m},$$

which is valid for $m = 1$ if k is continuous.

Exercises

3.1. A standard method for constructing examples of integral operators with special properties is as follows. Begin with an infinite orthonormal set $\{e_n(x)\}$ on $L_2^{(c)}(a,b)$ which may or may not be a basis, and with a sequence of real numbers $\{\lambda_n\}$ of decreasing modulus such that $\sum |\lambda_n|^2 < \infty$. Construct the function

$$k(x,\xi) = \sum_{n=1}^{\infty} \lambda_n e_n(x)\bar{e}_n(\xi),$$

which is clearly a symmetric kernel.

(a) Show that the corresponding integral operator K has the eigenpairs (λ_n, e_n) and $(0, u)$, where u is any element orthogonal to all the $\{e_n\}$. In particular, if $\{e_n\}$ is a basis, 0 is not an eigenvalue.

(b) Construct a symmetric kernel which has infinitely many nonzero eigenvalues and has 0 as an eigenvalue of multiplicity k (do both the case where k is finite and the case where k is infinite).

(c) Construct a symmetric kernel that has infinitely many positive and negative eigenvalues and for which 0 is not an eigenvalue. Show nevertheless that there are nontrivial elements u for which $\langle Ku, u \rangle = 0$.

3.2. The expansion theorem (3.11) no longer holds if K is not symmetric (in fact, K may have no eigenfunctions, yet $\bar{R}_K = H$). Something can be rescued, however, by introducing the left and right iterates: $L = K^*K, R = KK^*$.

(a) Show that these are both symmetric, nonnegative integral operators and that $N_L = N_K, N_R = N_{K^*}$.

(b) Show that L and R have the same positive eigenvalues, λ_n^2 (indexed in decreasing order with due respect to multiplicity). If $\{v_n\}$ is a corresponding orthonormal set of eigenfunctions for L, show that $\{u_n = Kv_n/\lambda_n\}$ is an orthonormal set of eigenfunctions for R. If $\{u_n\}$ is an orthonormal set of eigenfunctions for R, show that $\{v_n = K^*u_n/\lambda_n\}$ is an orthonormal set of eigenfunctions for L.

From the theory of symmetric operators we know that $\{v_n\}$ is an orthonormal basis for $\overline{R}_L$ and hence for $(N_L)^\perp = (N_K)^\perp = \overline{R}_{K^*}$ [see (5.5), Chapter 5]. Similarly $\{u_n\}$ is a basis for $\overline{R}_K$. Thus, if $f \in \overline{R}_K$,

$$(3.20) \qquad f = \Sigma\langle f, u_n\rangle u_n,$$

and, if f is arbitrary,

$$(3.21) \qquad f = f_0 + \Sigma\langle f, v_n\rangle v_n,$$

where f_0 is an element in N_K.

3.3. If K is a symmetric, positive H.-S. operator, show that

$$\lambda_1 = \lim_{m\to\infty}\left[\int_a^b k_m(x,x)\,dx\right]^{1/m},$$

where $k_m(x,\xi)$ is the mth iterated kernel of K.

4. THE INHOMOGENEOUS EQUATION

Neumann Series

The equation $Ku - \lambda u = f$ can be viewed as a fixed point problem in a suitable metric space. With $\lambda \neq 0$ the equation can be rewritten as

$$(4.1) \qquad Tu = u \qquad \text{with } Tu \doteq \frac{1}{\lambda}Ku - \frac{f}{\lambda}.$$

In Chapter 4 we showed that, for λ sufficiently large, T was a contraction mapping on the space of continuous functions with the uniform metric [on the assumption that the kernel $k(x,\xi)$ was a continuous function of x and ξ]. It is easy to reformulate the problem in an L_2 setting.

Assume that $k(x,\xi)$ is a kernel that generates a bounded integral operator K on $L_2^{(c)}(a,b)$. From

$$d_2(Tu, Tv) = \|Tu - Tv\| \leqslant \frac{\|K\|}{|\lambda|}\|u - v\| = \frac{\|K\|}{|\lambda|}d_2(u,v),$$

it follows that T is a contraction on $L_2^{(c)}(a,b)$ for $|\lambda| > \|K\|$. Thus (4.1) has one and only one solution whenever $|\lambda| > \|K\|$. (This is of course already known from the fact that the spectrum of K is confined to $|\lambda| \leqslant \|K\|$.) Moreover, the solution u can be determined by iteration from an arbitrary initial element v_0:

$$u = \lim_{n\to\infty} v_n, \qquad v_n = T^n v_0.$$

Thus

(4.2)

$$u = \lim_{n\to\infty}\left[\frac{1}{\lambda^n}K^n v_0 - \sum_{m=1}^{n-1}\frac{K^m f}{\lambda^{m+1}} - \frac{f}{\lambda}\right] = -\frac{f}{\lambda} - \sum_{m=1}^{\infty}\frac{K^m f}{\lambda^{m+1}} = -\sum_{m=0}^{\infty}\frac{K^m f}{\lambda^{m+1}},$$

where K^0 is the identity operator. The operator K^m in (4.2) is an integral operator whose kernel is the iterated kernel (1.26). These iterated kernels are not always easy to calculate explicitly. Although we have shown only that the solution (4.2) is valid for $|\lambda| > \|K\|$, it is actually valid for $|\lambda| > |\lambda_c|$, where λ_c is the complex number of largest modulus in the spectrum of K.

Example 1. Consider the case of a one-term separable kernel

$$k(x,\xi) = p(x)\bar{q}(\xi), \qquad 0 \leqslant x, \xi \leqslant 1,$$

for which

$$v \doteq Ku \doteq \int_0^1 p(x)\bar{q}(\xi)u(\xi)\,d\xi,$$

so that

$$\|v\|^2 = \|p\|^2|\langle u, q\rangle|^2 \leqslant \|p\|^2\|q\|^2\|u\|^2,$$

which becomes an equality if u is proportional to q. Thus

$$\|K\| = \|p\|\,\|q\|.$$

The iterated kernel k_2 is

$$k_2(x,\xi) = \int_0^1 p(x)\bar{q}(\xi_1)p(\xi_1)\bar{q}(\xi)\,d\xi_1$$

$$= p(x)\bar{q}(\xi)\langle p,q\rangle,$$

and, similarly,

$$k_n(x,\xi) = \langle p,q\rangle^{n-1}p(x)\bar{q}(\xi).$$

Thus (4.2) becomes

(4.3) $$u = -\frac{f}{\lambda} - \frac{1}{\lambda^2}p(x)\langle f,q\rangle \sum_{k=1}^\infty \left(\frac{\langle p,q\rangle}{\lambda}\right)^{k-1}.$$

The series converges for $|\lambda| > |\langle p,q\rangle|$, and, using the formula for a geometric series, we find that

(4.4) $$u = -\frac{f(x)}{\lambda} - \frac{\langle f,q\rangle}{\lambda^2}\left(1 - \frac{\langle p,q\rangle}{\lambda}\right)^{-1}p(x).$$

We observe that (4.3) is the one and only solution of (4.1) for $|\lambda| > |\langle p,q\rangle|$. Since $\langle p,q\rangle$ is the only nonzero eigenvalue of K, we have confirmed that the Neumann series converges not only for $|\lambda| > \|K\|$ but even for $|\lambda| > |\lambda_c|$, where λ_c is the point in the spectrum having the largest modulus (of course, if K is self-adjoint, $|\lambda_c| = \|K\|$, as is easily checked in this particular problem since then $p = q$ and $|\langle p,q\rangle| = \|p\|\|q\| = \|K\|$). Expression (4.4) provides the unique solution of $Ku - \lambda u = f$ whenever $\lambda \neq 0$ and $\lambda \neq \langle p,q\rangle$, but, in general, it is not possible to perform explicitly the analytic continuation which enabled us to go from (4.3) to (4.4) in our very simple special case.

Solution by Eigenfunction Expansion

Let the nonzero eigenvalues of the self-adjoint compact operator K be indexed with due regard to multiplicity as

$$|\lambda_1| \geqslant |\lambda_2| \geqslant \cdots \geqslant |\lambda_n| \geqslant \cdots,$$

the corresponding *orthonormal* set of eigenfunctions being denoted by $e_1(x), e_2(x), \ldots, e_n(x), \ldots$. Although occasionally there may be only a finite number of such eigenvalues (if the kernel is separable), in most cases of interest the sequence $\{\lambda_k\}$ will be infinite and, as was shown earlier, $\lim_{k\to\infty} \lambda_k = 0$.

Consider the inhomogeneous equation

(4.5) $$Ku - \lambda u = f,$$

where λ is a given complex number and f is an element in H. For a solution to exist, the element $z \doteq f + \lambda u$ must be in the range of K and so can be represented by its Fourier series in the set $\{e_k\}$:

$$z = Ku = \sum_k \gamma_k e_k, \qquad \gamma_k = \langle z, e_k \rangle = \langle Ku, e_k \rangle.$$

Since $\langle z, e_k \rangle = \langle f, e_k \rangle + \lambda \langle u, e_k \rangle$, and

$$\langle Ku, e_k \rangle = \langle u, Ke_k \rangle = \lambda_k \langle u, e_k \rangle,$$

we find that a solution u of (4.5) must satisfy

(4.6) $$(\lambda_k - \lambda)\langle u, e_k \rangle = \langle f, e_k \rangle, \qquad k = 1, 2, \ldots .$$

We must now consider a number of different cases.

1. If $\lambda \neq 0$ and λ is *not* one of the eigenvalues $\lambda_1, \lambda_2, \ldots$, then (4.6) gives

(4.7) $$\langle u, e_k \rangle = \frac{\langle f, e_k \rangle}{\lambda_k - \lambda}, \qquad \gamma_k = \lambda_k \frac{\langle f, e_k \rangle}{\lambda_k - \lambda},$$

so that

(4.8) $$u = \frac{z - f}{\lambda} = -\frac{f}{\lambda} + \sum_k \frac{\lambda_k}{\lambda(\lambda_k - \lambda)} \langle f, e_k \rangle e_k,$$

which is our candidate for the solution of (4.5). The coefficients of the series appearing in (4.8) tend to 0 more rapidly than $\langle f, e_k \rangle$ since the only possible limit-point of $\{\lambda_k\}$ is 0. In view of the convergence of $\sum_k |\langle f, e_k \rangle|^2$, the series

$$\sum_k \left| \frac{\lambda_k}{\lambda(\lambda_k - \lambda)} \right|^2 |\langle f, e_k \rangle|^2$$

must also converge—in fact, more rapidly—and hence the series in (4.8) converges in L_2. One can therefore insert (4.8) in the integral equation and apply K term by term to verify that the equation is satisfied. Thus (4.8) is the one and only solution of $Ku - \lambda u = f$. By setting $f = \Sigma \langle f, e_k \rangle e_k + f_0$, where f_0 is the projection of f on N_K, we can rewrite (4.8) as

$$(4.9) \qquad u = -\frac{f_0}{\lambda} + \sum_k \frac{1}{\lambda_k - \lambda} \langle f, e_k \rangle e_k,$$

which has a simpler appearance than does (4.8) but converges more slowly. By interchanging integration and summation, (4.8) can also be put in the form

$$(4.10) \qquad u = -\frac{f}{\lambda} + \frac{1}{\lambda} \int_a^b R(x, \xi, \lambda) f(\xi) d\xi,$$

where

$$(4.11) \qquad R(x, \xi, \lambda) = \sum_k \frac{\lambda_k}{\lambda_k - \lambda} e_k(x) \bar{e}_k(\xi)$$

is known as the *resolvent kernel* of K.

2. Let $\lambda = \lambda_m$, one of the nonzero eigenvalues of K. If λ_m is not degenerate, the first factor in (4.6) vanishes only for $k = m$. If $\langle f, e_m \rangle \neq 0$, (4.6) cannot be solved for $\langle u, e_m \rangle$ and $Ku - \lambda_m u = f$ has *no* solution. If $\langle f, e_m \rangle = 0$, $\langle u, e_m \rangle$ is arbitrary, but for $k \neq m$, $\langle u, e_k \rangle$ is still given by (4.7), so that $Ku - \lambda_m u = f$ has the infinite number of solutions

$$(4.12) \qquad u = -\frac{f}{\lambda_m} + \sum_{k \neq m} \frac{\lambda_k}{\lambda_m(\lambda_k - \lambda_m)} \langle f, e_k \rangle e_k + c e_m, \qquad \langle f, e_m \rangle = 0.$$

If λ_m is degenerate, a number of eigenvalues with successive indices will be equal to λ_m. Let these resonance indices be $m_1, \ldots, m_j$, where j is the multiplicity of λ_m; one of these indices will of course be m. Then, unless $\langle f, e_{m_i} \rangle = 0$ for each resonance index, $Ku - \lambda_m u = f$ will have *no* solution. If, however, these j *solvability conditions* are satisfied, the solutions of the inhomogeneous equation are given by

$$(4.13) \qquad u = -\frac{f}{\lambda_m} + \sum_{k \neq m_i} \frac{\lambda_k}{\lambda_m(\lambda_k - \lambda_m)} \langle f, e_k \rangle e_k + \sum_{i=1}^{j} c_i e_{m_i},$$

$$\langle f, e_{m_i} \rangle = 0, \quad i = 1, \ldots, j.$$

3. If $\lambda=0$, the inhomogeneous equation becomes $Ku=f$, and we need to characterize R_K. With u an arbitrary element of H, we have

$$Ku = \sum_k \lambda_k \langle u, e_k \rangle e_k.$$

Since the set $\{e_k\}$ is orthogonal to N_K, clearly f will have to be orthogonal to N_K. Then $f = \Sigma \langle f, e_k \rangle e_k$, and $Ku=f$ reduces to

$$\langle u, e_k \rangle = \frac{1}{\lambda_k} \langle f, e_k \rangle, \qquad k=1,2,\ldots .$$

The series $\Sigma_k \langle u, e_k \rangle e_k$ will converge if and only if

$$(4.14) \qquad\qquad \sum_k \frac{1}{|\lambda_k|^2} |\langle f, e_k \rangle|^2 < \infty.$$

To recapitulate, f is in R_K if and only if f is orthogonal to N_K and satisfies (4.14). The general solution of $Ku=f$ will then be

$$(4.15) \qquad\qquad u = u_0 + \sum_k \frac{1}{\lambda_k} \langle f, e_k \rangle e_k,$$

where u_0 is an arbitrary element in N_K (that is, an arbitrary eigenfunction corresponding to $\lambda=0$). Of course, if $\lambda=0$ is not an eigenvalue of K, the term u_0 does not appear in (4.15). Note that, if the sequence $\{\lambda_k\}$ is in fact an infinite sequence, condition (4.14) means that R_K is not closed; the closure of R_K is the set of all elements orthogonal to N_K.

Example 2.   Let

$$k(x,\xi) = p(x)\bar{p}(\xi), \qquad 0 \leqslant x, \xi \leqslant 1.$$

Then there is only one nonzero eigenvalue $\lambda_1 = \int_0^1 |p|^2 dx = \|p\|^2$ with normalized eigenfunction $e_1 = p(x)/\|p\|$.

If $\lambda \neq 0, \lambda \neq \lambda_1$, the one and only solution of $Ku - \lambda u = f$ is given by (4.8):

$$u = -\frac{f(x)}{\lambda} + \frac{\lambda_1 \langle f, p \rangle}{\lambda(\lambda_1 - \lambda)\|p\|^2} p(x) = -\frac{f(x)}{\lambda} + \frac{\langle f, p \rangle}{\lambda(\|p\|^2 - \lambda)} p(x),$$

which coincides with (4.4) when $p=q$.

The cases $\lambda=0$ and $\lambda=\lambda_1$ are left as exercises.

Example 3. Consider the integral operator with difference kernel

$$(4.16) \qquad Ku \doteq \int_{-\pi}^{\pi} k(\theta - \phi) u(\phi) \, d\phi, \qquad -\pi \leqslant \theta \leqslant \pi.$$

In order for the function Ku to be defined on $-\pi \leqslant \theta \leqslant \pi$, values of $k(x)$ will be required for $-2\pi \leqslant x \leqslant 2\pi$. Not much can be done for general k, but *if k has period 2π*, the operator K simplifies considerably by using the convolution formula for Fourier series (see Exercise 3.8, Chapter 2). Writing

$$u(\theta) = \sum \gamma_n e^{in\theta}, \qquad k(\theta) = \sum k_n e^{in\theta},$$

we find that

$$Ku = 2\pi \sum k_n \gamma_n e^{in\theta}.$$

The eigenvalue problem $Ku - \lambda u = 0$ reduces to

$$(4.17) \qquad (2\pi k_n - \lambda)\gamma_n = 0,$$

which must be satisfied for all integers, $-\infty < n < \infty$. This, in turn, requires that λ be one of the numbers $2\pi k_n$, the corresponding eigenfunction being proportional to $e^{in\theta}$. Of course, it may happen that some number $2\pi k_m$ occurs for a set of indices $m_1, \ldots, m_p$, which means that the eigenvalue $2\pi k_m$ has multiplicity p with eigenfunctions $c_1 e^{im_1\theta} + \cdots + c_p e^{im_p\theta}$.

As a particular case consider the Poisson kernel that occurs in potential theory:

$$k(\theta) = \frac{1}{2\pi} \frac{1 - r^2}{1 + r^2 - 2r\cos\theta}, \qquad \text{where } r \text{ is fixed,} \quad 0 < r < 1.$$

We have already studied this kernel in Section 2, Chapter 2, particularly in Exercise 2.9, where we obtained the Fourier expansion

$$k(\theta) = \frac{1}{2\pi} + \frac{1}{\pi} \sum_{n=1}^{\infty} r^n \cos n\theta = \frac{1}{2\pi} \sum_{n=-\infty}^{\infty} r^{|n|} e^{in\theta}.$$

The eigenvalues of K are therefore

$$\lambda_n = r^n, \qquad n = 0, 1, 2, \ldots.$$

The eigenvalue $\lambda_0 = 1$ has the eigenfunction $e_0(\theta) = A_0$, whereas each of the other eigenvalues λ_n is degenerate with the independent eigenfunctions $A_n \cos n\theta, B_n \sin n\theta$. For $\lambda \neq r^n$ the inhomogeneous equation $Ku - \lambda u = f$ has the one and only solution

$$u(\theta) = -\frac{f(\theta)}{\lambda} + \sum_{n=-\infty}^{\infty} \frac{r^{|n|} e^{in\theta}}{2\pi\lambda(r^{|n|} - \lambda)} \int_{-\pi}^{\pi} f(\theta) e^{-in\theta} d\theta.$$

If $\lambda = r^n$, a solution is possible only if $0 = \int_{-\pi}^{\pi} f(\theta) \cos n\theta \, d\theta = \int_{-\pi}^{\pi} f(\theta) \sin n\theta \, d\theta$.

Example 4. Consider again Example 3, Section 1. We are searching for an Ω-band-limited function $f(t)$ of unit energy which maximizes the energy in the time interval $|t| < T$:

(4.18) $$E_T = \int_{-T}^{T} |f(t)|^2 dt.$$

Of course the maximum value of E_T is $\leqslant 1$, but we shall see below that it is strictly smaller than 1.

We shall set the problem in the functional framework of the Hilbert space $H = L_2^{(c)}(-\infty, \infty)$, on which we define the two closed linear manifolds:

M_T: the set of elements $f(t)$ vanishing for $|t| \geqslant T$,
S_Ω: the set of elements $f(t)$ which are Ω-band limited.
The operator P_T, defined by

(4.19) $$P_T f = \begin{cases} f(t), & |t| \leqslant T, \\ 0, & |t| > T, \end{cases}$$

is clearly the orthogonal projection on M_T. The operator Q_Ω, defined by

(4.20) $$Q_\Omega f = \frac{1}{2\pi} \int_{-\Omega}^{\Omega} f^\wedge(\omega) e^{-i\omega t} d\omega,$$

is the orthogonal projection on S_Ω [as can be seen by using a slight modification of Parseval's identity (4.8), Chapter 2]. Both P_T and Q_Ω are symmetric nonnegative operators on H.

An easy calculation gives, for each f in H,

(4.21)

$$Q_\Omega P_T f = \int_{-T}^{T} \frac{\sin \Omega(t-s)}{\pi(t-s)} f(s) \, ds \doteq \int_{-\infty}^{\infty} k(t,s) f(s) \, ds, \qquad -\infty < t < \infty,$$

where

$$(4.22) \qquad k(t,s) = \begin{cases} \dfrac{\sin \Omega(t-s)}{\pi(t-s)}, & |s| \leqslant T, \\[2mm] 0, & |s| > T, \end{cases}$$

For f in S_Ω expression (4.18) for E_T becomes

$$(4.23) \qquad E_T = \langle P_T f, f \rangle = \langle P_T f, Q_\Omega f \rangle = \langle Q_\Omega P_T f, f \rangle.$$

Since $Q_\Omega P_T$ transforms elements of S_Ω into elements of S_Ω, it can be viewed as an operator on the infinite-dimensional Hilbert space S_Ω. On S_Ω the operator $Q_\Omega P_T$ is symmetric and positive. Indeed, for f and g in S_Ω, we have

$$\langle Q_\Omega P_T f, g \rangle = \langle P_T f, Q_\Omega g \rangle = \langle P_T f, g \rangle = \langle f, P_T g \rangle$$
$$= \langle Q_\Omega f, P_T g \rangle = \langle P_T Q_\Omega f, g \rangle.$$

Clearly (4.23) shows that $Q_\Omega P_T$ is a nonnegative operator on S_Ω; it is also known that a nontrivial band-limited function cannot vanish on any time interval, so that $Q_\Omega P_T$ is actually a positive operator.

When $Q_\Omega P_T$ is regarded as an operator on $L_2^{(c)}(-\infty, \infty)$, it is an integral operator generated by the kernel (4.22). The calculation [see the Féjer kernel of Example 3(d), Section 2, Chapter 2]

$$(4.24) \qquad \int\!\!\int_{-\infty}^{\infty} |k|^2 \, ds \, dt = \int_{-T}^{T} ds \int_{-\infty}^{\infty} \frac{\sin^2 \Omega w}{\pi^2 w^2} \, dw = \frac{2T\Omega}{\pi}$$

shows that k is Hilbert-Schmidt and hence that $Q_\Omega P_T$ is compact on H and, a fortiori, on S_Ω. Thus $Q_\Omega P_T$ is a positive compact operator on S_Ω, therefore generating a sequence of eigenpairs (λ_n, e_n) with $\lambda_n > 0$ and $\{e_n\}$ an orthonormal basis. By the theory of Section 3 [see (3.14)] the largest eigenvalue λ_1 is the maximum of $\langle Q_\Omega P_T f, f \rangle$ over unit elements f in S_Ω. We see from (4.23) that λ_1 *is therefore just the desired maximal value of E_T and $e_1(t)$ is the optimal signal $f(t)$.* The pair (λ_1, e_1) therefore satisfies the integral equation

$$\lambda_1 e_1(t) = \int_{-\infty}^{\infty} k(t,s) e_1(s) \, ds, \qquad -\infty < t < \infty,$$

that is,

$$(4.25) \qquad \lambda_1 e_1(t) = \int_{-T}^{T} \frac{\sin \Omega(t-s)}{t-s} e_1(s) \, ds, \qquad -\infty < t < \infty.$$

Note that λ_1 and $P_T e_1$ are determined by looking at (4.25) as an integral equation on $-T \leqslant t \leqslant T$ (more precisely: $P_T e_1$ is determined up to a multiplicative constant). The right side then has meaning for all t and, when divided by λ_1, provides the values of $e_1(t)$ for $|t| > T$; of course $e_1(t)$ is determined up to a multiplicative constant, which is in turn calculated by the normalization requirement

$$\int_{-\infty}^{\infty} |e_1(t)|^2 \, dt = 1.$$

Let us calculate some rough bounds for λ_1. By the trace inequality (3.19) with $m = 1$, we find that

$$\lambda_1 \leqslant \int_{-T}^{T} k(t,t) \, dt = \frac{2T\Omega}{\pi}.$$

Of course this upper bound yields new information only if $\Omega T < \pi/2$. A lower bound is obtained from

$$\lambda_1 \geqslant \int_{-T}^{T} |f(t)|^2 \, dt, \qquad \|f\| = 1, \quad f \in S_\Omega.$$

The simplest trial function is the function $f(t)$ of unit energy whose Fourier transform is constant for $|\omega| \leqslant \Omega$ and vanishes for $|\omega| > \Omega$. Then

$$f^\wedge(\omega) = \begin{cases} \left(\dfrac{\pi}{\Omega}\right)^{1/2}, & |\omega| \leqslant \Omega, \\ 0, & |\omega| > \Omega, \end{cases}$$

and we find that

$$f(t) = \frac{1}{2(\pi\Omega)^{1/2}} \int_{-\Omega}^{\Omega} e^{-i\omega t} \, d\omega = \frac{\sin \Omega t}{(\pi\Omega)^{1/2} t},$$

so that

$$\lambda_1 \geqslant \int_{-T}^{T} |f|^2 \, dt = \int_{-\Omega T}^{\Omega T} \frac{\sin^2 z}{\pi z^2} \, dz.$$

Most of these results are due to Landau, Pollak, and Slepian (see the book by Dym and McKean).

The Volterra Equation

Consider the integral equation

$$(4.26) \qquad \int_a^x k(x,\xi)u(\xi)\,d\xi = \lambda u(x) + f(x), \qquad a \leqslant x \leqslant b,$$

where the upper limit of integration is x instead of b. Such an equation can be considered as a special case of a Fredholm equation on $a \leqslant x \leqslant b$ with kernel

$$\tilde{k}(x,\xi) = \begin{cases} 0, & x < \xi, \\ k(x,\xi), & \xi < x. \end{cases}$$

From this point of view $\tilde{k}$ will *never* be symmetric, and therefore the results on the existence of eigenfunctions cannot be applied. However, the situation is quite a bit simpler than that for Fredholm equations, and it is easier to proceed directly from (4.26). Problem (4.26) has already been discussed in Example 4(b), Section 4, Chapter 4, in the functional framework of $C(a,b)$ with the uniform metric. We now merely assume that

$$\int_a^b \int_a^b |\tilde{k}^2|\,dx\,d\xi = M < \infty$$

and that f is in $L_2^{(c)}(a,b)$. We then look for solutions u in $L_2^{(c)}(a,b)$. Using the same fixed-point techniques as in Chapter 4, we can show that no value of $\lambda \neq 0$ can be an eigenvalue; if $k(x,x)$ vanishes for some x, 0 can be an eigenvalue. For instance, $u = x^2$ is a solution of $\int_0^x (x - \frac{4}{3}\xi)u(\xi)\,d\xi = 0$, a phenomenon related to singular points of differential equations.

Since the initial value problem for a differential equation of order p with $a_p(x) \neq 0$ can be transformed into a Volterra equation, we now have the existence and uniqueness of solution for such an IVP in an L_2 framework.

Exercises

4.1. The function $k(x) = \log(1 - \cos x)$ has period 2π and mild singularities at $x = 2n\pi$. Equation (3.34), Chapter 2, shows that its Fourier series is

$$-\log 2 - 2 \sum_{n=1}^{\infty} \frac{\cos nx}{n}.$$

Consider the integral operator on $L_2^{(c)}(-\pi,\pi)$:

$$Ku \doteq \int_{-\pi}^{\pi} k(x - \xi)u(\xi)\,d\xi,$$

and show that K is Hilbert-Schmidt and self-adjoint. Find all eigenfunctions and eigenvalues of K. Discuss the equation $Ku = f$.

4.2. Let $k(x)$ be an even, real function of period 2π such that
$\int_{-\pi}^{\pi} k^2(x)\,dx < \infty$. The function k has the Fourier cosine series

$$k(x) = \tfrac{1}{2}k_0 + \sum_{n=1}^{\infty} k_n \cos nx.$$

Consider the integral operator on $L_2^{(c)}(-\infty, \infty)$:

$$Ku \doteq \int_{-\pi}^{\pi} k(x+\xi)u(\xi)\,d\xi,$$

and find all its eigenvalues and eigenfunctions.

4.3. Let $k(x)$ be an odd, real function of period 2π such that
$\int_{\pi}^{\pi} k^2(x)\,dx < \infty$. Find the eigenvalues and eigenfunctions of the integral equation

$$\int_{-\pi}^{\pi} k(x+\xi)u(\xi)\,d\xi = \lambda u(x), \qquad -\pi \leqslant x \leqslant \pi.$$

(Note that $\lambda = 0$ is always an eigenvalue in Exercise 4.3 but not
necessarily in 4.2.)

4.4. For heat conduction (without sources or sinks) in a homogeneous
infinite rod, the temperature $U(x,t)$ at time $t > 0$ is related to the
initial temperature $u(x)$ by

(4.27) $$U(x,t) = \frac{1}{\sqrt{4\pi t}} \int_{-\infty}^{\infty} e^{-(x-\xi)^2/4t} u(\xi)\,d\xi \doteq Ku.$$

Suppose we want to determine the initial temperature from observations on the temperature U at some given time $t > 0$, say $t = 1$ for
definiteness. We then view (4.27) as an integral equation for $u(x)$
with $U(x, 1)$ given on $-\infty < x < \infty$. By applying a Fourier transform
on the space variable, show that it is possible to solve (4.27) *only* if
$U(x, 1)$ satisfies very severe restrictions. In particular, show that a
solution is possible if $U(x, 1)$ is band limited (that is, $U\hat{} \equiv 0$ for
$|\omega| > \omega_0$). Show that K^{-1} is unbounded but that $\overline{R}_K = L_2^{(c)}(-\infty, \infty)$.
The problem just considered is typical of ill-posed problems (see
Payne).

4.5. Let $k(x,\xi)$ be a continuous, symmetric H.-S. kernel that generates a nonnegative operator K. Show that $k(x,x)$ is pointwise nonnegative. (Note that k can be negative off the diagonal.)

4.6. Let A be a bounded operator on H. Show that, if $\langle Au, u \rangle$ is real for all u, then A is symmetric. [Use the polar identity (5.18), Chapter 4, with $b(u,v) \doteq \langle Au, v \rangle$.]

4.7. Consider a Volterra equation with a difference kernel:

$$\int_0^x k(x-y)u(y)\,dy - \lambda u(x) = f(x), \qquad x > 0.$$

Apply a Laplace transform to both sides, and express u as an inversion integral. Specialize to the Abel equation (1.1), and by bringing the Laplace transform of f' into the picture show that

$$u(x) = \frac{1}{\pi} \int_0^x \frac{f'(y)}{\sqrt{x-y}}\,dy,$$

which can be rewritten as

$$u(x) = \frac{1}{\pi} \frac{d}{dx} \int_0^x \frac{f(y)}{\sqrt{x-y}}\,dy.$$

What restrictions are placed on f in these two formulas?

4.8. Let t be fixed, $0 < t < 1$, and consider the integral equation

$$(4.28) \qquad Ku \doteq \int_{-\infty}^{\infty} k(x,y)u(y)\,dy = \lambda u(x), \qquad -\infty < x < \infty,$$

where

$$(4.29) \quad k(x,y) = (1-t^2)^{-1/2} \exp\left(\frac{x^2+y^2}{2} \right) \exp\left(-\frac{x^2+y^2-2xyt}{1-t^2} \right).$$

(a) Show directly that

$$u_0(x) = \exp\left(-\frac{x^2}{2} \right)$$

is an eigenfunction corresponding to $\lambda_0 = \sqrt{\pi}$.

(b) Let

(4.30) $$u_n(x) = \exp\left(-\frac{x^2}{2}\right) H_n(x),$$

where the functions

(4.31) $\quad H_n(x) = (-1)^n \exp(x^2) \dfrac{d^n \exp(-x^2)}{dx^n}, \qquad n = 0, 1, 2, \ldots,$

are clearly polynomials (the *Hermite polynomials*). *Assume* that $Ku_n = \lambda_n u_n$, and show that $Ku_{n+1} = \lambda_{n+1} u_{n+1}$, where $\lambda_{n+1} = t\lambda_n$. Together with (a), this shows that (4.28) has eigenvalues

$$\lambda_n = t^n \sqrt{\pi}, \qquad n = 0, 1, 2, \ldots,$$

with corresponding eigenfunctions $u_n(x)$.

(c) Prove that

$$\int_{-\infty}^{\infty} u_n^2 \, dx = 2^n n! \sqrt{\pi},$$

and, by using the bilinear series for k, that

$$k(x,y) = \sum_{n=0}^{\infty} \frac{u_n(x) u_n(y)}{2^n n!} t^n.$$

4.9. Suppose that K is a compact, nonsymmetric integral operator. Show that $Ku = f$ can have solutions only if $f \in N_{K^*}$ and

$$\sum_n \frac{1}{\lambda_n^2} |\langle f, u_n \rangle|^2 < \infty,$$

where (λ_n^2, u_n) are the eigenpairs of $R = KK^*$ as in Exercise 3.2. If these conditions are satisfied, show that the general solution of $Ku = f$ is given by

$$u = u_0 + \sum \frac{\langle f, u_n \rangle}{\lambda_n} v_n,$$

where u_0 is an arbitrary element in N_K, and v_n is defined in Exercise 3.2.

5. VARIATIONAL PRINCIPLES AND RELATED APPROXIMATION METHODS

We want to characterize the eigenvalues of a self-adjoint operator by extremal principles which can be used for approximate calculations. Our interest in this chapter is in compact operators, but the method applies with slight modifications to bounded operators (or even operators that are only bounded below or bounded above) as long as either the upper or lower part of the spectrum consists of eigenvalues.

To avoid the notational difficulties associated with the general case, let us confine the analysis to a *self-adjoint, nonnegative, compact* operator K. We will make remarks later about possible extensions. The spectrum of K consists of positive eigenvalues

$$(5.1) \qquad \lambda_1 \geqslant \lambda_2 \geqslant \cdots \geqslant \lambda_n \geqslant \cdots > 0,$$

listed with due regard to multiplicity, and of $\lambda = 0$, which either is an eigenvalue or is in the continuous spectrum. We shall be concerned only with the positive eigenvalues of K and their corresponding orthonormal eigenfunctions $e_1, \ldots, e_n, \ldots$. The span of $\{e_1, \ldots, e_n\}$ is denoted by M_n (thus M_n is an n-dimensional manifold in H). The set (5.1) of eigenvalues can be finite or infinite; *in the former case we agree to define $\lambda_n = 0$ for $n > m$, where m is the number of nonzero eigenvalues*. In this way we always have an infinite sequence $\{\lambda_n\}$, which proves convenient in the statement of theorems. If $\lambda_n = 0$, the corresponding e_n is understood to be an element of N_K.

The *Rayleigh quotient* is defined as

$$(5.2) \qquad R(u) = \frac{\langle Ku, u \rangle}{\|u\|^2},$$

where it is *always* assumed that $\|u\| \neq 0$. Theorem 3, Section 3, can be restated as follows.

Theorem 1.

$$(5.3) \qquad \max_{u \in M_n^{\perp}} R(u) = \lambda_{n+1}.$$

Proof. By (3.11) we have, for u in H,

$$\langle Ku, u \rangle = \sum_k \lambda_k |\langle u, e_k \rangle|^2.$$

If $u \in M_n^{\perp}$, $\langle u, e_1 \rangle = \cdots = \langle u, e_n \rangle = 0$ and

$$\langle Ku, u \rangle = \sum_{k>n} \lambda_k |\langle u, e_k \rangle|^2 \leqslant \lambda_{n+1} \sum_{k>n} |\langle u, e_k \rangle|^2 \leqslant \lambda_{n+1} \|u\|^2,$$

so that $R(u) \leqslant \lambda_{n+1}$. On the other hand, $e_{n+1} \in M_n^{\perp}$ and $R(e_{n+1}) = \lambda_{n+1}$, which proves the theorem. (If $n = 0$, the condition $u \in M_n^{\perp}$ is understood to be absent and the proof is the same.)

It is possible to characterize λ_{n+1} by an extremal principle that does not refer to the eigenfunctions $e_1, \ldots, e_n$, which, after all, are unknown. The idea is to first maximize $R(u)$ subject to orthogonalization with respect to a "wrong" set of functions $\{v_1, \ldots, v_n\}$ whose span is E_n rather than M_n. Next we choose E_n to minimize this maximum. We find that the minimax is just λ_{n+1}.

Theorem 2   (Weyl-Courant minimax theorem).   Set

$$\nu(E_n) = \max_{u \in E_n^{\perp}} R(u),$$

where, as indicated, ν will depend on the choice of E_n. Then

$$\lambda_{n+1} = \min_{\substack{\text{over all} \\ \text{choices of } E_n}} \nu(E_n),$$

or, combining the two statements, we obtain

$$\lambda_{n+1} = \min_{E_n \in S_n} \max_{u \in E_n^{\perp}} R(u),$$

where S_n is the set of all n-dimensional manifolds in H.

Proof.   First we note that, by Theorem 1, $\nu(M_n) = \lambda_{n+1}$ so that clearly $\min \nu(E_n) \leqslant \lambda_{n+1}$. To prove the reverse inequality, and hence the theorem, it is sufficient to exhibit for each choice of E_n an element w in $E_n^{\perp}$ such that $R(w) \geqslant \lambda_{n+1}$. Let us try an element w of the form

$$w = c_1 e_1 + \cdots + c_{n+1} e_{n+1},$$

with $c_1, \ldots, c_{n+1}$ chosen so that $\|w\| \neq 0$ and $0 = \langle w, v_1 \rangle = \cdots = \langle w, v_n \rangle$, where $\{v_1, \ldots, v_n\}$ is a basis for E_n. Such a choice is always possible since these conditions reduce to finding a nontrivial solution to a homogeneous system of n equations in $n+1$ unknowns. Now

$$R(w) = \frac{\langle Kw, w \rangle}{\|w\|^2} = \frac{\sum\limits_{k=1}^{n+1} \lambda_k |c_k|^2}{\sum\limits_{k=1}^{n+1} |c_k|^2} \geqslant \lambda_{n+1},$$

which completes the proof.

Ritz-Rayleigh Procedure

The simplest practical procedure for estimating eigenvalues starts from Theorem 1 with $n=0$:

(5.4) $\lambda_1 = \max R(u),$

and instead of taking the maximum of R over all u we take the maximum only for elements of the form $c_1 v_1 + \cdots + c_k v_k$, where $v_1, \ldots, v_k$ is a fixed, judiciously chosen set of independent functions in H. Denoting the span of $\{v_1, \ldots, v_k\}$ by E_k, we then try to calculate

(5.5) $\max\limits_{u \in E_k} R(u) = \max\limits_{c_1, \ldots, c_k} R(c_1 v_1 + \cdots + c_k v_k).$

Since the functional in (5.5) is the same as that in (5.4) but the maximum is taken over a subspace rather than the whole of H, it follows that (5.5) yields a value that cannot exceed λ_1. Let us try to calculate the maximum of (5.5) as explicitly as possible. We have

$$(5.6) \quad R(c_1 v_1 + \cdots + c_k v_k) = \frac{\left\langle \sum\limits_{i=1}^{k} c_i K v_i, \sum\limits_{j=1}^{k} c_j v_j \right\rangle}{\left\langle \sum\limits_{i=1}^{k} c_i v_i, \sum\limits_{j=1}^{k} c_j v_j \right\rangle} = \frac{\sum\limits_{i,j=1}^{k} c_i \bar{c}_j k_{ij}}{\sum\limits_{i,j=1}^{k} c_i \bar{c}_j \alpha_{ij}},$$

where

$$\alpha_{ij} = \langle v_i, v_j \rangle, \qquad k_{ij} = \langle K v_i, v_j \rangle$$

are regarded as known quantities that can be calculated before any maximization procedure. Note that $\alpha_{ij} = \bar{\alpha}_{ji}, k_{ij} = \bar{k}_{ji}$, so that both matrices are symmetric. We obtain greater geometrical insight by rewriting (5.6) in terms of the operator P (defined on H) which associates with each u in H its orthogonal projection Pu on E_k. If u is in E_k, Ku will not usually be in

E_k but PKu will be; for u in E_k we have $u = Pu$ and, by the symmetry of P,

$$(5.7) \qquad R(u) = \frac{\langle Ku, u \rangle}{\|u\|^2} = \frac{\langle Ku, Pu \rangle}{\|u\|^2} = \frac{\langle PKu, u \rangle}{\|u\|^2}, \qquad u \in E_k.$$

Since PK transforms elements of E_k into elements of E_k, it can be regarded as an operator on E_k; PK is known as the *part of K in E_k* and is easily seen to be symmetric and nonnegative. Thus (5.6) is just the Rayleigh quotient for the nonnegative symmetric operator PK on E_k.

Since such an operator is certainly compact, the usual extremal principles apply and the maximum of (5.7) and (5.6) is the largest eigenvalue Λ_1 of the algebraic eigenvalue problem

$$(5.8) \qquad\qquad PKw - \Lambda w = 0, \qquad w \in E_k,$$

which we can easily write in coordinate form. Since $\{v_1, \dots, v_k\}$ is a basis for E_k, (5.8) will hold if and only if $\langle PKw - \Lambda w, v_j \rangle = 0$, $j = 1, \dots, k$. By using $\langle PKw, v_j \rangle = \langle Kw, Pv_j \rangle = \langle Kw, v_j \rangle$, we find that

$$\langle Kw, v_j \rangle = \Lambda \langle w, v_j \rangle, \qquad j = 1, \dots, k,$$

and, setting $w = \sum_{i=1}^k c_i v_i$, we obtain

$$(5.9) \qquad \sum_{i=1}^k \langle Kv_i, v_j \rangle c_i = \Lambda \sum_{i=1}^k \langle v_i, v_j \rangle c_i, \qquad j = 1, \dots, k,$$

which is just the set of algebraic equations that would be obtained by maximizing (5.6) using the calculus (keeping in mind, however, that the c_i's might be complex).

Equation (5.8) or its equivalent coordinate form (5.9) is known as the *Galerkin equation*. Although we were led to the equation from an extremal principle, the equation has a simple, intrinsic, geometrical meaning independent of any variational principle. To see this let us examine again the original problem

$$(5.10) \qquad\qquad Ku - \lambda u = 0, \qquad u \in H,$$

for which we try to find approximate solutions lying in E_k. Such an approximate solution would be of the form $\sum_{i=1}^k c_i v_i$. Substitution in (5.10) would lead to an inconsistent equation since $K(\sum_{i=1}^k c_i v_i)$ is not usually in E_k. Setting our sights lower, we merely require that the projection of the left side of (5.10) on E_k vanish. This gives the Galerkin equation (5.8) or (5.9).

Remark.   Even if a variational principle is not available (for instance, if K is not symmetric or even if K is not linear), we can regard (5.8) and (5.9) as approximations to (5.10). The advantage of also having a variational principle is that we can make more precise statements about the relationship between the eigenvalues of (5.8) and (5.10).

For the class of problems we are considering here, PK is a symmetric nonnegative operator on E_k. Therefore PK has k nonnegative eigenvalues (not necessarily distinct):

$$(5.11) \qquad \Lambda_1 \geqslant \Lambda_2 \geqslant \cdots \geqslant \Lambda_k \geqslant 0.$$

The corresponding eigenvectors $\{w_1, \ldots, w_k\}$ are chosen to form an orthonormal basis for E_k. Of course in dealing with finite-dimensional problems the largest eigenvalue Λ_1 is characterized by a maximum principle, whereas the lowest eigenvalue Λ_k is characterized by the minimum principle

$$(5.12) \qquad \Lambda_k = \min_{u \in E_k} \frac{\langle PKu, u \rangle}{\|u\|^2} = \min_{u \in E_k} R(u).$$

We now compare the eigenvalues (5.11) of (5.8) with those of (5.10).

Theorem 3   (Poincaré).

$$(5.13) \qquad \Lambda_1 \leqslant \lambda_1, \ldots, \Lambda_k \leqslant \lambda_k.$$

Proof.   Let us show that $\Lambda_j \leqslant \lambda_j$ by applying Theorem 1 for $n = j - 1$. Our trial element w is a linear combination $d_1 w_1 + \cdots + d_j w_j$ of the first j eigenvectors of PK. We require that $\|w\| \neq 0$ and that w satisfy the $j - 1$ orthogonality conditions $\langle w, e_1 \rangle = \cdots = \langle w, e_{j-1} \rangle = 0$, where $e_1, \ldots, e_{j-1}$ are the familiar eigenfunctions of K. Since these orthogonality conditions reduce to $j - 1$ homogeneous equations in j unknowns, it is always possible to construct the desired element w. From Theorem 1 it follows that $R(w) \leqslant \lambda_j$, and from (5.7)

$$R(w) = \frac{\langle PKw, w \rangle}{\|w\|^2} = \frac{\displaystyle\sum_{i=1}^{j} \Lambda_i |d_i|^2}{\displaystyle\sum_{i=1}^{j} |d_i|^2} \geqslant \Lambda_j,$$

so that $\Lambda_j \leqslant \lambda_j$ as claimed.

Thus the Galerkin equation provides lower bounds to the largest k eigenvalues of K.

An immediate consequence of Theorem 3 is the maximin theorem below. Like Theorem 2, it gives a characterization of the eigenvalues of K without reference to eigenfunctions of lower index, but has the advantage of not requiring an extremum over an infinite-dimensional subspace such as $E_n^{\perp}$.

Theorem 4 (**Poincaré maximin theorem**).   Let S_k be the set of all linear manifolds of dimension k lying in H, and let E_k be a particular member of S_k. If we define

$$(5.14) \qquad \mu(E_k) = \min_{u \in E_k} R(u),$$

then

$$(5.15) \qquad \max_{E_k \in S_k} \mu(E_k) = \lambda_k,$$

or, combining these statements,

$$(5.16) \qquad \max_{E_k \in S_k} \min_{u \in E_k} R(u) = \lambda_k.$$

Proof.   From (5.12) we see that $\mu(E_k) = \Lambda_k$, the lowest eigenvalue of the part of K in E_k. By Theorem 3 $\Lambda_k \leqslant \lambda_k$, and therefore the maximum of Λ_k over all possible choices of E_k does not exceed λ_k. On the other hand, $\mu(M_k) = \lambda_k$, so Theorem 4 is proved.

Remarks on the Ritz-Rayleigh Procedure

1.   If K is nonpositive, Theorems 1, 2, 3, and 4 hold in revised form with the words "maximum" and "minimum" interchanged and all inequalities reversed. If K is indefinite, the original theorems hold for the upper end of the spectrum and the revised theorems for the lower end. In general, if A is merely bounded from one side and it is known that the spectrum at the bounded end is discrete, there are appropriate extremal principles for the discrete end of the spectrum.

2.   With a limited set of trial functions $v_1(x), \ldots, v_k(x)$ one cannot expect a very good approximation to the eigenfunction $e_1(x)$, but the approximation to λ_1 is much better. The reason is that the functional $R(u)$ is stationary (has a maximum) at $u = e_1$ and that the maximum value is λ_1;

roughly speaking, a first-order change in u about e_1 leads to a second-order change in R.

3.   With k fixed, the approximation Λ_1 to λ_1 is usually better than the approximation Λ_2 to λ_2, and so on.

4.   If we increase k, we improve the approximations. In theory we could start with a basis $\{v_1, \ldots, v_n, \ldots\}$ for H and define E_k as the span of $\{v_1, \ldots, v_k\}$; as $k \to \infty$, the eigenvalues of (5.8) would tend to those of K, since the compact operator PK tends to K in the operator norm.

5.   Equation (5.8) provides us with lower bounds to the eigenvalues of K. How do we obtain complementary bounds (that is, upper bounds)? The trace inequality (3.19) is one such bound. Another, the Kohn-Kato method, is presented in the following section. A third, due to Weinstein and Arondszajn and later modified by Bazley and Fox, is described in the article by Fox and Rheinboldt.

6.   For numerical purposes we must avoid ill-conditioned matrices. This means that the set $v_1, \ldots, v_n$ used as a basis for E_n should be either orthogonal or nearly so (in a real space the cosine of the angle between v_i and v_j is $\langle v_i, v_j \rangle / \|v_i\| \|v_j\|$, and we require that this number not be close to $+1$ or -1). In particular, if we are dealing with a problem originally in $L_2(-1, 1)$, the set $v_k(x) = x^{k-1}$ is unsuitable and should be first orthogonalized by the Gram-Schmidt process. For many purposes it is useful to employ a set $v_1, \ldots, v_n$ with narrow support. These functions are usually constructed by piecing together polynomials and requiring some continuity

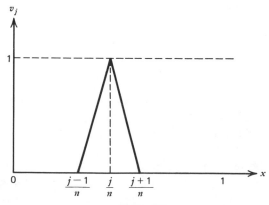

Figure 5.1

and perhaps smoothness at the junctions. A simple special example is the set of *roof* functions, one of which is shown in Figure 5.1. We divide the interval $0 \leqslant x \leqslant 1$ into n equal parts, and for $j = 1, \ldots, n-1$ we let

$$
v_j(x) = \begin{cases} 0, & x < \dfrac{j-1}{n}, x > \dfrac{j+1}{n}, \\[2mm] nx - (j-1), & \dfrac{j-1}{n} < x < \dfrac{j}{n}, \\[2mm] -nx + (j+1), & \dfrac{j}{n} < x < \dfrac{j+1}{n}. \end{cases}
$$

Clearly v_j and v_k are orthogonal if $|j - k| \geqslant 2$, but v_j is not orthogonal to v_{j-1} and v_{j+1}.

Eigenvalue Estimation Based on Spectral Theory

Although the methods described below are applicable for more general symmetric operators, we shall confine ourselves for simplicity to *positive compact operators*. Let K be such an operator. Its eigenvalues $\{\lambda_n\}$ are all positive, and the corresponding eigenfunctions $\{e_n\}$ form an orthonormal basis. Let z be an arbitrary unit vector which we call a trial element. We associate with z a mass distribution on the λ axis by placing at each point λ_n the concentrated mass $|\langle z, e_n \rangle|^2$, which is of course nonnegative. The total mass is $\sum_n |\langle z, e_n \rangle|^2 = \|z\|^2 = 1$, so that this distribution of mass can also be regarded as the probability distribution of a discrete random variable X. If S is a set on the λ axis, the mass in S is the probability $P(X \in S)$. As we change z, we merely redistribute our unit of mass over the same points $\{\lambda_n\}$.

The mean or center of mass is given by

$$
(5.17) \qquad m = E(X) = \sum \lambda_n |\langle z, e_n \rangle|^2 = \langle Kz, z \rangle.
$$

If $f(X)$ is any function of the random variable X, we have

$$
(5.18) \qquad E(f(X)) = \sum f(\lambda_n) |\langle z, e_n \rangle|^2,
$$

and, in particular,

$$
E(X^m) = \sum \lambda_n^m |\langle z, e_n \rangle|^2 = \langle K^m z, z \rangle
$$

and

$$
(5.19) \qquad \sigma^2 = E\big[(X - m)^2\big] = \sum (\lambda_n - m)^2 |\langle z, e_n \rangle|^2 = \|Kz - mz\|^2,
$$

where σ^2 is the *variance* of X (that is, the moment of inertia of the mass distribution about its center of mass). Note that m and σ^2 can be calculated directly from the trial element z by integration. Of course higher moments require iterated integrals which may be hard to compute. Our goal is to estimate an eigenvalue in terms of m and σ alone.

Since our mass is distributed on the part of the λ axis between $\lambda = 0$ and $\lambda = \lambda_1$, it is clear that the center of mass also lies in that interval, so that

$$(5.20) \qquad\qquad m = \langle Kz, z \rangle \leqslant \lambda_1,$$

which is just the familiar upper bound for the Rayleigh quotient with $\|z\| = 1$. The inequality will be useful if we can cleverly choose z so that most of the associated unit mass is at λ_1.

If we can show that a certain interval contains mass, then there must be at least one eigenvalue in that interval. We have immediately the following theorem.

Theorem 5. The closed interval $[m - \sigma, m + \sigma]$ contains an eigenvalue.

Proof. If all the mass is outside the interval in question, $(\lambda_n - m)^2 > \sigma^2$ for each n having $\langle z, e_n \rangle \neq 0$, and (5.19) gives an immediate contradiction.

Remark. The theorem is nothing more than the simplest application of the famous Chebyshev inequality of probability theory.

To use the theorem to estimate, say, the largest eigenvalue λ_1, we must know that the interval $[m - \sigma, m + \sigma]$ contains only the eigenvalue λ_1. This will be the case if $m - \sigma > \lambda_2$. Obviously this requires that the trial element z be chosen judiciously enough so that its mass center is larger than λ_2 by at least one standard deviation. If the initial trial element does not do the trick, the Schwarz iteration described in Exercise 5.2 will enable us to construct a suitable z. We then have

$$(5.21) \qquad\qquad m - \sigma \leqslant \lambda_1 \leqslant m + \sigma$$

or, in view of (5.20),

$$(5.22) \qquad\qquad m \leqslant \lambda_1 \leqslant m + \sigma,$$

where $m = \langle Kz, z \rangle, \sigma = \|Kz - mz\|$.

We can improve on these bounds by the following procedure, based on Chebyshev inequalities (see Marshall and Olkin, Rosenbloom). Let S be an interval on the λ axis which we want to contain mass, and let T be the complement of S. We would like to estimate $P(X \in S)$ and to show that it is positive or, equivalently, that $P(X \in T) < 1$. Let $f(\lambda)$ be a real-valued function satisfying

(5.23) $f(\lambda) \leqslant 0$ for all λ; $f(\lambda) \geqslant 1$, $\lambda \in T$.

We would like to choose $f(\lambda)$ as close as possible to the indicator function (see Section 1, Chapter 2) of the set T. We have

$$P(X \in T) = \sum_{\lambda_n \in T} |\langle z, e_n \rangle|^2 \leqslant \sum_{\lambda_n \in T} f(\lambda_n) |\langle z, e_n \rangle|^2$$

$$\leqslant \sum_{n=1}^{\infty} f(\lambda_n) |\langle z, e_n \rangle|^2 = E(f(X)),$$

and therefore

(5.24) $P(X \in S) \geqslant 1 - E(f(X)),$

which will be positive if

(5.25) $E(f(X)) < 1.$

We shall consider functions f, which also depend on certain parameters adjusted so that (5.23) and (5.25) are satisfied and the interval S containing mass is as narrow as possible. Suppose, for instance, that $m > A$ and we want to find the narrowest interval $(A, A + \Delta)$ which can be guaranteed to contain mass (note there must be some mass for $\lambda \geqslant m$ since m is the center of mass). We would like to determine this interval by using only the first two moments of X. Let $h > A$, and consider the function

$$f(\lambda, h) = \frac{(\lambda - h)^2}{(A - h)^2},$$

which satisfies (5.23) when $|\lambda - h| \geqslant |A - h|$. Thus we take S as the interval $(A, 2h - A)$, of which h is the midpoint.

A simple calculation gives

$$E(f(X, h)) = \frac{(h - m)^2 + \sigma^2}{(h - A)^2},$$

which satisfies (5.25) if

$$h > \frac{m + A}{2} + \frac{\sigma^2}{2(m - A)}.$$

Therefore the open interval

$$\left(A, m + \frac{\sigma^2}{m-A} + \varepsilon\right)$$

contains mass for every $\varepsilon > 0$. In view of the discrete nature of the spectrum and the fact that $m + \sigma^2/(m-A) > 0$, we conclude that the half-closed interval

$$\left(A, m + \frac{\sigma^2}{m-A}\right]$$

must contain mass. Similarly, if $m < B$, the interval

$$\left[m - \frac{\sigma^2}{B-m}, B\right)$$

contains mass.

Now suppose that we are trying to estimate some particular eigenvalue λ^* in the spectrum and that we know (perhaps by previous rough calculations) that the interval (A, B) contains only the eigenvalue λ^*. In this case we find the *Kohn-Kato inclusion interval*

$$(5.26) \qquad m - \frac{\sigma^2}{B-m} \leqslant \lambda^* \leqslant m + \frac{\sigma^2}{m-A}$$

if $A < m < B$ (as should be the case if z is a reasonable trial element). In fact, for a reasonable z both $\sigma/(m-A)$ and $\sigma/(B-m)$ will be much smaller than 1 and the inclusion interval (5.26) will be narrower than the one given by Theorem 5. If we apply (5.26) to estimate the kth eigenvalue λ_k, we need to set A equal to an upper bound U_{k+1} to λ_{k+1} and B to a lower bound L_{k-1} to λ_{k-1}. Then (5.26) gives, for a trial element z such that $U_{k+1} < m < L_{k-1}$,

$$(5.27) \qquad m - \frac{\sigma^2}{L_{k-1}-m} \leqslant \lambda_k \leqslant m + \frac{\sigma^2}{m-U_{k+1}}.$$

The method gives an improved upper bound to λ_1:

$$(5.28) \qquad \lambda_1 \leqslant m + \frac{\sigma^2}{m-U_2},$$

where U_2 is an upper bound (perhaps fairly crude) to λ_2 and $m > U_2$.

Remark. If an eigenvalue is degenerate or if some eigenvalues are clustered together, the method can still be applied but gives information only about the overall location of the cluster.

Eigenfunction Approximation

Consider a trial element z intended to approximate the eigenfunction e^* corresponding to the simple eigenvalue λ^*. Note that $-e^*$ and, more generally, $e^* \exp(i\psi)$ for real ψ are also unit eigenfunctions corresponding to λ^*. We are perfectly satisfied if z is close to any of these unit eigenvectors. Obviously z will be a good approximation to a unit eigenvector on the one-dimensional manifold generated by e^* if the magnitude of the projection $\langle z,e^* \rangle e^*$ is close to 1. By the projection theorem we can decompose z as

$$(5.29) \qquad z = u + v = \langle z,e^* \rangle e^* + v, \qquad \langle v,e^* \rangle = 0.$$

We will have $|\langle z,e^* \rangle|$ near 1 if

$$(5.30) \qquad \|v\|^2 = 1 - |\langle z,e^* \rangle|^2$$

is near 0.

If z is supposed to be an approximation to e_1, we can easily find a bound for $1 - |\langle z,e_1 \rangle|^2$. Indeed, we have

$$m = \langle Kz,z \rangle = \sum_k \lambda_k |\langle z,e_k \rangle|^2 \le \lambda_1 |\langle z,e_1 \rangle|^2 + \lambda_2 \sum_{k>2} |\langle z,e_k \rangle|^2$$

or

$$(5.31) \qquad m \le \lambda_1 |\langle z,e_1 \rangle|^2 + \lambda_2 \left(1 - |\langle z,e_1 \rangle|^2\right),$$

so that

$$(5.32) \qquad 1 - |\langle z,e_1 \rangle|^2 \le \frac{\lambda_1 - m}{\lambda_1 - \lambda_2}.$$

Of course neither λ_1 nor λ_2 is generally known. However, we may use upper bounds U_1, U_2 for λ_1, λ_2, respectively, in (5.31) to obtain

$$(5.33) \qquad 1 - |\langle z,e_1 \rangle|^2 \le \frac{U_1 - m}{U_1 - U_2}.$$

A somewhat better bound for $1 - |\langle z,e^* \rangle|^2$ can be found by using the decomposition (5.29). For any real c we have

$$Kz - cz = Ku - cu + Kv - cv,$$

where $Ku - cu$ is seen to be proportional to e^* and $Kv - cv$ is orthogonal to e^*. Therefore we find that

$$\|Kv - cv\|^2 \leqslant \|Kz - cz\|^2 = \sigma^2 + (m - c)^2,$$

so that

$$\|v\|^2 \leqslant \frac{\sigma^2 + (m - c)^2}{\|Kv - cv\|^2 / \|v\|^2}.$$

Since

$$\min_{\langle v, e^* \rangle = 0} \frac{\|Kv - cv\|^2}{\|v\|^2} = (\tilde{\lambda} - c)^2,$$

where $\tilde{\lambda}$ is the eigenvalue different from λ^* nearest c, we have

$$\|v\|^2 \leqslant \frac{\sigma^2 + (m - c)^2}{(\tilde{\lambda} - c)^2}.$$

If the interval (A, B) contains only the eigenvalue λ^* and $A < m < B$, we find, by choosing $c = m$, that

(5.34)
$$\|v\|^2 \leqslant \frac{\sigma^2}{(B - m)^2}, \qquad m \geqslant \frac{A + B}{2},$$

$$\|v\|^2 \leqslant \frac{\sigma^2}{(m - A)^2}, \qquad m \leqslant \frac{A + B}{2}.$$

The optimal choice of c is discussed in Exercise 5.9. If U_2 is an upper bound to λ_2 and $m > U_2$, we find from (5.34) that

(5.35)
$$1 - |\langle z, e_1 \rangle|^2 \leqslant \frac{\sigma^2}{(m - U_2)^2},$$

which is usually better than (5.33).

Exercises

5.1. Let A be a symmetric (perhaps indefinite) operator on $E_n^{(c)}$ whose eigenvalues listed in decreasing order, as follows: $\lambda_1 \geqslant \lambda_2 \geqslant \cdots \geqslant \lambda_n$. Show that λ_k can be characterized by two theorems of the

Weyl-Courant type (see Theorem 2), one requiring orthogonality to a $(k-1)$-dimensional space and the other to an $(n-k)$-dimensional space. The second of these is a maximin theorem, but is it the same as the Poincaré maximin theorem?

5.2. Let K be a real, nonnegative, symmetric, H.-S. operator defined on $L_2(a,b)$. Starting with an arbitrary real function f_0 such that $Kf_0 \neq 0$, let us define

$$f_n = Kf_{n-1} = \cdots = K^n f_0$$

and

$$a_k = \langle f_i, f_{k-i} \rangle,$$

where, as the notation suggests, the definition is independent of i. By expanding in the eigenfunctions of K, show that $a_n > 0$ and that the sequence

$$\theta_{k+1} \doteq \frac{a_{k+1}}{a_k}$$

is monotonically increasing and bounded above by the largest eigenvalue λ_1 of K. Show that, if f_0 is *not* orthogonal to the fundamental eigenfunction e_1, then $\theta_k \to \lambda_1$.

5.3. *An integrodifferential operator.* Let A be the operator defined by

$$(5.36) \qquad Au = -\frac{d^2 u}{dx^2} + \int_0^1 xyu(y)\,dy = -u''(x) + x\langle x, u \rangle,$$

with domain D_A consisting of all functions $u(x)$ on $0 < x < 1$ with a continuous second derivative and satisfying the boundary conditions

$$(5.37) \qquad\qquad u(0) = 0, \qquad u'(1) = 0.$$

(a) Show that A on D_A is a symmetric, positive operator. Since A is real, we can (and shall) restrict the domain of A to real-valued functions.

(b) Consider the inhomogeneous integrodifferential equation

(5.38)

$$-u'' + \int_0^1 xyu(y)\,dy = f(x), \quad 0 < x < 1; \qquad u(0) = u'(1) = 0.$$

A function $u(x)$ satisfies (5.38) if and only if $u(x)$ and α satisfy simultaneously

$$-u'' = f - \alpha x, \quad 0 < x < 1; \qquad u(0) = u'(1) = 0;$$

and

$$\alpha = \int_0^1 xu(x)\,dx = \langle x, u \rangle.$$

Using Green's function $g(x,\xi)$ for $-D^2$ with the boundary conditions (5.37), and letting G be the corresponding integral operator, show that

$$\alpha = \frac{\langle f, Gx \rangle}{1 + \langle Gx, x \rangle}$$

and

$$(5.39) \quad u(x) = \int_0^1 \left[g(x,\xi) - \frac{5}{204}(3x - x^3)(3\xi - \xi^3) \right] f(\xi)\,d\xi.$$

5.4 Consider the eigenvalue problem $Au = \lambda u$, where A is the operator of Exercise 5.3 with the boundary conditions (5.37). Since A on D_A is symmetric and positive, all eigenvalues are positive and eigenfunctions corresponding to different eigenvalues are orthogonal. In view of the fact that A is a real operator, we may restrict ourselves to real-valued eigenfunctions.

(a) Show that the problem can be reduced to an eigenvalue problem for a pure integral operator; hence show that the eigenfunctions form a basis.

(b) By observing that the problem $Au = \lambda u$ has the form $u'' + \lambda u = cx$, show that the eigenvalues $\lambda = \alpha^2$ are obtained from the positive solutions of

$$\tan \alpha = \alpha + \frac{\alpha^3}{3} - \alpha^5.$$

Sketch the functions $\tan \alpha$ and $\alpha + (\alpha^3/3) - \alpha^5$. Use tables to find an approximate value for the smallest eigenvalue, λ_1.

(c) In the extremal principle

$$\lambda_1 = \min_{v \in D_A} \frac{\int_0^1 (v')^2 \, dx + \left(\int_0^1 xv \, dx \right)^2}{\int_0^1 v^2 \, dx}$$

use the trial element $v = x(2-x)$ to find an upper bound to λ_1. Use the trace inequality for k_2 [the iterate of the kernel of part (a)] to find a lower bound to λ_1.

5.5. The following is more of an exposition than an exercise. The interested reader may wish to supply any missing proofs (see Catchpole). Consider the integrodifferential equation

(5.40) $-u'' + r(x)\langle r, u \rangle = \lambda u,$

where $r(x)$ is a given real-valued function, λ a given complex number, and

$$\langle r, u \rangle \doteq \int_0^1 r(x)u(x) \, dx.$$

Some solutions of (5.40) can be found by inspection: any solution u of the differential equation

(5.41) $-u'' = \lambda u$

for which $\langle r, u \rangle = 0$ is automatically a solution of (5.40). Since the set of solutions of (5.41) is two-dimensional, it is always possible to find at least a one-dimensional subspace for which $\langle r, u \rangle = 0$, but not all solutions of (5.40) are obtained in this way. We shall be interested in finding the general solution of (5.40) and in associated initial and boundary value problems.

Let $u_1(x, \lambda), u_2(x, \lambda)$ be the solutions of (5.41) with initial data $(1, 0)$ and $(0, 1)$, respectively, at $x = 0$, that is, $u_1(0, \lambda) = 1$, $u_1'(0, \lambda) = 0$ and $u_2(0, \lambda) = 0$, $u_2'(0, \lambda) = 1$. Let $P(x, \lambda)$ be the solution of the inhomogeneous problem

(5.42) $-u'' - \lambda u = r, \qquad u(0) = u'(0) = 0.$

The necessary and sufficient condition for u to be a solution of (5.40) with initial data (α, β) at $x = 0$ is that u satisfies the integral equation

(5.43) $$u = \alpha u_1 + \beta u_2 - \langle r, u \rangle P.$$

In view of the independence of u_1, u_2, P, we have shown that there are at most three independent solutions to (5.40).

If u is a solution of (5.43), we find that

(5.44) $$C \langle r, u \rangle = \alpha \langle r, u_1 \rangle + \beta \langle r, u_2 \rangle,$$

where

(5.45) $$C(\lambda) \doteq 1 + \langle r(x), P(x, \lambda) \rangle.$$

The analysis of (5.40) differs according to whether or not $C(\lambda)$ vanishes.

Case 1. For any value of λ such that $C(\lambda) \neq 0$, we have

$$\langle r, u \rangle = \frac{\alpha}{C} \langle r, u_1 \rangle + \frac{\beta}{C} \langle r, u_2 \rangle,$$

so that

(5.46) $$u = \alpha u_1 + \beta u_2 - \frac{\alpha}{C} \langle r, u_1 \rangle P - \frac{\beta}{C} \langle r, u_2 \rangle P.$$

Theorem. If $C \neq 0$, every solution of (5.40) is of the form (5.46) for some α and β, and, conversely, for any α, β (5.46) provides a solution of (5.40). Moreover, (5.46) is the one and only solution of (5.40) with initial data (α, β). In particular, the solutions with initial data $(1, 0)$ and $(0, 1)$ are the independent functions $u_1 - (\langle r, u_1 \rangle / C) P$ and $u_2 - (\langle r, u_2 \rangle / C) P$, respectively.

Case 2. If λ is such that $C(\lambda) = 0$, then P is itself a solution of (5.40). Moreover, (5.44) shows that (5.40) can be solved only if

(5.47) $$\alpha \langle r, u_1 \rangle + \beta \langle r, u_2 \rangle = 0.$$

(a). We suppose first that

(5.48) $$\langle r, u_1 \rangle = \langle r, u_2 \rangle = 0.$$

Then $\alpha u_1 + \beta u_2 + \gamma P$ is a solution of (5.40) for any α, β, γ. The IVP for (5.40) with initial data (α, β) is always solvable and has the class of solutions $\alpha u_1 + \beta u_2 + \gamma P$, where γ is arbitrary.

(b). If (5.48) does not hold, there is only one independent solution v of $-u'' - \lambda u = 0$ for which $\langle r, v \rangle = 0$, namely, $v = u_1 \langle r, u_2 \rangle - u_2 \langle r, u_1 \rangle$. Then the general solution of (5.40) is $\delta v + \gamma P$, where δ and γ are arbitrary. However, the IVP for (5.40) has solutions if and only if the initial data is of the form $(\alpha \langle r, u_2 \rangle, -\alpha \langle r, u_1 \rangle)$, in which case there is a family of solutions $\alpha v + \gamma P$, where γ is arbitrary.

5.6. Consider the special case of (5.40) when $r(x)$ is the real positive constant a. Show that

$$u_1 = \cos \sqrt{\lambda}\, x, \qquad u_2 = \frac{\sin \sqrt{\lambda}\, x}{\sqrt{\lambda}}, \qquad P = \frac{a}{\lambda}(\cos \sqrt{\lambda}\, x - 1),$$

$$C(\lambda) = 1 + \frac{a^2}{\lambda}\left(\frac{\sin \sqrt{\lambda}}{\sqrt{\lambda}} - 1 \right).$$

Show therefore that, if $a = 2\pi$ and $\lambda = 4\pi^2$, the integrodifferential equation has three independent solutions and that the corresponding BVP

(5.49) $\displaystyle -u'' + 4\pi^2 \int_0^1 u(x)\, dx = \lambda u, \ 0 < x < 1; \ u(0) = u(1) = 0$

has $\lambda = 4\pi^2$ as a double eigenvalue with independent eigenfunctions $\sin 2\pi x$ and $\cos 2\pi x - 1$.

5.7. Show that the BVP

(5.50)

$$-u'' + 4\pi^2 \int_0^1 u(x)\, dx = \lambda u, \quad 0 < x < 1; \qquad u(0) = u(1), \quad u'(0) = u'(1)$$

has $\lambda = 4\pi^2$ as an eigenvalue of *multiplicity 3*.

5.8. Find all eigenvalues and eigenfunctions of (5.50). Does the set of eigenfunctions form a basis?

5.9. For $A < c < B$ the inequality preceding (5.34) yields

$$\|v\|^2 \leqslant \frac{\sigma^2 + (m-c)^2}{\alpha^2},$$

where $\alpha = \min(c - A, B - c)$. Find the optimal value of c if $m \leqslant (A + B)/2$.

REFERENCES AND ADDITONAL READING

Carrier, G. F., Krook, M., and Pearson, C. E., *Functions of a complex variable*, McGraw-Hill, New York, 1966.

Catchpole, E. A., A Cauchy problem for an ordinary differential equation, *Proc. Royal Soc. Edinburgh* (A) **72** (1972-73), 4.

Dym, H. and McKean, H. P., *Fourier series and integrals*, Academic Press, New York, 1972.

Fox, D. W. and Rheinboldt, W. C., Computational methods for determining lower bounds for eigenvalues of operators in Hilbert space, *SIAM Rev.*, **8**, No. 4 (1966).

Hochstadt, H., *Integral equations*, Wiley-Interscience, New York, 1973.

Marshall, A. W. and Olkin, I., Multivariate Chebyshev inequalities, *Numerische Mathematik*, **6** (1964).

Noble, B., *The Wiener-Hopf technique*, Pergamon Press, London, 1958.

Payne, L. E., *Improperly posed problems in partial differential equations*, SIAM, Philadelphia, 1975.

Rosenbloom, P.C., Inequalities for moments and means, *J. Approx Th.*, **23** (1978).

Stakgold, I., *Boundary value problems of mathematical physics*, Vol. II, Macmillan, New York, 1968.

Weinstein, A. and Stenger, W., *Methods of intermediate problems for eigenvalues*, Academic Press, New York, 1972.

7

Spectral Theory
of Second-Order
Differential
Operators

1. INTRODUCTION; THE REGULAR PROBLEM

The main purpose of the chapter is to study the relationship between two methods for solving boundary problems, the one based on eigenfunction expansion and the other on Green's function. We first deal with operators whose spectrum is discrete; in this case the eigenfunction expansion is an infinite series, and we show how Green's function can be expressed as a series of eigenfunctions and how, conversely, the eigenfunctions can be generated from the closed-form expression for Green's function. If the spectrum is continuous, an integral expansion over a continuous parameter plays the role of the infinite series, but it is still possible to generate this expansion from a knowledge of Green's function.

We shall develop the theory for self-adjoint differential operators of the second order. The corresponding differential equations, which usually arise when separating variables in curvilinear coordinates for Laplace's equation or related equations, have the form

(1.1) $$-\frac{d}{dx}\left(p(x)\frac{du}{dx}\right) + q(x)u - \lambda s(x)u = 0, \qquad a < x < b,$$

where λ, which started life as a separation constant, is now viewed as an eigenvalue parameter. We can write (1.1) as

(1.2) $$Lu - \lambda u = 0,$$

where L is the operator defined by

(1.3)
$$Lu \doteq \frac{1}{s}\left[-(pu')' + qu\right].$$

Note that we have placed a $1/s$ factor in L, so that (1.2) will have the standard eigenvalue form. This causes the slight complication that L is no longer formally self-adjoint in the usual inner product $\langle u, v \rangle = \int_a^b u\bar{v}\,dx$. As we shall see, an appropriate remedy is to introduce the new inner product

$$\langle u, v \rangle_s = \int_a^b su\bar{v}\,dx.$$

Since $s(x) > 0$ in $a < x < b$, $\langle u, u \rangle_s$ is positive for $u \neq 0$ and can be used to define a norm

$$\|u\|_s = \langle u, u \rangle_s^{1/2} = \left(\int_a^b s|u|^2\,dx\right)^{1/2}.$$

Most of our work will take place in the Hilbert space H_s, consisting of all functions $u(x)$ for which $\int_a^b s|u|^2\,dx$ is finite. Orthogonality in H_s means that $\langle u, v \rangle_s = 0$, that is, $\int_a^b su\bar{v}\,dx = 0$. Let $\{e_n\}$ be an orthonormal basis in H_s; then, for each f in H_s,

(1.4)
$$f = \sum_n \langle f, e_n \rangle_s e_n = \sum_n \left[\int_a^b s(x)f(x)\bar{e}_n(x)\,dx\right]e_n(x).$$

If f is in $L_2(a, b)$, then f/s is in H_s and

(1.5)
$$\frac{f}{s} = \sum_n \langle \frac{f}{s}, e_n \rangle_s e_n = \sum_n \langle f, e_n \rangle e_n = \sum_n \left(\int_a^b f\bar{e}_n\,dx\right)e_n(x).$$

We make the following permanent assumptions regarding the coefficients in (1.1): p, q, s are *real-valued* functions on $a < x < b$; p, p', q, s are *continuous* on $a < x < b$; p and s are *positive* on $a < x < b$. If the interval is *finite* and all assumptions on the coefficients hold for the *closed* interval $a \leqslant x \leqslant b$, the problem is said to be *regular*; otherwise it is *singular*. In a singular problem solutions of (1.1) need not lie in H_s, and a more delicate analysis is required. In the regular case we associate with (1.1) two homogeneous boundary conditions of the *unmixed* type [see (2.3), Chapter 3]:

(1.6)
$$0 = B_a u \doteq \cos\alpha\, u(a) - \sin\alpha\, u'(a),$$
$$0 = B_b u \doteq \cos\beta\, u(b) + \sin\beta\, u'(b),$$

where α and β are given real numbers, $0 \leqslant \alpha < \pi$, $0 \leqslant \beta < \pi$. As β, for instance, takes on different values, the boundary condition at b specifies the ratio $u'(b)/u(b)$ as any positive or negative number [the cases $\beta = 0$ and $\beta = \pi/2$ correspond to $u(b) = 0$ and $u'(b) = 0$]. A similar statement holds for the endpoint a. The signs in (1.6) have been chosen so that the eigenvalues decrease as α or β increases.

We note the following properties of L and the boundary operators. Let u and v be arbitrary twice-differentiable functions:

(1.7)   (a) $L\bar{u} = \overline{(Lu)}, \qquad B_a \bar{u} = \overline{(B_a u)}, \qquad B_b \bar{u} = \overline{(B_b u)}.$

(1.8)   (b) $\bar{v} L u - u L \bar{v} = \dfrac{1}{s} [p(u\bar{v}' - u'\bar{v})]' = \dfrac{1}{s} [p(x) W(u, \bar{v}; x)]',$

where $W(u, \bar{v}; x) = u\bar{v}' - \bar{v}u'$ is the Wronskian of u and $\bar{v}$ [see (1.6), Chapter 3].

(1.9)   (c) $\langle Lu, v \rangle_s - \langle u, Lv \rangle_s = p(b) W(u, \bar{v}; b) - p(a) W(u, \bar{v}; a).$

(d)   If u and v both satisfy the boundary conditions (1.6), then $W(u, \bar{v}; b) = W(u, \bar{v}; a) = 0$, so that

(1.10) $$\langle Lu, v \rangle_s = \langle u, Lv \rangle_s$$

and

(1.11) $$\langle Lu, u \rangle_s \text{ is real.}$$

Thus, if we consider L as an operator whose domain D consists of functions having a continuous second derivative and satisfying (1.6), then L is symmetric. If we are willing to enlarge D slightly by relaxing the smoothness condition (to include functions whose first derivative is absolutely continuous and whose second derivative is in H_s), then L is actually self-adjoint on this new domain D_L.

We now investigate some of the properties of the eigenvalue problem (1.1)-(1.6):

(1.12a) $\begin{aligned} &-(pu')' + qu - \lambda su = 0, \quad a < x < b; \\ &\cos\alpha\, u(a) - \sin\alpha\, u'(a) = 0, \quad \cos\beta\, u(b) + \sin\beta\, u'(b) = 0, \end{aligned}$

or, equivalently, using definition (1.3),

(1.12b) $$Lu - \lambda u = 0, \quad a < x < b; \qquad B_a u = B_b u = 0.$$

1.   *The eigenvalues are real.* Multiply (1.12) by $s\bar{u}$ and integrate to obtain

$$\langle Lu, u \rangle_s = \lambda \langle u, u \rangle_s = \lambda \int_a^b s(x) |u(x)|^2 \, dx.$$

Since u is an eigenfunction, $\int_a^b s|u|^2\,dx > 0$, and it follows from (1.11) that λ is real.

2. *Eigenfunctions corresponding to different eigenvalues are orthogonal in H_s.* Let u and v satisfy (1.12) with respective eigenvalues λ and μ. Then

$$\langle Lu,v\rangle_s = \lambda\langle u,v\rangle_s = \lambda\int_a^b su\bar{v}\,dx,$$

$$\langle u,Lv\rangle_s = \langle u,\mu v\rangle_s = \bar{\mu}\int_a^b su\bar{v}\,dx = \mu\int_a^b su\bar{v}\,dx.$$

By (1.10), $\langle Lu,v\rangle_s = \langle u,Lv\rangle_s$; hence, for $\lambda\neq\mu$,

(1.13)
$$0 = \int_a^b su\bar{v}\,dx = \langle u,v\rangle_s,$$

so that u and v are orthogonal in H_s [other ways of expressing this property: u and v are orthogonal with weight s; $\sqrt{s}\,u$ and $\sqrt{s}\,v$ are orthogonal in $L_2(a,b)$].

3. The eigenvalues of (1.12) are simple and can be listed as the sequence

$$\lambda_1 < \lambda_2 < \lambda_3 < \dots < \lambda_n < \dots,$$

with $\lim_{n\to\infty}\lambda_n = +\infty$ (*thus there are at most finitely many negative eigenvalues*). *The corresponding normalized eigenfunctions $\{u_n\}$ form an orthonormal basis in H_s.* Assume without loss of generality that $\lambda=0$ is not an eigenvalue of (1.12), and introduce Green's function g_0 satisfying $-(pg_0')' + qg_0 = \delta(x-\xi)$, $B_a g_0 = B_b g_0 = 0$. Then (1.12) is equivalent to the integral equation

$$u(x) = \lambda\int_a^b g_0(x,\xi)s(\xi)u(\xi)\,d\xi,$$

or, setting $k(x,\xi) = \sqrt{s(x)}\,g_0(x,\xi)\sqrt{s(\xi)}$ and $\sqrt{s(x)}\,u(x) = v(x)$,

(1.14)
$$v(x) = \lambda\int_a^b k(x,\xi)v(\xi)\,d\xi \doteq \lambda Kv.$$

Since $g_0(x,\xi)$ is symmetric and Hilbert-Schmidt, so is $k(x,\xi)$ and the theory of self-adjoint H.-S. operators applies. Thus the eigenvalues of (1.12) are the *reciprocals* of the eigenvalues of K (recall that an eigenvalue of K is a number γ such that $Kv = \gamma v$). Since $Kv = 0$ implies that $v = 0$, the eigenfunctions $\{v_n\}$ form an orthonormal basis; now $v_n = \sqrt{s}\, u_n$, where u_n is the corresponding eigenfunction of (1.12); thus $\{u_n\}$ is an orthonormal basis in H_s. The differential equation (1.12) has two independent solutions, so that no eigenvalue can have multiplicity greater than 2. In fact, it can be shown that for unmixed boundary conditions the eigenvalues are simple. The theory of compact operators tells us that 0 is the only limit-point of the eigenvalues of K. Therefore $|\lambda_n| \to \infty$; in addition one can show (Exercise 1.1) that there are only finitely many negative eigenvalues, and hence it is possible to index the eigenvalues λ_n so that $\lambda_1 < \lambda_2 < \dots$ and $\lambda_n \to +\infty$.

Relation between Green's Function and the Eigenfunctions

We now study the relation between the eigenfunctions of (1.12) and Green's function $g(x, \xi; \lambda)$ satisfying

$$(1.15) \quad -(pg')' + qg - \lambda sg = \delta(x - \xi), \quad a < x, \xi < b; \quad B_a g = B_b g = 0.$$

It is an easy matter (and not our primary concern) to determine the Fourier expansion of $g(x, \xi; \lambda)$ in the eigenfunctions $u_n(x)$ of (1.12). We have, by (1.4),

$$g(x, \xi; \lambda) = \sum_n g_n(\xi, \lambda) u_n(x), \qquad g_n = \langle g, u_n \rangle_s = \int_a^b s(x) g(x, \xi; \lambda) \bar{u}_n(x)\, dx.$$

To find g_n, multiply (1.15) by $\bar{u}_n(x)$ and integrate from a to b to obtain

$$\langle Lg, u_n \rangle_s - \lambda \langle g, u_n \rangle_s = \bar{u}_n(\xi),$$

or, using (1.10),

$$\langle g, u_n \rangle_s (\lambda_n - \lambda) = \bar{u}_n(\xi),$$

so that

$$(1.16) \qquad g(x, \xi; \lambda) = \sum_n \frac{u_n(x) \bar{u}_n(\xi)}{\lambda_n - \lambda},$$

which is the *bilinear series* for g. As expected, g has singularities at $\lambda = \lambda_n$; in any event g can be constructed if $\{u_n\}$ and $\{\lambda_n\}$ are known. We can also find the Fourier expansion of the solution $w(x, \lambda)$ of the inhomogeneous equation

$$(1.17) \qquad -(pw')' + qw - \lambda sw = f, \quad a < x < b; \quad B_a w = B_b w = 0,$$

either by imitating the steps leading to (1.16) or by recalling that

(1.18)

$$w(x,\lambda) = \int_a^b g(x,\xi;\lambda) f(\xi) \, d\xi = \sum_n u_n(x) \frac{\int_a^b f(\xi) \bar{u}_n(\xi) \, d\xi}{\lambda_n - \lambda} = \sum_n \frac{\langle f, u_n \rangle}{\lambda_n - \lambda} u_n.$$

We now ask the question that concerns us most in the present chapter: given an explicit formula for $g(x,\xi;\lambda)$ as obtained, say, by the methods of Section 2, Chapter 3, how do we determine the eigenfunctions $\{u_n\}$ and eigenvalues $\{\lambda_n\}$ of (1.12)? Representation (1.16) shows that, as a function of the complex parameter λ, $g(x,\xi;\lambda)$ has simple poles at the real points $\lambda = \lambda_n$ with corresponding residues $-u_n(x)\bar{u}_n(\xi)$. Thus all we have to do is to examine $g(x,\xi;\lambda)$ in the complex λ plane and to pick out the singularities (which will be the eigenvalues) and the residues at these singularities (which are related to the eigenfunctions). This information is formally contained in the compact formula

(1.19)
$$\frac{1}{2\pi i} \int_{C_\infty} g(x,\xi;\lambda) \, d\lambda = -\sum_n u_n(x) \bar{u}_n(\xi),$$

where the integral in the λ plane is taken counterclockwise around the infinitely large circle C_∞, thus containing all the poles of g in (1.16). Admittedly the series in (1.19) fails to converge in the ordinary sense but has a distributional meaning (see Section 3, Chapter 2); in fact, using (1.5) with $f = \delta(x - \xi)$ and $e_n = u_n$, we have

(1.20)
$$\frac{\delta(x - \xi)}{s(x)} = \sum_n u_n(x) \bar{u}_n(\xi).$$

If one feels uncomfortable with the series in (1.19), it is possible instead to focus attention on (1.18). By integrating around C_∞, we find that

(1.21)
$$\frac{1}{2\pi i} \int_{C_\infty} w(x,\lambda) \, d\lambda = -\sum_n \langle f, u_n \rangle u_n = -\frac{f(x)}{s(x)},$$

which can be used as a basis for a more rigorous theory; we prefer, however, to use (1.19) to generate the eigenfunctions. Even in problems with a continuous spectrum it will be possible to use (1.19) with the summation replaced by an integral over a continuous parameter (Green's function then has a branch instead of poles).

The first step in using (1.19) is to construct $g(x,\xi;\lambda)$ in such a way that we can keep track of its dependence on λ. For the regular problem (1.15) we know that, for $x < \xi$, g is a solution of the homogeneous equation satisfying $0 = B_a g = (\cos \alpha g - \sin \alpha g')_{x=a}$. Therefore g is just a constant multiple of the solution $v(x, \lambda)$ of an appropriate initial value problem for the homogeneous equation; for instance, we can require v to satisfy $v(a, \lambda) = \sin \alpha$, $v'(a, \lambda) = \cos \alpha$. Then v obviously satisfies $B_a v = 0$. The advantage of proceeding in this way is that v is an *analytic* function of λ in the whole λ plane (a so-called entire function). This follows from the fact that v satisfies *initial* conditions independent of λ and a differential equation where λ appears analytically. Such a problem can be translated into a Volterra integral equation whose Neumann series converges in the whole λ plane (see Example 4b, Section 4, Chapter 4). Similarly, let $z(x, \lambda)$ be the unique solution of the homogeneous equation with the initial conditions $z(b, \lambda) = \sin \beta, z'(b, \lambda) = -\cos \beta$. Then $B_b z = 0$ and g is proportional to z for $x > \xi$. Thus we have

$$g(x, \xi; \lambda) = A v(x_<, \lambda) z(x_>, \lambda),$$

where $x_< = \min(x, \xi)$ and $x_> = \max(x, \xi)$. The jump condition on g' is

$$\frac{dg}{dx}\bigg|_{x=\xi+} - \frac{dg}{dx}\bigg|_{x=\zeta} = -\frac{1}{p(\xi)},$$

which becomes

$$A[v(\xi, \lambda) z'(\xi, \lambda) - v'(\xi, \lambda) z(\xi, \lambda)] = -\frac{1}{p(\xi)}.$$

The quantity in brackets is the Wronskian of v and z, which, by (1.8), Chapter 3, is of the form $C/p(\xi)$, where C is independent of ξ but may depend on λ. Therefore

$$(1.22) \qquad g(x, \xi; \lambda) = -\frac{v(x_<, \lambda) z(x_>, \lambda)}{C(\lambda)},$$

where $C(\lambda)$ is unambiguously determined from

$$(1.23) \qquad W[v(x, \lambda), z(x, \lambda); x] = vz' - zv' = \frac{C(\lambda)}{p(x)}.$$

Since v and z are entire functions, so are v', z', W, and $C(\lambda)$. Let $\lambda = \mu$ be a zero of C; that is, $C(\mu) = 0$. Then the Wronskian of $v(x, \mu)$ and $z(x, \mu)$ vanishes, and these functions are linearly dependent. In view of their initial values neither function can vanish identically in x. Therefore $v(x, \mu)$ is a nontrivial constant multiple of $z(x,\mu)$, and both functions satisfy the two boundary conditions and the differential equation in (1.12). Thus μ is an eigenvalue of (1.12) with eigenfunction (not normalized) $v(x,\mu)$. From (1.16) it is clear that at an eigenvalue g has a singularity, and therefore C must vanish. We conclude that the zeros of $C(\lambda)$ coincide with the eigenvalues of (1.12); we therefore label these zeros (or eigenvalues) $\lambda_1 < \lambda_2 < \ldots < \lambda_n < \ldots$, with $\lambda_n \to \infty$, and

$$(1.24) \qquad v(x,\lambda_n) = k_n z(x,\lambda_n),$$

where k_n is a real nonzero constant. We also know from (1.16) that g has only simple poles, so that the zeros of C are likewise simple. To simplify the notation a little, we let

$$v_n(x) \doteq v(x,\lambda_n), \qquad z_n(x) \doteq z(x,\lambda_n).$$

Thus $v_n(x)$ (or z_n) is a real eigenfunction corresponding to the simple eigenvalue λ_n and $v_n(x) = k_n z_n(x)$. Neither v_n nor z_n is normalized.

The residue of g at $\lambda = \lambda_n$ is

$$-\frac{v_n(x_<)z_n(x_>)}{C'(\lambda_n)} = -k_n \frac{z_n(x_<)z_n(x_>)}{C'(\lambda_n)} = -\frac{v_n(x_<)v_n(x_>)}{k_n C'(\lambda_n)}.$$

The quantity $z_n(x_<)z_n(x_>)$ is equal to $z_n(x)z_n(\xi)$ whether $x < \xi$ or $x > \xi$. It is remarkable that the discontinuity in the first derivative of g at $x = \xi$ has left no trace. The residue of g at $\lambda = \lambda_n$ becomes

$$-k_n \frac{z_n(x)z_n(\xi)}{C'(\lambda_n)} = -\frac{v_n(x)v_n(\xi)}{k_n C'(\lambda_n)} = -u_n(x)\bar{u}_n(\xi),$$

from which we recognize that the *real normalized eigenfunction* $u_n(x)$ is given by

$$(1.25) \qquad u_n(x) = \pm \frac{v_n(x)}{\left[k_n C'(\lambda_n) \right]^{1/2}} = \pm \left[\frac{k_n}{C'(\lambda_n)} \right]^{1/2} z_n(x).$$

Another method of obtaining the normalization constant is described in Exercise 1.2.

Example 1.   Consider one-dimensional heat conduction without sources in a rod $0 < x < 1$. The initial temperature is given, the left end is kept at temperature 0, and at the right end the temperature gradient is proportional to the temperature. The temperature $\Theta(x,t)$ in the rod satisfies

(1.26)

$$\frac{\partial\Theta}{\partial t} - \frac{\partial\Theta}{\partial x^2} = 0, \quad 0<x<1, \quad t>0; \quad \Theta(x,0)=f(x),$$

$$\Theta(0,t)=0, \quad \cos\beta\,\Theta(1,t) + \sin\beta\frac{\partial\Theta}{\partial x}(1,t)=0,$$

where β is a given real number in $[0,\pi]$. One tries to construct the solution of this problem by using as building blocks functions that are separable in x and t [that is, of the form $u(x)T(t)$]. These functions are chosen to satisfy the differential equation and the boundary conditions but *not* the arbitrary initial condition, which is ultimately satisfied by an appropriate sum of such separable functions. The steps in the procedure are purely formal, and one must check afterwards that the proposed solution actually meets all the requirements of (1.26).

Substituting $\Theta = u(x)T(t)$ in the differential equation, we find that

$$\frac{1}{T}\frac{dT}{dt} = \frac{1}{u}\frac{d^2u}{dx^2} = -\lambda,$$

where λ is an arbitrary complex parameter whose values will be determined from the x equation with its boundary conditions:

(1.27) $-u'' - \lambda u = 0, \quad 0<x<1; \quad u(0)=0, \quad \cos\beta\,u(1) + \sin\beta\,u'(1)=0.$

Let the eigenvalues of (1.27) be arranged in increasing order $\lambda_1<\lambda_2<\dots$, with the corresponding eigenfunctions $u_1(x),\dots,u_n(x),\dots$ forming an orthonormal basis. For $\lambda=\lambda_k$ the solution of the t equation is $T=e^{-\lambda_k t}$ and $\Theta = e^{-\lambda_k t}u_k(x)$. The sum $\sum_{k=1}^\infty c_k e^{-\lambda_k t}u_k(x)$ formally satisfies the homogeneous differential equation and the boundary conditions; it will also take on the correct initial value if the $\{c_k\}$ are chosen so that $f(x)=\sum_{k=1}^\infty c_k u_k$. Thus c_k must be chosen to be the Fourier coefficient $\langle f, u_k\rangle$.

For large values of time the behavior of the temperature is controlled by λ_1. According as $\lambda_1>0$ or $\lambda_1<0$, the temperature decays exponentially at the rate $e^{-\lambda_1 t}$ or blows up exponentially at the rate $e^{|\lambda_1|t}$ if $\langle f, u_1\rangle\neq 0$. In the case $\lambda_1<0$ we would say that the zero solution of the completely homogeneous problem ($f=0$) is *unstable*, for a slight change in the initial condition would create a disturbance that would grow in time.

Before analyzing (1.27) in detail we note that the eigenvalues $\{\lambda_n\}$ decrease as β increases in $0\leqslant\beta<\pi$. Mathematically this is a consequence of the Courant-Weyl variational principle (see Exercise 1.3). Physically, as

β increases, less and less heat is removed at the right end; $\beta=0$ means that the right end is kept at 0 temperature; if $0<\beta<\pi/2$, we have radiation into a surrounding medium at 0 temperature, heat being removed from the right end of the rod at a rate proportional to the temperature (the proportionality constant decreasing with increasing β); if $\beta=\pi/2$, the right end is insulated; if $\pi/2<\beta<\pi$, heat is being fed into the rod from its right end at a rate proportional to the temperature. Obviously the last case is one of potential instability, but since heat is removed from the left end (to keep it at 0 temperature), it is not clear how large β has to be before a negative eigenvalue λ_1 of (1.27) appears. In any event it seems physically obvious that at least the lowest eigenvalue $\lambda_1(\beta)$ is a decreasing function of β.

Green's function $g(x,\xi;\lambda)$ corresponding to (1.27) satisfies

(1.28)
$$-g''-\lambda g=\delta(x-\xi), \quad 0<x,\xi<1;$$

$$g|_{x=0}=0, \quad (\cos\beta g+\sin\beta g')_{x=1}=0.$$

Let $v(x,\lambda)$, $z(x,\lambda)$ be the solutions of the homogeneous equation satisfying the initial conditions

$$v(0,\lambda)=0, \quad v'(0,\lambda)=1; \qquad z(1,\lambda)=\sin\beta, \quad z'(1,\lambda)=-\cos\beta.$$

Then

(1.29)
$$v(x,\lambda)=\frac{\sin\sqrt{\lambda}\,x}{\sqrt{\lambda}},$$

$$z(x,\lambda)=\sin\beta\cos\sqrt{\lambda}\,(x-1)-\frac{\cos\beta}{\sqrt{\lambda}}\sin\sqrt{\lambda}\,(x-1),$$

where for definiteness we have chosen $\sqrt{\lambda}$ unambiguously as follows: for $\lambda=0$, $\sqrt{\lambda}=0$; for $\lambda\neq0$, λ has the unique polar representation $\lambda=|\lambda|e^{i\theta}$, $0\leqslant\theta<2\pi$, and we define

(1.30)
$$\sqrt{\lambda}=|\lambda|^{1/2}e^{i\theta/2}, \qquad 0\leqslant\theta<2\pi,$$

where $|\lambda|^{1/2}$ is the positive square root of the positive number $|\lambda|$. In this way each complex number λ has a well-defined square root $\sqrt{\lambda}$, which is analytic in the complex plane except on the positive real axis (for λ directly above the positive real axis, $\sqrt{\lambda}=|\lambda|^{1/2}$, whereas directly below, $\sqrt{\lambda}=-|\lambda|^{1/2}$). This is just the *principal value* of $\sqrt{\lambda}$ as defined in Exercise 2.3, Chapter 1. Note that $\sqrt{\lambda}$ has a *positive* imaginary part as long as λ is not on the positive real axis. Despite the branch singularity of $\sqrt{\lambda}$, the functions defined by (1.29) are analytic in the whole λ plane (both v and z are even functions of $\sqrt{\lambda}$, so that, whereas $\sqrt{\lambda}$ abruptly changes sign across the positive real axis, v and z do not; note also that there is no singularity at $\lambda=0$ if we use the limiting value of (1.29) as $\lambda\to0$).

Next we calculate, from (1.23) with $p=1$ and (1.29),

$$(1.31) \qquad C(\lambda) = vz' - zv' = -\sin\beta\cos\sqrt{\lambda} - \frac{\cos\beta}{\sqrt{\lambda}}\sin\sqrt{\lambda},$$

which is also an analytic function of λ. The eigenvalues of (1.27), which are necessarily real, are the zeros of $C(\lambda)$.

We first look for negative eigenvalues. Setting $\lambda=-r^2, r>0$, we find that the equation $C(\lambda)=0$ becomes (see Figure 1.1)

$$(1.32) \qquad \tanh r = -r\tan\beta, \qquad r>0,$$

which has a single root $r_1(\beta)$ if and only if $3\pi/4<\beta<\pi$. The corresponding negative eigenvalue $\lambda_1(\beta) = -r_1^2(\beta)$ decreases from

$$\lambda_1\left(\frac{3\pi}{4}+\right)=0 \qquad \text{to} \qquad \lambda_1(\pi-)=-\infty$$

as β increases from $(3\pi/4)+$ to $\pi-$.

If $\beta=3\pi/4$, then $\lambda_1=0$ is an eigenvalue with an eigenfunction $u_1(x)$ proportional to x.

Turning next to positive eigenvalues, we set $\lambda=r^2, r>0$. The equation $C(\lambda)=0$ now becomes

$$(1.33) \qquad \tan r = -r\tan\beta, \qquad r>0.$$

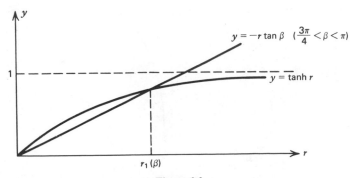

Figure 1.1

The values of r at the intersections of the curve $y = \tan r$ with the straight line $y = -r\tan\beta$ then give the square roots of the desired positive eigenvalues. Figure 1.2 shows these intersections for different values of β. For a fixed β there are infinitely many intersections. Since for β in $0 \le \beta < 3\pi/4$ all eigenvalues are positive, we label the r values at the intersections as $r_1(\beta), r_2(\beta), \ldots$; then $\lambda_k(\beta) = r_k^2(\beta)$ gives the sequence of eigenvalues in increasing order. In particular, we have $r_k(0) = k\pi$ and $\lambda_k(0) = k^2\pi^2$. As β increases in $0 \le \beta < 3\pi/4$, it is clear that $r_k(\beta)$ decreases and hence so does $\lambda_k(\beta)$. When β is slightly smaller than $3\pi/4$, the first intersection occurs close to $r = 0$; when $\beta = 3\pi/4$, the line $y = -r\tan\beta$ (that is, $y = r$) is tangent to the curve $y = \tan r$ at $r = 0$ and 0 now becomes an eigenvalue (as we saw earlier). For $3\pi/4 < \beta < \pi$ the smallest eigenvalue λ_1 is the negative eigenvalue $-r_1^2(\beta)$, found from Figure 1.1; the intersections on Figure 1.2 are then labeled $r_2(\beta), r_3(\beta), \ldots$ in increasing order, and the corresponding eigenvalues are $\lambda_2(\beta) = r_2^2(\beta), \ldots$.

With this notation we see that, as β increases from 0 to $\pi-$,

$\lambda_1(\beta)$ decreases from π^2 to $-\infty$,

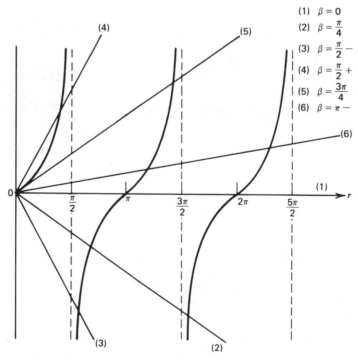

(1) $\beta = 0$

(2) $\beta = \dfrac{\pi}{4}$

(3) $\beta = \dfrac{\pi}{2} -$

(4) $\beta = \dfrac{\pi}{2} +$

(5) $\beta = \dfrac{3\pi}{4}$

(6) $\beta = \pi -$

Figure 1.2

$\lambda_k(\beta)$ decreases from $k^2\pi^2$ to $(k-1)^2\pi^2, k \geq 2$.

In all cases the dependence on β is continuous.

We now use (1.25) to find the normalized eigenfunctions (for the case $\beta < 3\pi/4$). We have, with $\lambda_n = r_n^2$,

$$v_n(x) = \frac{\sin r_n x}{r_n}, \qquad z_n(x) = -\frac{\cos\beta}{r_n \cos r_n} \sin r_n x,$$

where we have made use of (1.32), which is satisfied for $r = r_n$. From (1.24) we see that

$$k_n = \frac{-\cos r_n}{\cos\beta}$$

and, from (1.31),

$$C'(\lambda_n) = \frac{1}{2r_n}\left(\sin\beta \sin r_n - \frac{\cos\beta}{r_n}\cos r_n + \cos\beta \frac{\sin r_n}{r_n^2}\right).$$

A straightforward calculation now gives

$$k_n C'(\lambda_n) = \frac{1}{r_n^2}\left(\frac{1}{2} - \frac{\sin 2r_n}{4r_n}\right).$$

Therefore the normalized eigenfunctions can be found from (1.25):

$$u_n(x) = \frac{1}{N_n}\sin\sqrt{\lambda_n}\,x, \qquad N_n^2 = \frac{1}{2} - \frac{\sin 2\sqrt{\lambda_n}}{4\sqrt{\lambda_n}}.$$

Of course in our case the normalization could have been calculated by elementary integration of $\sin^2\sqrt{\lambda_n}\,x$, but this will usually be impossible in more difficult cases. Alternatively the method of Exercise 1.2 can be used.

Example 2. Any circularly symmetric natural mode u and the corresponding natural frequency λ of an annular membrane with fixed boundary satisfy

(1.34) $-(xu')' = \lambda xu, \quad 0 < a < x < b; \quad u(a) = u(b) = 0,$

where x is the *radial* coordinate measured from the center of the annulus.

Green's function $g(x,\xi;\lambda)$ is the solution of

(1.35) $-(xg')' - \lambda xg = \delta(x-\xi), \quad a < x, \xi < b; \quad g = 0 \text{ at } x = a, b.$

The homogeneous equation has the independent solutions $J_0(\sqrt{\lambda}\,x)$, $N_0(\sqrt{\lambda}\,x)$. The functions

(1.36)
$$v(x,\lambda) = J_0(\sqrt{\lambda}\,a)N_0(\sqrt{\lambda}\,x) - J_0(\sqrt{\lambda}\,x)N_0(\sqrt{\lambda}\,a),$$
$$z(x,\lambda) = J_0(\sqrt{\lambda}\,b)N_0(\sqrt{\lambda}\,x) - J_0(\sqrt{\lambda}\,x)N_0(\sqrt{\lambda}\,b)$$

satisfy the homogeneous equation and the respective initial conditions

$$v(a,\lambda) = 0, \qquad v'(a,\lambda) = \sqrt{\lambda}\,W\big(J_0, N_0; \sqrt{\lambda}\,a\big) = \frac{2}{\pi a},$$

$$z(b,\lambda) = 0, \qquad z'(b,\lambda) = \sqrt{\lambda}\,W\big(J_0, N_0; \sqrt{\lambda}\,b\big) = \frac{2}{\pi b},$$

where use has been made of the Wronskian relation

(1.37)
$$W(J_0, N_0; x) = \frac{2}{\pi x}.$$

Clearly $v(x,\lambda)$ and $z(x,\lambda)$ are entire functions, and

$$g = Av(x_<,\lambda)z(x_>,\lambda).$$

The jump condition on dg/dx is $g'|_{\xi_-}^{\xi_+} = -(1/\xi)$, so that

$$AW(v,z;\xi) = -\frac{1}{\xi}.$$

From (1.36) and (1.37) we have

$$W(v,z;x) = \frac{2}{\pi x}\Big[J_0(\sqrt{\lambda}\,a)N_0(\sqrt{\lambda}\,b) - J_0(\sqrt{\lambda}\,b)N_0(\sqrt{\lambda}\,a)\Big],$$

so that

(1.38)
$$g = -\frac{\pi}{2}\,\frac{v(x_<,\lambda)z(x_>,\lambda)}{J_0(\sqrt{\lambda}\,a)N_0(\sqrt{\lambda}\,b) - J_0(\sqrt{\lambda}\,b)N_0(\sqrt{\lambda}\,a)}.$$

The eigenvalues of (1.34) are the zeros of the denominator $D(\lambda)$ in (1.38). These are known from (1.34) to be real and positive. Setting $\lambda = r^2$, we find that the eigenvalues $\lambda_n = r_n^2$ are determined from

(1.39) $$D(r^2) \doteq J_0(ra)N_0(rb) - J_0(rb)N_0(ra) = 0 \quad \text{or} \quad \frac{J_0(ra)}{J_0(rb)} = \frac{N_0(ra)}{N_0(rb)}.$$

This equation gives rise to a sequence of positive simple roots

$$r_1 < r_2 < \cdots,$$

and we set

(1.40)
$$R_n = \frac{J_0(r_n a)}{J_0(r_n b)} = \frac{N_0(r_n a)}{N_0(r_n b)}.$$

We then find that

$$v_n(x) \doteq v(x, r_n^2) = J_0(r_n a) N_0(r_n x) - J_0(r_n x) N_0(r_n a)$$
$$= R_n J_0(r_n b) N_0(r_n x) - R_n N_0(r_n b) J_0(r_n x) = R_n z(x, r_n^2)$$
$$\doteq R_n z_n(x).$$

The residue of g at $\lambda = \lambda_n = r_n^2$ is

$$-\frac{\pi}{2} \frac{v_n(x_<) z_n(x_>)}{D'(\lambda_n)} = -\frac{\pi}{2}\left(\frac{2r_n}{R_n}\right) \frac{v_n(x) v_n(\xi)}{\left[dD(r^2)/dr\right]_{r=r_n}},$$

and, on using (1.39) and (1.40), we obtain

$$\frac{dD(r^2)}{dr}\bigg|_{r=r_n} = -\frac{a}{R_n}\frac{2}{\pi r_n a} + bR_n\frac{2}{\pi r_n b} = \frac{2}{\pi r_n}\left(\frac{R_n^2 - 1}{R_n}\right).$$

The normalized eigenfunctions are therefore given by

(1.41) $$u_n(x) = \frac{\pi r_n}{\sqrt{2}\sqrt{R_n^2 - 1}}\left[J_0(r_n a) N_0(r_n x) - J_0(r_n x) N_0(r_n a)\right],$$

where r_n is determined from (1.39) and R_n from (1.40).

Example 3. (See Eastham). (a) In propagation through periodic lattices one encounters the equation

(1.42) $$-u'' + Q(x)u = 0, \qquad -\infty < x < \infty,$$

where $Q(x)$ is periodic in x. Such an equation does not necessarily have a nontrivial periodic solution (for instance, $-u'' + u = 0$), but any periodic solution u has a period which coincides with one of the periods of Q. Suppose, without loss of generality, that

(1.43) $$Q(x + 1) = Q(x),$$

and let us look not only for solutions of (1.42) with period 1 but also, more generally, for solutions satisfying

$$(1.44) \qquad u(x+1)=\rho u(x),$$

where ρ is an undetermined constant known as a *multiplier*.

Let $v(x)$ and $z(x)$ be the solutions of the homogeneous equation (1.42) satisfying, respectively, the initial conditions

$$(1.45) \qquad v(0)=1, \quad v'(0)=0; \qquad z(0)=0, \quad z'(0)=1.$$

In view of (1.43) the functions $v(x+1)$ and $z(x+1)$ are also solutions of (1.42); moreover, these functions are independent. We can therefore write $v(x+1)$ and $z(x+1)$ as linear combinations of $v(x)$ and $z(x)$:

$$(1.46) \qquad \begin{aligned} v(x+1)&=v(1)v(x)+v'(1)z(x),\\ z(x+1)&=z(1)v(x)+z'(1)z(x). \end{aligned}$$

The general solution of (1.42) is $u=Av(x)+Bz(x)$; requiring (1.44) and using (1.46), we find that nontrivial solutions are possible if and only if the multiplier ρ satisfies the quadratic equation

$$\rho^2-\rho[v(1)+z'(1)]+v(1)z'(1)-z(1)v'(1)=0.$$

The usual Wronskian relation gives $v(1)z'(1)-z(1)v'(1)=v(0)z'(0)-z(0)v'(0)=1$, so that

$$(1.47) \qquad \rho^2-\rho[v(1)+z'(1)]+1=0.$$

If (1.47) has $\rho=1$ as a solution, (1.42) will have solution(s) with period 1. If (1.47) has $\rho=-1$ as a solution, (1.42) will have solution(s) with period 2.

In many applications $Q(x)=q(x)-\lambda$, where $q(x)$ has period 1 and λ is an eigenvalue parameter. We look for values of λ which give rise to periodic solutions of (1.42). Then the corresponding functions v and z depend on λ, and so does the coefficient of ρ in (1.47). We see below how to generate periodic solutions by considering eigenvalue problems on the basic interval $0<x<1$.

(b) We look at two related problems.

(i) Given q such that $q(x+1)=q(x)$, find λ and $u(x)$ with $u(x+1)=u(x)$ satisfying

$$-u''+q(x)u-\lambda u=0, \qquad -\infty<x<\infty.$$

(ii) Find the eigenvalues and eigenfunctions of

$$-u'' + q(x)u - \lambda u = 0, \quad 0 < x < 1; \qquad u(0) = u(1), \quad u'(0) = u'(1).$$

If (λ, u) is a solution of (i), then (λ, u) is clearly an eigenpair of (ii). If (λ, u) is an eigenpair of (ii), we can, by virtue of the boundary conditions, extend $u(x)$ to $-\infty < x < \infty$ as a continuously differentiable function with period 1. On $-\infty < x < \infty$ this periodic extension satisfies problem (i), where $q(x)$ stands for the periodic extension of the $q(x)$ that appears in (ii). *Thus problems (i) and (ii) are equivalent.*

If instead of the boundary conditions in (ii) we take $u(0) = -u(1), u'(0) = -u'(1)$, another eigenvalue problem results. Any eigenfunction of this new problem can be extended as a continuously differentiable function in $1 \leqslant x \leqslant 2$ by setting $u(x+1) = -u(x)$. This new function satisfies $u(0) = u(2), u'(0) = u'(2)$ and can therefore be extended as a function $u(x)$ *with period 2* on $-\infty < x < \infty$. The extended u satisfies $-u'' + qu - \lambda u = 0$, where q is the periodic extension referred to earlier: $q(x+1) = q(x)$. Let us now examine problem (ii) more closely.

Nontrivial solutions of (ii) are possible if and only if (1.47) is satisfied for $\mu = 1$. Since v and z are now functions of x and λ, this means that λ is determined from

$$(1.48) \qquad 2 - v(1, \lambda) - z'(1, \lambda) = 0.$$

Simple roots of (1.48) lead to simple eigenvalues of (ii), and double roots to eigenvalues of multiplicity 2. If the conditions $v'(1, \lambda) = 0$ and $z(1, \lambda) = 0$ hold simultaneously in addition to (1.48), the eigenvalue is of multiplicity 2. Indeed, from the Wronskian relation we then have $v(1, \lambda)z'(1, \lambda) = 1$, which together with (1.48) gives $v(1, \lambda) = z'(1, \lambda) = 1$. Thus both v and z satisfy the boundary conditions in (ii) and are therefore (independent) eigenfunctions corresponding to that value of λ. Since the differential equation is of the second order, every nontrivial solution is then an eigenfunction.

(c) We consider a special simple case of problem (ii):

$$(1.49) \qquad -u'' - \lambda u = 0, \quad 0 < x < 1; \qquad u(0) = u(1), \quad u'(0) = u'(1).$$

It is easy to find the eigenvalues and eigenfunctions explicitly. Let $v(x, \lambda), z(x, \lambda)$ be the solutions of the homogeneous equation satisfying, respectively, the initial conditions

$$v(0, \lambda) = 1, \quad v'(0, \lambda) = 0; \qquad z(0, \lambda) = 0, \quad z'(0, \lambda) = 1.$$

We find that

$$v(x,\lambda) = \cos\sqrt{\lambda}\,x, \qquad z(x,\lambda) = \frac{\sin\sqrt{\lambda}\,x}{\sqrt{\lambda}},$$

which are correct even for $\lambda = 0$ if the limiting value is used when needed. Therefore (1.48) becomes

$$(1.50) \qquad\qquad\qquad \cos\sqrt{\lambda} = 1.$$

Equation (1.50) has a simple zero at $\lambda = 0$ and double zeros at $\lambda = 4n^2\pi^2$, $n = 1, 2, \ldots$. Only $\lambda = 0$ is a simple eigenvalue of (1.49); all the others have multiplicity 2. To index all eigenvalues with regard to multiplicity, we set

$$(1.51) \qquad \lambda_0 = 0, \qquad \lambda_{2n-1} = \lambda_{2n} = 4n^2\pi^2, \quad n = 1, 2, \ldots .$$

The corresponding orthonormal basis of eigenfunctions is

$$(1.52) \qquad u_0 = 1, \qquad u_{2n-1} = \sqrt{2}\cos n\pi x, \qquad u_{2n} = \sqrt{2}\sin n\pi x.$$

Let us now see how we can generate the spectrum (1.51)-(1.52) from the residues at the singularities of Green's function $g(x,\xi;\lambda)$ satisfying

$$-g'' - \lambda g = \delta(x-\xi), \quad 0 < x, \xi < 1; \qquad g\big|_{x=0} = g\big|_{x=1}, \quad \big|g'_{x=0} = g'\big|_{x=1}.$$

We can write

$$(1.53) \qquad\qquad g(x,\xi;\lambda) = h(x,\xi;\lambda) + Av(x,\lambda) + Bz(x,\lambda),$$

where h is the causal Green's function given by

$$h(x,\xi;\lambda) = -\frac{\sin\sqrt{\lambda}\,(x-\xi)}{\sqrt{\lambda}}\,H(x-\xi).$$

On applying the boundary conditions, we find that

$$A(1 - \cos\sqrt{\lambda}\,) - B\frac{\sin\sqrt{\lambda}}{\sqrt{\lambda}} = -\frac{\sin\sqrt{\lambda}\,(1-\xi)}{\sqrt{\lambda}},$$

$$A\sqrt{\lambda}\sin\sqrt{\lambda} + B(1 - \cos\sqrt{\lambda}\,) = -\cos\sqrt{\lambda}\,(1-\xi),$$

from which follow

$$(1.54) \quad A = -\frac{\sin\sqrt{\lambda}\,(1-\xi)}{2\sqrt{\lambda}} - \frac{\sin\sqrt{\lambda}}{2\sqrt{\lambda}\,(1-\cos\sqrt{\lambda})}\cos\sqrt{\lambda}\,(1-\xi),$$

$$(1.55) \quad B = -\frac{\cos\sqrt{\lambda}\,(1-\xi)}{2} + \frac{\sin\sqrt{\lambda}}{2(1-\cos\sqrt{\lambda})}\sin\sqrt{\lambda}\,(1-\xi).$$

Expansion (1.16) is still valid, and we need only to keep track of the singularities of g. Since h, v, z, and the first term on the right of each of (1.54) and (1.55) are all analytic, the residues all stem from the singular part of $Av(x,\lambda) + Bz(x,\lambda)$, that is, from

$$R = -\frac{\sin\sqrt{\lambda}\,\cos\sqrt{\lambda}\,(1-\xi)\cos\sqrt{\lambda}\,x}{2(1-\cos\sqrt{\lambda})} + \frac{\sin\sqrt{\lambda}\,\sin\sqrt{\lambda}\,(1-\xi)\sin\sqrt{\lambda}\,x}{2(1-\cos\sqrt{\lambda})\sqrt{\lambda}}$$

$$-\frac{\cos\sqrt{\lambda}\,(x-\xi+1)}{2\sqrt{\lambda}}\left(\frac{\sin\sqrt{\lambda}}{1-\cos\sqrt{\lambda}}\right) = -\frac{\cos\sqrt{\lambda}\,(x-\xi+1)}{2\sqrt{\lambda}}\frac{\cos\left(\sqrt{\lambda}/2\right)}{\sin\left(\sqrt{\lambda}/2\right)}.$$

At $\lambda = 0$ we have a simple pole from the simple zero of $1 - \cos\sqrt{\lambda}$; this gives the residue

$$-\frac{1}{2}\frac{1}{\left[d/d\lambda(1-\cos\sqrt{\lambda})\right]_{\lambda=0}} = -1.$$

At $\lambda = 4n^2\pi^2$, $1 - \cos\sqrt{\lambda}$ has a double zero but $\sin\sqrt{\lambda}$ has a simple zero, so that we again have a simple pole whose residue is

$$-\frac{\cos 2n\pi(x-\xi+1)}{4n\pi}\frac{\cos n\pi}{d/d\lambda\left[\sin\left(\sqrt{\lambda}/2\right)\right]_{\lambda=2n\pi}} = -2\cos 2n\pi(x-\xi+1)$$

$$= -2\cos 2n\pi(x-\xi) = -2(\cos 2n\pi x \cos 2n\pi\xi + \sin 2n\pi x \sin 2n\pi\xi).$$

Using (1.19), we can set the sum of these residues equal to $-\sum_n u_n(x)\bar{u}_n(\xi)$, and we recover (1.52). Note that (1.20) becomes the familiar

(1.56)

$$\delta(x-\xi) = 1 + 2\sum_{n=1}^{\infty}(\cos 2n\pi x \cos 2n\pi\xi + \sin 2n\pi x \sin 2n\pi\xi), \qquad 0 < x, \xi < 1.$$

Transition to a Singular Problem for First-Order Equations

Consider the eigenvalue problem

$$(1.57) \qquad u' = i\lambda s(x)u, \quad 0 < x < b; \qquad u(b) = u(0),$$

where $s(x) > 0$ for $x \geqslant 0$. Letting $H_s(0, b)$ be the space of complex-valued functions such that $\int_0^b s|u|^2 dx < \infty$, we can rewrite our problem in the form

$$(1.58) \qquad\qquad\qquad Lu = \lambda u,$$

where

$$Lu \doteq -\frac{i}{s(x)}u',$$

and the domain D_L consists of absolutely continuous functions with Lu in $H_s(0, b)$ satisfying $u(b) = u(0)$. We proved in Chapter 5 that L is self-adjoint. Problem (1.57) or (1.58) is regular for finite b, and we shall be interested in the limit as $b \to \infty$ when the problem becomes singular.

For finite b we can solve (1.57) explicitly. Setting $S(x) = \int_0^x s(t) dt$, we see that the general solution of the differential equation is $A\phi(x, \lambda)$, where

$$(1.59) \qquad\qquad\qquad \phi(x, \lambda) = e^{i\lambda S(x)}.$$

Applying the boundary condition, we find that

$$e^{i\lambda S(b)} = 1,$$

so that the eigenvalues are given by

$$(1.60) \qquad\qquad\qquad \lambda_n = \frac{2n\pi}{S(b)}, \qquad n = 0, \pm 1, \pm 2 \ldots .$$

Note that the eigenvalues are uniformly spaced along the whole real axis.

The eigenfunctions are $Ae^{i\lambda_n S}$, which we now proceed to normalize. Setting $\phi_n = \phi(x, \lambda_n)$, we find that

$$\langle \phi_n, \phi_m \rangle_s = \int_0^b s(x) e^{i(\lambda_n - \lambda_m) S(x)} dx.$$

Since S is increasing and differentiable, we may make the change of variable

$$y = \frac{2\pi S(x)}{S(b)}$$

to obtain

$$\langle \phi_n, \phi_m \rangle_s = \frac{S(b)}{2\pi} \int_0^{2\pi} e^{i(n-m)y} \, dy = \begin{cases} 0, & n \neq m, \\ S(b), & n = m. \end{cases}$$

Thus (1.57) leads to the orthonormal basis of eigenfunctions in H_s,

(1.61) $$u_n(x) = [S(b)]^{-1/2} e^{i\lambda_n S(x)}.$$

It now becomes easy to analyze the behavior as $b \to \infty$. Since $S(x)$ is an increasing continuous function, there are only two possibilities: (1) $\lim_{b \to \infty} S(b)$ is a finite number denoted by $S(\infty)$, or (2) $S(b) \to +\infty$ as $b \to +\infty$. If $s(x) = 1$, for instance, case 2 obtains; but if $s(x) = 1/(1+x^2)$, we have case 1.

In case 1 everything is pretty much the same as for the bounded interval. We see from (1.60) that the spectrum remains discrete. It suffices to replace $S(b)$ by $S(\infty)$ in (1.60) and (1.61) to obtain the normalized eigenpairs for the interval $0 < x < \infty$. In fact, even the boundary condition in (1.57) is satisfied in the form $u(\infty) = u(0)$. We should also note that the general solution of the differential equation (1.57) without boundary condition is in H_s for all λ in the complex plane. Indeed, we have from (1.59)

(1.62) $$\|\phi\|_s^2 = \int_0^\infty s e^{-2\beta S} \, dx = \int_0^{S(\infty)} e^{-2\beta z} \, dz,$$

where $\beta = \operatorname{Im}\lambda$. Since $S(\infty)$ is finite, so is the integral in (1.62) for all β.

In case 2, $\lim_{b \to \infty} S(b) = \infty$, so that the eigenvalues (1.60) coalesce into the whole real axis. The solution $\phi(x, \lambda)$ is no longer in H_s when λ is real. In fact, (1.62) shows that $\phi(x, \lambda)$ is in H_s if and only if $\operatorname{Im}\lambda > 0$. Thus our limiting spectral problem has no eigenvalues. What happens is that the real λ axis is now in the continuous spectrum.

Additional information can be obtained from a study of Green's function $g_b(x, s; \lambda)$ satisfying

(1.63) $$-ig_b' - \lambda s g_b = \delta(x - \xi), \quad 0 < x, \xi < b; \qquad g_b|_{x=b} = g_b|_{x=0}.$$

The bilinear series (1.16), as well as (1.19) and (1.20), are again valid for regular self-adjoint boundary value problems involving equations of the first order. The functions $\{u_n\}$ are now given explicitly by (1.61). Of course Green's function can also be constructed in closed form from (1.63) when λ is not one of the eigenvalues (1.60). Certainly this construction is valid for $\operatorname{Im}\lambda \neq 0$. The function g_b is proportional to ϕ for $x < \xi$ and for $x > \xi$ but with different proportionality constants. At $x = \xi$, g_b satisfies the jump condition

$$g_b(\xi+,\xi;\lambda)-g_b(\xi-,\xi;\lambda)=i.$$

A straightforward calculation gives

(1.64) $$g_b(x,\xi;\lambda)=ie^{i\lambda[S(x)-S(\xi)]}\left[\frac{1}{e^{-i\lambda S(b)}-1}+H(x-\xi)\right],$$

where H is the Heaviside function.

As expected, g_b has simple poles at the eigenvalues λ_n given by (1.60), and the residue of g_b at $\lambda=\lambda_n$ is easily verified to be

$$\frac{-e^{i\lambda_n S(x)}e^{-i\lambda_n S(\xi)}}{S(b)},$$

which agrees with $-u_n(x)\bar{u}_n(\xi)$ from (1.61), thereby confirming (1.19).

We can also determine the behavior of g_b as $b\to\infty$. In case 1 when $S(b)$ tends to the finite limit $S(\infty)$, g_b itself has a limiting value with simple poles at $\lambda_n=2\pi/S(\infty)$. The spectrum therefore remains discrete. If, however, $S(b)\to\infty$, the simple poles in (1.64) disappear. Indeed, we have, on setting $\lambda=\alpha+i\beta$,

$$\frac{1}{e^{-i\lambda S(b)}-1}=\frac{1}{e^{-i\alpha S(b)}e^{\beta S(b)}-1},$$

which tends, as $b\to\infty$, to 0 for $\beta>0$ and to -1 for $\beta<0$ (there is no limit for $\beta=0$). The limiting Green's function g has therefore different analytic expressions in the upper and lower halves of the λ plane. We have the explicit forms

(1.65) $$g(x,\xi;\lambda)=ie^{i\lambda[S(x)-S(\xi)]}H(x-\xi) \qquad \text{for } \text{Im}\lambda>0,$$
$$g(x,\xi;\lambda)=ie^{i\lambda[S(x)-S(\xi)]}\left[-1+H(x-\xi)\right] \quad \text{for } \text{Im}\lambda<0.$$

It turns out that the same limiting form for g would result even if we had used in (1.63) the more general self-adjoint boundary condition $u(b)=e^{i\gamma}u(0)$.

Since g is analytic in both half-planes, we reduce the integral over a large circle C_∞ in the complex plane to an integral along the real axis:

(1.66) $$\int_{C_\infty}g(x,\xi;\lambda)\,d\lambda=-\int_{-\infty}^{\infty}\left[g(x,\xi;\lambda_+)-g(x,\xi;\lambda_-)\right]d\lambda,$$

where λ_+ and λ_- denote the values just above the real axis and just below, respectively. Using (1.65), we find that

$$(1.67) \qquad \frac{1}{2\pi i} \int_{C_\infty} g(x,\xi;\lambda)\,d\lambda = -\frac{1}{2\pi} \int_{-\infty}^{\infty} e^{i\lambda[S(x)-S(\xi)]}\,d\lambda.$$

Thus the integral over the real λ axis now plays the role of the sum appearing in (1.19). The equivalent form of (1.20) is then

$$(1.68) \qquad \frac{\delta(x-\xi)}{s(x)} = \frac{1}{2\pi}\int_{-\infty}^{\infty} e^{i\lambda[S(x)-S(\xi)]}\,d\lambda,$$

which contains, for the case $s(x)=1$, the famous Fourier integral formula

$$(1.69) \qquad \delta(x-\xi) = \frac{1}{2\pi}\int_{-\infty}^{\infty} e^{i\lambda(x-\xi)}\,d\lambda,$$

which has been derived for $0<x,\xi<\infty$, but is clearly valid for all x,ξ, since $x-\xi$ takes on all possible real values as x and ξ vary between 0 and infinity.

We shall pursue the transition to singular problems for second-order DVPs in Sections 2 and 3.

Exercises

1.1. An operator L defined on a dense domain D_L in H_s is said to be *bounded below* if there exists a real constant c (possibly negative) such that

$$(1.70) \qquad \langle Lu,u\rangle_s \geqslant c\|u\|_s^2 \qquad \text{for all } u\in D_L.$$

Show that the operator L given by (1.3) on the domain D_L of functions having a continuous second derivative and satisfying (1.6) is bounded below. (It follows that L on D_L has only a finite number of negative eigenvalues since we have shown elsewhere that the eigenvalues cannot have a finite limit-point.) *Hint*: Follow these steps with $\|u\|^2 = \int_a^b |u|^2\,dx$:

(a) There exists x_0 such that $\|u\|^2 = (b-a)|u(x_0)|^2$.

(b) $\big||u(x_0)|^2 - |u(a)|^2\big| = \big|\int_a^{x_0}(d/dx)(u\bar{u})\,dx\big| \leqslant 2\|u\|\,\|u'\|$.

(c) $|u(a)|^2 \leqslant [\|u\|^2/(b-a)] + 2\|u\|\,\|u'\|$. This is also true for $|u(b)|^2$.

(d) $\langle Lu,u\rangle_s \geqslant m_1\|u'\|^2 + m_2\|u\|^2 + m_3\|u\|\,\|u'\|$, $m_1>0$.

(e) $\langle Lu,u\rangle_s \geqslant d\|u\|^2$.

(f) $\langle Lu,u\rangle_s \geqslant c\|u\|_s^2$.

1.2.   Consider the eigenvalue problem

$$-(pu')' + qu - \lambda su = 0, \quad a < x < b; \qquad B_a u = 0, \quad B_b u = 0,$$

where B_a and B_b are defined by (1.6). Let $v(x, \lambda)$ be the solution of the homogeneous differential equation satisfying $v(a, \lambda) = \sin \alpha$, $v'(a, \lambda) = \cos \alpha$. Note that $v(x, \lambda)$ is real for λ real. The nontrivial function $v(x, \lambda)$ satisfies the boundary condition at the left end; an eigenvalue (necessarily real) λ_n is a number such that $v(x, \lambda_n)$ also satisfies the boundary condition at the right end. The real function $v_n(x) = v(x, \lambda_n)$ is the corresponding eigenfunction but is not normalized. We would like to calculate $\int_a^b s v_n^2(x) \, dx$. From the respective BVPs satisfied by $v_n(x)$ and $v(x, \lambda)$, where λ is a real number near λ_n, show that

$$(\lambda - \lambda_n) \int_a^b s v_n v \, dx = p(b) W(v, v_n; b),$$

and, by taking the limit as $\lambda \to \lambda_n$,

$$\int_a^b s v_n^2 \, dx = p(b) \left[v_n'(b) \frac{dv(b, \lambda)}{d\lambda} \bigg|_{\lambda = \lambda_n} - v_n(b) \frac{dv'(b, \lambda)}{d\lambda} \bigg|_{\lambda = \lambda_n} \right].$$

Prove the equivalence of this result and (1.25).

1.3.   Consider the eigenvalue problem

(1.71)
$$-(pu')' + qu - \lambda su = 0, \quad a < x < b;$$
$$\cos \alpha \, u(a) - \sin \alpha \, u'(a) = 0, \qquad \cos \beta \, u(b) + \sin \beta \, u'(b) = 0,$$

where $0 \leqslant \alpha < \pi, 0 \leqslant \beta < \pi$. The eigenvalues (which depend on α, β) are indexed in the usual order, $\lambda_1 < \lambda_2 < \ldots < \lambda_n < \ldots$. Show that $\lambda_n(\alpha, \beta)$ is a decreasing function of α and of β. *Hint*: Use the Weyl-Courant principle,

$$\lambda_{n+1}(\alpha, \beta) = \max_{E_n \in S_n} \min_{\substack{u \in E_n \\ u \in D(\alpha, \beta)}} R(u),$$

where

$$R(u) = \frac{\int_a^b (pu'^2 + qu^2) \, dx + p(b) u^2(b) \cot \beta + p(a) u^2(a) \cot \alpha}{\int_a^b su^2 \, dx},$$

$D(\alpha,\beta)$ is the set of piecewise differentiable real functions $u(x)$ (if α or $\beta=0$, u is also required to vanish at the corresponding endpoint, and the term involving α or β in R is omitted), and S_n is the class of all n-dimensional subspaces of $D(\alpha,\beta)$.

1.4. Consider the eigenvalue problem (1.71) and the related eigenvalue problem obtained by replacing $p(x)$ by $P(x)$ and $q(x)$ by $Q(x)$, but keeping s,α,β the same. Denote the eigenvalues of the related problem by Λ_n. Show that, if $P \geqslant p, Q \geqslant q$, then

$$\Lambda_n \geqslant \lambda_n, \qquad n=1,2,\ldots .$$

1.5. Show that the eigenvalues of

$$-(x^2u')' -\lambda u=0, \qquad 1<x<e; \qquad u(1)=u(e)=0$$

are positive. By trying solutions of the form x^μ, find the general solution of the differential equation. Show that the eigenvalues are $\lambda_n=n^2\pi^2+\frac{1}{4}$, $n=1,2,\ldots$, with corresponding (nonnormalized) eigenfunctions

$$v_n(x)=x^{-1/2}\sin(n\pi\log x).$$

By considering the differential equations satisfied by $v_n(x)$ and by $v=x^{-1/2}\sin(\alpha\log x)$, show that

$$\int_1^e v_n^2(x)\,dx=\tfrac{1}{2},$$

a result that could have also been obtained by elementary integration.

1.6. Consider the eigenvalue problem

$$-u'' -\lambda u=0, \qquad -1<x<1;$$

$$u'(-1)=\cot\alpha\, u(-1), \qquad u'(1)=\cot\beta\, u(1),$$

where $0\leqslant\alpha<\pi, 0\leqslant\beta<\pi$, with the understanding that the cases $\alpha=0,\beta=0$ are to be interpreted as $u(-1)=0, u(1)=0$, respectively. Find Green's function $g(x,\xi;\lambda)$, and use (1.19) to generate the normalized eigenfunctions. Show that there exists a negative eigenvalue if and only if $\cot\alpha+\cot\beta<0$.

1.7. Follow the procedure of Example 3(c) to find the eigenvalues and normalized eigenfunctions of

$$-u'' - \lambda u = 0, \qquad 0 < x < 1; \qquad u(0) = -u(1), \qquad u'(0) = -u'(1).$$

1.8. Consider the eigenvalue problem

(1.72) $-(pu')' - \lambda u = 0, \quad -1 < x < 1; \quad u(-1) = u(1) = 0,$

where we are interested in the limiting case $p(x) = p_1, x < 0, p(x) = p_2, x > 0$, where p_1 and p_2 are positive constants (for discussion of such interface problems, see Section 4, Chapter 1). Even though p is discontinuous at $x = 0$, pu' and u remain continuous at $x = 0$, as noted in (4.35) and (4.36), Chapter 1. One can therefore carry out an analysis similar to the one leading to (1.22). Let $v(x, \lambda)$ be the solution of the homogeneous equation satisfying $v(-1, \lambda) = 0$, $v'(-1, \lambda) = 1$. Show that the eigenvalues of (1.72) are the zeros of $v(1, \lambda)$, where

$$v(1, \lambda) = \sqrt{\frac{p_1}{\lambda}} \left(\sin \sqrt{\frac{\lambda}{p_1}} \cos \sqrt{\frac{\lambda}{p_2}} + \sqrt{\frac{p_1}{p_2}} \cos \sqrt{\frac{\lambda}{p_1}} \sin \sqrt{\frac{\lambda}{p_2}} \right).$$

1.9. Let $g(x)$ have period 1. Show that the function

$$u(x) = e^{G(x)},$$

where $G(x) = \int_0^x g(y)\, dy$, is a solution of the equation with periodic coefficient

$$-u'' + (g' + g^2)u = 0.$$

What is the multiplier [see (1.44)] in this case? When is this solution periodic?

1.10. Consider the eigenvalue problem

$$u' = i\lambda s(x)u, \quad a < x < b; \qquad u(b) = e^{i\gamma}u(a),$$

where $s > 0$ on the real axis and γ is a real constant. Find the eigenvalues and normalized eigenfunctions. Discuss the transition to a singular problem as $b \rightarrow \infty, a \rightarrow -\infty$. Construct Green's function for the regular problem, and calculate the limit as $b \rightarrow \infty, a \rightarrow \infty$. Find the corresponding forms of (1.67) and (1.68).

1.11. Consider the equation

$$(1.73) \qquad -iu' + qu - \lambda su = 0,$$

where $q(x)$, $s(x)$ are real and $s(x) > 0$ for $x \geqslant 0$. Define

$$S(x) \doteq \int_0^x s(t)\, dt, \qquad Q(x) \doteq \int_0^x q(t)\, dt, \qquad P(x) = \lambda S - Q.$$

Show that, if $S(\infty)$ is finite, the solutions of $(1.73) \in H_s(0, \infty)$ for all λ, whereas, if $S(\infty) = \infty$, the nontrivial solutions are in $H_s(0, \infty)$ only for $\operatorname{Im}\lambda > 0$.

Consider next (1.73) with the general boundary condition $u(b) = e^{i\gamma} u(0)$, where γ is a real number. Find all eigenvalues of this self-adjoint problem.

Show that Green's function $g_b(x, \xi; \lambda)$ satisfying

$$-ig_b' + qg_b - \lambda s g_b = \delta(x - \xi), \quad 0 < x, \xi < b; \qquad g_b|_{x=b} = e^{i\gamma} g_b|_{x=0}$$

is given by

$$(1.74) \qquad g_b(x, \xi; \lambda) = ie^{i[P(x) - P(\xi)]}\left[H(x - \xi) + C \right],$$

where

$$C = \left[e^{i\gamma} e^{-iP(b)} - 1 \right]^{-1}.$$

Show that for fixed b and β the locus of C is a circle in the complex C plane. For β fixed and b increasing, each circle is enclosed in the previous one, so that as $b \rightarrow \infty$ the locus of C is either a point (*limit-point*) or a circle (*limit-circle*). If $S(\infty) = \infty$, show that the limit-point case obtains and that $g_b \rightarrow g$, g being given by (1.65) with $S(x)$ replaced by $P(x)$. If $S(\infty)$ is finite, we have the limit-circle case. Points on the limit-circle correspond to boundary conditions of the form $|u(\infty)| = |u(0)|$. Then $g_b \rightarrow g$, g being given by (1.74) with

$$C = \left[e^{i\theta} e^{-i\lambda S(\infty)} - 1 \right]^{-1},$$

where θ is a real constant.

2. WEYL'S CLASSIFICATION OF SINGULAR PROBLEMS

Suppose (1.1) has *one* endpoint singular; that is, either the interval is semi-infinite or, if finite, $p(x)$ vanishes at one of the endpoints (this may also be accompanied by the unboundedness of q or s at the same endpoint). The solutions of the differential equation are no longer necessarily in the space H_s of functions such that $\int_a^b s|u|^2\, dx < \infty$. The following theorem of Weyl enables us to classify the singular endpoint according to the number of solutions in H_s.

Weyl's Theorem.   Consider

$$(2.1) \qquad\qquad -(pu')' + qu - \lambda su = 0, \qquad a < x < b,$$

when one of the endpoints is regular and the other singular. The coefficients are fixed except for the parameter λ. Then the following hold:

1. If for some particular value of λ every solution of (2.1) is in H_s, for any other value of λ every solution is again in H_s.

2. For every λ with $\mathrm{Im}\,\lambda \neq 0$, there exists at least one solution of (2.1) in H_s.

Remarks

1. The theorem (whose proof we omit) tells us that the singular point must fall into one of the following two mutually exclusive categories:

Limit-circle case. All solutions are in H_s for all λ.

Limit-point case. For λ with $\mathrm{Im}\,\lambda \neq 0$ there is exactly one (independent) solution in H_s, and then for $\mathrm{Im}\,\lambda = 0$ there may be one or no solution in H_s.

To determine which case applies it suffices to examine a *single* value of λ.

2. If both endpoints are singular, we introduce an intermediate point l, $a < l < b$, and then classify a according to the behavior of solutions in $a < x < l$ and classify b according to the behavior of solutions in $l < x < b$ (the classification is clearly independent of l).

Example 1.   Consider $-u'' - \lambda u = 0$ on three different intervals:

(a) The interval $a < x < \infty$, where a is finite. Then $s = p = 1$ and $H_s = L_2^{(c)}(a, \infty)$. The point a is regular, but the right endpoint $b = \infty$ is singular. If $\lambda = 0$, the general solution is $Ax + B$, so that *no* solution is in $L_2^{(c)}(a, \infty)$, which means that we have the limit-point case at the singular point $b = \infty$. It is easy to check the prediction, contained in Weyl's theorem, that there is exactly one solution in $L_2^{(c)}(a, \infty)$ for $\mathrm{Im}\,\lambda \neq 0$. Indeed the general solution of $-u'' -\lambda u = 0$ is $A\exp(i\sqrt{\lambda}\,x) + B\exp(-i\sqrt{\lambda}\,x)$, where, as usual, $\sqrt{\lambda}$ is the principal value (1.30). We denote the real interval $0 \leqslant \lambda < \infty$ by $[0, \infty)$. If $\lambda \notin [0, \infty)$, $\sqrt{\lambda}$ has a positive imaginary part and $\exp(i\sqrt{\lambda}\,x)$ is in $L_2^{(c)}(a, \infty)$ while $\exp(-i\sqrt{\lambda}\,x)$ is not. If $\lambda \in [0, \infty)$, no solution is in $L_2^{(c)}(a, \infty)$. Thus when $\mathrm{Im}\,\lambda \neq 0$ there is exactly one solution in H_s, whereas when $\mathrm{Im}\,\lambda = 0$ there is one solution in H_s if $\lambda < 0$ and no solution in H_s if $\lambda \geqslant 0$.

(b) The interval $-\infty < x < b$ with b finite. Again we have the limit-point case at the singular endpoint (which is now the left endpoint $a = -\infty$).

(c) The interval $-\infty < x < \infty$. By Remark 2 and the results of (a) and (b) it follows that both endpoints are in the limit-point case.

Example 2.   Bessel's equation of order ν (fixed $\geqslant 0$) with parameter λ

(2.2) $$-(xu')' + \frac{\nu^2}{x}u - \lambda x u = 0.$$

(a) On $0 < x < b$, with b finite. Then 0 is a singular point since $p(0) = 0$ [and also $q(0) = \infty$ if $\nu \neq 0$]. The space H_s consists of functions u for which $\int_0^b x|u|^2\,dx < \infty$. For $\lambda = 0$, $\nu \neq 0$, (2.2) degenerates into a much simpler equation with the independent solutions x^ν and $x^{-\nu}$, which are both in H_s if $\nu < 1$; if $\nu \geqslant 1$, only $x^{-\nu}$ is in H_s. For $\lambda = 0$ and $\nu = 0$, both the independent solutions 1 and $\log x$ are in H_s. Thus if $\nu < 1$ we have the *limit-circle* case at $x = 0$, whereas for $\nu \geqslant 1$ we have the *limit-point* case at $x = 0$. We can easily confirm the prediction of Weyl's theorem that for $\nu < 1$ all solutions are in H_s for all λ. The functions $J_\nu(\sqrt{\lambda}\,x)$ and $J_{-\nu}(\sqrt{\lambda}\,x)$ are independent solutions of (2.2) for $\lambda \neq 0$ and $\nu \neq 0$. J_ν is finite at $x = 0$ and $J_{-\nu}$ has a singularity of order $x^{-\nu}$; thus J_ν and $J_{-\nu}$ are both in H_s. If $\nu = 0$, the independent solutions can be taken as $J_0(\sqrt{\lambda}\,x)$ and $N_0(\sqrt{\lambda}\,x)$, the first of which is finite at $x = 0$ while the second has a logarithmic singularity at $x = 0$; again both solutions are in H_s.

(b) On $a < x < \infty$, where $a > 0$. Then the right endpoint is singular, but the left endpoint is regular. The space $H_s : \int_a^\infty x|u|^2\,dx < \infty$. Taking $\lambda = 0$, we see that neither of the independent solutions x^ν, $x^{-\nu}(\nu > 0)$ is in H_s; the same is true for $\nu = 0$ when 1 and $\log x$ are independent solutions. Thus we have the *limit-point* case at ∞ for all ν.

(c) On $0 < x < \infty$ with both endpoints singular. Here we combine the results of parts (a) and (b).

Example 3.   The Hermite equation.   The quantum-mechanical problem of a particle in a quadratic potential field (the so-called harmonic oscillator) is described by the equation

(2.3) $-u'' + x^2 u - \lambda u = 0, \qquad -\infty < x < \infty.$

Each of the endpoints is singular. The substitution $u = z e^{x^2/2}$ transforms the differential equation into

$$-z'' - 2xz' - z = \lambda z,$$

which, for $\lambda = -1$, has the independent solutions 1 and $\int_0^x e^{-t^2} dt$. Thus, for $\lambda = -1$, (2.3) has the solutions

(2.4) $u_1(x) = e^{x^2/2}, \qquad u_2(x) = e^{x^2/2} \int_0^x e^{-t^2} dt.$

Since $u_1(x)$ is of infinite norm in both $(-\infty, l)$ and (l, ∞), both endpoints are in the *limit-point* case.

Example 4.   The Legendre equation

(2.5) $-\left[(1 - x^2) u'\right]' + \lambda u = 0, \quad -1 < x < 1; \qquad p(x) = 1 - x^2, \quad s(x) = 1,$

has both of the endpoints -1 and 1 as singular points since $p(-1) = p(1) = 0$. For $\lambda = 0$ the equation has the independent solutions $u_1(x) = 1$ and $u_2(x) = \log(1 + x) - \log(1 - x)$. For any l, $-1 < l < 1$, each of the integrals $\int_{-1}^l |u_1|^2 dx$, $\int_l^1 |u_1|^2 dx$, $\int_{-1}^l |u_2|^2 dx$, $\int_l^1 |u_2|^2 dx$ is finite. Thus we have the *limit-circle* case at both endpoints.

Construction of Green's Function in Limit-Point Case

If both endpoints in (2.1) are regular, then, after associating the boundary conditions (1.6), Green's function can be constructed whenever λ is not an eigenvalue. Since these eigenvalues form a discrete set of real numbers $\{\lambda_n\}$ with $\lim_{n \to \infty} \lambda_n = \infty$, we know a priori that $g(x, \xi; \lambda)$ exists whenever $\mathrm{Im}\,\lambda \neq 0$ and even when λ is real negative with sufficiently large modulus. If the endpoint $x = a$ is regular but b is singular in the *limit-point* case, we can proceed in the same way for $\mathrm{Im}\,\lambda \neq 0$ by attaching a condition of the type (1.6) at $x = a$ but *no* condition at b (it suffices to require that $g \in H_s$). The condition at a selects a solution of the homogeneous equation for $a < x < \xi$, and the requirement g in H_s picks out a solution for $x > \xi$. The arbitrary multiplicative constants are then determined by the continuity of g and the jump condition in dg/dx at $x = \xi$. Consider, for instance, Green's function for Example 1(a), where we have set $a=0$,

$$(2.6) \quad -\frac{d^2g}{dx^2} -\lambda g = \delta(x-\xi), \quad 0<x,\xi<\infty; \quad g|_{x=0}=0, \quad g\in L_2^{(c)}(0,\infty).$$

For $x<\xi$ we have $g=A\sin\sqrt{\lambda}\,x$; for $x>\xi$ the only solution in $L_2^{(c)}(0,\infty)$ is $B\exp(i\sqrt{\lambda}\,x)$, an observation which is true not only when $\mathrm{Im}\lambda\neq0$ but also when λ is real and negative. The matching conditions then give

$$(2.7) \quad g(x,\xi;\lambda)=\frac{1}{\sqrt{\lambda}}\sin\sqrt{\lambda}\,x_<\exp(i\sqrt{\lambda}\,x_>), \quad \lambda\notin[0,\infty)$$

Regarded as a function of the complex variable λ, g has no poles (not even at $\lambda=0$) but has a branch on the positive real axis. In fact, letting $[g]$ stand for the jump in g across the positive real axis, we find that

$$(2.8) \qquad [g]=g(x,\xi;|\lambda|e^{i0+})-g(x,\xi;|\lambda|e^{i2\pi-})$$

$$=\frac{2i\sin|\lambda|^{1/2}x_<\sin|\lambda|^{1/2}x_>}{|\lambda|^{1/2}}$$

$$=\frac{2i}{|\lambda|^{1/2}}\sin|\lambda|^{1/2}x\sin|\lambda|^{1/2}\xi.$$

It is perhaps worth interrupting the discussion to compare (2.7) with Green's function g_l for the regular problem

$$(2.9) \quad -\frac{d^2g_l}{dx^2}-\lambda g_l=\delta(x-\xi), \quad 0<x,\xi<l; \quad g_l|_{x=0}=g_l|_{x=l}=0.$$

An easy calculation gives

(2.10)

$$g_l(x,\xi;\lambda)=\frac{1}{\sqrt{\lambda}\,\sin\sqrt{\lambda}\,l}\left[\sin\sqrt{\lambda}\,x_<\sin\sqrt{\lambda}\,(l-x_>)\right], \quad 0<x,\xi<l,$$

which has simple poles at the eigenvalues $\lambda_n=n^2\pi^2/l^2$, $n=1,2,\ldots$, of the problem $-u_l''-\lambda u_l=0$, $u_l(0)=u_l(l)=0$. Setting $\sqrt{\lambda}=\alpha+i\beta(\beta\geqslant0)$, we find that

$$g_l=\frac{\sin\sqrt{\lambda}\,x_<}{\sqrt{\lambda}}\left[\frac{\exp(i\alpha l-\beta l-i\sqrt{\lambda}\,x_>)-\exp(-i\alpha l+\beta l+i\sqrt{\lambda}\,x_>)}{\exp(i\alpha l-\beta l)-\exp(-i\alpha l+\beta l)}\right].$$

For $\beta > 0$ the limit as $l \to \infty$ is easily calculated since $e^{\beta l}$ dominates $e^{-\beta l}$ while $e^{i\alpha l}$ and $e^{-i\alpha l}$ remain bounded. Thus the term in brackets tends to $\exp(i\sqrt{\lambda}\, x_>)$ when $\beta > 0$ and has no limit for $\beta = 0$. Therefore, if $\lambda \notin [0, \infty)$,

$$\lim_{l \to \infty} g_l(x, \xi; \lambda) = \frac{\sin \sqrt{\lambda}\, x_< \exp(i\sqrt{\lambda}\, x_>)}{\sqrt{\lambda}},$$

which is just (2.7). We have shown explicitly how the simple poles of (2.10) coalesce into the branch singularity in (2.7). Even if we had used a different boundary condition of type (1.6) at $x = l$ to define g, we would still find that g_l tends to (2.7) as $l \to \infty$. Therefore (2.7) is the common limit as $l \to \infty$ of solutions of (2.9), whatever boundary condition is imposed at $x = l$.

The inhomogeneous problem related to (2.9) is

$$(2.11) \qquad -w_l'' - \lambda w_l = f(x), \quad 0 < x < l; \qquad w_l|_{x=0} = w_l|_{x=l} = 0.$$

We know that, if $\lambda \neq n^2 \pi^2 / l^2$, (2.11) has one and only one solution, given by

$$w_l(x) = \int_0^l g_l(x, \xi; \lambda) f(\xi) \, d\xi.$$

If λ is an eigenvalue, g_l does not exist and (2.11) has solutions only if f satisfies a solvability condition. What is the corresponding statement for the inhomogeneous equation associated with (2.6)? Consider the problem

$$(2.12) \qquad -w'' - \lambda w = f(x), \quad 0 < x < \infty; \qquad w(0) = 0,$$

where $f(x) \in L_2^{(c)}(0, \infty)$, and we are looking for solutions also in $L_2^{(c)}(0, \infty)$. Then, if $\lambda \notin [0, \infty)$, (2.12) has one and only one solution in $L_2^{(c)}(0, \infty)$, given by

$$(2.13) \qquad w(x) = \int_0^\infty g(x, \xi; \lambda) f(\xi) \, d\xi,$$

where g is Green's function (2.7). The proof is simple: for fixed λ, $\lambda \notin [0, \infty)$, let G be the integral operator with kernel g. Then G is a Holmgren operator; indeed we have the estimate

$$\int_0^\infty |g(x, \xi; \lambda)| \, d\xi = \int_0^\infty |g(\xi, x; \lambda)| \, d\xi \leqslant \frac{2}{\beta |\lambda|^{1/2}}; \qquad \sqrt{\lambda} = \alpha + i\beta, \quad \beta > 0,$$

which implies condition (1.11), Chapter 5, characterizing Holmgren kernels. Thus G is a bounded operator on $L_2^{(c)}(0, \infty)$, although G is, of course, not compact. It is now easy to verify by differentiation that (2.13) satisfies (2.12). Let us state the results in spectral terminology. Consider the operator $L = -(d^2/dx^2)$ defined on a domain D_L consisting of complex-valued functions $u(x)$ in $L_2^{(c)}(0, \infty)$, with u continuous, u' absolutely continuous, u'' in $L_2^{(c)}(0, \infty)$ and $u(0) = 0$. Then every value of λ not in $[0, \infty)$ is in the resolvent set. Moreover, in Exercises 2.1 and 2.2 we show that every value of λ in $[0, \infty)$ is in the continuous spectrum.

Next let us consider a case where *both endpoints* are singular in the limit-point case. It is still easy to construct $g(x, \xi; \lambda)$ for $\mathrm{Im}\,\lambda \neq 0$. Take, for instance, the specific problem

$$(2.14) \qquad -\frac{d^2 g}{dx^2} - \lambda g = \delta(x - \xi), \qquad -\infty < x, \xi < \infty,$$

where both endpoints are in the limit-point case $\left[\exp(i\sqrt{\lambda}\,x) \text{ is in } L_2^{(c)}(l, \infty), \text{ and } \exp(-i\sqrt{\lambda}\,x) \text{ is in } L_2^{(c)}(-\infty, l)\right]$. We impose no boundary condition at either endpoint, but merely require square integrability. Thus $g = A \exp(-i\sqrt{\lambda}\,x)_< \exp i\sqrt{\lambda}\,x)_>$, and an easy calculation gives $A = i/2\sqrt{\lambda}$, so that

$$(2.15) \qquad g(x, \xi; \lambda) = \frac{i}{2\sqrt{\lambda}} \exp\left(i\sqrt{\lambda}\,|x - \xi|\right).$$

We note that $g \in L_2^{(c)}(-\infty, \infty)$ for $\lambda \notin [0, \infty)$. The equation

$$(2.16) \qquad -w'' - \lambda w = f, \qquad -\infty < x < \infty; \qquad f \in L_2^{(c)}(-\infty, \infty)$$

has one and only one solution in $L_2^{(c)}(-\infty, \infty)$ for $\lambda \notin [0, \infty)$. That solution is given by

$$w(x) = \int_{-\infty}^{\infty} \frac{i}{2\sqrt{\lambda}} \exp\left(i\sqrt{\lambda}\,|x - \xi|\right) f(\xi) \, d\xi.$$

There are instances in applications when one actually wants to calculate g (or w) when λ is in the continuous spectrum, that is, when λ is real and $0 \leq \lambda < \infty$. There is no longer a natural mathematical criterion to select the solutions of (2.14) for $x < \xi$ and $x > \xi$ since no solution is square integrable at either $+\infty$ or $-\infty$. Obviously this means that an L_2 theory is really insufficient for our purposes. Fortunately, however, there will usually be a physical criterion to guide us. The case where λ is real positive often arises in nondissipative wave propagation, and the natural requirement is that the problem be considered as the limit of vanishing dissipation. It turns out that small dissipation means that λ has a small, positive imaginary part; thus the limit of vanishing dissipation is obtained by letting λ approach real values from above the real axis. The corresponding limit for g is

$$\frac{i}{2|\lambda|^{1/2}} e^{i|\lambda|^{1/2}|x-\xi|},$$

which will then be the appropriate Green's function for a nondissipative problem with $0<\lambda<\infty$.

Green's Function in the Limit-Circle Case

Suppose $x=a$ is a regular point and $x=b$ is a singular point in the limit-circle case. With the regular point we associate a boundary condition of the form $0=B_a u \doteq \cos\alpha \, u(a)-\sin\alpha \, u'(a)$ with *real* coefficients. In constructing Green's function this boundary condition determines the form of g for $x<\xi$; for $x>\xi$ the criterion of belonging to H_s is not sufficient to specify g since all solutions of the homogeneous equation are in H_s. Thus we will have to *impose a boundary condition* at the singular point $x=b$. In trying to imitate the regular case we encounter the difficulty that $u(b)$ and $u'(b)$ may be infinite, so that the usual condition relating $u(b)$ and $u'(b)$ may not make sense. We shall, however, reformulate below the boundary condition in the regular case to allow its extension to the singular limit-circle problem.

If u and v have continuous second derivatives, and $a<b_0<b$, then, the point b_0 being regular, we have from Green's theorem

$$(2.17) \qquad \int_a^{b_0} s(\bar{v}Lu - uL\bar{v}) \, dx = p(b_0) W(u,\bar{v};b_0) - p(a) W(u,\bar{v};a).$$

Let D be the class of functions u such that both u and Lu are in H_s (that is, $\int_a^b s|u|^2 \, dx < \infty$ and $\int_a^b s|Lu|^2 \, dx < \infty$). If u and v both belong to D, the left side of (2.17) has a limit as $b_0 \to b$ and therefore

$$(2.18) \qquad \lim_{b_0 \to b} p(b_0) W(u,\bar{v};b_0) \doteq [u,\bar{v}]_b$$

exists. At a regular point we can of course use this same notation for $p(a) W(u,\bar{v};a)$, and (2.17) becomes

$$(2.19) \qquad \int_a^b s(\bar{v}Lu - uL\bar{v}) \, dx = [u,\bar{v}]_b - [u,\bar{v}]_a, \qquad u,v \in D.$$

If u is also a solution of $Lu - \lambda u = 0$, then $L\bar{u} - \bar{\lambda}\bar{u} = 0$ because the coefficients in L are real, and therefore (2.19) with $v = u$ gives

$$(2.20) \qquad 2i(\text{Im}\lambda)\int_a^b s|u|^2 dx = [u,\bar{u}]_b - [u,\bar{u}]_a.$$

Now if, at the regular point $x = a$, u satisfies a condition of the type $\cos\alpha u(a) - \sin\alpha u'(a) = 0$ with *real* coefficients, *then*

$$[u,\bar{u}]_a = \{u(a)\bar{u}'(a) - \bar{u}(a)u'(a)\} p(a) = 0;$$

and, conversely, if $[u,\bar{u}]_a = 0$, then u satisfies $\cos\alpha u(a) - \sin\alpha u'(a) = 0$ for some real α. At the singular point b, $[u,\bar{u}]_b$ exists if $u \in D$; it is therefore natural to regard

$$(2.21) \qquad [u,\bar{u}]_b = 0$$

as a *boundary condition* at b equivalent to the boundary condition with real coefficients that applies in the regular case. It follows from (2.20) that no solution of $Lu - \lambda u = 0$ except the trivial solution can satisfy both $[u,\bar{u}]_a = 0$ and $[u,\bar{u}]_b = 0$ when $\text{Im}\lambda \neq 0$.

Another important feature of the regular case is that the boundary condition is independent of the parameter λ; this is not taken into account by (2.21), which merely makes sure that the coefficients in the boundary condition are real. To see how to deal with the other question, let us begin with Green's function $g(x,\xi;\lambda_0)$ for a particular $\lambda = \lambda_0$ with $\text{Im}\lambda_0 \neq 0$:

$$(2.22) \quad Lg - \lambda_0 g = \frac{\delta(x-\xi)}{s(x)}, \quad a<x,\xi<b; \qquad B_a g = 0, \qquad [g,\bar{g}]_b = 0.$$

For $x < \xi$ let $v(x,\lambda_0)$ be the solution of the homogeneous equation satisfying $B_a v = 0$, that is, $\cos\alpha v(a,\lambda_0) - \sin\alpha v'(a,\lambda_0) = 0$ for a particular α. Thus v is determined up to a multiplicative constant, and $[v,\bar{v}]_a = 0$. For $x > \xi$ let $z(x,\lambda_0)$ be a solution of the homogeneous equation satisfying $[z,\bar{z}]_b = 0$, which is regarded as a boundary condition at the singular point b. Unlike v, z is not determined up to a multiplicative constant; nevertheless the functions z and v are independent since there is no nontrivial solution of the homogeneous equation satisfying both $[u,\bar{u}]_a = 0$ and $[u,\bar{u}]_b = 0$. Thus, if we choose the multiplicative constant in v so that

$$(2.23) \qquad p(x)W(v,z;x) = -1,$$

then

$$(2.24) \qquad g(x,\xi;\lambda_0) = v(x_<,\lambda_0)z(x_>,\lambda_0).$$

Obviously g as a function of either x or ξ is in H_s, but we have the even more powerful result (not always true in the limit-point case)

$$(2.25) \qquad \int_a^b \int_a^b s(x)s(\xi)|g(x,\xi;\lambda_0)|^2 \, dx \, d\xi < \infty,$$

which will pave the way for the application of the Hilbert-Schmidt theory. To prove (2.25), we observe from (2.24) that

$$\int_a^b |g|^2 s(\xi) \, d\xi = |z(x)|^2 \int_a^x s(\xi)|v(\xi)|^2 \, d\xi + |v(x)|^2 \int_x^b s(\xi)|z(\xi)|^2 \, d\xi$$

$$\leqslant |z(x)|^2 \|v\|_s^2 + |v(x)|^2 \|z\|_s^2,$$

where we have suppressed the dependence on λ_0. Hence

$$\int_a^b \int_a^b |g(x,\xi;\lambda_0)|^2 s(x)s(\xi) \, dx \, d\xi \leqslant 2\|v\|_s^2 \|z\|_s^2,$$

which is finite since we are in the limit-circle case. If $\tilde{z}$ is a solution of $L\tilde{z} - \lambda_0 s\tilde{z} = 0$ which satisfies $[\tilde{z},\tilde{z}]_b = 0$ and is not a multiple of z, we will obtain a different Green's function in (2.24) corresponding to a "different" boundary condition at the singular point b.

Consider next the inhomogeneous problem

$$(2.26) \qquad Lw - \lambda_0 w = f, \qquad a < x < b; \quad B_a w = 0,$$

where we have not yet specified the boundary condition at the singular point b. We assume $f \in H_s$, and we look for solutions in H_s. Green's function (2.24) enables us to write one solution of (2.26) as

(2.27)

$$w(x,\lambda_0) = \int_a^b g(x,\xi;\lambda_0)s(\xi)f(\xi) \, d\xi$$

$$= z(x,\lambda_0) \int_a^x v(\xi,\lambda_0)s(\xi)f(\xi) \, d\xi + v(x,\lambda_0) \int_x^b z(\xi,\lambda_0)s(\xi)f(\xi) \, d\xi,$$

from which follow

$$w'(x,\lambda_0) = z'(x,\lambda_0) \int_a^x v(\xi,\lambda_0)s(\xi)f(\xi) \, d\xi + v'(x,\lambda_0) \int_x^b z(\xi,\lambda_0)s(\xi)f(\xi) \, d\xi,$$

$$[w,\tilde{z}]_x = [z,\tilde{z}]_x \int_a^x v(\xi,\lambda_0)s(\xi)f(\xi) \, d\xi + [v,\tilde{z}]_x \int_x^b z(\xi,\lambda_0)s(\xi)f(\xi) \, d\xi.$$

Now let $x \to b$, and use the fact that $[z, \bar{z}]_b = 0$ to obtain

(2.28) $$[w, \bar{z}]_b = 0.$$

From (2.7) we find that

(2.29) $$\|w\|_s^2 \leqslant \left[\int_a^b \int_a^b |g(x, \xi; \lambda_0)|^2 s(x) s(\xi) \, dx \, d\xi \right] \|f\|_s^2.$$

Thus the BVP

(2.30) $$Lw - \lambda_0 w = f, \quad a < x < b; \quad B_a w = 0, \quad [w, \bar{z}]_b = 0$$

has one and only one solution, given by (2.27); moreover from (2.29) and (2.25) we conclude that the inverse operator is bounded.

We now formulate a spectral problem for the operator L:

(2.31) $$Lu - \lambda u = 0, \quad a < x < b; \quad B_a u = 0, \quad [u, \bar{z}]_b = 0,$$

where the same auxiliary function $z(x, \lambda_0)$ satisfying $Lz - \lambda_0 z = 0$ and $[z, \bar{z}]_b = 0$ is used for all λ. We rewrite the differential equation in (2.31) as $Lu - \lambda_0 u = (\lambda - \lambda_0) u$; from (2.27), (2.28), and (2.30) it then follows that the BVP (2.31) is equivalent to the integral equation

(2.32) $$u(x, \lambda) = (\lambda - \lambda_0) \int_a^b g(x, \xi; \lambda_0) s(\xi) u(\xi, \lambda) \, d\xi,$$

which can be written as

(2.33) $$\mu y(x) = \int_a^b k(x, \xi) y(\xi) \, d\xi \quad \text{or} \quad \mu y = Ky,$$

where

$$\mu = \frac{1}{\lambda - \lambda_0}, \quad y = \sqrt{s} \, u, \quad k = \sqrt{s(x)} \, \sqrt{s(\xi)} \, g(x, \xi; \lambda_0).$$

From (2.25) k is a Hilbert-Schmidt kernel, so that (2.33) has at most denumerably many eigenvalues and there must therefore exist a *real* number θ which is *not* an eigenvalue of (2.31). If we then repeat our derivation using $\lambda_0 = \theta$, k will be a *real symmetric* H.-S. kernel. The eigenfunctions of (2.31) must therefore form an orthonormal basis in H_s.

We can therefore conclude that problems with singular points in the limit-circle case (at one end or both) are quite similar to regular problems. The boundary condition at the singular point(s) is specified in a somewhat indirect way as in (2.31), but the problem then generates a discrete spectrum and the usual type of eigenfunction. The only difference from the regular case is that the eigenvalues λ_n need only satisfy $|\lambda_n| \to \infty$ rather than $\lambda_n \to \infty$.

Example 5. Setting $\nu = 0$ in (2.2), we have Bessel's equation of order 0:

$$(2.34) \qquad -(xu')' - \lambda xu = 0.$$

We shall consider spectral problems for this equation on the interval $0 < x < 1$. At the regular endpoint $x = 1$ we impose the boundary condition $u(1) = 0$. At the singular (limit-circle) endpoint $x = 0$, we impose, in keeping with (2.31), a boundary condition of the form $[u, \bar{z}]_0 = 0$, where the auxiliary function z remains to be specified. From our previous theory we know that z can be chosen to be a solution of $Lz - \lambda_0 z = 0$ with $\text{Im}\,\lambda_0 \neq 0$ and $[z, \bar{z}]_0 = 0$. It is permissible, however, to use instead a real value of λ_0 as long as the solution z is chosen independently of the solution v of the same equation satisfying the boundary condition at the regular endpoint.

With $\lambda_0 = 0$, (2.34) has independent solutions 1 and $\log x$. Thus $v(x) = \log x$, and every independent solution is a multiple of $z = -1 + A \log x$, where A must be *real* to satisfy $[z, \bar{z}]_0 = 0$. The particular form chosen for z has the property $[\bar{z}, v]_x = x(\bar{z}v' - v\bar{z}') = -1$. The condition $[u, \bar{z}]_0 = 0$, with A fixed, becomes

$$\lim_{x \to 0} Au - (A \log x - 1)xu' = 0.$$

For each real A we can consider the spectral problem

$$(2.35)$$
$$-(xu')' - \lambda xu = 0, \quad 0 < x < 1; \qquad u(1) = 0, \quad \lim_{x \to 0} Au - (A \log x - 1)xu' = 0,$$

and each problem yields a different set of eigenvalues and eigenfunctions. By far the most important problem in applications occurs for $A = 0$, when the boundary condition at $x = 0$ is

$$(2.36) \qquad \lim_{x \to 0} xu' = 0.$$

The independent solutions of (2.34) for $\lambda \neq 0$ are $J_0(\sqrt{\lambda}\,x)$ and $N_0(\sqrt{\lambda}\,x)$, of which only $J_0(\sqrt{\lambda}\,x)$ satisfies (2.36). In applications to heat conduction x is the radial cylindrical coordinate, and (2.36) states that the total heat flux through a small circle surrounding the origin vanishes, that is, there is no heat source at the origin. It is usual to replace (2.36) by the condition that the solution be *finite* at the origin, which has the same effect as far as selecting $J_0(\sqrt{\lambda}x)$ as the admissible solution. I would argue, however, that it is both mathematically *and* physically more natural to impose condition (2.36).

Let us consider in more detail problem (2.35) with $A=0$ [that is, with boundary condition (2.36)]. Then $J_0(\sqrt{\lambda}\,x)$ is the only solution satisfying (2.36). The boundary condition at $x=1$ implies $J_0(\sqrt{\lambda}\,)=0$, which gives rise to a sequence $\{\lambda_k\}$ of positive eigenvalues. The corresponding, nonnormalized eigenfunctions are $J_0(\sqrt{\lambda_k}\,x)$. It is instructive to calculate the normalization factor by the method of Section 1 [see, for instance, (1.25)]. We first need Green's function $g(x,\xi;\lambda)$ satisfying, for $\mathrm{Im}\lambda\neq0$,

$$-(xg')' - \lambda xg = \delta(x-\xi), \quad 0<x,\xi<1; \quad xg'|_{x=0}=0, \quad g|_{x=1}=0.$$

We find, on using the Wronskian relationship (1.37), that

$$g = J_0(\sqrt{\lambda}\,x_<)v(x_>,\lambda),$$

where

$$v(x,\lambda) = \frac{\pi}{2J_0(\sqrt{\lambda}\,)}\left[J_0(\sqrt{\lambda}x)N_0(\sqrt{\lambda}\,) - J_0(\sqrt{\lambda}\,)N_0(\sqrt{\lambda}x)\right].$$

Thus g has simple poles at $\lambda=\lambda_k$, where $J_0(\sqrt{\lambda_k}\,)=0$. The residue at $\lambda=\lambda_k$ is

$$\frac{\pi\sqrt{\lambda_k}}{J_0'(\sqrt{\lambda_k}\,)}J_0(\sqrt{\lambda_k}\,x)J_0(\sqrt{\lambda_k}\,\xi)N_0(\sqrt{\lambda_k}\,).$$

From the Wronskian relation we have $N_0(\sqrt{\lambda_k}\,)J_0'(\sqrt{\lambda_k}\,) = -2/(\pi\sqrt{\lambda_k}\,)$, and the residue can be rewritten as

$$-\frac{2}{\left[J_0'(\sqrt{\lambda_k}\,)\right]^2}J_0(\sqrt{\lambda_k}\,x)J_0(\sqrt{\lambda_k}\,\xi),$$

so that from (1.19) we find that

(2.37) $$u_k(x) = \frac{\sqrt{2}}{J_0'(\sqrt{\lambda_k}\,)}J_0(\sqrt{\lambda_k}\,x)$$

is an orthonormal basis of eigenfunctions in H_s, that is,

$$\int_0^1 x u_k(x) \bar{u}_j(x)\, dx = \begin{cases} 0, & k \neq j, \\ 1, & k = j. \end{cases}$$

The spectral relation (1.20) takes the form

$$(2.38) \qquad \frac{\delta(x-\xi)}{x} = \sum_{k=1}^{\infty} \frac{2}{\left[J_0'\left(\sqrt{\lambda_k}\right)\right]^2} J_0\left(\sqrt{\lambda_k}\, x\right) J_0\left(\sqrt{\lambda_k}\, \xi\right),$$

which gives rise to the usual Fourier-Bessel series:

$$(2.39) \qquad F(\xi) = 2 \sum_{k=1}^{\infty} \frac{J_0\left(\sqrt{\lambda_k}\, \xi\right)}{\left[J_0'\left(\sqrt{\lambda_k}\right)\right]^2} \int_0^1 x J_0\left(\sqrt{\lambda_k}\, x\right) f(x)\, dx.$$

The question remains open whether (2.35) might occur in a physical setting with $A \neq 0$. The eigenfunctions would then contain a combination of a source term $N_0(\sqrt{\lambda}\, x)$ and of a term $J_0(\sqrt{\lambda}\, x)$. Only the ratio of J_0 and N_0 would be determined. No simple physical interpretation seems to be available.

Exercises

In Exercises 2.1 and 2.2 we complete the discussion of the spectral properties of the operator $L \doteq -d^2/dx^2$ defined on the domain D_L of complex-valued functions $u(x)$ in $L_2^{(c)}(0, \infty)$, with u continuous, u' absolutely continuous, u'' in $L_2^{(c)}(0, \infty)$ and $u(0) = 0$. We want to prove that every value of λ in $[0, \infty)$ is in the continuous spectrum. Setting $\lambda = \alpha^2$, $\alpha \geqslant 0$, we must do two things

1. Show that

$$(2.40) \qquad -u'' - \alpha^2 u = f(x), \quad 0 < x < \infty; \quad u(0) = 0,$$

has an L_2 solution for $f \in M$, where M is a dense set in L_2. This means that the range of $L - \alpha^2 I$ is dense in L_2.

2. Show that the inverse of $L - \alpha^2 I$ is unbounded, that is, for each α exhibit a sequence $\{u_n \neq 0\}$ such that

$$(2.41) \qquad \frac{\| L u_n - \alpha^2 u_n \|}{\| u_n \|} \to 0.$$

2.1. Clearly there are some $f \in L_2$ for which (2.40) has an L_2 solution: start with any function U of *compact support* satisfying $U(0)=0$ and having a piecewise continuous second derivative. Then $f \doteq -U'' - \alpha^2 U$ will be in L_2, so that for that particular f (2.40) has the solution U.

Show that, if f is any L_2 function with compact support, (2.40) has a solution in L_2 if and only if

$$(2.42) \qquad \int_0^\infty (\sin \alpha x) f(x) \, dx = 0, \quad \alpha > 0; \qquad \int_0^\infty x f(x) \, dx = 0, \quad \alpha = 0.$$

It can also be shown that the set M_α of L_2 functions with compact support satisfying condition (2.42) for fixed α is dense in L_2.

2.2. To exhibit a sequence with property (2.41), we note first that $\sin \alpha x$ satisfies the homogeneous equation $Lu - \alpha^2 u = 0$, $u(0)=0$ but fails to be in $L_2^{(c)}(0, \infty)$. Nevertheless $\sin \alpha x$ is "nearly" an eigenfunction and, suitably modified into an L_2 function, will become the desired u_n.

Let $F(x)$ be a fixed twice-differentiable function on $0 \leqslant x \leqslant 1$, satisfying $F(0)=0$, $F'(0)=1$, $F(1)=0$, $F'(1)=0$. Using the definition

$$u_n(x) = \begin{cases} \sin \alpha x, & 0 \leqslant x \leqslant l_n, \\ F(x - l_n), & l_n \leqslant x \leqslant l_n + 1, \\ 0, & x \geqslant l_n + 1, \end{cases}$$

where $l_n = 2n\pi/\alpha$, show that $\{u_n\}$ satisfies (2.41).

2.3. Consider the spectral problem (2.3). Construct Green's function for the case $\lambda = -1$ and translate the problem into an integral equation. Show that the kernel is symmetric and H.S. with eigenfunctions forming an orthonormal basis. Verify that $u_n(x)$ given by (4.30), Chapter 6, is an eigenfunction of (2.3) corresponding to $\lambda_n = 2n + 1$.

3. SPECTRAL PROBLEMS WITH A CONTINUOUS SPECTRUM

In regular problems (or problems in the limit-circle case) we can use (1.19) to generate the orthonormal eigenfunctions. Relation (1.20) further tells us that the eigenfunctions form a basis; indeed, multiplying (1.20) by $f(\xi)s(\xi)$ and integrating from $\xi = a$ to $\xi = b$, we obtain

$$(3.1) \qquad f(x) = \sum_n \langle f, u_n \rangle_s u_n(x),$$

which is the Fourier expansion of f in the set $\{u_n\}$. Of course this approach only guarantees the validity of (3.1) for test functions f with compact support in (a,b), but this is enough to tell us that the set $\{u_n\}$ is a basis.

If we have a singular problem with at least one endpoint in the *limit-point* case, it is possible (but not necessary) for the spectrum to be in part continuous. This continuous portion arises as a consequence of a branch in $g(x,\xi;\lambda)$, as we saw in (2.7), for instance. We shall still accept the validity of the formula

$$(3.2) \qquad -\frac{\delta(x-\xi)}{s(x)} = \frac{1}{2\pi i}\int_{C_\infty} g(x,\xi;\lambda)\,d\lambda, \qquad a<x,\xi<b.$$

The integral around the large circle in the complex plane can now be reduced to a sum of residues (the point spectrum) plus a branch-cut integral over a portion of the real axis (the continuous spectrum). Instead of (1.19) and (1.20) one then obtains for $a<x,\xi<b$

$$(3.3)$$

$$-\frac{\delta(x-\xi)}{s(x)} = \frac{1}{2\pi i}\int_{C_\infty} g(x,\xi;\lambda)\,d\lambda = -\sum_n u_n(x)\bar u_n(\xi) - \int u_\nu(x)\bar u_\nu(\xi)\,d\nu.$$

If the sum is not present, one has an expansion over a continuous index, leading to transform pairs rather than series expansions. The ideas will perhaps be clearer through examples.

Sine Transform

Consider the spectral problem

$$(3.4) \qquad -u'' - \lambda u = 0, \quad 0<x<\infty; \qquad u(0)=0, \quad u\in L_2^{(c)}(0,\infty).$$

To apply (3.3) we must first construct Green's function $g(x,\xi;\lambda)$. This was done in (2.7), and we observed there that g was analytic in the λ plane except for a branch on the positive real axis. It therefore follows from Cauchy's integral theorem that

$$\frac{1}{2\pi i}\int_{C_\infty} g(x,\xi;\lambda)\,d\lambda = -\frac{1}{2\pi i}\int_0^\infty [g]\,d|\lambda|,$$

where $[g]$ stands for the jump of g across the positive real axis. Thus we can write, since $s(x)=1$,

$$(3.5) \qquad \delta(x-\xi) = \frac{1}{2\pi i}\int_0^\infty [g]\,d|\lambda|, \qquad 0<x,\xi<\infty,$$

where, from (2.8),

$$[g] = \frac{2i}{|\lambda|^{1/2}} \sin|\lambda|^{1/2}x \sin|\lambda|^{1/2}\xi.$$

Letting $\lambda = \nu^2$, we find that (3.5) becomes

(3.6) $$\delta(x-\xi) = \frac{2}{\pi} \int_0^\infty \sin\nu x \sin\nu\xi \, d\nu, \qquad 0 < x, \xi < \infty.$$

The spectral relation (3.6) contains the formula for the Fourier sine transform. Indeed, if we multiply (3.6) by $f(x)$ and integrate from $x = 0$ to ∞, we find that

(3.7) $$f(\xi) = \frac{2}{\pi} \int_0^\infty \sin\nu\xi \, F_s(\nu) \, d\nu, \qquad 0 < \xi < \infty,$$

where

(3.8) $$F_s(\nu) \doteq \int_0^\infty \sin\nu x \, f(x) \, dx, \qquad 0 < \nu < \infty.$$

Equation (3.8) defines the Fourier sine transform of $f(x)$, and (3.7) is an inversion integral or, if one prefers, an expansion of f in the set $\sin\nu x$ over the continuous parameter ν.

We can now use (3.7) and (3.8) to solve the inhomogeneous problem

(3.9) $$-w'' - \lambda w = f(x), \quad 0 < x < \infty; \quad w(0) = \alpha, \quad w \in L_2^{(c)}(0, \infty)$$

on the assumption that f is given in $L_2^{(c)}(0, \infty)$ and λ is not in the spectrum, that is, $\lambda \notin [0, \infty)$. We multiply (3.9) by $\sin\nu x$ and integrate from 0 to ∞ to obtain

$$-\int_0^\infty w'' \sin\nu x \, dx - \lambda W_s(\nu) = F_s(\nu),$$

where F_s and W_s are the sine transforms of f and w, respectively. We integrate by parts to find

$$-\int_0^\infty w'' \sin\nu x \, dx = \nu^2 \int_0^\infty w \sin\nu x \, dx + (w\nu \cos\nu x - w' \sin\nu x)_0^\infty,$$

where the contribution from 0 is $-\alpha\nu$, while the terms from infinity are dropped (at this stage it usually does not pay to argue that w and w' vanish at infinity; instead proceed as if they did and then check that the solution obtained actually satisfies the original problem). Thus we have

$$(\nu^2 - \lambda) W_s(\nu) = \alpha \nu + F_s(\nu), \qquad W_s(\nu) = \frac{\alpha \nu}{\nu^2 - \lambda} + \frac{F_s(\nu)}{\nu^2 - \lambda}$$

and

$$w(x) = \frac{2}{\pi} \int_0^\infty \frac{\alpha \nu}{\nu^2 - \lambda} \sin \nu x \, d\nu + \frac{2}{\pi} \int_0^\infty \frac{F_s(\nu)}{\nu^2 - \lambda} \sin \nu x \, d\nu$$

$$= \frac{2i}{2\pi} \int_{-\infty}^\infty e^{-i\nu x} \left[\frac{\alpha \nu}{\nu^2 - \lambda} + \frac{F_s(\nu)}{\nu^2 - \lambda} \right] d\nu,$$

where we have used (3.8) to define $F_s(\nu)$ for negative ν. The problem has been reduced to ordinary Fourier inversion. The first integral is easily calculated by residues and gives (for $x > 0$) $\alpha \exp(i\sqrt{\lambda} x)$. The second integral is the convolution of the Fourier originals of $2iF_s(\nu)$ and $1/(\nu^2 - \lambda)$, which are, respectively, $f_0(x)$ and $i \exp(i\sqrt{\lambda} |x|)/2\sqrt{\lambda}$, where $f_0(x)$ is the odd extension of $f(x)$. Thus, for $x > 0$,

(3.10)

$$w(x) = \alpha \exp(i\sqrt{\lambda} x) + i \int_{-\infty}^\infty \frac{\exp(i\sqrt{\lambda} |x - \xi|)}{2\sqrt{\lambda}} f_0(\xi) \, d\xi$$

$$= \alpha \exp(i\sqrt{\lambda} x) + i \int_0^\infty \frac{\exp(i\sqrt{\lambda} |x - \xi|) - \exp(i\sqrt{\lambda} |x + \xi|)}{2\sqrt{\lambda}} f(\xi) \, d\xi,$$

which is just the same solution that would be obtained by using Green's function (2.7), which in turn can be deduced from (2.15) by images.

As an application of the Fourier sine transform in partial differential equations, consider the problem of finding Green's function (with Dirichlet boundary conditions) for the negative Laplacian in a two-dimensional semi-infinite strip. Let x, y be Cartesian coordinates in the plane, and let the strip occupy the domain Ω: $0 < x < \pi, 0 < y < \infty$, with the source at $Q = (\xi, \eta)$. Green's function $g(P, Q) = g(x, y; \xi, \eta)$ satisfies

(3.11)

$$-\Delta g = \delta(P - Q) = \delta(x - \xi)\delta(y - \eta), \quad P, Q \in \Omega; \quad g(P, Q)|_{P \in \Gamma} = 0,$$

where Γ is the boundary of the strip. The relation between delta functions can be found in Exercise 2.13, Chapter 2.

Multiply both sides of (3.11) by $\sin \nu y$, and integrate from $y=0$ to $y=\infty$ to obtain (after integration by parts of the term $\sin \nu y \, \partial^2 g / \partial y^2$)

$$(3.12) \quad -\frac{\partial^2 G_s}{\partial x^2} + \nu^2 G_s = \sin \nu \eta \, \delta(x-\xi), \quad 0<x,\xi<\pi; \qquad G_s|_{x=0} = G_s|_{x=\pi} = 0,$$

where

$$(3.13) \qquad\qquad \begin{aligned} G_s &= G_s(x,\nu;Q) = \int_0^\infty g(x,y;Q)\sin \nu y \, dy, \\ g(x,y;Q) &= \frac{2}{\pi} \int_0^\infty G_s(x,\nu;Q)\sin \nu y \, d\nu. \end{aligned}$$

In deriving (3.12) we dropped integrated terms of the form $g\cos \nu y$ and $g'\sin \nu y$ evaluated at $y=\infty$. This seems reasonable since g and g' can be expected to vanish at $y=\infty$, but full justification can come only after careful examination of the final solution (a step we shall leave to the conscientious reader).

The solution of (3.12) is easily obtained by the methods of Chapter 1. We find that

$$G_s = A \sinh \nu x_< \sinh \nu(x_> - \pi),$$

where A is determined from the jump condition

$$\frac{dG_s}{dx}(\xi+,\nu;Q) - \frac{dG_s}{dx}(\xi-,\nu;Q) = -\sin \nu \eta,$$

which gives $A = -\sin \nu \eta / (\nu \sinh \nu \pi)$, so that

$$(3.14) \qquad\qquad G_s = -\frac{\sin \nu \eta}{\nu \sinh \nu \pi} \sinh \nu x_< \sinh \nu(x_> - \pi).$$

We can calculate $g(P,Q)$ from (3.13), but the integration can be performed numerically with reasonable effort only if G_s decreases exponentially for large ν. Analysis of (3.14) shows that this behavior occurs as long as $|x-\xi|$ is not small. Thus the form (3.13) is useful whenever the x coordinate of the observation point is not too close to the x coordinate of the source.

Alternatively we can find an expression for g that is particularly useful if $|y-\eta|$ is not small. This requires an expansion in the x direction rather than in the y direction. On separating variables, the x problem is $-u'' - \lambda u = 0$ with $u(0) = u(\pi) = 0$ giving rise to the eigenfunctions $\sin nx$. Starting with (3.11), we multiply by $(2/\pi)\sin nx$ and integrate from 0 to π to obtain

$$(3.15) \quad n^2 g_n - \frac{\partial^2 g_n}{\partial y^2} = \frac{2}{\pi} \sin n\xi \, \delta(y - \eta), \quad 0 < y, \eta < \infty; \qquad g_n|_{y=0} = 0,$$

where

$$g_n = g_n(y; Q) = \frac{2}{\pi} \int_0^\pi g(x, y; Q) \sin nx \, dx,$$

(3.16)

$$g(x, y; Q) = \sum_{n=1}^\infty g_n(y; Q) \sin nx.$$

The solution of (3.15) which vanishes at $y = \infty$ is

$$(3.17) \qquad\qquad g_n = \frac{2}{\pi n} \sin n\xi \sinh ny_< \exp(-ny_>),$$

where $y_< = \min(y, \eta)$, $y_> = \max(y, \eta)$. We then substitute g_n in (3.16) to find g. The series for g converges rapidly if g_n tends to 0 exponentially as $n \to \infty$, as turns out to be the case whenever $|y - \eta|$ is not small.

By using contour integration or the Poisson sum formula it is possible to pass directly from (3.13)-(3.14) to (3.16)-(3.17) without reference to the differential equation. The solution of (3.11) could also have been found by images or by conformal mapping.

For another illustration of the use of the Fourier sine transformation, consider heat flow in a semi-infinite rod with 0 initial temperature and temperature 1 at the end. The temperature $u(x, t)$ satisfies

$$(3.18) \quad \frac{\partial u}{\partial t} - \frac{\partial^2 u}{\partial x^2} = 0, \quad 0 < x < \infty, \, t > 0; \qquad u(x, 0) = 0, \quad u(0, t) = 1.$$

Apply a sine transform to the x coordinate to obtain

$$(3.19) \qquad\qquad \frac{\partial U_s}{\partial t} + \nu^2 U_s = \nu, \quad t > 0; \qquad U_s|_{t=0} = 0,$$

where

$$U_s(\nu, t) = \int_0^\infty u(x, t) \sin \nu x \, dx,$$

(3.20)

$$u(x, t) = \frac{2}{\pi} \int_0^\infty U_s(\nu, t) \sin \nu x \, d\nu.$$

Again we dropped some integrated terms at infinity in the derivation of (3.19), whose solution is

$$(3.21) \qquad U_s(\nu,t) = \frac{1}{\nu}(1 - e^{-\nu^2 t})$$

and

$$u(x,t) = \frac{2}{\pi} \int_0^\infty \frac{\sin \nu x}{\nu} \, d\nu - \frac{2}{\pi} \int_0^\infty \frac{\sin \nu x}{\nu} e^{-\nu^2 t} \, d\nu.$$

The first term on the right is equal to 1 for all $x > 0$. To evaluate the second integral $I(x, t)$ note that $\partial I / \partial x = \int_0^\infty \cos \nu x \, e^{-\nu^2 t} \, d\nu = \frac{1}{2}(\pi/t)^{1/2} e^{-x^2/4t}$. Since $I(0, t) = 0$, we have

$$I(x,t) = \int_0^x \frac{1}{2}\sqrt{\frac{\pi}{t}} \, e^{-y^2/4t} \, dy = \int_0^{x/2\sqrt{t}} \sqrt{\pi} \, e^{-z^2} \, dz,$$

and hence

$$(3.22) \qquad u(x,t) = 1 - \frac{2}{\sqrt{\pi}} \int_0^{x/2\sqrt{t}} e^{-z^2} \, dz = 1 - \operatorname{erf}\left(\frac{x}{2\sqrt{t}}\right),$$

where "erf" stands for the error function

$$\operatorname{erf} x = \frac{2}{\sqrt{\pi}} \int_0^x e^{-z^2} \, dz.$$

The more general problem with boundary temperature $h(t)$ instead of 1 can be solved by superposition. First we note that, if the boundary temperature in (3.18) is 0 up to time τ and is then suddenly increased to 1, the solution will be 0 up to time τ and then $u(x, t - \tau)$, where $u(x, t)$ is given by (3.22). In other words, the solution for the displaced step temperature $H(t - \tau)$ is $H(t - \tau)u(x, t - \tau)$. Now we can write, for $t > 0$, as in (1.16), Chapter 1,

$$h(t) = \int_0^\infty \delta(t - \tau)h(\tau) \, d\tau = -\int_0^\infty \frac{dH(t - \tau)}{d\tau} h(\tau) \, d\tau$$

$$= \int_0^\infty H(t - \tau)h'(\tau) \, d\tau + h(0).$$

Therefore, by superposition

$$(3.23) \quad w(x,t) \doteq \int_0^\infty H(t-\tau)u(x,t-\tau)h'(\tau)\,d\tau + h(0)u(x,t)$$

$$= \int_0^t u(x,t-\tau)h'(\tau)\,d\tau + h(0)u(x,t)$$

$$= -\int_0^t \left[\frac{\partial}{\partial \tau} u(x,t-\tau) \right] h(\tau)\,d\tau = \int_0^t \frac{d}{dt} u(x,t-\tau)h(\tau)\,d\tau$$

is the solution of

$$(3.24) \quad \frac{\partial w}{\partial t} - \frac{\partial w}{\partial x^2} = 0, \quad 0 < x < \infty, t > 0; \qquad w(x,0) = 0, \quad w(0,t) = h(t).$$

The form (3.23) of the solution is known as *Duhamel's formula* (see also Exercise 1.3, Chapter 3). In view of (3.22) we can write explicitly

$$(3.25) \qquad w(x,t) = \int_0^t \frac{x}{2\sqrt{\pi}} \frac{e^{-x^2/4(t-\tau)}}{(t-\tau)^{3/2}} h(\tau)\,d\tau.$$

The general initial value problem for the semi-infinite rod is best attacked by finding Green's function $g(x,t;\xi,0)$ corresponding to an initial temperature $\delta(x-\xi)$ and a vanishing boundary temperature

$$(3.26) \quad \frac{\partial g}{\partial t} - \frac{\partial^2 g}{\partial x^2} = 0, \quad 0 < x < \infty, t > 0; \qquad g|_{x=0} = 0, \quad g|_{t=0} = \delta(x-\xi).$$

Applying a Fourier sine transform on x, we obtain

$$(3.27) \qquad \frac{\partial}{\partial t} G_s + \nu^2 G_s = 0, \quad t > 0; \qquad G_s|_{t=0} = \sin \nu \xi,$$

$$G_s = \int_0^\infty g(x,t;\xi,0)\sin \nu x\,dx, \qquad g = \frac{2}{\pi} \int_0^\infty G_s \sin \nu x\,d\nu.$$

Solving (3.27) and writing the inversion formula, we find that

$$g = \frac{2}{\pi} \int_0^\infty e^{-\nu^2 t} \sin \nu \xi \sin \nu x\,d\nu = \frac{1}{\pi} \int_0^\infty e^{-\nu^2 t} \left[\cos \nu(x-\xi) - \cos \nu(x+\xi) \right] d\nu,$$

an integral already calculated just above (3.22). We conclude that

$$(3.28) \qquad g(x,t;\xi,0) = \frac{1}{\sqrt{4\pi t}} \left[e^{-(x-\xi)^2/4t} - e^{-(x+\xi)^2/4t} \right],$$

which confirms the result obtained by the method of images.

The IVP

$$(3.29) \qquad \frac{\partial v}{\partial t} - \frac{\partial^2 v}{\partial x^2} = 0, \quad 0 < x < \infty, \, t > 0; \qquad v(0,t) = 0, \quad v(x,0) = f(x),$$

has the solution

$$(3.30) \qquad v(x,t) = \int_0^\infty g(x,t; \xi, 0) f(\xi) \, d\xi.$$

Hankel Transform of Order 0 and Applications

Consider Bessel's equation of order 0 on $0 < x < \infty$:

$$(3.31) \qquad -(xu')' - \lambda xu = 0, \qquad 0 < x < \infty.$$

Here both endpoints are singular, and we saw in Example 2, Section 2, that the left end $x = 0$ is in the limit-circle case and the right end is in the limit-point case. We shall therefore require no boundary condition at infinity; at $x = 0$ we shall ask that the boundary condition $\lim_{x \to 0} xu' = 0$ be satisfied. As we saw in (2.36), this is an appropriate condition for (3.31) at $x = 0$ and has the effect of selecting the solution which is bounded at the origin. We proceed with the construction of Green's function satisfying

$$(3.32) \qquad \begin{aligned} -(xg')' - \lambda xg &= \delta(x - \xi), \quad 0 < x, \xi < \infty; \\ (xg')_{x=0+} &= 0, \quad \int_0^\infty x |g|^2 \, dx < \infty. \end{aligned}$$

Whenever $\lambda \notin [0, \infty)$, we can actually meet these requirements and we find that

$$(3.33) \qquad g(x, \xi; \lambda) = \frac{\pi i}{2} J_0\left(\sqrt{\lambda} \, x_<\right) H_0^{(1)}\left(\sqrt{\lambda} \, x_>\right),$$

which has a branch on the positive real axis and no other singularity. Applying (3.2), we obtain

$$(3.34)$$

$$\frac{\delta(x - \xi)}{x} = -\frac{1}{2\pi i} \int_{C_\infty} g(x, \xi; \lambda) \, d\lambda = \frac{1}{2\pi i} \int_0^\infty [g] \, d|\lambda|, \qquad 0 < x, \xi < \infty,$$

where

$$(3.35)$$

$$[g] = \frac{\pi i}{2} \left\{ J_0\left(|\lambda|^{1/2} x_<\right) H_0^{(1)}\left(|\lambda|^{1/2} x_>\right) - J_0\left(-|\lambda|^{1/2} x_<\right) H_0^{(1)}\left(-|\lambda|^{1/2} x_>\right) \right\}$$

is the jump in g across the real axis.

By using the formulas

(3.36) $J_0(-a) = J_0(a), \qquad H_0^{(1)}(a) - H_0^{(1)}(-a) = 2J_0(a),$

we can simplify (3.35), and after setting $|\lambda| = \gamma^2$, (3.34) becomes

(3.37) $\dfrac{\delta(x-\xi)}{x} = \displaystyle\int_0^\infty \gamma J_0(\gamma x) J_0(\gamma \xi)\, d\gamma, \qquad 0 < x, \xi < \infty.$

The spectral relation (3.37) contains the pair of transform formulas

(3.38) $F_H(\gamma) = \displaystyle\int_0^\infty x J_0(\gamma x) f(x)\, dx, \qquad 0 < \gamma < \infty,$

(3.39) $f(x) = \displaystyle\int_0^\infty \gamma J_0(\gamma x) F_H(\gamma)\, d\gamma, \qquad 0 < x < \infty,$

where (3.38) defines the Hankel transform of order 0 of f, and (3.39) is an inversion formula enabling us to recover f from its Hankel transform.

We now consider a number of related applications of the Hankel transform of order 0. All of these problems are axisymmetric, with the z axis as the axis of symmetry. First, in free space let us place a ring of sources on the circle $r = a, z = 0$; here r is the radial cylindrical coordinate (that is, $r = \sqrt{x^2 + y^2}$). The total strength of this ring is taken to be unity, so that the source density is $\delta(r-a)\delta(z)/2\pi a = \delta(r-a)\delta(z)/2\pi r$. We shall suppose that we are dealing with a steady neutron source in a subcritical assembly. The resulting spatial distribution u of neutrons is clearly independent of polar angle and satisfies [see (2.27), Chapter 2]

$$-\Delta u + k^2 u = \frac{\delta(r-a)\delta(z)}{2\pi r}, \qquad 0 < r, \quad -\infty < z < \infty,$$

or

(3.40) $-\dfrac{\partial}{\partial r}\left(r\dfrac{\partial u}{\partial r} \right) - r\dfrac{\partial^2 u}{\partial z^2} + k^2 r u = \dfrac{\delta(r-a)\delta(z)}{2\pi}.$

If we separated variables for the homogeneous equation, the radial part would satisfy (3.31), so that a Hankel transform (with r playing the role of x) of order 0 should help in solving (3.40). Multiply the equation by $J_0(\gamma r)$ and integrate from $r = 0$ to ∞ to obtain, after integration by parts of the first term,

(3.41) $-\dfrac{\partial^2 U_H}{\partial z^2} + (k^2 + \gamma^2) U_H = \dfrac{\delta(z) J_0(\gamma a)}{2\pi},$

where U_H is the Hankel transform of u:

$$(3.42) \quad U_H(\gamma,z) = \int_0^\infty r J_0(\gamma r) u(r,z)\, dr; \quad u(r,z) = \int_0^\infty \gamma J_0(\gamma r) U_H(\gamma,z)\, d\gamma.$$

We solve the ordinary differential equation in the usual way, requiring that at $z = \pm\infty$ the neutron density tend to 0. This gives

$$U_H(\gamma,z) = \frac{\exp\left(-\sqrt{k^2+\gamma^2}\,|z|\right)}{4\pi\sqrt{k^2+\gamma^2}} J_0(\gamma a),$$

$$(3.43) \quad u(r,z) = \frac{1}{4\pi} \int_0^\infty \gamma J_0(\gamma a) J_0(\gamma r) \frac{\exp\left(-\sqrt{k^2+\gamma^2}\,|z|\right)}{\sqrt{k^2+\gamma^2}}\, d\gamma.$$

If $a\to 0$, the ring degenerates into a unit point source at the origin; the corresponding solution of (3.40) depends only on the distance from the origin and $u = (1/4\pi)\exp(-k\sqrt{r^2+z^2})/\sqrt{r^2+z^2}$. Comparing with (3.43) for $a=0$, we find that

$$(3.44) \quad \frac{\exp\left(-k\sqrt{r^2+z^2}\right)}{\sqrt{r^2+z^2}} = \int_0^\infty \gamma J_0(\gamma r) \frac{\exp\left(-\sqrt{k^2+\gamma^2}\,|z|\right)}{\sqrt{k^2+\gamma^2}}\, d\gamma,$$

an interesting relation due essentially to Sommerfeld, who studied the corresponding acoustical problem. For $k=0$, (3.44) remains valid and reduces to

$$(3.45) \quad \frac{1}{\sqrt{r^2+z^2}} = \int_0^\infty J_0(\gamma r) e^{-\gamma|z|}\, d\gamma.$$

Another expression for (3.43) can be obtained by applying a Fourier transform in the z direction to (3.40). Letting

$$u^\wedge(r,\alpha) = \int_{-\infty}^\infty e^{i\alpha z} u(r,z)\, dz,$$

we find that

$$-\frac{d}{dr}\left(r\frac{du^\wedge}{dr}\right) + (k^2+\alpha^2) r u^\wedge = \frac{\delta(r-a)}{2\pi}, \qquad 0 < r < \infty.$$

The corresponding homogeneous equation is the modified Bessel equation with independent solutions $I_0(\sqrt{k^2+\alpha^2}\,r)$, $K_0(\sqrt{k^2+\alpha^2}\,r)$, the first of which is bounded at $r=0$, and the second square integrable at $r=\infty$.

Thus, after using the Wronskian relations for I_0, K_0, we have

$$u^{\wedge}(r,\alpha) = \frac{1}{2\pi} I_0\left(\sqrt{k^2+\alpha^2}\; r_<\right) K_0\left(\sqrt{k^2+\alpha^2}\; r_>\right),$$

where $r_< = \min(r,a)$, $r_> = \max(r,a)$. The inversion formula then gives

$$u(r,z) = \frac{1}{4\pi^2} \int_{-\infty}^{\infty} e^{-i\alpha z} I_0\left(\sqrt{k^2+\alpha^2}\; r_<\right) K_0\left(\sqrt{k^2+\alpha^2}\; r_>\right) d\alpha,$$

which reduces, as $a \to 0$, to

$$\frac{\exp(-k\sqrt{r^2+z^2})}{\sqrt{r^2+z^2}} = \frac{1}{\pi} \int_{-\infty}^{\infty} e^{-i\alpha z} K_0\left(\sqrt{k^2+\alpha^2}\; r\right) d\alpha$$

or

$$K_0\left(\sqrt{k^2+\alpha^2}\; r\right) = \frac{1}{2} \int_{-\infty}^{\infty} e^{i\alpha z} \frac{\exp(-k\sqrt{r^2+z^2})}{\sqrt{r^2+z^2}} dz.$$

On setting $\alpha = 0$, we obtain the integral representation of the Macdonald function:

$$(3.46) \qquad K_0(r) = \frac{1}{2} \int_{-\infty}^{\infty} \frac{\exp(-\sqrt{r^2+z^2})}{\sqrt{r^2+z^2}} dz.$$

Exercises

3.1. *Fourier transform*

(a) Consider the spectral problem

$$-u'' - \lambda u = 0, \qquad -\infty < x < \infty,$$

where both endpoints are in the limit-point case. Show that when $\lambda \notin [0,\infty)$ Green's function is given by (2.15). By using (3.2) obtain the completeness relation

$$(3.47) \quad \delta(x-\xi) = \frac{1}{2\pi} \int_{-\infty}^{\infty} e^{i\nu x} e^{-i\nu\xi} d\nu, \qquad -\infty < x, \xi < \infty,$$

from which the usual Fourier transform and inversion follow.

(b) Consider the Dirichlet problem for a strip:

(3.48)
$$-\Delta u=0, \quad -\infty<x<\infty, 0<y<b; \quad u(x,0)=0, \quad u(x,b)=f(x),$$

where $f(x)$ is a given function. Apply a Fourier transform on the x coordinate to obtain

$$u^{\hat{}}(\omega,y)=f^{\hat{}}(\omega)\frac{\sinh\omega y}{\sinh\omega b}.$$

By the convolution theorem for Fourier transforms we then have

(3.49)
$$u(x,y)=\int_{-\infty}^{\infty} I(x-\xi,y)f(\xi)d\xi,$$

where $I(x,y)$ is the inverse transform of $(\sinh\omega y)/(\sinh\omega b)$.

(c) Show by the calculus of residues that

(3.50)
$$I(x,y)=\frac{1}{2b}\frac{\sin(\pi y/b)}{\cos(\pi y/b)+\cosh(\pi x/b)}.$$

3.2. The following problem can have a spectrum which is partly discrete and partly continuous:

$$-u''-\lambda u=0, \quad 0<x<\infty; \quad \cos\alpha\, u(0)-\sin\alpha\, u'(0)=0,$$

where α is a real number $0\leqslant\alpha<\pi$. The physical significance of this boundary condition was explained in Example 1, Section 1 (the change in sign in the boundary condition is a result of applying the physical condition at the left endpoint instead of the right).

(a) By elementary methods show that an eigenvalue is possible if and only if $(\pi/2)<\alpha<\pi$. Then there is exactly one eigenvalue, given by $\lambda=-\cot^2\alpha$ with normalized eigenfunction $\sqrt{-2\cot\alpha}\; e^{x\cot\alpha}$.

(b) Show that Green's function for $\text{Im}\lambda\neq0$ is given by

$$g(x,\xi;\lambda)=\frac{i\left[\exp\left(i\sqrt{\lambda}\,x_>\right)\right]}{\sqrt{\lambda}\,\sin\alpha+i\cos\alpha}\left(\sin\alpha\cos\sqrt{\lambda}\,x_<+\frac{\cos\alpha\sin\sqrt{\lambda}\,x_<}{\sqrt{\lambda}}\right).$$

For $0\leqslant\alpha\leqslant\pi/2$ the spectrum is purely continuous and consists of the real interval $0\leqslant\lambda<\infty$; for $\pi/2<\alpha<\pi$ the spectrum consists of the continuous part $0\leqslant\lambda<\infty$ and of the single eigenvalue

$\lambda = -\cot^2 \alpha$. Derive the expansion

$$\delta(x-\xi) = \frac{2}{\pi}\int_0^\infty \frac{\nu^2}{\cos^2\alpha + \nu^2\sin^2\alpha}\, u_\nu(x)u_\nu(\xi)\, d\nu, \qquad 0\leqslant\alpha\leqslant\frac{\pi}{2},$$

$$\delta(x-\xi) = \frac{2}{\pi}\int_0^\infty \frac{\nu^2}{\cos^2\alpha + \nu^2\sin^2\alpha}\, u_\nu(x)u_\nu(\xi)\, d\nu - 2\cot\alpha\, e^{(x+\xi)\cot\alpha},$$

$$\frac{\pi}{2} < \alpha < \pi,$$

where

$$u_\nu(x) = \sin\alpha\cos\nu x + \frac{\cos\alpha}{\nu}\sin\nu x.$$

(c) Show that the first of these spectral expansions leads to the transform pair

$$F(\nu) = \int_0^\infty \frac{\nu}{(\cos^2\alpha + \nu^2\sin^2\alpha)^{1/2}}\left(\sin\alpha\cos\nu x + \frac{\cos\alpha}{\nu}\sin\nu x\right)f(x)\, dx,$$

$$f(x) = \frac{2}{\pi}\int_0^\infty \frac{\nu}{(\cos^2\alpha + \nu^2\sin^2\alpha)^{1/2}}\left(\sin\alpha\cos\nu x + \frac{\cos\alpha}{\nu}\sin\nu x\right)F(\nu)\, d\nu.$$

These reduce to the ordinary sine and cosine transforms for $\alpha = 0$ and $\alpha = \pi/2$, respectively.

(d) Solve the problem

$$\frac{\partial u}{\partial t} - \frac{\partial^2 u}{\partial x^2} = 0, \quad t>0,\, x>0; \qquad u(x,0) = f(x),$$

$$(\cos\alpha)u(0,t) - (\sin\alpha)\frac{\partial u}{\partial x}(0,t) = 0,$$

by applying the transform in (c) when $0 \leqslant \alpha \leqslant \pi/2$.

3.3. *The Weber transform*

(a) Consider the spectral problem for Bessel's equation:

$$-(xu')' - \lambda xu = 0, \quad a<x<\infty; \qquad u(a) = 0,$$

where a is a given positive number. Clearly the left endpoint is regular, and the right endpoint is in the limit-point case. Show that for $\mathrm{Im}\,\lambda \neq 0$

$$g(x,\xi;\lambda) = \frac{\pi i}{2H_0^{(1)}(\sqrt{\lambda}\,a)}\,v(x_<,\lambda)H_0^{(1)}(\sqrt{\lambda}\,x_>),$$

where

$$v(x,\lambda) = J_0(\sqrt{\lambda}\,x)H_0^{(1)}(\sqrt{\lambda}\,a) - J_0(\sqrt{\lambda}\,a)H_0^{(1)}(\sqrt{\lambda}\,x)$$

is the solution of the homogeneous equation satisfying $v(a,\lambda)=0$, $v'(a,\lambda) = -2i/\pi a$. Show that the spectrum is purely continuous and consists of the real values $0 \leqslant \lambda < \infty$. Obtain the expansions

$$\frac{\delta(x-\xi)}{x} = -\int_0^\infty \frac{v(x,\mu^2)v(\xi,\mu^2)}{J_0^2(a\mu)+N_0^2(a\mu)}\,\mu\,d\mu, \qquad a<x, \xi<\infty,$$

$$\frac{\delta(x-\xi)}{x} = \int_0^\infty \frac{w(x,\mu^2)w(\xi,\mu^2)}{J_0^2(a\mu)+N_0^2(a\mu)}\,\mu\,d\mu, \qquad a<x, \xi<\infty,$$

where

$$v(x,\mu^2) = J_0(\mu x)H_0^{(1)}(\mu a) - J_0(\mu a)H_0^{(1)}(\mu x),$$
$$w(x,\mu^2) = J_0(\mu x)N_0(\mu a) - J_0(\mu a)N_0(\mu x).$$

The transform pair corresponding to the second spectral expansion is known as the *Weber transform* pair.

(b) Use a Weber transform to find the solution of the heat equation problem in two space dimensions (cylindrical hole):

$$\frac{\partial u}{\partial t} - \Delta u = \frac{\partial u}{\partial t} - \frac{1}{r}\frac{\partial}{\partial r}\left(r\frac{\partial u}{\partial r}\right) = 0, \qquad r>a, \quad t>0;$$

$$u(a,t)=1, \qquad u(r,0)=0.$$

3.4. *The Mellin transform.* Consider the spectral problem

$$-(rR')' - \frac{\lambda}{r}R = 0, \quad 0<r<\infty; \qquad p(r)=r, \quad s(r)=\frac{1}{r},$$

where the independent variable has been called r because the problem arises in polar coordinates in the *plane*.

(a) Show that the limit-point case holds at both singular points, $r=0$ and $r=\infty$.

(b) For $\lambda \notin [0, \infty)$ construct Green's function

$$g(r,r_0) = \frac{i}{2\sqrt{\lambda}} r_<^{-i\sqrt{\lambda}} \, r_>^{i\sqrt{\lambda}},$$

where $r_< = \min(r,r_0)$, $r_> = \max(r,r_0)$.

(c) Obtain formally the completeness relation

$$(3.51) \quad r\delta(r-r_0) = \frac{1}{2\pi} \int_{-\infty}^{\infty} r_0^{i\nu} r^{-i\nu} \, d\nu, \qquad 0<r, \quad r_0 < \infty,$$

and the *Mellin transform pair*

$$(3.52) \qquad F_M(\nu) \doteq \int_0^{\infty} f(r) r^{-i\nu - 1} \, dr \text{ (Mellin transform)},$$

$$(3.53) \qquad f(r) = \frac{1}{2\pi} \int_{-\infty}^{\infty} r^{i\nu} F_M(\nu) \, d\nu.$$

(d) Solve the BVP for the Laplacian in a wedge:

$$-\Delta u = 0, \quad 0 < r < \infty, \quad 0 < \phi < \alpha; \qquad u(r,0) = 0, \quad u(r,\alpha) = h(r),$$

where $h(r)$ is a prescribed function. Proceed formally by taking the Mellin transform to obtain the solution in the form of an inverse transform.

REFERENCES AND ADDITIONAL READING

Atkinson, F. V., *Discrete and continuous boundary problems*, Academic Press, New York, 1964.

Eastham, M. S. P., *The spectral theory of periodic differential equations*, Scottish Academic Press, Edinburgh, 1973.

Titchmarsh, E. C., *Eigenfunction expansions associated with second-order differential equations*, Oxford University Press, Oxford, 1946.

Yosida, K., *Lectures on differential and integral equations*, Interscience, New York, 1960.

8
Partial
Differential
Equations

1. CLASSIFICATION OF PARTIAL DIFFERENTIAL EQUATIONS

Introduction

The only difficulty that can arise for the initial value problem for an ordinary linear differential equation is the presence of a singular point. The equivalent concept for partial differential equations is more subtle. Not surprisingly, one must first recast the one-dimensional problem before finding a useful generalization to higher dimensions.

In the IVP for an ordinary linear differential equation of order p, we look for a solution $u(x)$ of

$$(1.1) \qquad Lu \doteq a_p(x)\frac{d^p u}{dx^p} + \cdots + a_1(x)\frac{du}{dx} + a_0(x)u = f(x),$$

with given initial data $\{u(x_0), u'(x_0), \ldots, u^{(p-1)}(x_0)\}$. We can rewrite (1.1) as

$$(1.2) \qquad a_p(x)u^{(p)}(x) = f(x) - Mu,$$

where M is a differential operator of order $p-1$.

If the coefficients $\{a_m(x)\}$ and $f(x)$ are continuous and if $a_p(x_0) \neq 0$, the IVP has one and only one solution in a neighborhood of x_0. If the leading coefficient $a_p(x)$ vanishes at x_0, one can try dividing by $a_p(x)$, but this will usually only cause some of the other coefficients to become unbounded at x_0.

In the case $a_p(x_0) \neq 0$, the initial data *together* with the differential equation enable us to construct the solution in a neighborhood of x_0 by a stepwise numerical procedure. As a first and crucial step we are able to calculate $u^{(p)}(x_0)$ unambiguously from the data and the differential equation since the right side of (1.2) is known at x_0 from the initial data and division by $a_p(x_0)$ gives $u^{(p)}(x_0)$. Armed with this additional piece of information, we can calculate new initial data $\{u(x_1), \ldots, u^{(p-1)}(x_1)\}$ at the neighboring point $x_1 = x_0 + \Delta x$ through the approximate formulas

$$u(x_1) = u(x_0) + u'(x_0)\Delta x, \qquad \ldots, \qquad u^{(p-1)}(x_1) = u^{(p-1)}(x_0) + u^{(p)}(x_0)\Delta x.$$

If x_1 is sufficiently close to x_0, $a_p(x_1)$ will not vanish and we can use the differential equation again to find $u^{(p)}(x_1)$. We now repeat the procedure to construct the solution of the original problem on an interval (x_0, b) as long as $a_p(x)$ does not vanish anywhere in $x_0 \leqslant x \leqslant b$.

If, however, $a_p(x_0) = 0$, we cannot calculate $u^{(p)}(x_0)$ unambiguously from (1.2). The right side is known from the initial data, and the left side vanishes; thus either $u^{(p)}(x_0)$ does not exist or it is indeterminate. In any event the earlier numerical construction fails, and so does the existence and uniqueness theorem. Another way of stating the difficulty is that Lu itself (rather than just Mu) is known at x_0 from the initial data alone.

Cauchy Problem

Consider a linear partial differential equation of order p in the n variables $x_1, \ldots, x_n$:

$$(1.3) \qquad\qquad Lu = f(x),$$

where, in the usual notation of (1.3), Chapter 2,

$$(1.4) \qquad\qquad L = \sum_{|k| \leqslant p} a_k(x) D^k,$$

with

$$k = (k_1, \ldots, k_n), \qquad |k| = k_1 + \cdots + k_n, \qquad D^k = D_1^{k_1} \cdots D_n^{k_n}, \qquad D_i = \frac{\partial}{\partial x_i}.$$

Anticipating that many of the qualitative properties of the solution of (1.3) will depend only on the terms of highest order in L, we define the *principal part* of the operator (1.4) as

$$(1.5) \qquad L_p = \sum_{|k|=p} a_k(x) D^k = \sum_{|k|=p} a_k(x) D_1^{k_1} \cdots D_n^{k_n}.$$

It is often convenient to regard L_p, for fixed x, as a pth degree polynomial in the n variables $D_1, \ldots, D_n$, the components of the n-dimensional vector D. When this point of view is taken, we shall write $L_p(x, D)$ for (1.5). On occasion we shall use P rather than x to denote a point in R_n. The coefficients are then $a_k(P)$, and $L_p(x, D)$ becomes $L_p(P, D)$.

The coefficients and the forcing term in (1.3) are assumed continuous, and a solution is sought in a specified domain in R_n (often this domain is the whole of R_n). We now wish to associate initial data with (1.3). The appropriate generalization of the one-dimensional case is to assign the values of u and its *normal* derivatives of order $\leqslant p-1$ on a smooth hypersurface Γ (a hypersurface being a manifold of dimension $n-1$). This type of initial data is known as *Cauchy data*, and the resulting initial value problem as the *Cauchy problem* for L. As we shall see, the Cauchy data actually determines on Γ all derivatives (with respect to $x_1, \ldots, x_n$, say) of order $\leqslant p-1$.

Let P be a point on Γ. In view of the smoothness of Γ we can introduce an $(n-1)$-dimensional coordinate system $\xi_2, \ldots, \xi_n$ to label points on Γ in some neighborhood of P which is taken as the origin. This can be done, for instance, by assigning to each point on Γ the Cartesian coordinates of its projection on the tangent plane to Γ at P. In any event we shall refer to $\xi_2, \ldots, \xi_n$ as *tangential* coordinates. Now let $\nu = \xi_1$ stand for a coordinate along the normal to Γ with origin on Γ. Then $(\nu, \xi_2, \ldots, \xi_n)$ is a normal-tangential coordinate system for all points in R_n sufficiently close to P. The coordinate surface $\nu = 0$ is Γ, and the point $(0, \ldots, 0)$ is P.

From the given Cauchy data on Γ we know the quantities

$$u(0, \xi_2, \ldots, \xi_n), \qquad \frac{\partial u}{\partial \nu}(0, \xi_2, \ldots, \xi_n), \qquad \ldots, \qquad \frac{\partial^{p-1} u}{\partial \nu^{p-1}}(0, \xi_2, \ldots, \xi_n).$$

Provided that there is enough smoothness, we can therefore calculate on Γ derivatives of any order with respect to the tangential coordinates and of order $\leqslant p-1$ with respect to the normal coordinate. In the coordinates $(\nu, \xi_2, \ldots, \xi_n)$ the only derivative of order p on Γ which is not known from the Cauchy data is $\partial^p u / \partial \nu^p$. To return to the original coordinates $x_1, \ldots, x_n$, we use the transformation law relating the two coordinate systems. We have

$$\frac{\partial u}{\partial x_i} = \sum_{j=1}^{n} \frac{\partial u}{\partial \xi_j} \frac{\partial \xi_j}{\partial x_i}, \qquad \nu \doteq \xi_1,$$

so that $\partial u / \partial x_i$ is a linear combination of $\partial u / \partial \xi_j$ with variable coefficients. Thus any mth derivative with respect to the x coordinates will involve only derivatives of order $\leqslant m$ with respect to the ξ coordinates. Since derivatives of order $\leqslant p-1$ with respect to the ξ coordinates are known from the

Cauchy data, so are derivatives of order $\leqslant p-1$ with respect to the x coordinates. In fact, we even have partial information about derivatives of order p.

To construct the solution numerically for points near P but away from Γ we will need to be able to calculate $\partial^p u/\partial \nu^p$ at P (or, equivalently, to calculate unambiguously the set of all pth derivatives with respect to $x_1,\ldots,x_n$ at P). This calculation will *not* be possible if Lu at P can be determined from the Cauchy data alone. We are therefore led to the following definition.

Definition. If Lu can be evaluated at a point P on Γ from the Cauchy data alone, the surface Γ is said *to be characteristic (for L) at P*. If the surface Γ is characteristic (for L) at every point P on Γ, we say that Γ is a *characteristic surface (for L)*.

Remark. Since the operator L is usually fixed throughout the discussion, the qualifier "for L" may be omitted.

If Γ is characteristic at P, then, when we express L in terms of the normal tangential coordinates $(\nu,\xi_2,\ldots,\xi_n)$, the coefficient of $\partial^p/\partial \nu^p$ will vanish at $P=(0,\ldots,0)$. In that sense the leading coefficient "vanishes" as for singular points of ordinary differential equations. We must note, however, that the new notion depends not only on the point P but also on Γ (or, more precisely, on the normal direction to Γ at P). This leads to the following theorem and corollaries, whose proofs we omit.

Theorem 1. Let L be a given operator of order p, and let P be a point on the smooth hypersurface Γ. The necessary and sufficient condition for Γ to be characteristic at P is that the coefficient of $\partial^p/\partial \nu^p$ vanishes when L is expressed in the coordinate system $(\nu,\xi_2,\ldots,\xi_n)$.

Corollary 1. Γ is characteristic at the point P if and only if

$$(1.6) \qquad\qquad L_p(P,\nu)=0,$$

where ν is the normal vector to Γ at P.

Corollary 2. Γ is *not* characteristic at the point P if and only if all pth order derivatives of u with respect to $x_1,\ldots,x_n$ are unambiguously determined at P from the Cauchy data on Γ together with the differential equation.

The results so far are essentially negative: if Γ is characteristic at P, there is no hope that the Cauchy problem is well posed, even in a neighborhood of P. Is the Cauchy problem well posed if Γ is nowhere characteristic? We shall find out that this is so for certain classes of partial differential equations but not for others.

Equations of the First Order in Two Independent Variables

Let x,y be Cartesian coordinates in R_2, and let Γ be a smooth *curve* in the plane. Since L is to be an operator of the first order, the Cauchy data will consist of specifying u on Γ.

We begin with the operator $L = \partial/\partial x$ and the simple Cauchy problem

$$(1.7) \qquad \frac{\partial u}{\partial x} = 0, \qquad u \text{ given on } \Gamma.$$

Let A be a point on Γ. The curve Γ will be characteristic at A if $\partial u/\partial x$ can be calculated at A from the Cauchy data alone. Obviously this is possible if and only if the tangent to Γ at A is in the x direction. The characteristic curves are therefore the *straight lines* $y =$ constant. The general solution of (1.7) is $u = F(y)$, where F is arbitrary. Thus u is constant on any horizontal line. If the initial curve Γ on which the Cauchy data is given is *nowhere characteristic*, as in Figure 1.1, the solution will be unambiguously determined in the whole strip by the formula $u_P = u_A$. [Note that, even if the Cauchy data has a discontinuity at some point B on Γ, we can still define the solution in the horizontal strip by $u_P = u_A$, but the solution is then discontinuous along the entire horizontal line through B. We showed in Example 5, Section 5, Chapter 2, that a discontinuous function $F(y)$ still satisfies (1.7) in the sense of distributions.] If the initial curve is characteristic even at a single point A, as in Figure 1.2, the Cauchy data will usually be incompatible with the differential equation; since u is constant along a horizontal line, a solution is possible only if the given values at points such as A_1 and A_2 are the same.

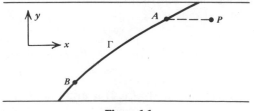

Figure 1.1

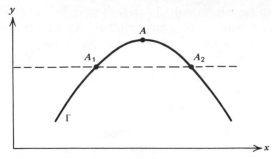

Figure 1.2

Let us now look at the equation of the first order:

$$(1.8) \qquad a(x,y)\frac{\partial u}{\partial x} + b(x,y)\frac{\partial u}{\partial y} + c(x,y,u) = 0,$$

which is a linear equation if $c(x,y,u)$ is of the form $c(x,y)u + d(x,y)$ and is said to be *semilinear* if $c(x,y,u)$ depends nonlinearly on u (note that in a semilinear equation the highest derivatives appear linearly). Since the theory of characteristics applies just as easily to semilinear equations, we may as well consider this more general case. Let Γ be a smooth curve in R_2, and let P be a point on Γ. The value of u is given on Γ. By Corollary 2, only if we can calculate $\partial u/\partial x$ and $\partial u/\partial y$ unambiguously at P from the data and the differential equation will Γ not be characteristic at P. Consider two neighboring points $P = (x,y)$ and $Q = (x+dx, y+dy)$ located on Γ. Let us try to calculate $\partial u/\partial x$ and $\partial u/\partial y$ at P. The following relations hold:

$$u(Q) - u(P) = dx\frac{\partial u}{\partial x}(P) + dy\frac{\partial u}{\partial y}(P),$$

$$c(P, u(P)) = a(P)\frac{\partial u}{\partial x}(P) + b(P)\frac{\partial u}{\partial y}(P).$$

These simultaneous linear equations for $\partial u/\partial x$ and $\partial u/\partial y$ at P will have a unique solution if and only if

$$(1.9) \qquad \begin{vmatrix} dx & dy \\ a(P) & b(P) \end{vmatrix} \neq 0 \qquad \text{or} \qquad b(P)\,dx - a(P)\,dy \neq 0.$$

The curve Γ is characteristic at $P = (x,y)$ if and only if

$$(1.10) \qquad b(x,y)\,dx - a(x,y)\,dy = 0.$$

Thus at every point in the plane a unique characteristic direction is defined by (1.9) as long as a and b do not vanish simultaneously. If the ordinary differential equation (1.10) is integrated, a "one-parameter" family of curves is obtained—the characteristic curves for (1.8). The same result could have been obtained by using (1.6). We have $p = 1$ and

$$L_1(P,v) = a(P)v_1 + b(P)v_2.$$

On setting $L_1 = 0$, we find that Γ is characteristic at P if its normal is in the direction $(-b(P), a(P))$, which means that its slope must be $b(P)/a(P)$, confirming (1.10).

We state without proof for the linear version of (1.8) a theorem similar to that obtained for $\partial u/\partial x = 0$: If Γ is *nowhere characteristic*, the solution of (1.8) exists and is unique in the characteristic "strip" bounded by the characteristic curves through the endpoints of Γ. Moreover, the solution depends continuously on the data. If the given value of u on Γ has a discontinuity at a point P, the discontinuity is propagated along the characteristic curve through P. The solution is then a weak solution in the sense of Section 5, Chapter 2.

As a simple illustration of (1.8) consider the equation

$$(1.11) \qquad \frac{\partial u}{\partial x} + \frac{\partial u}{\partial t} + \sigma u = 0,$$

where σ is a nonnegative constant, x is a space coordinate, and t is the time (corresponding to y in the earlier notation). The case $\sigma = 0$ governs the plug flow of a fluid [see (3.11), Chapter 0], whereas the case $\sigma > 0$ describes the absorption (without scattering) of neutrons streaming in the positive x direction [see (5.18), Chapter 0]. Since $a = b = 1$, the characteristics of (1.11) are the straight lines $x - t = $ constant. The usual Cauchy problem associated with (1.11) involves data given on the initial curve $t = 0$, which is *nowhere characteristic*. We then want to solve (1.11) for $t > 0$ and $-\infty < x < \infty$, subject to the initial condition $u(x,0) = f(x)$, where $f(x)$ is arbitrary. By making the change of variables $\alpha = x - t$, $\beta = x + t$, we reduce (1.11) to

$$\frac{\partial u}{\partial \beta} + \frac{\sigma}{2}u = 0,$$

so that $u(x,t) = F(x-t)e^{-\sigma(x+t)/2}$, and, on imposing the initial condition,

$$(1.12) \qquad u(x,t) = f(x-t)e^{-\sigma t}.$$

If $\sigma = 0$, the solution is $f(x-t)$, which for fixed $t > 0$ is just the initial function $f(x)$ displaced t units to the right along the x axis. We can therefore view $f(x-t)$ as a wave traveling to the right with unit velocity. If $\sigma > 0$, we also have an attenuation factor from the exponential in (1.12). This is in agreement with our intuitive ideas of absorption.

Equations of the Second Order in Two Variables

Consider the equation of the second order in the two independent variables x and y:

(1.13)

$$a(x,y)\frac{\partial^2 u}{\partial x^2} + 2b(x,y)\frac{\partial^2 u}{\partial x \, \partial y} + c(x,y)\frac{\partial^2 u}{\partial y^2} + F\left(x,y,u,\frac{\partial u}{\partial x},\frac{\partial u}{\partial y}\right) = 0.$$

The equation will be *linear* if

$$F\left(x,y,u,\frac{\partial u}{\partial x},\frac{\partial u}{\partial y}\right) = d\frac{\partial u}{\partial x} + e\frac{\partial u}{\partial y} + fu + g,$$

where d, e, f, g are arbitrary functions of x and y alone, and otherwise *semilinear*. We are given u and its normal derivative on a smooth curve Γ in the x-y plane. Thus both $\partial u/\partial x$ and $\partial u/\partial y$ are known on Γ.

If we can calculate all second derivatives at a point P on Γ from the data together with the differential equation, Γ is not characteristic at P. Again let $P = (x,y)$ and $Q = (x+dx, y+dy)$ be two neighboring points on Γ. Then the following simultaneous linear equations for $\partial^2 u/\partial x^2, \partial^2 u/\partial y^2, \partial^2 u/\partial x \, \partial y$, all evaluated at P, must hold:

$$\frac{\partial u}{\partial x}(Q) - \frac{\partial u}{\partial x}(P) = dx\frac{\partial^2 u}{\partial x^2}(P) + dy\frac{\partial^2 u}{\partial x \, \partial y}(P),$$

$$\frac{\partial u}{\partial y}(Q) - \frac{\partial u}{\partial y}(P) = dx\frac{\partial^2 u}{\partial x \, \partial y}(P) + dy\frac{\partial^2 u}{\partial y^2}(P),$$

$$-F_P = a(P)\frac{\partial^2 u}{\partial x^2}(P) + 2b(P)\frac{\partial^2 u}{\partial x \, \partial y}(P) + c(P)\frac{\partial^2 u}{\partial y^2}(P),$$

where $F_P \doteq F(P, u(P), \partial u/\partial x(P), \partial u/\partial y(P))$

The left sides are known from the Cauchy data, so that a necessary and sufficient condition for a unique solution is

(1.14) $\quad \begin{vmatrix} dx & dy & 0 \\ 0 & dx & dy \\ a(P) & 2b(P) & c(P) \end{vmatrix} \neq 0 \quad$ or $\quad a\,dy^2 - 2b\,dx\,dy + c\,dx^2 \neq 0.$

If $a(P) \neq 0$, Γ will be characteristic at $P = (x,y)$ if and only if the slope of Γ at P satisfies

(1.15)
$$\frac{dy}{dx} = \frac{b \pm \sqrt{b^2 - ac}}{a}.$$

We distinguish among three cases:

If $b^2 - ac < 0$ at P, no real curve Γ can satisfy (1.15) and no curve is characteristic at P. Equation (1.13) is said to be *elliptic* at P. For the Laplace equation $\partial^2 u / \partial x^2 + \partial^2 u / \partial y^2 = 0$, we have $b^2 - ac = -1$, so that the equation is elliptic in the whole plane. The Laplace equation is the prototype of elliptic equations.

If $b^2 - ac > 0$ at P, there are two characteristic directions at P and the equation is said to be *hyperbolic* at P. The equation $\partial^2 u / \partial x \, \partial y = 0$ and the wave equation $\partial^2 u / \partial t^2 - \partial^2 u / \partial x^2 = 0$ are hyperbolic in the entire x-y and x-t planes, respectively.

If $b^2 - ac = 0$ at P, there is only one characteristic direction at P and the equation is *parabolic* at P. The diffusion equation $\partial u / \partial t - a \partial^2 u / \partial x^2 = 0$ is parabolic in the whole x-t plane.

Remarks

1.   If the coefficients are not constant, the equation may be of different types in various parts of the plane. For instance, $\partial^2 u / \partial x^2 + x \partial^2 u / \partial y^2 = 0$ is elliptic in the half-plane $x > 0$ and hyperbolic in the half-plane $x < 0$.

2.   If the equation is hyperbolic in the whole plane, we can integrate (1.15) to find two families of characteristic curves, whereas for the parabolic case we can find only one such family.

3.   Characteristics can be a blessing or a curse. If the equation is hyperbolic and Cauchy data is given on a curve Γ, the problem is well posed when Γ is not a characteristic but ill posed if Γ is a characteristic. One might therefore think that in the elliptic case—when characteristics are totally absent—the Cauchy problem would be well posed. We shall see, however, that this is not the case and that, instead, boundary value problems are usually well posed for elliptic equations (but not for hyperbolic ones).

To show how characteristics can be used, we shall investigate a few problems related to the hyperbolic equation

(1.16)
$$\frac{\partial^2 u}{\partial x \, \partial y} = f(x, y, u),$$

where f is a continuous function of its arguments. If $f=0$, we have the simple equation

(1.17)
$$\frac{\partial^2 u}{\partial x \, \partial y} = 0,$$

which we study first. Since $(\partial/\partial x)(\partial u/\partial y)=0$, $\partial u/\partial y = G(y)$ and $u = F(x) + G(y)$ is the general solution of (1.17). The characteristics for (1.17) as well as for (1.16) are the straight lines $x=$ constant, $y=$ constant. Let us consider a few initial value problems for (1.17).

1. *The Cauchy problem.* Let Γ be a curve that is nowhere characteristic, as in Figure 1.3. We shall show how the solution can be determined within the large rectangle from the Cauchy data on Γ. Indeed, let $P=(x,y)$ be a point in the rectangle. Integrating $\partial u/\partial x$ from A to P, we obtain

$$u(P) = u(A) + \int_{x_A}^x \frac{\partial u}{\partial x}(\xi,y) \, d\xi.$$

Here x_A is the value of x at the point A, that is, $x_A = \psi^{-1}(y)$, where $\psi(x)$ is the equation of the curve Γ. The differential equation tells us that $\partial u/\partial x$ is independent of y, so that

$$\frac{\partial u}{\partial x}(\xi,y) = \frac{\partial u}{\partial x}(\xi,\psi(\xi)).$$

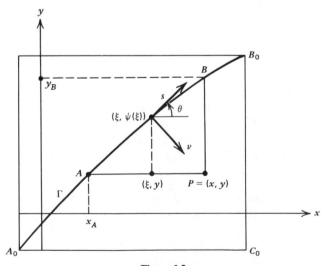

Figure 1.3

Thus we obtain

$$(1.18) \qquad u(P) = u(A) + \int_{x_A}^{x} \frac{\partial u}{\partial x}(\xi, \psi(\xi)) \, d\xi,$$

which expresses $u(x,y)$ in terms of known quantities on Γ. One can easily verify that (1.18) actually solves the Cauchy problem for (1.17). By integrating $\partial u / \partial y$ from P to B one also finds that

$$(1.19) \qquad u(P) = u(B) - \int_{y}^{y_B} \frac{\partial u}{\partial y}(\psi^{-1}(\eta), \eta) \, d\eta.$$

We can express all integrals in terms of the arclength s along Γ by using the relations $d\xi = \cos\theta \, ds$, $d\eta = \sin\theta \, ds$. Adding (1.18) and (1.19), we then find that

$$u(P) = \frac{u(A)}{2} + \frac{u(B)}{2} + \frac{1}{2} \int_{s_A}^{s_B} \left(\frac{\partial u}{\partial x} \cos\theta - \frac{\partial u}{\partial y} \sin\theta \right) ds.$$

Introducing the unit vector $q = (\cos\theta, -\sin\theta)$, known as the *transversal* to Γ, we can simplify the last formula to

$$(1.20) \qquad u(P) = \frac{u(A)}{2} + \frac{u(B)}{2} + \frac{1}{2} \int_{s_A}^{s_B} \frac{\partial u}{\partial q} \, ds,$$

where $\partial u / \partial q$ is known from the Cauchy data. The transversal will coincide with the unit normal $\nu = (\sin\theta, -\cos\theta)$ if $\theta = \pi/4$. Thus, if Γ is a 45° line, we have the useful result

$$(1.21) \qquad u(P) = \frac{u(A)}{2} + \frac{u(B)}{2} + \frac{1}{2} \int_{A}^{B} \frac{\partial u}{\partial \nu} \, ds.$$

Formulas such as (1.18), (1.19), and (1.20) establish the existence of a solution and its continuity with respect to the data. Uniqueness is easy. Suppose we have two solutions u_1 and u_2 corresponding to the same Cauchy data. Then the difference $u = u_1 - u_2$ satisfies (1.17) with vanishing Cauchy data on Γ. Since the general solution of (1.17) is $u = F(x) + G(y)$, it follows that

$$0 = F(x_A) + G(y), \qquad 0 = F(x) + G(y_B),$$

so that $u(x,y)$ is constant in the large rectangle of Figure 1.3. Since u vanishes on Γ, that constant is 0.

2. If Γ is characteristic at P, the Cauchy data and the differential equations are generally incompatible. This is clearly seen if Γ is a portion

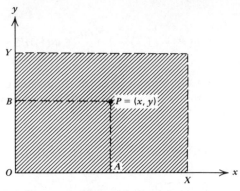

Figure 1.4

of the characteristic curve $y = $ constant. The Cauchy problem consists of assigning u and $\partial u/\partial y$ arbitrarily on Γ, but the differential equation tells us that $\partial u/\partial y$ is independent of x, so that $\partial u/\partial y$ must be constant on Γ.

3.   There are sensible problems that can be formulated with data on characteristics, but these are not strict Cauchy problems. Consider two intersecting characteristics which, for simplicity, we have taken along the coordinate axes, as in Figure 1.4. If we are given u *alone* on the segments OX and OY, we can easily show that the solution is uniquely determined in the shaded rectangle by the formula

$$(1.22) \qquad\qquad u(P) = u(A) + u(B) - u(O),$$

where $u(O)$ is the value of u at the origin (we have assumed that the given values of u on the segments OX and OY coincide at O). With enough smoothness in the data, it is easy to verify that (1.22) is a classical solution. Even if the data is discontinuous, we find that (1.22) still provides a weak solution to the Cauchy problem. Discontinuities in the data are propagated along characteristics.

4.   Another type of problem occurs when u is given on a characteristic segment and on an intersecting curve nowhere characteristic, as in Figure 1.5. We then find the solution in the shaded region as

$$(1.23) \qquad\qquad u(P) = u(A) + u(B) - u(C).$$

Let us now turn to the more difficult nonlinear problem (1.16) with Cauchy data on the noncharacteristic curve Γ. We shall still refer to Figure

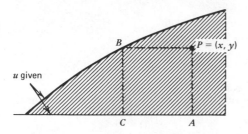

Figure 1.5

1.3. Integrating $\partial u / \partial x$ from A to P, we have

$$u(P) = u(A) + \int_{x_A}^x \frac{\partial u}{\partial x}(\xi, y) \, d\xi,$$

and, from the differential equation,

$$\frac{\partial u}{\partial x}(\xi, \psi(\xi)) - \frac{\partial u}{\partial x}(\xi, y) = \int_y^{\psi(\xi)} f(\xi, \eta, u(\xi, \eta)) \, d\eta,$$

so that

$$(1.24) \quad u(P) = u(A) + \int_{x_A}^x \frac{\partial u}{\partial x}(\xi, \psi(\xi)) \, d\xi - \iint_{D(x,y)} f(\xi, \eta, u(\xi, \eta)) \, d\xi \, d\eta,$$

where $D(x,y)$ is the triangular region bounded by AP, PB, and Γ. Although the first two terms on the right side of (1.24) are known from the Cauchy data, the last term is not. Thus (1.24) is an integral equation (nonlinear) for the unknown $u(x,y)$, completely equivalent to the Cauchy problem for (1.16).

By subtracting from the solution of the Cauchy problem for (1.16) the solution (previously calculated) of (1.17) with the same Cauchy data, we obtain a problem of type (1.16) with vanishing Cauchy data but different f. It is therefore with no loss of generality that we may analyze (1.24) with vanishing Cauchy data:

$$(1.25) \quad u(x,y) = - \iint_{D(x,y)} f(\xi, \eta, u(\xi, \eta)) \, d\xi \, d\eta.$$

We would like to show that (1.25) has one and only one solution, at least in some sufficiently small region to the right of Γ. We shall use the ideas developed in Section 4, Chapter 4, on contraction transformations. First we must make some restriction on the nature of the nonlinearity in f. We shall require that the Lipschitz condition

(1.26) $$|f(x,y,u_2-u_1)| \leqslant C|u_2-u_1|$$

hold for all u_1, u_2 and for all x, y in the closed region Ω bounded by the segments $A_0 C_0, C_0 B_0$, and Γ (see Figure 1.3).

Let M_Ω be the space of continuous functions on Ω with the uniform metric

$$d_\infty(u_1, u_2) = \int\int_\Omega |u_1 - u_2| \, dx \, dy.$$

The right side of (1.25) defines a transformation of M_Ω into itself. We shall denote this transformation by T. A solution of (1.25) is then just a fixed point of T and vice versa. Let u_1 and u_2 be arbitrary elements of M_Ω, and let v_1 and v_2 be their respective images under T. Then

$$|v_1(x,y) - v_2(x,y)| = |Tu_1 - Tu_2| \leqslant \int\int_{D(x,y)} |f(\xi,\eta,u_2) - f(\xi,\eta,u_1)| \, d\xi \, d\eta$$

$$\leqslant CS(x,y) d_\infty(u_1, u_2),$$

where $S(x,y)$ is the area of $D(x,y)$. Therefore

$$d_\infty(v_1, v_2) \leqslant CS d_\infty(u_1, u_2),$$

where S is the area of Ω. Clearly, if the constant C in (1.26) is smaller than $1/S$, then T is a contraction on M_Ω and therefore has a unique fixed point which is the required solution of (1.25). In any event we can always choose a smaller region Ω (of the same triangular form) so that $CS < 1$; then within this smaller region (1.25) and the equivalent Cauchy problem for (1.16) will have one and only one solution.

Remarks

1. Existence and uniqueness for the Cauchy problem (1.16) can be extended to the case where f depends on first derivatives, but again a growth condition is needed.

2. It may appear that we have studied only a very special hyperbolic equation. However, by a change of coordinates, any hyperbolic equation of the second order in two variables can be transformed into (1.16) with f also depending on $\partial u / \partial x$ and $\partial u / \partial y$. For instance, the linear wave

equation

$$\frac{\partial^2 u}{\partial t^2} - c^2 \frac{\partial^2 u}{\partial x^2} = f(x,t)$$

takes the form

$$\frac{\partial^2 u}{\partial \alpha \, \partial \beta} = \tilde{f}(\alpha,\beta)$$

with $\alpha = x - ct$, $\beta = x + ct$, and

$$\tilde{f}(\alpha,\beta) = -\frac{1}{4c^2} f\left(\frac{\alpha+\beta}{2}, \frac{\beta-\alpha}{2c}\right).$$

Exercises

1.1. The Cauchy problem for elliptic equations is ill posed despite the absence of characteristics. Consider Laplace's equation $\Delta u = 0$ in the half-plane $x > 0$, $-\infty < y < \infty$ with the Cauchy data $u(0,y) = 0$, $\partial u / \partial x = \varepsilon \sin \alpha y$, where ε and α are constants. A solution of this problem is $u(x,y) = \varepsilon \sin \alpha y \sinh \alpha x$, which reduces—as it should—to $u \equiv 0$ when $\varepsilon = 0$. If $|\varepsilon|$ is small, the data is small in the uniform norm, independently of α, that is,

$$\|u(0,y)\|_\infty \leqslant \varepsilon \qquad \text{and} \qquad \left\|\frac{\partial u}{\partial x}(0,y)\right\|_\infty \leqslant \varepsilon.$$

Show that for any $\varepsilon \neq 0$ and any preassigned $x > 0$, the solution $u(x,y)$ can be made arbitrarily large by choosing α large enough. Thus the solution does not depend continuously on the data.

1.2. The pure BVP with data on a noncharacteristic rectangle is ill posed for hyperbolic equations. Consider the homogeneous wave equation

$$\frac{\partial^2 u}{\partial t^2} - \frac{\partial^2 u}{\partial x^2} = 0$$

on the rectangle $0 < x < \pi$, $0 < t < \tau$ with $u(x,0) = u(0,t) = u(\pi,t) = 0$ and $u(x,\tau) = \varepsilon f(x)$, where $f(x)$ is a given function in $C^2(0,\pi)$ that satisfies $f(0) = f(\pi) = 0$ and all of whose Fourier sine coefficients are nonvanishing. Show that, if τ is rational, the solution obtained by separation of variables cannot be made to satisfy the condition at $t=\tau$; if τ is irrational, the solution does not depend continuously on the data.

2. TYPICAL WELL-POSED PROBLEMS FOR HYPERBOLIC AND PARABOLIC EQUATIONS

Hyperbolic Equations

Consider the small, transverse vibrations of a very long, taut string subject to a pressure $q(x,t)$ for $t>0$. At $t=0$ we give the initial "state" of the string, that is, the initial displacement $f_1(x)$ and the initial velocity $f_2(x)$. Our boundary value problem is

(2.1)

$$\frac{\partial^2 u}{\partial t^2} - \frac{\partial^2 u}{\partial x^2} = q(x,t), \quad t>0, \; -\infty<x<\infty;$$

$$u(x,0)=f_1(x), \quad \frac{\partial u}{\partial t}(x,0)=f_2(x).$$

We first consider the *homogeneous* equation ($q\equiv 0$) whose general solution is

(2.2) $$F(x-t)+G(x+t),$$

where F and G are arbitrary. As we have already seen in Section 1, and in Example 7, Section 5, Chapter 2, $F(x-t)$ can be regarded as a wave traveling to the right with unit velocity. Similarly $G(x+t)$ is a wave traveling to the left with unit velocity. To solve the initial value problem for the homogeneous equation it suffices to express F and G in terms of f_1 and f_2. This is easily done for the infinite string. Although it is no trouble to handle both initial conditions simultaneously, we prefer, with an eye to future applications, to treat them separately. We begin with the case where $f_1=0$, that is, with the BVP

(2.3)

$$\frac{\partial^2 u}{\partial t^2} - \frac{\partial^2 u}{\partial x^2}=0, \quad t>0, \; -\infty<x<\infty; \qquad u(x,0)=0, \quad \frac{\partial u}{\partial t}(x,0)=f(x).$$

Since $u(x,t)$ can be written as $F(x-t)+G(x+t)$, we immediately find that

$$F(x)+G(x)=0, \qquad -F'(x)+G'(x)=f(x).$$

The last equation gives

$$-F(x)+G(x)=\int_0^x f(\xi)\,d\xi+A,$$

so that, together with $F+G=0$, we obtain

$$G(x)=\frac{A}{2}+\frac{1}{2}\int_0^x f(\xi)\,d\xi, \qquad F(x)=-\frac{1}{2}\int_0^x f(\xi)\,d\xi-\frac{A}{2},$$

and

$$(2.4) \qquad u(x,t)=F(x-t)+G(x-t)=\frac{1}{2}\int_{x-t}^{x+t}f(\xi)\,d\xi,$$

which is the solution of (2.3).

If the initial displacement is $f(x)$ and the initial velocity 0, the corresponding solution of (2.3) is just the time derivative of (2.4). This is contained in the following theorem.

Theorem. If $u(x,t)$ is a solution of (2.3), then $v(x,t) \doteq (\partial u/\partial t)(x,t)$ is a solution of

$$(2.5) \qquad \frac{\partial^2 v}{\partial t^2}-\frac{\partial^2 v}{\partial x^2}=0, \quad t>0, \ -\infty<x<\infty; \quad v(x,0)=f(x), \quad \frac{\partial v}{\partial t}=0.$$

Proof. Since u is a solution of a homogeneous linear differential equation with constant coefficients, any derivative of u is also a solution of the same equation. Also

$$v(x,0+)=f(x) \qquad \text{and} \qquad \frac{\partial v}{\partial t}(x,0+)=\frac{\partial^2 u}{\partial t^2}(x,0+)=\frac{\partial^2 u}{\partial x^2}(x,0+)=0,$$

since $u(x,0+)=0$.

It therefore follows from (2.4) that the solution of (2.5) is

$$(2.6) \qquad v(x,t)=\tfrac{1}{2}f(x+t)+\tfrac{1}{2}f(x-t).$$

If the initial displacement is $\delta(x)$ and the initial velocity is 0 (that is, we are plucking the string at $x=0$), the corresponding solution (2.6) is $\tfrac{1}{2}\delta(x+t)+\tfrac{1}{2}\delta(x-t)$, which means that the initial displacement splits into two waves, one traveling to the left and the other to the right with unit velocity. If the initial displacement is 0 and the initial velocity is $\delta(x)$ (that is, we are striking the string at $x=0$), a well of constant depth forms under the blow and the well front spreads outward with unit velocity.

The solution of (2.3) with initial velocity $=\delta(x)$ coincides for $t>0$ with the causal fundamental solution with pole at $x=0, t=0$. A similar relationship has already been shown for ordinary differential equations. The causal fundamental solution $E(t,0)$ for the operator

$$L = a_p(t)\frac{d^p}{dt^p} + \cdots + a_1(t)\frac{d}{dt} + a_0(t)$$

satisfies

$$LE = \delta(t), \qquad -\infty < t < \infty; \quad E \equiv 0, \quad t < 0.$$

In (5.48), Chapter 2, we showed that E coincides for $t>0$ with the solution of the IVP (for the homogeneous equation)

$$Lu = 0, \quad t>0; \qquad u(0)=u'(0)=\cdots=u^{(p-2)}(0)=0, \quad u^{(p-1)}(0)=\frac{1}{a_p(0)}.$$

For our partial differential equation we want a fundamental solution that is causal with respect to time; thus $E(x,t;0,0)$ satisfies

$$(2.7) \qquad \frac{\partial^2 E}{\partial t^2} - \frac{\partial^2 E}{\partial x^2} = \delta(x)\delta(t), \quad -\infty < x, t < \infty; \quad E \equiv 0, \quad t < 0.$$

By analogy with ordinary differential equations we can equally well characterize E for $t>0$ as the solution of the IVP

$$(2.8)$$

$$\frac{\partial^2 u}{\partial t^2} - \frac{\partial^2 u}{\partial x^2} = 0, \quad -\infty < x < \infty, t > 0; \qquad u(x,0)=0, \quad \frac{\partial u}{\partial t}(x,0)=\delta(x).$$

As a check we recall that (2.7) was solved in Exercise 5.4, Chapter 2. The solution was found to be

$$(2.9) \qquad E(x,t) = \tfrac{1}{2}H(t-|x|) = \tfrac{1}{2}H(t)[H(x+t)-H(x-t)].$$

On the other hand, the solution of (2.8) follows from (2.4) with $f(x)=\delta(x)$,

$$(2.10) \quad u(x,t) = \frac{1}{2}\int_{x-t}^{x+t}\delta(\xi)\,d\xi = \tfrac{1}{2}[H(x+t)-H(x-t)], \qquad t>0,$$

and (2.10) coincides with (2.9) for $t>0$. We note that the function $\frac{1}{2}[H(x+t)-H(x-t)]$ is a solution of the homogeneous wave equation for $-\infty<x,t<\infty$. This is true despite the discontinuities along the characteristics $x=t$ and $x=-t$. Function (2.9), however, has a source at $x=0, t=0$ created by piecing together two different solutions of the homogeneous equation: the zero solution for $t<0$, and (2.10) for $t>0$.

If the source is introduced at $x=\xi, t=\tau$, the corresponding causal fundamental solution is

(2.11) $$E(x,t;\xi,\tau)=\tfrac{1}{2}H(t-\tau-|x-\xi|).$$

Consider now the inhomogeneous equation with vanishing initial data

(2.12)

$$\frac{\partial^2 w}{\partial t^2}-\frac{\partial^2 w}{\partial x^2}=q(x,t),\quad -\infty<x<\infty, t>0;\quad w(x,0)=\frac{\partial w}{\partial t}(x,0)=0.$$

We can immediately write the solution in terms of the fundamental solution (2.11) as

$$w(x,t)=\int_0^t d\tau\int_{-\infty}^{\infty}d\xi\,E(x,t;\xi,\tau)q(\xi,\tau).$$

Since $E=\tfrac{1}{2}$ for $x-(t-\tau)<\xi<x+(t-\tau)$ and vanishes elsewhere, we have

$$w(x,t)=\frac{1}{2}\int_0^t d\tau\int_{x-(t-\tau)}^{x+(t-\tau)}q(\xi,\tau)d\xi=\frac{1}{2}\iint_D q(\xi,\tau)d\xi d\tau,$$

where D is the triangular domain in Figure 2.1.

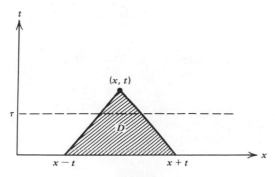

Figure 2.1

The solution of the general problem (2.1) is therefore

(2.13)

$$u(x,t) = \frac{1}{2} \iint_D q(\xi,\tau) d\xi d\tau + \tfrac{1}{2} f_1(x-t) + \tfrac{1}{2} f_1(x+t) + \frac{1}{2} \int_{x-t}^{x+t} f_2(\xi) d\xi.$$

It is clear that the solution depends continuously on the data $\{q, f_1, f_2\}$, although there are some related difficulties that should not be overlooked (see Exercise 2.10). Note that the solution at (x,t) depends only on the data in $\overline{D}$, which is called the *domain of dependence* of (x,t).

Let us now turn to the semi-infinite string. If the left end $x=0$ is kept fixed, the causal Green's function $g(x,t;\xi,\tau)$ satisfies

(2.14)

$$\frac{\partial^2 g}{\partial t^2} - \frac{\partial^2 g}{\partial x^2} = \delta(x-\xi)\delta(t-\tau), \quad 0<x,\xi<\infty, \; -\infty<t,\tau<\infty;$$

$$g|_{x=0} = 0; \quad g \equiv 0, \quad t > \tau.$$

It obviously suffices to find $g(x,t;\xi,0)$ and set $g(x,t;\xi,\tau) = g(x,t-\tau;\xi,0)$. Of course we cannot handle the space part in the same way because the geometry spoils the invariance under translations in space. To find $g(x,t;\xi,0)$ for $t>0$ we may choose to solve instead the IVP

$$\frac{\partial^2 g}{\partial t^2} - \frac{\partial^2 g}{\partial x^2} = 0, \quad 0<x<\infty, \, t>0;$$

(2.15)

$$g|_{x=0} = 0, \quad g|_{t=0} = 0, \quad \frac{\partial g}{\partial t}\bigg|_{t=0} = \delta(x-\xi).$$

Whether we deal with (2.14) or (2.15), the solution is most easily found by images as

(2.16) $$g(x,t;\xi,\tau) = \tfrac{1}{2} H(t-\tau-|x-\xi|) - \tfrac{1}{2} H(t-\tau-|x+\xi|).$$

The problem

$$\frac{\partial^2 u}{\partial t^2} - \frac{\partial^2 u}{\partial x^2} = q(x,t), \quad 0<x<\infty, \, t>0;$$

(2.17)

$$u(x,0) = f_1(x), \quad \frac{\partial u}{\partial t}(x,0) = f_2(x), \quad u(0,t)=0$$

has the solution

$$u(x,t) = \int_0^t d\tau \int_0^\infty d\xi\, g(x,t;\xi,\tau) q(\xi,\tau) + \int_0^\infty g(x,t;\xi,0) f_2(\xi)\, d\xi$$

$$+ \frac{d}{dt} \int_0^\infty g(x,t;\xi,0) f_1(\xi)\, d\xi.$$

Alternatively, we can extend $q(x,t), f_1(x), f_2(x)$ as *odd functions* for $x < 0$ and then use (2.13).

A more interesting feature of the semi-infinite string is the possibility of an inhomogeneous boundary condition at $x = 0$. Consider, for instance, the problem

(2.18)

$$\frac{\partial^2 v}{\partial t^2} - \frac{\partial^2 v}{\partial x^2} = 0, \quad 0 < x < \infty,\, t > 0; \qquad v(x,0) = \frac{\partial v}{\partial t}(x,0) = 0, \quad v(0,t) = h(t).$$

There are many interrelated methods for solving a problem of this type: reduction to a problem with a homogeneous boundary condition, Green's function, Duhamel's formula, Fourier sine transform on space, Laplace transform on time. We shall illustrate the last method and the method of characteristics.

First let us look at the method of characteristics. The solution for $x > t$ is determined from the Cauchy data on the noncharacteristic initial curve $t = 0, x > 0$, and therefore $v \equiv 0, x > t$. For $x < t$ we have a problem in which the value of v is given on the characteristic $x = t$ and on the noncharacteristic $x = 0$. Since the jump along a characteristic remains constant for this particular hyperbolic equation [see (5.54), Chapter 2], we have $v(x,x+) = h(0+)$. Therefore $v(x,t) = v(0,t-x) = h(t-x), x < t$. Thus the solution of (2.18) is

(2.19) $$v(x,t) = H(t-x) h(t-x).$$

Let us illustrate how to apply the Laplace transform to (2.18). We multiply both sides by e^{-st} and integrate from $t = 0$ to ∞. Integrating the first term by parts twice, and using the initial conditions and the fact that e^{-st} is exponentially small at $t = \infty$ for $\text{Re}\, s > 0$, we find that

(2.20) $$-\frac{\partial^2 \tilde{v}}{\partial x^2} + s^2 \tilde{v} = 0, \quad 0 < x < \infty; \qquad \tilde{v}(0,s) = \tilde{h}(s),$$

where

$$\tilde{v}(x,s) = \int_0^\infty e^{-st} v(x,t)\, dt, \qquad \tilde{h}(s) = \int_0^\infty e^{-st} h(t)\, dt$$

are the Laplace transforms of v and h, respectively.

If we solve (2.20) under the requirement that

$$\lim_{x \to \infty} \tilde{v}(x,s) = 0, \qquad \text{Re}\, s > 0,$$

we obtain

(2.21) $\tilde{v}(x,s) = \tilde{h}(s) e^{-sx},$

which we want to invert for fixed $x > 0$. The original of e^{-sx} is $\delta(t-x)$, so that by the convolution theorem for Laplace transforms [see (4.46), Chapter 2]

$$v(x,t) = \int_0^t \delta(t-\tau-x) h(\tau)\, d\tau = \begin{cases} h(t-x), & t > x, \\ 0, & t < x, \end{cases}$$

in agreement with (2.19). If this treatment using delta functions is deemed a bit cavalier, (2.21) can be written instead as

$$\tilde{v}(x,s) = \left[s\tilde{h} - h(0) \right] \frac{e^{-sx}}{s} + h(0) \frac{e^{-sx}}{s}.$$

The original of $s\tilde{h} - h(0)$ is $h'(t)$, and that of e^{-sx}/s is $H(t-x)$. Therefore

(2.22) $v(x,t) = \int_0^t H(t-\tau-x) h'(\tau)\, d\tau + h(0) H(t-x),$

which is easily reduced to (2.19). Note that (2.22) is just the version of Duhamel's formula (3.23), Chapter 7, appropriate for our problem since $H(t-x)$ is the solution of (2.18) with $h(t) \equiv 1$.

Parabolic Equations

The one-dimensional heat operator $\partial/\partial t - \partial^2/\partial x^2$ has the characteristics $t = \text{constant}$. Here these characteristics do not play the same significant role as they do for hyperbolic equations. A typical well-posed problem for heat conduction in an infinite rod is

(2.23) $\dfrac{\partial u}{\partial t} - \dfrac{\partial^2 u}{\partial x^2} = q(x,t), \quad -\infty < x < \infty,\ t > 0; \qquad u(x,0) = f(x),$

where only the initial value for u is required since the equation is first order in time. The causal Green's function $g(x,t;0,0)$ satisfies

$$(2.24) \qquad \frac{\partial g}{\partial t} - \frac{\partial^2 g}{\partial x^2} = \delta(x)\delta(t), \quad -\infty < x, t < \infty; \qquad g \equiv 0, \quad t < 0,$$

or, equivalently (for $t > 0$),

$$(2.25) \qquad \frac{\partial g}{\partial t} - \frac{\partial^2 g}{\partial x^2} = 0, \quad -\infty < x < \infty, t > 0; \qquad g|_{t=0} = \delta(x).$$

In the first of these characterizations we view g as a distribution on $C_0^\infty(R_2)$ with the properties

$$\phi(0,0) = \langle g, L^*\phi \rangle = \left\langle g, -\frac{\partial\phi}{\partial t} - \frac{\partial^2\phi}{\partial x^2} \right\rangle \qquad \text{for each } \phi(x,t) \in C_0^\infty(R_2),$$

$$\langle g, \phi \rangle = 0 \qquad \text{for each } \phi \in C_0^\infty(R_2) \text{ with support in } t < 0.$$

In (2.25) we regard g as a one-dimensional distribution with t appearing as a parameter. Thus

$$\frac{\partial}{\partial t} \langle g, \phi \rangle \quad \left\langle g, \frac{\partial^2\phi}{\partial x^2} \right\rangle = 0 \qquad \text{for } t > 0 \text{ and each } \psi(x) \subset C_0^\infty(R_1),$$

$$\lim_{t \to 0+} \langle g, \phi \rangle = \phi(0) \qquad \text{for each } \phi \in C_0^\infty(R_1).$$

We saw in (5.51), Chapter 2, that

$$(2.26) \qquad g(x,t;0,0) = H(t)\frac{e^{-x^2/4t}}{\sqrt{4\pi t}},$$

so that

$$(2.27) \qquad g(x,t;\xi,\tau) = H(t-\tau)\frac{e^{-(x-\xi)^2/4(t-\tau)}}{\sqrt{4\pi(t-\tau)}},$$

and the solution of (2.23) is

(2.28)

$$u(x,t) = \int_0^t d\tau \int_{-\infty}^\infty d\xi \frac{e^{-(x-\xi)^2/4(t-\tau)}}{\sqrt{4\pi(t-\tau)}} q(\xi,\tau) + \int_{-\infty}^\infty d\xi \frac{e^{-(x-\xi)^2/4t}}{\sqrt{4\pi t}} f(\xi).$$

It is clear that u depends continuously on the data.

In an IVP such as (2.23) we have not yet specified precisely how the initial values are to be taken on by the solution. A natural requirement would be

$$(2.29) \qquad \lim_{t \to 0+} u(x,t) = f(x) \qquad \text{for each } x, \quad -\infty < x < \infty,$$

but as we shall see the use of a pointwise limit is unsatisfactory both mathematically and physically to obtain uniqueness. To see what goes wrong let us look at the function

$$(2.30) \qquad h(x,t) = \frac{x}{4\pi^{1/2}t^{3/2}} e^{-x^2/4t}, \qquad t > 0,$$

which is just $-(\partial/\partial x)g(x,t;0,0)$. Since $g(x,t;0,0)$ satisfies the homogeneous heat equation for $t > 0$, so does any partial derivative of g (because the coefficients in the differential equation are constant). The function $h(x,t)$ therefore satisfies $\partial h/\partial t - \partial^2 h/\partial x^2 = 0$, $t > 0$; moreover,

$$\lim_{t \to 0+} h(x,t) = 0 \qquad \text{for each } x, \quad \infty < x < \infty.$$

Of course we expect the trivial solution to be the only solution of the homogeneous heat equation with vanishing initial data. The paradox is only apparent, for h really does not correspond to vanishing initial data. Let us see what $\lim_{t \to 0+} h(x,t)$ is in the distributional sense. We think of h as a distribution in x with parameter t so that

$$\langle h, \phi(x) \rangle = \int_{-\infty}^{\infty} h(x,t)\phi(x)\,dx, \qquad t > 0.$$

Hence

$$\langle h, \phi \rangle = \int_{-\infty}^{\infty} -\frac{\partial g}{\partial x}\phi\,dx = \int_{-\infty}^{\infty} g\frac{d\phi}{dx}\,dx = \int_{-\infty}^{\infty} \frac{e^{-x^2/4t}}{\sqrt{4\pi t}}\frac{d\phi}{dx}\,dx$$

and

$$\lim_{t \to 0+} \langle h, \phi \rangle = \phi'(0),$$

which is not the zero distribution. In fact, $h(x,t)$ is the temperature corresponding to an initial unit dipole at $x = 0$.

Requirement (2.29) is not sufficient for a proper formulation of the IVP. There are two ways of remedying the situation. The first works adequately if $f(x)$ and $q(x,t)$ are continuous functions. We then look for a solution

$u(x,t)$ of (2.23) for $t>0$ which, together with the initial values $f(x)$, defines a continuous function in $t \geqslant 0$. Note that the function $h(x,t)$ for $t>0$, together with the value $f=0$ for $t=0$, is not a continuous function in $t \geqslant 0$. In fact, if we approach the origin along the curve $x=2\sqrt{t}$, h tends to $+\infty$ (and along $x=-2\sqrt{t}$ to $-\infty$).

Another formulation is more suitable when f or g may be discontinuous (or perhaps distributions). Then we require that the initial values be taken in the distributional sense:

$$(2.31) \qquad \lim_{t \to 0+} \langle u, \phi \rangle = \langle f, \phi \rangle \qquad \text{for each } \phi(x) \in C_0^\infty(R_1).$$

The first formulation tells us only that $h(x,t)$ in (2.30) is not a solution which assumes zero initial values. The second formulation gives much more information: the initial value for $h(x,t)$ is $\delta'(x)$.

Let us turn next to some heat conduction problems for finite rods. Consider first a thin ring constructed by bending a rod of unit length so that its ends meet perfectly. If x measures position along the centerline of the ring, we may assume that the ring extends from $x=-\frac{1}{2}$ to $x=\frac{1}{2}$, these two values of x corresponding to the same *physical* point in the ring. If the ring is insulated, the flow of heat is essentially one-dimensional along the centerline of the ring. The novel feature is that the boundary conditions state that the temperature and any derivative of the temperature at $x=\frac{1}{2}$ must be equal to the corresponding quantity at $x=-\frac{1}{2}$. It actually suffices to require that the temperature and its first derivative have, respectively, the same value at $x=\pm\frac{1}{2}$; the relations for the higher derivatives then follow from successive differentiations of the differential equation.

The causal fundamental solution $E(x,t;\xi,\tau)$ for this ring is the temperature when a unit of heat is suddenly introduced at time τ at the poing ξ when the ring was at 0 temperature up to time τ. It is clear that E depends only on $x-\xi$ and $t-\tau$, so that we need only to calculate $E(x,t) \doteq E(x,t;0,0)$ satisfying

$$(2.32) \qquad \frac{\partial E}{\partial t} - \frac{\partial^2 E}{\partial x^2} = \delta(x)\delta(t), \qquad -\tfrac{1}{2}<x<\tfrac{1}{2}, \quad -\infty<t<\infty;$$

$$E \equiv 0, \quad t<0; \quad E\left(-\tfrac{1}{2},t\right)=E\left(\tfrac{1}{2},t\right), \quad \frac{\partial E}{\partial x}\left(-\tfrac{1}{2},t\right)=\frac{\partial E}{\partial x}\left(\tfrac{1}{2},t\right).$$

The solution is most easily obtained by images. Consider a fictitious infinite straight rod with positive unit sources at all integers. The resulting temperature is clearly periodic in x with period 1, so that the boundary conditions in (2.32) are satisfied; moreover, the part of the rod between $-\frac{1}{2}$ and $\frac{1}{2}$ contains only the single source at $x=0$. If we believe that the solution of (2.32) is unique, it must coincide with the temperature in the infinite rod between $-\frac{1}{2}$ and $\frac{1}{2}$. The temperature in the infinite rod is the *theta function*

$$(2.33) \qquad \theta(x,t) \doteq \frac{1}{\sqrt{4\pi t}} \sum_{n=-\infty}^{\infty} e^{-(x-n)^2/4t}, \qquad t>0,$$

and therefore $E(x,t)=\theta(x,t)$, $-\frac{1}{2}<x<\frac{1}{2}, t>0$. Formula (2.33) is useful for small t, and only the term corresponding to $n=0$ contributes appreciably to the temperature in the ring. Physically the interpretation is that the finiteness of the ring is not felt for small t, and it is as if we were dealing with a source in an infinite rod.

Another way of determining E is by an expansion in a series of spatial eigenfunctions $e^{i2n\pi x}$. We multiply (2.32) by $e^{-i2n\pi x}$ and integrate from $x=-\frac{1}{2}$ to $x=\frac{1}{2}$ to obtain

$$(2.34) \qquad \frac{\partial E_n}{\partial t} + 4n^2\pi^2 E_n = \delta(t); \qquad E_n = 0, \quad t<0,$$

where

$$E(x,t) = \sum_{n=-\infty}^{\infty} E_n(t) e^{i2n\pi x}, \qquad E_n(t) = \int_{-1/2}^{1/2} E(x,t) e^{-i2n\pi x} \, dx.$$

From (2.34) we find that $E_n = e^{-4n^2\pi^2 t}$ and

$$(2.35) \qquad E = \sum_{n=-\infty}^{\infty} e^{-4n^2\pi^2 t} e^{i2n\pi x} = 1 + 2 \sum_{n=1}^{\infty} e^{-4n^2\pi^2 t} \cos 2n\pi x.$$

The cosine series in (2.35) converges rapidly for large t since the ratio of two successive terms is exponentially small. It is possible to pass from (2.33) to (2.35) directly by using the Poisson summation formula (3.26), Chapter 2.

We shall now use the results for the ring to deal with the finite rod $0<x<l$ with vanishing end temperatures. The corresponding causal Green's function $g(x,t;\xi,0)$ satisfies

$$(2.36) \qquad \begin{aligned} &\frac{\partial g}{\partial t} - \frac{\partial^2 g}{\partial x^2} = \delta(x-\xi)\delta(t), \qquad 0<x,\xi<l, \quad \infty<t<\infty; \\ &g \equiv 0, \quad t<0; \qquad g|_{x=0} = g|_{x=l} = 0, \quad t>0. \end{aligned}$$

Consider an infinite rod with the array of unit sources and sinks shown in

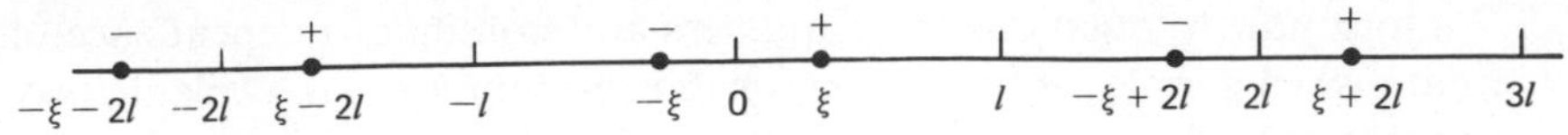

Figure 2.2

Figure 2.2. Thus we place a unit source at each of the points $\xi+2nl$ and a unit sink at each of the points $-\xi+2nl$, where n is an arbitrary integer ranging from $-\infty$ to ∞. The resulting temperature is an odd function about $x=0$ and about $x=l$, so that the temperatures at these points must vanish for $t>0$. Since there is only a unit source in the part of the rod between $x=0$ and $x=l$, the temperature for the infinite rod will satisfy in $0<x<l$ the BVP (2.36). We have explicitly

$$g(x,t;\xi,0)=\sum_{n=-\infty}^{\infty}\frac{1}{\sqrt{4\pi t}}\left[e^{-(x-\xi-2nl)^2/4t}-e^{-(x+\xi-2nl)^2/4t}\right]$$

$$=\frac{1}{2l}\left[\theta\left(\frac{x-\xi}{2l},\frac{t}{4l^2}\right)-\theta\left(\frac{x+\xi}{2l},\frac{t}{4l^2}\right)\right],$$

where θ is the theta function (2.33).

Either by using relation (2.35) for theta functions or by expanding (2.36) directly in the space eigenfunctions $\sin(n\pi x/l)$, we find the alternative expression

$$g(x,t;\xi,0)=\frac{2}{l}\sum_{n=1}^{\infty}\sin\frac{n\pi x}{l}\sin\frac{n\pi\xi}{l}e^{-n^2\pi^2 t/l^2}.$$

Using Green's function $g(x,t;\xi,\tau)=g(x,t-\tau;\xi,0)$, it is easy to solve the initial-boundary value problem

$$\frac{\partial u}{\partial t}-\frac{\partial^2 u}{\partial x^2}=q(x,t),\quad 0<x<l,\,t>0;\qquad u(x,0)=f(x),\quad u(0,t)=u(l,t)=0,$$

in the form

$$(2.37)\quad u(x,t)=\int_0^t d\tau\int_0^l d\xi\, g(x,t;\xi,\tau)q(\xi,\tau)+\int_0^l g(x,t;\xi,0)f(\xi)\,d\xi.$$

Continuous dependence on the data is an immediate consequence of (2.37). For the problem where the boundary values depend on t, one can use either a Green's function method or one of the many techniques mentioned in the discussion of the wave equation. Again continuity with respect to data follows from the explicit expression for the solution.

We turn now to questions of uniqueness and continuous dependence on the data when simple explicit forms for the solution are not available (say, for problems with more than one space dimension). A typical well-posed problem for the heat equation is a combined initial value and boundary value problem. Let Ω_x be a *bounded* domain in the space variables $x = (x_1, \ldots, x_n)$, and let its boundary be Γ_x. We shall want to solve the heat equation with given source density $q(x, t)$ in the cylindrical space-time domain Ω, the Cartesian product of Ω_x and the segment of the time axis $0 < t < T$. (See Figure 2.3.) Here T is a large value of time used for convenience instead of $t = \infty$. The boundary Γ of Ω consists of three portions: the bases $(x \in \Omega_x, t = 0)$ and $(x \in \Omega_x, t = T)$, and the lateral surface $(x \in \Gamma_x, 0 < t < T)$. Initial data is given on the lower base, and boundary data on the lateral surface. We shall denote the union of the lower base and the lateral surface by Γ_d. We assume that the data is continuous on Γ_d. Our BVP can be formulated as follows: to find $u(x, t)$ continuous in $\overline{\Omega}$, satisfying

(2.38)
$$\frac{\partial u}{\partial t} - \Delta u = q(x, t), \qquad (x, t) \in \Omega;$$

$$u(x, 0) = f(x), \quad x \in \Omega_x; \qquad u(x, t) = h(x, t), \quad x \in \Gamma_x, \quad 0 < t < T$$

(these two conditions mean that u is given on Γ_d).

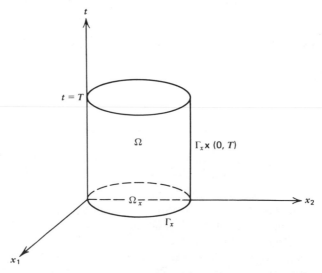

Figure 2.3

As always, the question of existence is delicate, and we shall concentrate on uniqueness and continuous dependence on the data. If (2.38) has two solutions u_1 and u_2, their difference must satisfy the completely homogeneous problem

$$(2.39) \qquad \frac{\partial u}{\partial t} - \Delta u = 0, \quad (x,t) \in \Omega; \qquad u = 0 \text{ on } \Gamma_d.$$

We now show that the only solution of (2.39) is the zero solution, so that (2.38) has at most one solution (uniqueness). Multiply the differential equation in (2.39) by $\bar{u}(x,t)$, and integrate over the space domain Ω_x to obtain

$$0 = \int_{\Omega_x} \left(\bar{u} \frac{\partial u}{\partial t} - \bar{u} \Delta u \right) dx = \frac{1}{2} \frac{d}{dt} \int_{\Omega_x} |u|^2 dx + \int_{\Omega_x} |\text{grad } u|^2 dx - \int_{\Gamma_x} \bar{u} \frac{\partial u}{\partial \nu} dS.$$

Since u vanishes on Γ_x for $t < T$, we find that

$$(2.40) \qquad 0 = \frac{1}{2} \frac{d}{dt} \int_{\Omega_x} |u|^2 dx + \int_{\Omega_x} \text{grad} |u|^2 dx \geqslant \frac{1}{2} \frac{d}{dt} \int_{\Omega_x} |u|^2 dx.$$

The initial condition implies that $\int_{\Omega_x} |u|^2 dx = 0$ at $t = 0$, and therefore (2.40) shows that $\int_{\Omega_x} |u|^2 dx = 0$ for all $t < T$. The continuity of u then yields $u(x,t) \equiv 0$, $(x,t) \in \Omega$.

A more powerful method of proving uniqueness (and also continuous dependence on data) utilizes the *maximum principle* for solutions of the homogeneous heat equation $(\partial u / \partial t) - \Delta u = 0$. Since any solution of this homogeneous equation also obeys a *minimum principle* of equal importance, it is somewhat misleading to give special status to the maximum principle, but we bow to tradition. When we deal with differential inequalities, however, the distinction between principles is essential.

Theorem 1 (Maximum principle). Let $u(x,t)$ be continuous in $\bar{\Omega}$ and satisfy the differential inequality

$$(2.41) \qquad \frac{\partial u}{\partial t} - \Delta u \leqslant 0, \qquad (x,t) \in \Omega.$$

If $u \leqslant M$ on Γ_d, then $u(x,t) \leqslant M$ on $\bar{\Omega}$.

The physical interpretation is simple. Equation (2.41) tells us that no heat is being added in the interior. It therefore follows that the interior temperature must be less than the maximum of the combined initial temperature and boundary temperature.

Corollary (Minimum principle). Suppose u is continuous in $\bar{\Omega}$ and

$$\frac{\partial u}{\partial t} - \Delta u \geqslant 0, \qquad (x,t) \in \Omega.$$

If $u \geqslant m$ on Γ_d, then $u(x,t) \geqslant m$ in $\bar{\Omega}$.

The proof of the corollary follows by setting $u = -v$ and applying Theorem 1. To prove Theorem 1 we consider first the simpler case where there is a sink at every point in Ω [strict inequality in (2.41)].

Lemma. Let v be continuous in $\bar{\Omega}$ and satisfy

(2.42) $$\frac{\partial v}{\partial t} - \Delta v < 0 \text{ in } \Omega;$$

then the maximum of v occurs on Γ_d.

Proof of Lemma. A maximum for v at an interior point (x,t) would imply that $\Delta v \leqslant 0$ and $\partial v / \partial t = 0$ at that point, which contradicts (2.42). If the maximum occurred at a point in Ω_x at time T, we would have $\Delta v \leqslant 0$ and $\partial v / \partial t \geqslant 0$, which again contradicts (2.42).

Proof of Theorem 1. Since Ω_x is bounded, it can be enclosed in an n-ball of radius a with center at the origin. Let $u(x,t)$ satisfy the hypothesis in Theorem 1, and define $v(x,t) = u(x,t) + \varepsilon |x|^2$. Then

$$\frac{\partial v}{\partial t} - \Delta v \leqslant -2n\varepsilon < 0,$$

so that v satisfies the inequality of the lemma. Thus

$$u(x,t) \leqslant v(x,t) \leqslant \max_{(x,t) \in \Gamma_d} v \leqslant \left[\varepsilon a^2 + \max_{(x,t) \in \Gamma_d} u \right] \leqslant \varepsilon a^2 + M,$$

and, by letting $\varepsilon \to 0$, we find that $u \leqslant M$.

For the *homogeneous* heat equation both Theorem 1 and its corollary apply, so that we have the following theorem.

Theorem 2. Let $u(x,t)$ be a continuous function on $\bar{\Omega}$ satisfying $(\partial u / \partial t) - \Delta u = 0$. If, on Γ_d, we have $m \leqslant u \leqslant M$, then, everywhere in Ω, $m \leqslant u(x,t) \leqslant M$.

Remark. Our proof does not exclude the possibility that at an interior point the temperature may equal the maximum of the initial and the boundary temperature. If, for instance, $f(x)=1$ and $h(x,t)=1$, $0<t<t_0$, $h(x,t)<1$, $t>t_0$, then $u(x,t)\equiv 1$ for $t<t_0$ but starts decreasing in the interior thereafter. The only way we can have an interior maximum or minimum is by having constant initial and boundary temperature up to the time at which the interior extremum occurs.

There are many important consequences of the maximum and minimum principles. First let us consider the inhomogeneous problem (2.38) for two different sets of data, $\{q_1,f_1,h_1\}$ and $\{q_2,f_2,h_2\}$. Assume that the second set *dominates* the first, that is, $q_2\geqslant q_1, f_2\geqslant f_1, h_2\geqslant h_1$. Physically this means that the sources and boundary data in the second set are larger than those in the first set, so that the solution should also be larger.

Theorem 3 (Comparison theorem). If the data $\{q_2,f_2,h_2\}$ dominates $\{q_1,f_1,h_1\}$, then $u_2(x,t)\geqslant u_1(x,t)$.

Proof. The function $v=u_1-u_2$ satisfies $(\partial v/\partial t)-\Delta v\leqslant 0$ in Ω, $v\leqslant 0$ on Γ_d. By Theorem 1, $v\leqslant 0$ in $\overline{\Omega}$.

Theorem 4 (Uniqueness). Problem (2.38) has at most one solution.

Proof. Let u_1 and u_2 satisfy (2.38) for the *same* data. Then $\{q_2,f_2,h_2\}$ dominates $\{q_1,f_1,h_1\}$ and vice versa. Therefore, by Theorem 3, $u_1\leqslant u_2$ and $u_2\leqslant u_1$.

Theorem 5 (Continuous dependence on data). Let u_1 and u_2 be solutions of (2.38) corresponding, respectively, to data $\{q_1,f_1,h_1\}$ and $\{q_2,f_2,h_2\}$. Let

$$|q_1-q_2|<\alpha, \quad (x,t)\in\overline{\Omega},$$
$$|u_1-u_2|<\beta, \quad (x,t)\in\Gamma_d;$$

then

$$|u_1(x,t)-u_2(x,t)|<\alpha T+\beta, \quad (x,t)\in\overline{\Omega}.$$

Proof. The function $u\doteq u_1-u_2$ satisfies the heat equation with data lying between $\{-\alpha,-\beta,-\beta\}$ and $\{\alpha,\beta,\beta\}$. Write $u=v+w$, where v has a vanishing inhomogeneous term and w has vanishing initial and boundary data. Then, by Theorem 2, $|v| < \beta$. The functions $w_1 = \alpha t, w_2 = -\alpha t$ satisfy the heat equation with inhomogeneous terms $+\alpha$ and $-\alpha$, respectively; on Γ_d, $w_1 \geqslant 0$ and $w_2 \leqslant 0$. Thus the data for w_1 dominates that for w; and the data for w dominates that for w_2. From Theorem 3 it follows that $w_2 \leqslant w \leqslant w_1$ and hence $-\alpha T \leqslant w \leqslant \alpha T$. Together with $|v| < \beta$, this gives the required result.

It is possible to extend these results with some modifications to unbounded domains and to more general parabolic equations (see Exercise 2.2, for instance).

Exercises

2.1. The concentration $C(x,t)$ of a pollutant subject to diffusion and convection along the x axis satisfies

$$(2.43) \qquad \frac{\partial C}{\partial t} + U(t)\frac{\partial C}{\partial x} - D\frac{\partial^2 C}{\partial x^2} = 0,$$

where D is the constant diffusivity and $U(t)$ is the wind velocity in the positive x direction. Both D and U are regarded as known.

Make a transformation to an appropriate moving coordinate system $t' = t$, $x' = x - V(t)$ to reduce (2.43) to

$$\frac{\partial C}{\partial t'} - D\frac{\partial^2 C}{\partial x'^2} = 0.$$

Find the free space causal fundamental solution for (2.43) with source at $x = 0$, $t = 0$. The curves $x - V(t) = $ constant are called *drift curves*. Let Γ be the drift curve through the origin. Solve the initial-boundary value problem for (2.43) with $C(x,0)$ given for $x > 0$ and C given on Γ for $t > 0$.

2.2. Let Ω be defined as in Figure 2.3. Suppose u is continuous in $\overline{\Omega}$ and satisfies

$$\frac{\partial u}{\partial t} - \Delta u + cu \leqslant 0, \quad (x,t) \in \Omega; \quad c(x) \geqslant 0.$$

Show that if, on Γ_d, $u \leqslant M$, where $M \geqslant 0$, then $u \leqslant M$ in $\overline{\Omega}$. What is the corresponding principle when the sign is reversed in the differential inequality? What is the appropriate principle if u satisfies

$$\frac{\partial u}{\partial t} - \Delta u + cu = 0?$$

2.3. For c *constant*, solve by separation of variables

$$\frac{\partial u}{\partial t} - \Delta u + cu = 0, \quad (x,t) \in \Omega; \quad u(x,0) = f(x); \quad u|_{\Gamma_x} = 0, \quad t > 0.$$

When can you guarantee that the solution tends to the steady state $u \equiv 0$ as $t \to \infty$?

2.4. Consider the wave equation on the domain of Figure 2.3:

(2.44)
$$\begin{cases} \dfrac{\partial^2 u}{\partial t^2} - \Delta u = q(x,t), & (x,t) \in \Omega, \\[2mm] u(x,0) = f(x), \quad \dfrac{\partial u}{\partial t}(x,0) = g(x), & x \in \Omega_x, \\[2mm] u(x,t) = h(x,t), & x \in \Gamma_x, \quad 0 < t < T. \end{cases}$$

To prove uniqueness, consider the completely homogeneous problem satisfied by the difference v of two solutions u_1 and u_2 of (2.44). Multiply the homogeneous differential equation by $\bar{v}$, and integrate over Ω_x to show that $v \equiv 0$ in Ω.

2.5. Use the Poisson summation formula of Chapter 2 to show the equality of (2.33) and (2.35).

2.6. Consider the heat conduction problem for a semi-infinite rod:

(2.45)
$$\frac{\partial w}{\partial t} - \frac{\partial^2 w}{\partial x^2} = 0, \quad 0 < x < \infty, \quad t > 0;$$

$$w(x,0) = 0, \quad w(0,t) = 1.$$

[We shall see in Exercise (2.7) that $w(x,t)$ enables us to solve the problem in which an arbitrary temperature is imposed on the left-hand end.] We now solve (2.45) in three ways:
(a) Write $w = 1 - z(x,t)$, and find $z(x,t)$ by using Green's function (obtained by images) for the semi-infinite rod with vanishing temperature at $x = 0$.
(b) Use a Laplace transform on the time variable to show that $\tilde{w}(x,s) = e^{-x\sqrt{s}}/s$, and invert.
(c) Use the method of similarity solution: show first that the equation and boundary conditions in (2.45) are invariant under the transformation $x^* = \alpha x$, $t^* = \alpha^2 t$, so that $w(\alpha x, \alpha^2 t) = w(x,t)$ for all

$\alpha > 0$. With x, t fixed, choose $\alpha = \frac{1}{2}\sqrt{t}$ to show that $w(x,t) = w(x/2\sqrt{t}, \frac{1}{4})$. Thus all points in the x-t plane with the same value of $x/2\sqrt{t}$ have the same temperature; hence

$$w(x,t) = h\left(\frac{x}{2\sqrt{t}}\right).$$

Substitution in (2.45) shows that $h(\xi)$ satisfies the ordinary differential equation

$$h'' + 2\xi h' = 0$$

with the boundary conditions $h(0) = 1$, $h(\infty) = 0$, whose solution is

$$h = 1 - \text{erf}\,\xi, \qquad \text{erf}\,\xi \doteq \frac{2}{\sqrt{\pi}} \int_0^\xi e^{-\eta^2}\,d\eta.$$

This proves that

(2.46) $$w(x,t) = 1 - \text{erf}\,\frac{x}{2\sqrt{t}},$$

as was shown by still another method in (3.22), Chapter 7.

2.7. Consider the heat conduction problem

$$\frac{\partial u}{\partial t} - \frac{\partial^2 u}{\partial x^2} = 0, \quad 0 < x < \infty, \quad t > 0; \qquad u(x,0) = 0, \quad u(0,t) = h(t).$$

Take a Laplace transform on time, and use the convolution theorem and Exercise 2.6 to show that

$$u(x,t) = \int_0^t w(x,t-\tau)h'(\tau)\,d\tau + h(0)w(x,t),$$

which can be reduced to

$$u(x,t) = -\int_0^t \frac{\partial w}{\partial \tau}(x,t-\tau)h(\tau)\,d\tau.$$

2.8. (a) Find a similarity solution (see Exercise 2.6) to the problem

$$\frac{\partial u}{\partial t} - \frac{\partial^2 u}{\partial x^2} = 0, \quad -\infty < x < \infty, \, t > 0; \qquad u(x,0) = H(x).$$

(b) Find Green's function $g(x,t;0,0)$ for an infinite rod by noting that its initial value is dH/dx.

2.9. Find a similarity solution to the nonlinear heat conduction problem

$$\frac{\partial u}{\partial t} - \frac{\partial}{\partial x}\left(k(u)\frac{\partial u}{\partial x}\right)=0, \quad 0<x<\infty, \, t>0; \qquad u(x,0)=0, \quad u(0,t)=1.$$

2.10. Let us examine (2.13) on the region

$$R_T: -\infty<x<\infty, \quad 0\leqslant t\leqslant T,$$

when $q=0$ and $f_1(x)$, $f_2(x)$ are continuous. Using the uniform norm on $-\infty<x<\infty$ for f_1 and f_2, and defining

$$\|u\|_{\infty,T}=\sup_{(x,t)\in R_T}|u(x,t)|,$$

we easily see that $\|u\|_{\infty,T}\to 0$ when both $\|f_1\|_\infty$ and $\|f_2\|_\infty$ tend to 0, so that we have continuous dependence on the data.

One may, however, rightly dispute the choice of norm for the solution u. It is perhaps more reasonable to regard the pair $(u, \partial u/\partial t)$ as characterizing the state of the string since that is the information needed at a particular time to calculate the solution for later times. This leads us to define a new norm,

$$\|u\|_{\infty,T}^1=\sup_{(x,t)\in R_T}\left\{|u(x,t)|+\left|\frac{\partial u}{\partial t}(x,t)\right|\right\}.$$

Show that, even if T is small, $\|u\|_{\infty,T}^1$ can be very large for small initial data.

3. ELLIPTIC EQUATIONS

Since elliptic equations have no characteristics, it might seem at first that they are ideally suited to the Cauchy problem. We saw, however, in Exercise 1.1 that this is *not* the case. In fact, one obtains existence and uniqueness only in the small and in situations when everything in sight is analytic; moreover, there is no continuous dependence on the data. It turns out that the pure boundary value problem is well posed for elliptic equations (a certain amount of preliminary grooming may be necessary). For instance, if Ω is a bounded domain in R_n, the BVP

$$(3.1) \qquad\qquad -\Delta u=q(x), \quad x\in\Omega; \qquad u=f, \quad x\in\Gamma,$$

is well posed if appropriate conditions (much less severe than analyticity) are placed on the nature of the boundary and the data. At the simplest level we would make the following assumptions: f continuous on Γ, q continuous in $\bar{\Omega}$, the desired solution u continuous in $\bar{\Omega}$. Even then some less than obvious assumptions must be made on the nature of Ω (or Γ). To see the difficulty, let Ω be the punctured disk $0<|x|<1$; then Γ consists of the unit circle and of the isolated point $x=0$ ($\bar{\Omega}$ is the closed unit disk). The problem $\Delta u=0$ in Ω with $u=0$ on the unit circle and $u=1$ at $x=0$ has no solution.

The Dirichlet problem for Laplace's equation in three dimensions may have no solution when Γ has very sharp spikes. An extreme example of this kind occurs when Ω consists of the open unit ball from which a radial line has been removed (in two dimensions this problem causes no difficulty). The physical interpretation is that, when Γ is viewed as a charged conductor, the spikes may enable the charge to leak off.

Let us look at a more typical problem where no difficulties arise.

Interior Dirichlet Problem for the Unit Disk

A function which satisfies Laplace's equation $\Delta u=0$ in a domain Ω is said to be *harmonic* in Ω. Let Ω be the unit disk; we wish to construct a function u harmonic in Ω and continuous in $\bar{\Omega}$, which takes on prescribed continuous boundary values on the unit circle Γ.

For purposes of calculations it will be advantageous to introduce in Ω polar coordinates (r,ϕ) with $0\leqslant r<1$, $-\pi\leqslant\phi\leqslant\pi$. Before proceeding, we should pause to reflect on the implications of using this coordinate system. The coordinate endpoints $r=0,\phi=\pi,\phi=-\pi$ do *not* correspond to any physical boundaries; $r=0$ is the center of the disk, whereas $\phi=\pi$ and $\phi=-\pi$ both represent the same radial line, a line which has no special distinction in the original formulation of the problem and which should ultimately be returned to the anonymity it deserves. We must therefore make sure that any proposed solution $u(r,\phi)$ behaves properly at $r=0$ and on the radial line whose equation is both $\phi=\pi$ and $\phi=-\pi$. With this warning we now express the Laplacian in polar coordinates, so that the problem has the form

$$(3.2)\qquad 0=\Delta u=\frac{1}{r}\frac{\partial}{\partial r}\left(r\frac{\partial u}{\partial r}\right)+\frac{1}{r^2}\frac{\partial^2 u}{\partial\phi^2},\quad (r,\phi)\in\Omega;\quad u(1,\phi)=f(\phi),$$

where $f(\phi)$ is continuous, $-\pi\leqslant\phi\leqslant\pi$, and $f(-\pi)=f(\pi)$.

The simplest way of solving (3.2) is by separation of variables. We shall construct u from basic building blocks u^*, which are harmonic functions of the special form $R(r)\Phi(\phi)$. Substitution in $\Delta u=0$ yields, after division by

$R\Phi$,

$$\frac{r}{R}\frac{d}{dr}\left(r\frac{dR}{dr}\right) = -\frac{1}{\Phi}\frac{d^2\Phi}{d\phi^2}.$$

Each of the two sides depends on a single coordinate. Since these coordinates are independent, the two sides must be equal to the same constant, labeled λ and known as a *separation constant*. We are thus led to the two ordinary differential equations

(3.3)
$$-\frac{d^2\Phi}{d\phi^2} = \lambda\Phi, \qquad -\pi < \phi < \pi,$$

(3.4)
$$r\frac{d}{dr}\left(r\frac{dR}{dr}\right) = \lambda R, \qquad 0 < r < 1.$$

For $u^* = R\Phi$ to be harmonic in Ω, its second derivatives must exist, so that both u^* and grad u^* certainly need to be continuous. To ensure the continuity of u^* and grad u^* on the radial line whose equation is both $\phi = \pi$ and $\phi = -\pi$, we must associate with (3.3) the so-called *periodic boundary conditions*

(3.5)
$$\Phi(-\pi) = \Phi(\pi), \qquad \Phi'(-\pi) = \Phi'(\pi).$$

For the radial equation (3.4) we will require $R(0)$ finite (or, equivalently, $\lim_{r \to 0} rR' = 0$, which tells us that there is no source at $r = 0$). Equation (3.3) subject to (3.5) is an eigenvalue problem with eigenvalues $n^2, n = 0, 1, 2, \ldots$, and eigenfunctions $\Phi_n = C\sin n\phi + D\cos n\phi$ (or $Ae^{in\phi} + Be^{-in\phi}$) for $n > 0$ and $\Phi_0 = A$. For our purposes it is slightly more convenient to let n also assume negative values and to regard $e^{in\phi}$ as the single eigenfunction corresponding to n.

Setting $\lambda = n^2$, we obtain from (3.4)

$$R_n(r) = Cr^n + Dr^{-n}, \qquad n \neq 0,$$
$$R_0(r) = E + F\log r,$$

where we set $F = 0$ and either D or $C = 0$ according to whether $n > 0$ or $n < 0$. This guarantees that R is bounded near the origin. In all cases we can write

$$R_n(r) = Cr^{|n|}.$$

For any integer n the function $u_n^*(r, \phi) = r^{|n|}e^{in\phi}$ is harmonic in the whole plane, and so is any finite linear combination of such functions. To satisfy the boundary condition at $r=1$ it will be necessary, however, to consider an infinite linear combination

$$(3.6) \qquad\qquad u(r,\phi) = \sum_{n=-\infty}^{\infty} a_n r^{|n|} e^{in\phi}, \qquad r < 1,$$

which must reduce to $f(\phi)$ at $r=1$, so that

$$(3.7) \quad f(\phi) = \sum_{n=-\infty}^{\infty} a_n e^{in\phi} \quad \text{and} \quad a_n = \frac{1}{2\pi} \int_{-\pi}^{\pi} f(\phi) e^{-in\phi} d\phi.$$

With these values for a_n substituted in (3.6) we presumably have the solution of (3.2). Rather than verifying this directly, we first transform (3.6) into an expression which exhibits more clearly the dependence of u on the boundary data. From (3.6) and the expression for a_n, we have

$$u(r,\phi) = \int_{-\pi}^{\pi} \left[\frac{1}{2\pi} \sum_{n=-\infty}^{\infty} r^{|n|} e^{in(\phi-\psi)} \right] f(\psi) \, d\psi, \qquad r < 1.$$

Now, for any complex z with $|z| < 1$ and for any real a,

$$\sum_{n=-\infty}^{\infty} z^{|n|} e^{ina} = 1 + \sum_{n=1}^{\infty} z^n e^{ina} + \sum_{n=1}^{\infty} z^n e^{-ina}$$

$$= 1 + \frac{ze^{ia}}{1 - ze^{ia}} + \frac{ze^{-ia}}{1 - ze^{-ia}} = \frac{1-z^2}{1+z^2-2z\cos a},$$

so that we obtain

$$(3.8) \quad u(r,\phi) = \frac{1}{2\pi} \int_{-\pi}^{\pi} \frac{1-r^2}{1+r^2-2r\cos(\phi-\psi)} f(\psi) \, d\psi, \qquad r < 1.$$

The function

$$(3.9) \qquad\qquad k(r,\phi) = \frac{1}{2\pi} \frac{1-r^2}{1+r^2-2r\cos\phi}$$

is known as *Poisson's kernel* and can be interpreted as a boundary influence function. Thus (3.9) is the temperature at the interior point (r,ϕ) when the boundary temperature is a delta function concentrated at $\phi=0$.

In view of the fact that (3.9) is harmonic for $r<1$ it is easy to see that (3.8) represents a harmonic function for $r<1$. Moreover, we showed in

Exercise 2.9, Chapter 2, that

$$\lim_{r \to 1^-} k(r,\phi) = \delta(\phi), \qquad |\phi| \leqslant \pi.$$

Application of the methods of Chapter 2 shows that $u(r,\phi)$ tends to $f(\phi)$ uniformly in $|\phi| \leqslant \pi$ and that the boundary values are therefore assumed in the sense of continuity. Continuous dependence on the boundary data is a fairly easy consequence of the explicit formula (3.8).

We now turn to the inhomogeneous equation (Poisson's equation) in the unit disk. It suffices to consider only 0 boundary data since we can always add the solution of a problem of type (3.2). We shall therefore consider

$$(3.10) \qquad -\Delta v = q(r,\phi), \quad (r,\phi) \in \Omega; \qquad v(1,\phi) = 0.$$

Although we can solve (3.10) directly by a Fourier expansion in ϕ, it is perhaps more in keeping with the spirit of our work to find Green's function and then use superposition. Green's function g for a unit source at $\phi = 0, r = r_0$ satisfies

$$(3.11) \qquad -\frac{1}{r}\frac{\partial}{\partial r}\left(r\frac{\partial g}{\partial r}\right) - \frac{1}{r^2}\frac{\partial^2 g}{\partial \phi^2} = \frac{\delta(r-r_0)\delta(\phi)}{r}, \quad r < 1; \qquad g|_{r=1} = 0.$$

The right side of the differential equation is the delta function expressed in polar coordinates; see (2.26), Chapter 2. Although the solution depends on r, ϕ, and r_0, we shall sometimes suppress the dependence on r_0 for notational convenience.

We multiply (3.11) by $(1/2\pi)e^{-in\phi}$ and integrate from $\phi = -\pi$ to $\phi = \pi$. Integrating the second term by parts and using the fact that both g and $e^{-in\phi}$ satisfy (3.5), we find that

$$(3.12) \qquad -\frac{d}{dr}\left(r\frac{dg_n}{dr}\right) + \frac{n^2}{r}g_n = \frac{1}{2\pi}\delta(r-r_0), \quad 0 < r, r_0 < 1; \qquad g_n|_{r=1} = 0,$$

where

$$g_n(r,r_0) = \frac{1}{2\pi}\int_{-\pi}^{\pi} g e^{-in\phi}d\phi, \qquad g = \sum_{n=-\infty}^{\infty} g_n(r,r_0)e^{in\phi}.$$

The general solution of the homogeneous equation corresponding to (3.12) is $Ar^n + Br^{-n}$ for $n \neq 0$ and $A + B\log r$ for $n = 0$. We require that g_n be bounded at $r = 0$ and vanish at $r = 1$. On applying the usual matching conditions at $r = r_0$, we find that

$$g_n(r, r_0) = -\frac{1}{4\pi|n|}(r_<^{|n|})(r_>^{|n|} - r_>^{-|n|}), \qquad n \neq 0,$$

$$g_0(r, r_0) = -\frac{1}{2\pi}\log r_>,$$

where $r_> = \max(r, r_0)$, $r_< = \min(r, r_0)$. We then conclude that

$$g = -\frac{1}{2\pi}\log r_> - \frac{1}{4\pi}\sum_{\substack{n=-\infty \\ n \neq 0}}^{\infty} \frac{e^{in\phi}}{|n|}(r_<^{|n|})(r_>^{|n|} - r_>^{-|n|})$$

$$= -\frac{1}{2\pi}\log r_> - \frac{1}{2\pi}\sum_{n=1}^{\infty} \frac{\cos n\phi}{n}(r_<^n)(r_>^n - r_>^{-n}).$$

If the source is at $\phi = \phi_0$, we have instead

(3.13)

$$g(r, \phi; r_0, \phi_0) = -\frac{1}{2\pi}\log r_> - \frac{1}{2\pi}\sum_{n=1}^{\infty}\frac{\cos n(\phi - \phi_0)}{n}(r_<^n)(r_>^n - r_>^{-n}).$$

The solution of (3.10) can therefore be written as

(3.14) $$v(r, \phi) = \int_{-\pi}^{\pi}d\phi_0\int_0^1 r_0\,dr_0\,g(r, \phi; r_0, \phi_0)q(r_0, \phi_0),$$

from which one can deduce the continuous dependence of v on the forcing term q.

If we prefer to express the solution as a series in trigonometric exponentials, we set

$$q(r, \phi) = \sum_{n=-\infty}^{\infty}q_n(r)e^{in\phi}, \qquad v(r, \phi) = \sum_{n=-\infty}^{\infty}v_n(r)e^{in\phi},$$

and, by substitution in either (3.14) or (3.10), we obtain

$$v_n(r) = 2\pi\int_0^1 r_0 q_n(r_0)g_n(r, r_0)\,dr_0.$$

The solution of the Dirichlet problem (3.2) can also be expressed in terms of Green's function as

(3.15) $$u(r, \phi) = -\int_{-\pi}^{\pi}\frac{\partial g}{\partial r_0}(r, \phi; r_0, \phi_0)f(\phi_0)r_0\,d\phi_0,$$

which, on substitution of (3.13), reduces to (3.6) and hence to (3.8).

Of course, for most domains, we cannot find an explicit formula for Green's function. We must then resort to other techniques such as that of integral equations, described later in the section.

Mean Value Theorem and Maximum Principle

An immediate consequence of (3.8) or (3.6) is that the *value of a harmonic function at the center of a disk is the average of its values on the boundary* (the extension to a disk of radius a instead of 1 is easily accomplished by a change of variables). By splitting the disk into concentric annuli, we conclude that the value at the center is also the average over the area of the disk. Similar properties hold in n dimensions. Thus, if a function is harmonic in a ball Ω (and continuous in $\bar{\Omega}$), its value at the center is equal to its average on any sphere or any ball having the same center and lying within $\bar{\Omega}$. This mean value theorem enables us to prove a strong form of the maximum principle for harmonic functions.

Theorem. Let u be harmonic in the bounded domain Ω and continuous in $\bar{\Omega}$. If $m \leqslant u \leqslant M$ on the boundary Γ of Ω, then $m < u < M$ in Ω unless $m = M$, in which case u is constant in $\bar{\Omega}$.

Proof. Suppose that u has a maximum at the interior point x_0. Then $u(x) \leqslant u(x_0)$ for all $x \in \bar{\Omega}$. Let B_0 be the largest ball, with center at x_0, that lies entirely in $\bar{\Omega}$. If at some point of B_0 we had $u(x) < u(x_0)$, the continuity of u and the fact that $u(x) \leqslant u(x_0)$ for all x in B_0 would imply that $u(x_0)$ is greater than its average on B_0. Since this contradicts the mean value theorem, we must have $u(x) \equiv u(x_0)$ in B_0. We now show that u must have this same value everywhere in Ω (and hence in $\bar{\Omega}$, by continuity). Let ξ be an arbitrary point in Ω, and connect ξ with x_0 by a curve lying in Ω. The intersection of this curve and B_0 is denoted by x_1; next, we construct the largest ball B_1, with center at x_1, lying entirely in $\bar{\Omega}$. Repeating the previous argument, we find that $u(x) \equiv u(x_0)$ in B_1. Continuing in this manner, we finally cover the point ξ by some ball B_k and therefore $u(\xi) = u(x_0)$. We conclude that, if u is harmonic in Ω, continuous in $\bar{\Omega}$, and not identically constant, its maximum occurs on the boundary and not in the interior. The same reasoning applied to the harmonic function $-u(x)$ shows that the minimum of u occurs on the boundary.

Remark. A weaker version of the maximum principle was proved in Section 3, Chapter 1. We were then able to prove uniqueness and continuous dependence on the data for (3.1).

An alternative method for proving the uniqueness of (3.1) is based on "energy" integrals. Suppose u_1 and u_2 are two solutions of (3.1). Their difference $v(x)$ satisfies the homogeneous problem $-\Delta v = 0, x \in \Omega; v = 0$ on Γ. Therefore

$$0 = \int_\Omega v \Delta v \, dx = -\int_\Omega |\mathrm{grad}\, v|^2 dx + \int_\Gamma v \frac{\partial v}{\partial \nu} \, dS = -\int_\Omega |\mathrm{grad}\, v|^2 dx,$$

and $v = $ constant in Ω. Since $v = 0$ on Γ, this constant must be 0. Thus $v \equiv 0$ and $u_1 = u_2$.

Surface Layers

It will be convenient in this section to use the language of electrostatics, although the results, being of a purely mathematical nature, can be applied whenever the physical laws lead to potential theory.

If a unit positive charge is located at the point ξ in R_3, the corresponding potential (free space fundamental solution) $E(x, \xi)$ satisfies

$$(3.16) \qquad\qquad -\Delta E = \delta(x - \xi), \qquad x, \xi \in R_3.$$

The solution is determined only up to an arbitrary solution of the homogeneous equation. If we also require that E be spherically symmetric about the source and vanish at infinity, the unique solution of (3.16) becomes

$$(3.17) \qquad\qquad E(x, \xi) = \frac{1}{4\pi |x - \xi|},$$

as was shown in Chapter 2. Equation (3.16) is interpreted in its distributional sense:

$$\langle E, \Delta \phi \rangle = -\phi(\xi) \qquad \text{for each } \phi \in C_0^\infty(R_3).$$

Now suppose that charge is distributed in free space with density $q(x)$, where, for simplicity, we shall assume $q \equiv 0$ outside some bounded ball. The element of volume $d\xi$ at ξ carries a charge $q(\xi) d\xi$ whose potential is $(1/4\pi |x - \xi|) q(\xi) d\xi$. By superposition the potential $u(x)$ of the entire charge distribution is

$$(3.18) \qquad u(x) = \int_{R_3} \frac{1}{4\pi |x - \xi|} q(\xi) d\xi = \int_\Omega \frac{1}{4\pi |x - \xi|} q(\xi) d\xi,$$

where Ω is the domain where q does not vanish. If x is outside Ω, the integrand is not singular as ξ traverses Ω. We can therefore calculate derivatives with respect to x by differentiating under the integral sign, and we see that $\Delta u = 0$. If x is in the charge region Ω, the integrand in (3.18) has a very mild singularity which causes little difficulty. The integral (3.18) still converges, as can easily be seen by passing to spherical coordinates with origin at x, the element of volume $d\xi$ furnishing a factor proportional to $|x - \xi|^2$ which more than cancels the denominator. Differentiating formally under the integral sign, we obtain

$$(3.19) \qquad -\Delta u = \int_{R_3} \delta(x - \xi) q(\xi) \, d\xi = q(x),$$

a result which is rigorous in the sense of the theory of distributions (and also holds classically if we assume that q is slightly better than just continuous—Hölder continuity suffices; differentiability is more than enough).

A *unit dipole at ξ with axis l* is a singular charge distribution obtained by taking the limit as $h \to 0$ of charges $-(1/h)$ and $1/h$ located at ξ and $\xi + hl$, respectively, where l is a given unit vector. The corresponding potential $D_l(x, \xi)$ satisfies

$$-\Delta D_l = \lim_{h \to 0} \frac{1}{h} \left\{ \delta[x - (\xi + hl)] - \delta(x - \xi) \right\} = \frac{d}{dl_\xi} \delta(x - \xi),$$

where d/dl_ξ refers to differentiation with respect to ξ in the l direction. By the principle of superposition, we can find D_l by performing the differentiation on the solution of $-\Delta E = \delta(x - \xi)$. A simple calculation then gives

$$(3.20) \qquad D_l(x, \xi) = \frac{d}{dl_\xi} \frac{1}{4\pi |x - \xi|} = \frac{\cos(x - \xi, l)}{4\pi |x - \xi|^2},$$

where $(x - \xi, l)$ is the angle between the vectors $x - \xi$ and l.

In (3.18) we considered the potential of a volume distribution of charges with density q. The potential $u(x)$ not only is continuous within the charge region but even satisfies the differential equation $-\Delta u = q$. We now turn to surface distributions of charge. In the hierarchy of charge configurations, a surface distribution is less singular than a point charge but not as regular as a volume distribution. (A line charge is less singular than a point charge but more so than a surface charge.) Let the surface Γ carry a charge with surface density $a(x)$. Such a charge configuration is known as a *simple layer*. This means that the portion Γ' of Γ has the charge $\int_{\Gamma'} a(\xi) \, dS_\xi$. One can think of the surface element dS_ξ at ξ as carrying the small concentrated charge $a(\xi) \, dS_\xi$. The potential of the simple layer is

$$(3.21) \qquad u(x) = \int_\Gamma \frac{1}{4\pi |x - \xi|} a(\xi) \, dS_\xi, \qquad x \in R_3.$$

Consider next a surface distribution of dipoles of density $b(x)$ with axis *normal* to Γ. This configuration is known as a *double layer*. The element of surface dS_ξ at ξ carries a dipole of strength $b(\xi)\,dS_\xi$ with axis n, where n is the normal to Γ at ξ. The potential $v(x)$ of the double layer follows from (3.20) by superposition:

$$(3.22) \quad v(x) = \int_\Gamma \frac{\cos(x-\xi,n)}{4\pi|x-\xi|^2} b(\xi)\,dS_\xi = \int_\Gamma \frac{\partial}{\partial n_\xi} \frac{1}{4\pi|x-\xi|} b(\xi)\,dS_\xi.$$

The integrals in (3.21) and (3.22) are obviously well defined for x not on Γ. Even if x is a point on Γ, the integral in (3.21) converges. Any difficulty would arise from points ξ near x, and then the corresponding integration is over a flat portion of Γ; introducing polar coordinates with origin at x, we see that dS_ξ contributes a factor proportional to $|x-\xi|$ which cancels the denominator. Similarly it can be shown that (3.22) converges when x is on Γ because $\cos(x-\xi,n)$ tends to 0 as ξ approaches x. Of course the fact that (3.21) and (3.22) make sense even when x is on Γ does not tell us how these integrals, evaluated for x near Γ, are related to the values on Γ. We study this question next.

The potential of a point source located at the origin is

$$(3.23) \qquad\qquad E(x_1,x_2,x_3) = \frac{1}{4\pi\left(x_1^2 + x_2^2 + x_3^2\right)^{1/2}}.$$

By regarding x_3 as a parameter, we would like to relate the behavior of E as x_3 tends to 0 with its behavior at $x_3 = 0$. In effect we have selected a set of parallel planes ($x_3 = $ constant) on which to study the potential (3.23). For each $x_3 \neq 0$, E is a continuous function of x_1, x_2 and so defines a distribution on the space of test functions $\phi(x_1, x_2)$ of compact support in R_2. When $x_3 = 0$, E becomes unbounded at $x_1 = x_2 = 0$ but remains locally integrable in the x_1-x_2 plane and therefore still defines a distribution. Setting $\rho^2 = x_1^2 + x_2^2$, we have the following theorem.

Theorem 1. In the sense of distributions,

$$(3.24) \quad \lim_{x_3 \to 0\pm} E(x_1,x_2,x_3) = \lim_{x_3 \to 0\pm} \frac{1}{4\pi\left(\rho^2 + x_3^2\right)^{1/2}} = \frac{1}{4\pi\rho} = E(x_1,x_2,0),$$

$$(3.25) \quad \lim_{x_3 \to 0\pm} \frac{\partial E}{\partial x_3}(x_1,x_2,x_3) = \lim_{x_3 \to 0\pm} \frac{-x_3}{4\pi\left(\rho^2 + x^2\right)^{3/2}} = \mp \tfrac{1}{2}\delta(x_1)\delta(x_2).$$

Proof. For all x_3, $1/4\pi(\rho^2+x_3^2)^{1/2}$ is dominated by the locally integrable function $1/4\pi\rho$. In the pointwise sense, as $x_3\to 0$, $1/4\pi(\rho^2+x_3^2)^{1/2}$ tends to $1/4\pi\rho$ everywhere except at $\rho=0$. It follows from the Lebesgue dominated convergence theorem (see Section 7, Chapter 0) that

$$(3.26) \qquad \lim_{x_3\to 0\pm} \int_{R_2} \frac{\phi(x_1,x_2)}{4\pi(\rho^2+x_3^2)^{1/2}}\, dx_1\, dx_2 = \int_{R_2} \frac{\phi(x_1,x_2)}{4\pi\rho}\, dx_1\, dx_2$$

for every integrable function ϕ (hence certainly for test functions).

Next we turn to $\partial E/\partial x_3$, which is the derivative normal to the planes $x_3=$ constant. For $x_3\neq 0$ we have

$$(3.27) \qquad \frac{\partial E}{\partial x_3}(x_1,x_2,x_3) = -\frac{x_3}{4\pi(\rho^2+x_3^2)^{3/2}},$$

whereas, on $x_3=0$, $\partial E/\partial x_3=0$ except at $\rho=0$. The Lebesgue theorem is not applicable because we cannot find an integrable function which dominates (3.27) in a neighborhood of $x_3=0$. However, (3.25) follows from Example 3(h), Section 2, Chapter 2. Thus, if ϕ is a test function, or even just a function continuous at the origin and integrable, we have

$$(3.28) \qquad \lim_{x_3\to 0\pm} \int_{R_2} \phi(x_1,x_2)\frac{\partial E}{\partial x_3}(x_1,x_2,x_3)\, dx_1\, dx_2 = \mp \tfrac{1}{2}\phi(0,0).$$

Remark. Results (3.26) and (3.28) remain valid even if the integral is taken only over a portion σ of the plane $x_3=0$. Indeed, it suffices to extend ϕ to be identically 0 outside σ and then apply (3.26) and (3.28), which hold for piecewise continuous integrands.

Now suppose Γ is a *smooth* surface, one side of which is taken as the positive side. By definition the *normal to* Γ is the normal on the positive side. If Γ is a closed surface without boundary, such as a sphere, it is customary to label the exterior side as positive. We shall investigate the behavior of potentials of layers in the neighborhood of a *fixed point s* on Γ. The normal at s will be called ν; of course the direction ν is then also determined by parallelism at every point in space.

Let $u(x)$ be the potential of a simple layer of surface density $a(x)$ spread on Γ; then

$$(3.29) \qquad u(x) = \int_\Gamma \frac{a(\xi)}{4\pi|x-\xi|}\, dS_\xi,$$

which is well defined not only when x is outside Γ but also when x is on Γ. Furthermore, if we imitate the arguments that led to (3.26), we can show that

$$(3.30) \qquad \lim_{x \to s\pm} u(x) = \int_\Gamma \frac{a(\xi)}{4\pi |s - \xi|} \, dS_\xi,$$

so that the limits from either side coincide with the potential right on the surface. Let us now turn to the derivative $\partial u / \partial \nu$ in the direction ν, which coincides with the normal at s. If x is not on Γ, we can calculate $\partial u / \partial \nu$ by differentiating (3.29) under the integral sign to obtain

$$\frac{\partial u}{\partial \nu}(x) = \int_\Gamma a(\xi) \frac{\partial}{\partial \nu} \frac{1}{4\pi |x - \xi|} \, dS_\xi = \int_\Gamma a(\xi) \frac{\cos(\xi - x, \nu)}{4\pi |x - \xi|^2} \, dS_\xi.$$

To analyze the behavior as $x \to s$, we divide Γ into the complementary parts Γ_ε and $\Gamma - \Gamma_\varepsilon$, where Γ_ε is the portion of Γ within a ball of radius ε centered at s. Thus we have

$$(3.31) \qquad \frac{\partial u}{\partial \nu}(x) = \int_{\Gamma - \Gamma_\varepsilon} a(\xi) \frac{\cos(\xi - x, \nu)}{4\pi |x - \xi|^2} \, dS_\xi + \int_{\Gamma_\varepsilon} a(x) \frac{\cos(\xi - x, \nu)}{4\pi |x - \xi|^2} \, dS_\xi.$$

The first integral is clearly continuous at any point x not lying on $\Gamma - \Gamma_\varepsilon$, and will therefore be continuous as $x \to s$. For ε sufficiently small, Γ_ε is nearly a flat surface and we can regard Γ_ε as lying in the plane tangent to Γ at s. We introduce a Cartesian coordinate system (x_1, x_2, x_3) with origin at s; x_1, x_2 lie in the tangent plane, and the positive x_3 axis is in the ν direction. If we let x lie on the normal to Γ at s, the second integral in (3.31) may be rewritten as

$$\int_{\Gamma_\varepsilon} a(\xi_1, \xi_2) \left[\frac{-x_3}{4\pi (\xi_1^2 + \xi_2^2 + x_3^2)^{3/2}} \right] d\xi_1 d\xi_2.$$

To let $x \to s\pm$ is the same as allowing x_3 to tend to $0\pm$. Hence from (3.28) and the remark following Theorem 1, the integral on Γ_ε tends to $\mp \frac{1}{2} a(s)$. The remaining integral in (3.31) approaches the integral

$$\int_{\Gamma - \Gamma_\varepsilon} a(\xi) \frac{\cos(\xi - s, \nu)}{4\pi |s - \xi|^2} \, dS_\xi,$$

which in the limit as $\varepsilon \to 0$ is equal to the convergent integral

$$\int_\Gamma a(\xi) \frac{\cos(\xi - s, \nu)}{4\pi |s - \xi|^2} \, dS_\xi.$$

We therefore conclude that

$$(3.32) \qquad \lim_{x \to s\pm} \frac{\partial u}{\partial \nu}(x) = \mp \tfrac{1}{2}a(s) + \int_\Gamma a(\xi) \frac{\cos(\xi - s, \nu)}{4\pi|x - \xi|^2} dS_\xi,$$

which contains the well-known result of electrostatics that, at a charged surface, the normal component of the electric field [that is, $-(\partial u/\partial \nu)$] jumps by an amount equal to the surface charge density. This result is usually derived by elementary arguments from Gauss's theorem, but (3.32) also yields deeper information about one-sided limits.

The potential (3.22) of a normally oriented *double layer* of surface density $b(x)$ spread on Γ can be written as

$$v(x) = \int_{\Gamma - \Gamma_\epsilon} b(\xi) \frac{\cos(x - \xi, n)}{4\pi|x - \xi|^2} dS_\xi + \int_{\Gamma_\epsilon} b(\xi) \frac{\cos(x - \xi, n)}{4\pi|x - \xi|^2} dS_\xi,$$

where *n is the normal to* Γ *at* ξ. The first integral is continuous as $x \to s$, and the second is over a nearly flat surface if ϵ is sufficiently small. Using the same Cartesian coordinate system introduced earlier, and with x lying on the normal to Γ at s, we have

$$\frac{\cos(x - \xi, n)}{4\pi|x - \xi|^2} = \frac{x_3}{4\pi(\xi_1^2 + \xi_2^2 + x_3^3)^{3/2}},$$

and hence

$$(3.33) \qquad \lim_{x \to s\pm} v(x) = \pm \tfrac{1}{2}b(s) + \int_\Gamma b(\xi) \frac{\cos(s - \xi, n)}{4\pi|s - \xi|^2} dS_\xi.$$

Thus the potential of a dipole layer is *discontinuous* on crossing the layer. We shall not need to investigate the behavior of the normal derivative of the potential of a double layer.

Interior Dirichlet Problem—Integral Equation of the Second Kind

Let Ω be a bounded domain in R_3 with smooth boundary Γ. Consider the Dirichlet problem for Laplace's equation

$$(3.34) \qquad \Delta w = 0, \quad x \in \Omega; \quad w = f, \quad x \in \Gamma,$$

where we assume f continuous on Γ, and we are looking for a solution w continuous on $\overline{\Omega}$.

The classical approach, which at the same time provides an existence proof, attempts to find the solution of (3.34) as a double layer on Γ. Thus our candidate for the solution of (3.34) is

$$(3.35) \qquad w(x) = \int_\Gamma b(\xi) \frac{\cos(x - \xi, n)}{4\pi |x - \xi|^2} \, dS_\xi,$$

where b is to be determined from

$$(3.36) \qquad \lim_{x \to s-} w(x) = f(s) \qquad \text{for every } s \text{ on } \Gamma.$$

In view of (3.33), condition (3.36) becomes the Fredholm integral equation of the second kind:

$$(3.37) \qquad f(s) = -\tfrac{1}{2} b(s) + \int_\Gamma k(s, \xi) b(\xi) \, dS_\xi,$$

where the kernel

$$(3.38) \qquad k(s, \xi) = \frac{\cos(s - \xi, n)}{4\pi |s - \xi|^2}$$

is usually not symmetric.

The singularity of the kernel at $x = \xi$ is only of order $1/|x - \xi|$ because of the mitigating effect of the numerator. The corresponding integral operator K is Hilbert-Schmidt (the proof is similar to Exercise 1.3, Chapter 6). It can be shown that $\tfrac{1}{2}$ is not an eigenvalue of K, so that (3.37) has one and only one solution b for each f. The corresponding double layer potential (3.35) is then a solution of Laplace's equation in Ω satisfying (3.36). With a little more work it can be shown that the boundary values are taken on in the sense of continuity. We have therefore shown the existence of a solution to (3.34). Existence under less restrictive conditions on Γ and f will be obtained in Section 4 by variational methods. Uniqueness was proved earlier.

Interior and Exterior Dirichlet Problems by an Integral Equation of the First Kind

Correct formulation of the Dirichlet problem for an unbounded domain requires a boundary condition at infinity. Consider, for instance, the domain exterior to the unit ball in R_3. The functions $u_1(x) = 1$ and $u_2(x) = 1/|x|$ are both harmonic for $|x| > 1$ and take on the same value on the unit sphere. In many (but not all) physical applications the appropriate requirement is that u vanish as $|x| \to \infty$.

Definition. Let Ω_e be the exterior of a bounded domain Ω_i in R_3. The common boundary of Ω_i and Ω_e is denoted by Γ. The *exterior Dirichlet problem* is the BVP

$$(3.39) \qquad \Delta u_e = 0, \quad x \in \Omega_e; \qquad u_e|_\Gamma = f; \qquad \lim_{|x|\to\infty} u_e = 0.$$

Expansion in a series of spherical harmonics shows that the solution of (3.39) has the properties

$$u_e = 0\left(\frac{1}{|x|}\right), \qquad |\operatorname{grad} u_e| = 0\left(\frac{1}{|x|^2}\right) \quad \text{as } |x| \to \infty.$$

An immediate consequence is that

$$(3.40) \qquad \lim_{R\to\infty} \int_{\Gamma_R} \left(E \frac{\partial u_e}{\partial R} - u_e \frac{\partial E}{\partial R} \right) dS_x = 0,$$

where $E = 1/4\pi|x-\xi|$, ξ is a fixed point, and Γ_R is a sphere of radius R with center at ξ.

Simultaneously with (3.39) we consider the *interior* Dirichlet problem with the same boundary values on Γ:

$$(3.41) \qquad \Delta u_i = 0, \quad x \in \Omega_i; \qquad u_i|_\Gamma = f.$$

Note that the solutions of the interior and exterior problems are not related in any obvious manner. For instance, if $f = 1$, then $u_i \equiv 1$ but u_e can be quite complicated (if Ω_i is the unit ball, $u_e = 1/|x|$). Nevertheless we shall be able to reduce both problems to a single integral equation on Γ whose solution can then be used to calculate both $u_i(x)$ and $u_e(x)$.

Recall that $E(x,\xi) = 1/4\pi|x-\xi|$ satisfies

$$(3.42) \qquad -\Delta E = \delta(x-\xi), \qquad x, \xi \in R_3.$$

We regard ξ as a fixed point in either Ω_i or Ω_e. Multiply (3.41) by E, (3.42) by u_i, add, and integrate over Ω_i to obtain

$$\int_{\Omega_i} (E\Delta u_i - u_i \Delta E)\, dx = \begin{cases} u_i(\xi), & \xi \in \Omega_i, \\ 0, & \xi \in \Omega_e. \end{cases}$$

On applying Green's theorem, this becomes

$$(3.43) \qquad \int_\Gamma \left(E \frac{\partial u_i}{\partial \nu} - u_i \frac{\partial E}{\partial \nu} \right) dS_\xi = \begin{cases} u_i(\xi), & \xi \in \Omega_i, \\ 0, & \xi \in \Omega_e, \end{cases}$$

where ν is the outward normal to Γ (that is, pointing away from Ω_i into Ω_e).

Similarly, we multiply (3.39) by E, (3.42) by u_e, add, and integrate over the domain bounded internally by Γ and externally by a sphere Γ_R of large radius R with center at ξ (such a large sphere will contain Γ if R is large enough). After applying Green's theorem, we note that the contribution from Γ_R tends to 0 as $R \to \infty$ by (3.40). With respect to Ω_e the outward normal on Γ is the negative of the previous ν. Therefore we obtain

$$(3.44) \qquad \int_\Gamma \left(-E \frac{\partial u_e}{\partial \nu} + u_e \frac{\partial E}{\partial \nu} \right) dS_x = \begin{cases} 0, & \xi \in \Omega_i, \\ u_e(\xi), & \xi \in \Omega_e. \end{cases}$$

Adding (3.43) and (3.44), and noting that $u_i = u_e = f$ on Γ, we find that

$$\int_\Gamma E(x, \xi) \left[\frac{\partial u_i}{\partial \nu}(x) - \frac{\partial u_e}{\partial \nu}(x) \right] dS_x = \begin{cases} u_i(\xi), & \xi \in \Omega_i, \\ u_e(\xi), & \xi \in \Omega_e. \end{cases}$$

Interchanging the labels of the variables and invoking the symmetry of E, we obtain

$$(3.45) \qquad \int_\Gamma E(x, \xi) I(\xi) dS_\xi = \begin{cases} u_i(x), & x \in \Omega_i, \\ u_e(x), & x \in \Omega_e, \end{cases}$$

where

$$(3.46) \qquad I(\xi) \doteq \frac{\partial u_i}{\partial \nu}(\xi) - \frac{\partial u_e}{\partial \nu}(\xi), \qquad \xi \in \Gamma.$$

We have succeeded in expressing u_i and u_e within their respective domains as the potential of the same simple layer with unknown surface density $I(\xi)$. Letting x tend to a point s on Γ, we obtain the integral equation for I:

$$(3.47) \qquad \int_\Gamma \frac{1}{4\pi|s - \xi|} I(\xi) dS_\xi = f(s), \qquad s \in \Gamma,$$

where we have used the continuity of the simple layer potential. After solving (3.47) for I, we would substitute in (3.45) to find both $u_i(x)$ and $u_e(x)$.

Equation (3.47) is a Fredholm equation of the first kind with a symmetric H.-S. kernel. We know that 0 is in the continuous spectrum of the corresponding integral operator, so that (3.47) cannot have an L_2 solution for each $f \in L_2(\Gamma)$. Nevertheless, even in these cases one can interpret the "solution" in a distributional sense, and, although the formal eigenfunction expansion for I may diverge, the corresponding expressions for u_i and u_e obtained from (3.45) will be well behaved (see Hsiao and Wendland, and the references quoted there for some recent results on this problem).

Another advantage of (3.47) is that the corresponding integral operator is *positive*. To show this let $I(\xi)$ be an arbitrary function in $L_2(\Gamma)$, and define $f(s)$ from (3.47). What we wish to prove is that

$$(3.48) \qquad \int_\Gamma f(s)I(s)\,dS > 0 \qquad \text{for } I \not\equiv 0.$$

Let us define the simple layer potential

$$u(x) = \int_\Gamma E(x,\xi)I(\xi)\,dS_\xi,$$

which is harmonic in both Ω_i and Ω_e; moreover, u vanishes at infinity. Thus we obtain

$$0 = \int_{\Omega_i} u\,\Delta u\,dx = \int_\Gamma u(s-)\frac{\partial u}{\partial \nu}(s-)\,dS - \int_{\Omega_i} |\operatorname{grad} u|^2\,dx,$$

$$0 = \int_{\Omega_e} u\,\Delta u\,dx = -\int_\Gamma u(s+)\frac{\partial u}{\partial \nu}(s+)\,dS - \int_{\Omega_e} |\operatorname{grad} u|^2\,dx.$$

Adding these equations and recalling the simple layer properties

$$u(s-) = u(s+) = f(s), \qquad \frac{\partial u}{\partial \nu}(s-) - \frac{\partial u}{\partial \nu}(s+) = I(s),$$

we find that

$$0 = \int_\Gamma f(s)I(s)\,dS - \int_{R_3} |\operatorname{grad} u|^2\,dx.$$

It follows that $\int_\Gamma fI\,dS > 0$ unless u is constant, in which case the behavior at infinity tells us that $u \equiv 0$ and therefore $I \equiv 0$. Hence we have proved (3.48).

Exercises

3.1. Green's function for the unit disk Ω can also be obtained by images. We set

$$(3.49) \qquad g(x,\xi) = \frac{1}{2\pi}\log\frac{1}{|x-\xi|} + v(x,\xi), \qquad \xi\in\Omega,$$

where $v(x,\xi)$ is required to be harmonic in Ω and g must vanish on Γ. The idea of images is to express v as the potential due to sources lying outside Ω. Let ξ^* be the inverse point of ξ with respect to the unit circle, that is, the polar coordinates of ξ^* are $(1/r,\phi)$ if those of ξ are (r,ϕ).

(a) Show that

$$(3.50) \quad g(x,\xi) = \frac{1}{2\pi}\log\frac{1}{|x-\xi|} - \frac{1}{2\pi}\log\frac{1}{|x-\xi^*|} + \frac{1}{2\pi}\log|\xi|.$$

(b) Show that using (3.50) in (3.15) reduces the latter to (3.8).

3.2. Let Ω be an arbitrary bounded domain in R_3, and let Γ be its boundary. Green's function $g(x,\xi)$ satisfies

$$(3.51) \qquad -\Delta g = \delta(x-\xi), \quad x,\xi \in \Omega; \qquad g=0, \quad x \in \Gamma.$$

It will sometimes be useful to write

$$(3.52) \qquad\qquad g = \frac{1}{4\pi|x-\xi|} + v(x,\xi),$$

where v is harmonic in Ω.

Demonstrate the following properties of g:

(a) g exists and is unique.

(b) $g(x,\xi) = g(\xi,x)$.

(c) g is positive in Ω. *Hint*: Draw a small ball about ξ, and use the maximum principle outside the ball.

(d) $g < 1/4\pi|x-\xi|, \quad x,\xi \in \Omega$.

(e) The integral operator G defined by

$$(3.53) \qquad\qquad Gu \doteq \int_\Omega g(x,\xi)u(\xi)\,d\xi, \qquad x \in \Omega,$$

is a H.-S. operator on $L_2(\Omega)$.

3.3. Consider the eigenvalue problem for $-\Delta$ in a *bounded domain* Ω in R_3:

$$(3.54) \qquad\qquad -\Delta u = \lambda u, \quad x \in \Omega; \qquad u=0, \quad x \in \Gamma.$$

We set the problem in the functional framework of the Hilbert space $L_2^{(c)}(\Omega)$ in which the inner product $\langle u,v \rangle$ and norm $\|u\|$ are given, respectively, by

$$\int_\Omega u\bar{v}\,dx \quad \text{and} \quad \left[\int_\Omega |u|^2\,dx\right]^{1/2}.$$

Prove the following statements:

(a) All eigenvalues are real and positive.

(b) Eigenfunctions corresponding to different eigenvalues are orthogonal.

(c) Since G is Hilbert-Schmidt (see Exercise 3.2), the eigenfunctions of (3.54) form an orthonormal basis $\{e_i\}$.

3.4. Let $\{e_i\}$ be the orthonormal basis of eigenfunctions of (3.54). Write the solution of

$$\frac{\partial u}{\partial t} - \Delta u = q(x,t), \quad x \in \Omega, t > 0;$$

$$u(x,0) = f(x); \quad u(x,t) = 0, \quad x \in \Gamma, t > 0,$$

as an eigenfunction expansion. Do the same for the wave equation

$$\frac{\partial^2 u}{\partial t^2} - \Delta u = q(x,t), \quad x \in \Omega, t > 0;$$

$$u(x,0) = f(x), \quad \frac{\partial u}{\partial t}(x,0) = g(x); \quad u(x,t) = 0, \quad x \in \Gamma, \quad t > 0.$$

3.5. Let a unit source in free space in R_2 be located at the point with polar coordinates $(r_0, 0)$. By taking a Mellin transform on the radial coordinate, show that

$$g(r,\phi;r_0,0) = \frac{1}{2\pi}\int_{-\infty}^{\infty}\frac{r^{iv}r_0^{-iv}}{2v\sinh v\pi}\cosh v(\pi - |\phi|)\,dv.$$

3.6. (a) Show that the Neumann problem

(3.55) $-\Delta u = q(x), \quad x \text{ in } \Omega; \qquad \frac{\partial u}{\partial n} = f, \quad x \in \Gamma,$

can have a solution only if f and g satisfy the solvability condition

(3.56) $$\int_\Omega q(x)\,dx = -\int_\Gamma f(x)\,dS_x,$$

and give a physical interpretation.

(b) The modified Green's function for the Neumann problem satisfies

(3.57) $\quad -\Delta g_M = \delta(x-\xi) - \dfrac{1}{V}, \quad x,\xi \in \Omega; \qquad \dfrac{\partial g_M}{\partial n} = 0, \quad x \in \Gamma,$

where V is the volume of Ω. Construct g_M when Ω is the unit disk in R_2 by a Fourier expansion in the angle ϕ. Show that

$$g_M(r,\phi;r_0,0) = A + \frac{r^2}{4\pi} - \frac{1}{2\pi}\log r_> + \sum_{n=1}^{\infty} \frac{\cos n\phi}{2\pi n} r_<^n (r_>^n + r_>^{-n}),$$

where the notation is the same as in (3.13).

3.7. Calculate explicitly the potentials of the following charge distributions in R_3:

(a) Line density 1 along the segment $-1 \leqslant z \leqslant 1$.

(b) Surface density 1 on the unit disk $z=0$, $r \leqslant 1$.

(c) A normally oriented double layer of uniform density 1 on the unit disk $z=0$, $r \leqslant 1$.

Show in (a) that the potential itself becomes singular as the observation point approaches the charge distribution. For (b) verify (3.30) and (3.32) at the center of the charge distribution. For (c) verify (3.33) at the center of the dipole layer.

4. VARIATIONAL PRINCIPLES FOR INHOMOGENEOUS PROBLEMS

Introduction

The solution $u(x)$ of a boundary value problem can also be characterized as the element for which a related functional J is stationary. Such a stationary characterization is a *variational principle* whose Euler-Lagrange equation is the differential equation of the original BVP. The functional J will usually have physical significance as an energy or equivalent quantity such as capacity or torsional rigidity. In many instances the stationary value of J turns out to be a minimum or maximum rather than a saddle-point. When the BVP is linear, the corresponding J is of second degree.

The equivalence between the BVP and the variational principle can be exploited numerically and theoretically. The variational principle is the basis for the Ritz-Rayleigh method for finding an approximation, say $\tilde{u}$, to the exact solution u of the BVP, and the corresponding approximation $J(\tilde{u})$ to the stationary value $J(u)$. Since J is "flat" at $\tilde{u}$, it is not surprising that $J(\tilde{u})$ furnishes a better estimate of $J(u)$ than $\tilde{u}$ does of u.

The variational principle may also provide guidance to existence questions for the BVP. We shall analyze the case of an inhomogeneous linear differential equation (ordinary or partial) subject to homogeneous boundary conditions. The appropriate modifications for inhomogeneous boundary conditions will be taken up later.

To fix ideas, let us examine the concrete BVP

$$(4.1) \qquad -u'' + ku = f(x), \quad 0 < x < 1; \qquad u(0) = u(1) = 0,$$

where $f(x)$ is a given real-valued function and k a *nonnegative* constant. To translate this problem into an operator equation we must specify the domain of the differential operator on the left of (4.1). The simplest approach is to define the operator A by

$$(4.2) \qquad Au \doteq -u'' + ku$$

on the domain D_A consisting of all functions $u(x)$ in $C^2(0, 1)$ with $u(0) = u(1) = 0$. Here we shall view D_A as a linear manifold in $H = L_2^{(r)}(0, 1)$; clearly D_A is dense in H. Thus (4.1) will be replaced by the operator equation

$$(4.3) \qquad Au = f, \qquad u \in D_A,$$

where we realize that we can hope for solutions only if $f(x)$ is a continuous function. The reader who has thoroughly mastered Chapter 5 might start more ambitiously with (4.2) defined on the larger domain consisting of functions $u(x)$ satisfying the boundary conditions and such that u' is absolutely continuous with u'' in L_2, but we shall not take this as a starting point.

One physical interpretation of (4.1) is that $u(x)$ is the transverse deflection of a taut string, fixed at its ends, under an applied, transverse, distributed load $f(x)$ and subject to a restoring force proportional to the displacement u (so that k is a "spring constant"). The three terms

$$\frac{1}{2} \int_0^1 (u')^2 \, dx, \qquad \frac{k}{2} \int_0^1 u^2 \, dx; \qquad \int_0^1 fu \, dx$$

represent, respectively, the strain energy of the string, the energy stored in the spring, and the work done by the applied load to take the string from its reference configuration to its deflected state. We shall find it convenient to deal with *twice* the total potential energy in the deflected state:

$$(4.4) \qquad \int_0^1 (u')^2 \, dx + k \int_0^1 u^2 \, dx - 2 \int_0^1 fu \, dx.$$

If u is replaced in (4.4) by an arbitrary element v in D_A, a functional J is defined on D_A:

$$(4.5) \qquad J(v) = \int_0^1 (v')^2 \, dx + k \int_0^1 v^2 \, dx - 2 \int_0^1 fv \, dx, \qquad v \in D_A.$$

We can regard $J(v)$ as the total potential energy associated with the virtual deflection v, whereas $J(u)$ is the energy associated with the correct deflection u. For any element $v \in D_A$ we have

$$(4.6) \qquad \langle Av, v \rangle = \int_0^1 (-v'' + kv)v \, dx = \int_0^1 \left[(v')^2 + kv^2 \right] dx,$$

so that we can rewrite J as

$$(4.7) \qquad J(v) = \langle Av, v \rangle - 2 \langle f, v \rangle, \qquad v \in D_A.$$

We note that, if u is a solution of (4.3), then $\langle Au, u \rangle = \langle f, u \rangle$, so that

$$(4.8) \qquad J(u) = -\langle f, u \rangle = -\langle Au, u \rangle.$$

We now consider the problem of finding the minimum of $J(v)$ among all functions $v \in D_A$. We write this minimal problem in the schematic form

$$(4.9) \qquad J(v) \to \min, \qquad v \in D_A.$$

The following theorem of *minimum potential energy* shows that problems (4.3) and (4.9) are equivalent.

Theorem. (a) Let u be a solution of (4.3); then $J(u) \leqslant J(v)$ for all $v \in D_A$, so that u is a solution of the minimum problem (4.9).

(b) Let u be a solution of the minimum problem (4.9); then u satisfies (4.3).

Proof. Let v and w be arbitrary elements of D_A, and set $v = w + h$. Then from (4.7) we find that

$$(4.10) \qquad \begin{aligned} J(v) - J(w) &= \langle A(w+h), w+h \rangle - \langle Aw, w \rangle - 2\langle f, h \rangle \\ &= 2\langle Aw - f, h \rangle + \langle Ah, h \rangle. \end{aligned}$$

Now let (4.1) have a solution u, and set $w = u$ in (4.10). Then, since $\langle Ah, h \rangle \geqslant 0$ for $h \in D_A$, we conclude that $J(v) \geqslant J(u)$, which proves part (a).

Next, assume that there exists an element $u \in D_A$ for which the minimum in (4.9) is attained. Then, from (4.10) with $w = u$, we find that

$$2\langle Au - f, h \rangle + \langle Ah, h \rangle \geqslant 0 \qquad \text{for every } h \in D_A.$$

Considering elements h of the form $\varepsilon\eta$, where ε is a real number and η is an element of D_A, we obtain

$$2\langle Au - f, \eta \rangle + \varepsilon\langle A\eta, \eta \rangle \geqslant 0, \qquad \varepsilon > 0, \quad \eta \in D_A,$$
$$2\langle Au - f, \eta \rangle + \varepsilon\langle A\eta, \eta \rangle \leqslant 0, \qquad \varepsilon < 0, \quad \eta \in D_A.$$

By taking the limit as $\varepsilon \to 0$, we conclude that $\langle Au - f, \eta \rangle = 0$ for all $\eta \in D_A$. Since the set D_A is dense in H, it follows that $Au - f = 0$, completing the proof of part (b).

Remark. We have *not* proved existence in (4.3) or (4.9), but have proved only that a solution to either problem is necessarily a solution to the other. If we are willing to accept (or prove independently) that (4.3) has a solution, that solution is also the minimizing function in (4.9). If, however, we want to use variational methods to establish that (4.3) or (4.1) has a solution, we must prove directly the existence of a minimum for J.

We shall give the existence proof for the general case shortly, but first let us outline the steps as related to the present problem (4.1) or (4.3). Our starting point is $J(v)$ as given by (4.5). It is relatively easy to show that J is bounded below on D_A, so that J has an infimum, say d, on D_A. If there is an element $u \in D_A$ for which $J(u) = d$, the infimum is attained on D_A, so that the minimum problem for J on D_A has the solution u. There are two reasons for considering the minimum problem for J on a larger domain M (as yet unspecified). First, J may not have a minimum on D_A; if f is only piecewise continuous, the solution of (4.1) is not C^2, so that (4.3) has no solution and hence, by the theorem of minimum potential energy, J does not have a minimum on D_A. The trouble here is simply that D_A is not a suitable domain on which to analyze the BVP when f is not continuous. The second reason is more subtle: even if J has a minimum on D_A (and therefore the BVP has a classical solution in D_A), it may be hard to prove this fact directly; it turns out to be easier to prove the existence of a minimum on the larger domain M and *then* to show that the minimizing element lies in D_A.

Thus we shall try to enlarge D_A so that a minimum for J can be guaranteed on the new domain. It is reasonable to hope for such an extension since the expression for J involves only the first derivative of v, whereas an element of D_A has a continuous derivative of order 2.

We write (4.5) as

$$J(v) = a(v,v) - 2l(v),$$

where $a(v,w)$ is the bilinear form

$$a(v,w) \doteq \int_0^1 (v'w' + kvw) \, dx,$$

and $l(v)$ is the linear functional

$$l(v) \doteq \int_0^1 fv \, dx.$$

Since $a(v,v) > 0$ for all nonzero elements v in D_A, it is clear that $a(v,w)$ is an acceptable inner product on D_A, known as the *energy inner product*. The corresponding *energy norm* $\| \ \ \|_a$ is given by

(4.11) $$\|v\|_a^2 = a(v,v) = \int_0^1 \left[(v')^2 + kv^2 \right] dx.$$

We now complete the set D_A in the energy norm. This gives us the larger domain M, on which $J(v)$ given by (4.11) can be guaranteed to have a minimum. The corresponding minimizing function $u \in M$ satisfies the equivalent variational equation

(4.12) $$a(u,\eta) = l(\eta) \qquad \text{for every } \eta \in M.$$

Equation (4.12) is a weak form of the original BVP (4.1) and is the integral form of the Euler-Lagrange equation for the variational principle. It is worth noting that (4.12) is known in mechanics as the principle of virtual work. If f is given as a continuous function, one can show that the solution of (4.12) is a classical solution of (4.1). When f is only in L_2, the solution of (4.12) is regarded as a generalized solution of (4.1). In our problem, which involves an ordinary differential equation, this generalized solution is in the class of functions u with u' absolutely continuous and u'' in L_2. However, for partial differential equations the generalized solution is not characterized so easily.

To avoid too much repetition we now proceed to the general case, after which we shall return to our special BVP (4.1) and to other examples.

Quadratic Functionals

Let H be a real Hilbert space with inner product $<,>$ and norm $\| \ \|$. We warn the reader that a second inner product will be introduced later. Let J be a real-valued functional defined on a linear manifold D in H. We say that J is *stationary* at $u \in D$ if

$$(4.13) \qquad \left[\frac{d}{d\varepsilon} J(u + \varepsilon \eta) \right]_{\varepsilon = 0} = 0 \qquad \text{for every } \eta \in D,$$

so that, at u, the directional derivative of J vanishes in every direction lying in D. Similar ideas, for operators as well as functionals, are discussed in Chapter 9.

Since our interest is in linear BVPs, we shall need to consider only quadratic functionals as in (4.5). Throughout we take

$$(4.14) \qquad J(v) = a(v,v) - 2l(v),$$

where l is a linear functional and $a(v,v)$ is the quadratic form associated with a bilinear form $a(v,w)$ defined for $v,w \in D$. The following definitions are given in order of increasing stringency.

Definition. The bilinear form $a(v,w)$ defined on D (that is, for $v,w \in D$) is said to be

$$(4.15) \quad \begin{cases} \text{symmetric:} & \text{if } a(v,w) = a(w,v) \text{ for all } v,w \in D, \\ \text{nonnegative:} & \text{if symmetric and } a(v,v) \geqslant 0 \text{ for all } v \in D, \\ \text{positive:} & \text{if symmetric and } a(v,v) > 0 \text{ for all } v \in D, v \neq 0, \\ \text{coercive:} & \text{if symmetric and there exists a positive constant } c \\ & \text{such that } a(v,v) \geqslant c\|v\|^2 \text{ for all } v \in D. \end{cases}$$

The same terminology is used for an operator A on D if the bilinear form $a(v,w) \doteq \langle Av,w \rangle$ has the corresponding property (see also Section 8, Chapter 5).

If the form is nonnegative, we have the Schwarz inequality

$$(4.16a) \qquad a^2(v,w) \leqslant a(v,v) a(w,w), \qquad v,w \in D.$$

If the form is positive, $a(v,w)$ can serve as an *energy inner product* with corresponding *energy norm* given by

$$\|v\|_a^2 = a(v,v),$$

and the Schwarz inequality can be rewritten as

(4.16b) $|a(v,w)| \leqslant \|v\|_a \|w\|_a, \qquad v, w \in D.$

Assuming only that the bilinear form $a(v,w)$ is symmetric, we note from (4.14) that

(4.17) $J(v+h) - J(v) = 2a(v,h) - 2l(h) + a(h,h) \qquad$ for all $v, h \in D,$

so that

(4.18) $\left[\dfrac{d}{d\varepsilon} J(v+\varepsilon\eta)\right]_{\varepsilon=0} = 2a(v,\eta) - 2l(\eta) \qquad$ for all $v, \eta \in D.$

We are therefore led to the following theorem.

Theorem 1.   Let $a(v,w)$ be a symmetric bilinear form for $v, w \in D$. Then the necessary and sufficient condition for the problem

(4.19) $J(v) \rightarrow$ stationary, $v \in D,$

to have a solution $u \in D$ is that this element satisfy the *variational equation*

(4.20) $a(u,\eta) = l(\eta) \qquad$ for every $\eta \in D.$

For any solution u of these equivalent problems we have the *reciprocity relation*

(4.21) $a(u,u) = l(u),$

so that

(4.22) $J(u) = -a(u,u) = -l(u).$

Proof.   If u is an element of D for which the variational equation (4.20) is satisfied, (4.18) shows that

$$\left[\frac{d}{d\varepsilon} J(u+\varepsilon\eta)\right]_{\varepsilon=0} = 0,$$

so that, by definition (4.13), J is stationary at u. If J is stationary at u, (4.18) and the definition of stationarity show that (4.20) is satisfied. Thus the equivalence of problems (4.19) and (4.20) has been proved. If u is a solution of these equivalent problems, we may substitute $\eta = u$ in (4.20) to obtain (4.21) and (4.22).

For positive bilinear forms we can replace "stationary" by "minimum," and we can also guarantee uniqueness (but of course not existence as yet).

Theorem 2. Let $a(v,w)$ be a positive bilinear form on D. Then the problem

$$(4.23) \qquad\qquad J(v) \to \min, \qquad v \in D,$$

is equivalent to the variational equation (4.20). There is at most one solution to these equivalent problems.

Proof. If J has a minimum at u, J is certainly stationary at u, so that (4.20) holds. If (4.20) holds, then (4.17) shows that $J(u + h) - J(u) = a(h,h)$. Since $a(h,h) > 0$ for $h \neq 0$, J clearly has an absolute minimum at u and there is at most one minimizing element.

For an existence proof we need two additional hypotheses: l must be bounded in the energy norm, and D must be complete in energy. The first hypothesis postulates the existence of a constant c such that

$$(4.24) \qquad\qquad |l(v)| \leqslant c\|v\|_a \qquad \text{for every } v \in D,$$

and the second means that, whenever a sequence $\{v_n\}$ in D satisfies $\|v_n - v_m\|_a \to 0$, there exists $v \in D$ such that $\|v_n - v\|_a \to 0$.

Theorem 3. Let $a(v,w)$ be a positive bilinear form on D which is complete in energy. Let l satisfy (4.24). Then the minimum in (4.23) is actually attained for an element (necessarily unique) in D.

Proof. Since D is complete in energy, it is a Hilbert space in its own right under the energy inner product. The boundedness of l allows us to use the Riesz representation theorem [see (7.2), Chapter 4] to guarantee the existence of a unique element $u \in D$ such that

$$(4.25) \qquad\qquad l(v) = a(v,u) \qquad \text{for all } v \in D.$$

We can therefore write

$$(4.26) \qquad\qquad J(v) = \|v\|_a^2 - 2a(v,u) = \|v - u\|_a^2 - \|u\|_a^2,$$

so that it is clear that J attains its minimum at the element u and nowhere else. Hence u is the solution of the minimum problem and of the variational equation. Moreover, the minimal value of J is

$$(4.27) \qquad\qquad J(u) = -\|u\|_a^2 = -l(u).$$

Theorem 3 still begs the question since the original domain D on which $a(v,w)$ is defined may not satisfy the additional hypotheses. Condition (4.24) is usually easy to verify in applications, but we are still left with the problem of constructing a manifold that is complete in energy.

We now show how to add elements to D to make it complete in energy. *We shall require that $a(v,w)$ be coercive [see (4.15)] on D.* Let $\{v_n\}$ be a sequence in D which is fundamental in energy. Thus $\|v_n - v_m\|_a \to 0$, and because of coercivity we also have $\|v_n - v_m\| \to 0$, so that $\{v_n\}$ is also fundamental in the ordinary norm. Since H is complete in the ordinary norm, there exists a unique element $v \in H$ such that $\|v_n - v\| \to 0$. The set of all such elements in H is the space of elements of finite energy.

Definition. The space M of *elements of finite energy* is the set of elements $v \in H$ with the simultaneous properties

$$(4.28) \qquad\qquad \|v_n - v_m\|_a \to 0, \qquad \|v_n - v\| \to 0,$$

and the sequence $\{v_n\} \in D$ is said to be *representative* for v.

We note that all elements of D belong to M since we can take $v_n = v$ in the definition. The energy norm and inner product can be extended to the whole of M by the obvious definitions

$$\|v\|_a = \lim_{n\to\infty} \|v_n\|_a, \qquad a(v,w) = \lim_{n\to\infty} a(v_n, w_n),$$

where $\{v_n\}$ and $\{w_n\}$ are representative for v and w, respectively. One must show of course that the limits in these definitions exist and are independent of the representative sequences used, but we omit the proofs. Both (4.15) and (4.24) are then easily extended to the whole of M. It can be shown that M is complete in energy, so that the representative sequence $\{v_n\}$ in (4.28) also has the property

$$(4.29) \qquad\qquad \|v_n - v\|_a \to 0.$$

Much of what we have done is related to the standard completion procedure described in Exercise 3.1, Chapter 4.

Theorem 4. Let $a(v,w)$ be coercive on D, and let M be the set of elements of finite energy (that is, the completion of D in energy). Then the problem

$$(4.30) \qquad\qquad J(v) \to \min, \qquad v \in M,$$

has one and only one solution $u \in M$ which is also the solution of the variational problem

$$(4.31) \qquad\qquad a(u,\eta) = l(\eta) \qquad \text{for every } \eta \in M.$$

Proof. Since $a(v,w)$ is coercive on M and (4.24) holds for $\eta \in M$, we can apply Theorem 3.

In most applications we are interested in the inhomogeneous problem

$$(4.32) \qquad\qquad Au = f, \qquad u \in D_A,$$

where A is a symmetric operator on the linear manifold D_A dense in H. We associate with A the symmetric bilinear form

$$(4.33) \qquad\qquad a(v,w) \doteq \langle Av, w \rangle, \qquad v, w \in D_A.$$

The linear functional l is defined from

$$(4.34) \qquad\qquad l(v) \doteq \langle f, v \rangle,$$

so that

$$(4.35) \qquad\qquad J(v) = \langle Av, v \rangle - 2\langle f, v \rangle, \qquad v \in D_A.$$

The variational equation on D_A becomes

$$\langle Au - f, \eta \rangle = 0 \qquad \text{for all } \eta \in D_A,$$

which is equivalent (since D_A is dense in H) to

$$(4.36) \qquad\qquad Au = f, \qquad u \in D_A.$$

Theorems 1 and 2 now show the equivalence between a variational principle for (4.35) and problem (4.36), whereas Theorem 3 gives existence when D_A is complete in energy. Of course when we construct M by completing D_A in energy, relation (4.33) between a and A is lost, so that

Theorem 4 cannot be stated in terms of A. However, if it happens that the solution u of (4.31), and hence of (4.30), lies in the subset D_A of M, then u satisfies $Au = f$; if not, we view u as a generalized solution of $Au = f$.

Suppose that A is coercive and that (4.32) has a solution u (necessarily unique) in D_A. We can then obtain bounds for $J(u)$ *from both sides* in terms of an arbitrary element $v \in D_A$. With $h = v - u$, we have

$$J(u) = J(v) - \langle Ah, h \rangle.$$

Since $\langle Ah, h \rangle > 0$ for $h \neq 0$, we have the usual upper bound $J(u) \leqslant J(v)$. To find a lower bound we first recall that, from coercivity,

$$\langle Ah, h \rangle \geqslant L_A \|h\|^2, \qquad \text{where } L_A > 0.$$

Schwarz's inequality then shows that

$$\|Ah\| \geqslant L_A \|h\|.$$

It follows that

$$\langle Ah, h \rangle \leqslant \|Ah\| \, \|h\| \leqslant \frac{1}{L_A} \|Ah\|^2 = \frac{1}{L_A} \|Av - f\|^2,$$

which gives the explicit lower bound

$$(4.37) \qquad\qquad J(u) \geqslant J(v) - \frac{1}{L_A} \|Av - f\|^2.$$

An improved upper bound for $J(u)$ can be obtained on the additional assumption that A is bounded above:

$$\langle Ah, h \rangle \leqslant U_A \|h\|^2.$$

It is known that there exists an unambiguous positive symmetric operator $A^{1/2}$ (the *square root* of A) with the property $A^{1/2} A^{1/2} = A$. We then find that

$$\|Ah\|^2 = \langle Ah, Ah \rangle = \langle A(A^{1/2}h), A^{1/2}h \rangle$$
$$\leqslant U_A \|A^{1/2}h\|^2 = U_A \langle A^{1/2}h, A^{1/2}h \rangle = U_A \langle Ah, h \rangle,$$

so that

$$\langle Ah, h \rangle \geqslant \frac{1}{U_A} \|Ah\|^2 = \frac{1}{U_A} \|Av - f\|^2,$$

and, together with (4.37), we have

$$(4.38) \qquad J(v) - \frac{1}{L_A} \|Av - f\|^2 \leqslant J(u) \leqslant J(v) - \frac{1}{U_A} \|Av - f\|^2.$$

A few examples may serve to clarify some of the ideas just presented, particularly the construction of the space M complete in energy.

Example 1. For purposes of comparison we shall simultaneously study the BVP (4.1) and a second BVP obtained from (4.1) by changing the boundary condition at the right end to $u'(1) = 0$. In operator form we have

$$(4.39) \qquad A_1 u = f, \qquad u \in D_1,$$
$$(4.40) \qquad A_2 u = f, \qquad u \in D_2,$$

where $A_1 u = A_2 u = -u'' + ku$, D_1 is the domain consisting of all functions $u(x)$ in $C^2(0,1)$ with $u(0) = u(1) = 0$, and D_2 is the domain of all functions $u(x)$ in $C^2(0,1)$ with $u(0) = u'(1) = 0$. Both D_1 and D_2 are linear manifolds dense in $H = L_2(0,1)$.

If $v, w \in D_1$, we find, on integration by parts, that

$$\langle A_1 v, w \rangle = \int_0^1 (v'w' + kvw) \, dx.$$

If $v, w \in D_2$, we obtain exactly the same expression for $\langle A_2 v, w \rangle$, so that for both problems we define the same bilinear form:

$$a(v, w) \doteq \int_0^1 (v'w' + kvw) \, dx$$

with the associated quadratic form

$$a(v, v) = \int_0^1 \left[(v')^2 + kv^2 \right] dx.$$

It is easy to see that $a(v,v) > 0$ for all nontrivial v in D_1 and D_2. Thus $a(v,w)$ can serve as the energy inner product with corresponding energy norm given by $\|v\|_a^2 = a(v,v)$. Note that the energy norm can also be written in terms of the L_2 norms of v and its first derivative,

$$(4.41) \qquad \|v\|_a^2 = a(v,v) = \|v'\|^2 + k\|v\|^2.$$

It is clear that for $k > 0$ the form is coercive [it suffices to set $c = k$ in (4.15)]. Consider therefore the case $k = 0$ (see also the Example following (8.6), Chapter 5). Then, if $v \in D_1$ or D_2, we have $v(0) = 0$ and hence

$v(x) = \int_0^x v'(t)\,dt$, so that, by the Schwarz inequality,

$$v^2(x) \leqslant \int_0^x 1^2\,dt \int_0^x (v')^2\,dt \leqslant x \int_0^1 (v')^2\,dt,$$

from which we infer the desired inequality

$$\int_0^1 (v')^2\,dt \geqslant 2\|v\|^2, \qquad v \in D_1 \quad \text{or} \quad v \in D_2.$$

The constant 2 that appears on the right side can be improved by appealing to the minimum characterization of eigenvalues. For instance, the best possible inequality for $v \in D_1$ is

$$\int_0^1 (v')^2\,dt \geqslant \pi^2 \|v\|^2.$$

Let M_1 and M_2 be the sets of functions of finite energy corresponding to the BVPs (4.39) and (4.40), respectively. A function $v \in M_1$ if there exists a sequence $\{v_n\}$ in D_1 such that

(4.42) $$\|v_n - v_m\|_a \to 0 \qquad \text{and} \qquad \|v_n - v\| \to 0.$$

The same definition applies to M_2 except that the sequence $\{v_n\}$ must then be chosen from D_2. In either case we see from (4.41) that conditions (4.42) are equivalent to

$$\|v_n - v\| \to 0 \qquad \text{and} \qquad \|v_n' - v_m'\| \to 0,$$

from the second of which we infer that v_n' converges in L_2 to some element w. Conditions (4.42) are therefore equivalent to the existence of elements $v, w \in L_2$, for which, simultaneously,

(4.43) $$\|v_n - v\| \to 0 \qquad \text{and} \qquad \|v_n' - w\| \to 0.$$

Whenever (4.43) holds with a sequence $\{v_n\}$ in $C^1(0,1)$, we refer to w as the *strong L_2 derivative* of v and write $w = v'$. In one dimension this notion is identical with the following property: $v(x)$ is absolutely continuous with the first derivative in L_2 (see Chapter 5 for the definition and discussion of absolute continuity). The space of elements v having a strong L_2 derivative is known as the *Sobolev space* $H^1(0,1)$. Thus both M_1 and M_2 are subsets of H^1, M_1 being formed by sequences $\{v_n\}$ in D_1 and M_2 being formed by sequences $\{v_n\}$ in D_2.

An immediate consequence of (4.43) is that $v_n(x)$ tends to $v(x)$ *uniformly* on $0 \leqslant x \leqslant 1$. We can verify this easily for elements in M_1 and M_2. From

Schwarz's inequality we have

$$\left| \int_0^x v_n' \, dt - \int_0^x w \, dt \right|^2 \leqslant \int_0^x 1^2 \, dt \int_0^x |v_n' - w|^2 \, dt \leqslant \|v_n' - w\|^2,$$

so that $\int_0^x v_n' \, dt$ converges uniformly to $\int_0^x w \, dt$ on $0 \leqslant x \leqslant 1$. Since $v_n(0) = 0$, we find that $v_n(x)$ converges uniformly to $\int_0^x w \, dt$, which must therefore coincide with $v(x)$. Hence

$$v(x) = \int_0^x w \, dt,$$

which confirms our earlier statement that $v(x)$ is absolutely continuous with its derivative in L_2. For elements in M_1 the same procedure, using integrals from x to 1 instead of from 0 to x, yields $v(x) = -\int_x^1 w \, dt$. Thus elements in M_1 satisfy the two boundary conditions $v(0) = v(1) = 0$, but all we can show for elements in M_2 is that $v(0) = 0$.

One can also show geometrically that M_1 and M_2 must be different. We claim that the function $v = x$ belongs to M_2 but not to M_1. If v were to belong to M_1, there would be a sequence $\{v_n\}$ in D_1 with the simultaneous properties $\|v_n - x\| \to 0$, $\|v_n' - 1\| \to 0$. In Figure 4.1 we have drawn a candidate for such a sequence $\{v_n\}$. The sequence satisfies $\|v_n - x\| \to 0$ but not $\|v_n' - 1\| \to 0$. The latter condition is violated because v_n' becomes infinite too rapidly near $x = 1$; more precisely, if we had $\|v_n' - 1\| \to 0$ and $\|v_n - x\| \to 0$, then v_n would converge uniformly to x on $0 \leqslant x \leqslant 1$, but this is impossible as long as v_n vanishes at $x = 1$, while v does not. On the other hand, $v \in M_2$; the sequence $\{z_n\}$ of Figure 4.2 satisfies $\|z_n - x\| \to 0$ and

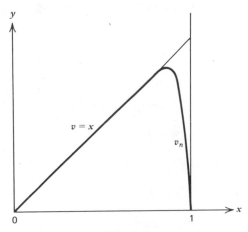

Figure 4.1

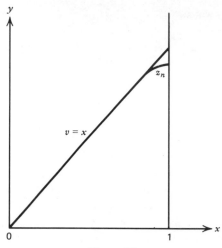

Figure 4.2

$\|z_n' - 1\| \to 0$. To prove the second condition, note that

$$|z_n' - 1| \leqslant 1, \quad x_n \leqslant x \leqslant 1,$$
$$z_n' = 1, \quad\quad 0 \leqslant x < x_n.$$

Since $x_n \to 1$ as $n \to \infty$, $\|z_n' - 1\| \to 0$, as claimed.

To recapitulate:

M_1 is the set of elements $v(x)$ in $H^1(0, 1)$ satisfying $v(0) = v(1) = 0$. M_2 is the set of elements $v(x)$ in $H^1(0, 1)$ satisfying $v(0) = 0$.

Theorem 4 then gives the following results. Setting

$$J(v) = a(v, v) - 2l(v) = \int_0^1 \left[(v')^2 + kv^2 \right] dx - 2 \int_0^1 fv \, dx,$$

we find that the problem

$$J(v) \to \min, \quad v \in M_1,$$

has a unique solution u which satisfies

(4.44) $a(u, \eta) = l(\eta) \quad$ for every $\eta \in M_1$.

The problem

$$J(v) \to \min, \quad v \in M_2,$$

has a unique solution u which satisfies

(4.45) $a(u, \eta) = l(\eta)$ for every $\eta \in M_2$.

We would now like to show that when f is continuous the solutions of (4.44) and (4.45) satisfy (4.39) and (4.40), respectively. This means that we must show that our minimizing functions are twice differentiable (functions in M_1 and M_2 need have only one derivative) and that in (4.45) the minimizing function satisfies the boundary condition $u'(1) = 0$. Any boundary condition in the BVP which need not be imposed on the set of admissible functions in the variational principle is said to be a *natural boundary condition*. Other boundary conditions are *essential*. Thus the boundary condition $u(0) = 0$ is essential, as is $u(1) = 0$, but the boundary condition $u'(1) = 0$ is natural. We shall carry out the proofs for the more interesting case, that of (4.45).

Equation (4.45) characterizes a function $u \in M_2$ such that

(4.46) $\displaystyle\int_0^1 (u'\eta' + ku\eta - f\eta)\,dx = 0$ for every $\eta \subset M_2$.

Since we are not allowed to assume that u'' exists, we shall use integration by parts on the term $(ku - f)\eta$. Setting

(4.47) $G(x) = \displaystyle\int_0^x (ku - f)\,dt,$

we obtain from (4.46)

(4.48) $\displaystyle\int_0^1 \eta'(u' - G)\,dx + \eta G]_0^1 = 0$ for every $\eta \in M_2$.

Since (4.48) holds for all $\eta \in M_2$, it certainly holds for those $\eta \in M_2$ that also satisfy $\eta(1) = 0$. This happens to be the set $\eta \in M_1$. Thus we find that

$$\int_0^1 \eta'(u' - G)\,dx = 0 \qquad \text{for every } \eta \in M_1,$$

from which we conclude (see Exercise 4.1) that

$$u'(x) = G(x) + C. \quad \cdot$$

Hence from (4.47) we see that u'' exists and $u'' = ku - f$. Now that the existence of u'' has been established, we return to (4.46) and integrate the first term by parts to obtain

$$0 = \int_0^1 \eta(-u'' + ku - f)\,dx + \eta(1)u'(1) - \eta(0)u'(0), \qquad \eta \in M_2.$$

Since we have shown that $-u'' + ku - f = 0$, and since $\eta(0) = 0$, we have

$$\eta(1)u'(1) = 0 \qquad \text{for every } \eta \in M_2.$$

But when $\eta \in M_2$, the value of $\eta(1)$ is arbitrary, so that $u'(1) = 0$. We have shown that the solution of (4.40) or, equivalently, the function that minimizes J in M_2 is the solution of

$$-u'' + ku = f, \quad 0 < x < 1; \qquad u(0) = u'(1) = 0.$$

Example 2. Let Ω be a bounded domain in R_n with boundary Γ, and consider the BVP

$$(4.49) \qquad\qquad -\Delta u + ku = f, \quad x \in \Omega; \qquad u|_\Gamma = 0,$$

where k is a nonnegative constant and f is a given element of the Hilbert space $H = L_2(\Omega)$, whose inner product and norm are denoted by $\langle\,,\rangle$ and $\|\ \|$, respectively. We translate the BVP (4.49) into the operator equation

$$(4.50) \qquad\qquad Au = f, \qquad u \in D_A,$$

where $A = -\Delta + k$ and D_A is the set of functions $u(x)$ in $C^2(\overline{\Omega})$ which vanish on Γ. Clearly D_A is dense in H.

For $v, w \in D_A$ we have

$$\langle Av, w \rangle = \int_\Omega (-\Delta v + kv) w \, dx = \int_\Omega (\operatorname{grad} v \cdot \operatorname{grad} w + kvw) \, dx;$$

this suggests introducing the bilinear form

$$(4.51) \qquad\qquad a(v, w) \doteq \int_\Omega (\operatorname{grad} v \cdot \operatorname{grad} w + kvw) \, dx,$$

which is clearly symmetric. We observe that

$$(4.52) \qquad\qquad a(v, v) = \int_\Omega (|\operatorname{grad} v|^2 + kv^2) \, dx$$

is positive for every nontrivial v in D_A. The statement is obvious for $k>0$; if $k=0$, then $a(v,v)=0$ implies that $v=$ constant, but since elements in D_A vanish on Γ, the constant must be 0.

Hence $a(v,w)$ can serve as energy inner product with $a(v,v)$ the square of the energy norm. Coercivity follows immediately from (4.52) for $k>0$. If $k=0$, it is necessary to prove the existence of a positive constant c such that

$$\int_\Omega |\operatorname{grad} v|^2 \, dx \geqslant c \int_\Omega v^2 \, dx \qquad \text{for every } v \in D_A.$$

The inequality is known in the literature as Friedrichs' inequality. We shall omit its proof, but we note that the best possible value of c is just the lowest eigenvalue of $-\Delta u = \lambda u$ in Ω, with u vanishing on Γ.

The completion of D_A in the energy norm is the Sobolev space $H_0^1(\Omega)$, defined independently as the closure of $C_0^\infty(\Omega)$ in the Sobolev norm $\| \ \|_1$, given by

$$\|u\|_1^2 = \|u\|^2 + \|\operatorname{grad} u\|^2.$$

Thus $M = H_0^1(\Omega)$. Again it can be proved that the solution of the minimum problem is somewhat smoother. However, even if f is continuous and Γ smooth, (4.49) may not have a classical solution; slightly more is needed: f must be Hölder continuous. We do not attempt any discussion of these ideas.

Our minimum principle is as follows:

$$(4.53) \qquad J(v) \doteq \int_\Omega (|\operatorname{grad} v|^2 + kv^2) \, dx - 2 \int_\Omega fv \, dx \to \min, \qquad v \in H_0^1(\Omega),$$

has a unique solution which is regarded as the generalized solution of (4.49).

Example 3. Let A be a bounded, symmetric operator defined on the whole Hilbert space H. We shall also assume that A is coercive, that is, there exists a positive constant L_A such that

$$\langle Au, u \rangle \geqslant L_A \|u\|^2 \qquad \text{for every } u \in H.$$

The bilinear form

$$a(v,w) \doteq \langle Av, w \rangle$$

is clearly symmetric, positive, and coercive. Thus $a(v,w)$ can serve as an energy inner product with energy norm

$$\|v\|_a^2 = a(v,v) = \langle Av, v \rangle.$$

Since A is both bounded and coercive, we have

$$L_A \|v\|^2 \leqslant \|v\|_a^2 \leqslant \|A\| \|v\|^2,$$

so that the ordinary norm and the energy norm are equivalent. Thus the completion of H in the energy norm is just H, and any linear functional $\langle f, v \rangle$ is also bounded in energy. We therefore have the following minimum principle:

$$(4.54) \qquad\qquad J(v) \doteq \langle Av, v \rangle - 2\langle f, v \rangle, \qquad v \in H,$$

has one and only one minimizing element u which is the solution of the inhomogeneous equation

$$(4.55) \qquad\qquad\qquad\qquad Au = f.$$

This completes the proof of the existence of a solution to (4.55). This result should be compared with those of Chapter 5.

One particular case of importance occurs when $A = K - \mu I$, where K is an integral operator on $L_2(a,b)$:

$$Ku \doteq \int_a^b k(x,\xi) u(\xi) \, d\xi.$$

We assume that K is symmetric, nonnegative, and Hilbert-Schmidt and that the real number μ is negative. Our previous results apply, and the problem

$$J(v) \doteq \int_a^b \int_a^b k(x,\xi) v(x) v(\xi) \, dx \, d\xi - \mu \int_a^b v^2 \, dx - 2 \int_a^b fv \, dx \rightarrow \min$$

has one and only one solution $u(x)$, which is also the solution of the integral equation

$$\int_a^b k(x,\xi) u(\xi) \, d\xi - \mu u(x) = f(x), \qquad a < x < b.$$

Example 4 (Schwinger-Levine principle). Let u be the solution of the problem

$$J(v) \doteq a(v,v) - 2l(v) \to \min, \qquad v \in M.$$

Then the problem

(4.56) $$R(v) \doteq -\frac{l^2(v)}{a(v,v)} \to \min, \qquad v \in M, \quad v \neq 0,$$

has the same minimum value as J, and the minimum is attained for any element of the form $cu, c \neq 0$. The proof is easy. First we note by the reciprocity relation (4.21) that

$$R(cu) = -\frac{l^2(u)}{a(u,u)} = J(u),$$

so that $\inf R(v) \leqslant \min J(v)$. If $\inf R < \min J$, there exists an element $z \in M$ such that $R(z) < \min J$. With $c = l(z)/a(z,z)$ we find that $J(cz) = R(z)$, which is a contradiction. Therefore $\inf R = \min J$. Clearly the infimum is achieved for elements of the form cu and for no others.

The Ritz-Rayleigh Approximation

We want to approximate the solution u of the minimum problem

$$J(v) \doteq a(v,v) - 2l(v), \qquad v \in M,$$

by minimizing J on an n-dimensional subspace E_n of M. We then have

$$\min_{v \in E_n} J(v) \geqslant \min_{v \in M} J(v) = J(u).$$

Since J is bounded below, it has an infimum on E_n and that infimum must be attained because E_n is finite-dimensional. We recall that if $v \in M$ we can write, from (4.26),

$$J(v) = \|v - u\|_a^2 - \|u\|_a^2,$$

so that, since u is a fixed (though unknown) element,

(4.57) $$\min_{v \in E_n} J(v) = -\|u\|_a^2 + \min_{v \in E_n} \|v - u\|_a^2.$$

The minimum appearing on the right side characterizes the element in E_n which is nearest in energy to the fixed element u, that is, the orthogonal projection (in the energy inner product) of u on E_n. We shall denote this projection by $u^{(n)}$, the superscript anticipating that the dimension of E_n may vary. Introducing a basis $\{h_i\}$ in E_n, we can write

$$(4.58) \qquad u^{(n)} = \sum_{j=1}^{n} \alpha_j^{(n)} h_j,$$

and since $u^{(n)} - u$ is orthogonal to E_n in energy,

$$(4.59) \qquad a(u^{(n)} - u, h_i) = 0, \qquad i = 1, \ldots, n.$$

A consequence of the variational equation (4.31) satisfied by u is that $a(u, h_i) = l(h_i), i = 1, \ldots, n$. Thus we can write (4.59) as

$$(4.60) \qquad a(u^{(n)}, h_i) = l(h_i), \qquad i = 1, \ldots, n,$$

or as the set of n linear equations for $\alpha_1^{(n)}, \ldots, \alpha_n^{(n)}$,

$$(4.61) \qquad \sum_{j=1}^{n} a_{ij} \alpha_j^{(n)} = l(h_i), \qquad i = 1, \ldots, n,$$

where

$$(4.62) \qquad a_{ij} \doteq a(h_i, h_j).$$

Equations (4.61) are known as the *Galerkin equations*, which determine un- ambiguously the *Ritz-Rayleigh approximation* $u^{(n)}$.

If we multiply (4.52) by α_i and sum over n, we note that our Ritz-Rayleigh approximation $u^{(n)}$ satisfies the reciprocity principle

$$(4.63) \qquad a(u^{(n)}, u^{(n)}) = l(u^{(n)}).$$

In view of the orthogonality of $u^{(n)}$ and u, we see from (4.57) that

$$(4.64) \qquad J(u^{(n)}) = -\|u^{(n)}\|_a^2 = -a(u^{(n)}, u^{(n)}) = -l(u^{(n)}),$$

which is an upper bound to $J(u)$.

If the elements in the basis for E_n are chosen to be orthonormal in energy, some simplifications take place. Denoting the elements of the basis by $\{e_i\}$ instead of $\{h_i\}$, we have

$$a(e_i, e_j) = \begin{cases} 0, & i \neq j, \\ 1, & i = j. \end{cases}$$

Then (4.61) immediately gives

(4.65) $$\alpha_i^{(n)} = l(e_i),$$

the Ritz-Rayleigh approximation is

(4.66) $$u^{(n)} = \sum_{j=1}^{n} l(e_j) e_j,$$

and from (4.64) the corresponding value of J is

(4.67) $$J(u^{(n)}) = - \sum_{j=1}^{n} l^2(e_j),$$

which provides an explicit upper bound to $J(u)$.

Now suppose that the set $\{e_i\}$ is an infinite orthonormal set (in energy) which is also a basis (in energy) for M. Then every element $v \in M$ can be written as

$$v = \sum_{j=1}^{\infty} a(v, e_j) e_j,$$

and, in particular, the solution u of the original minimum problem is given by

(4.68) $$u = \sum_{j=1}^{\infty} a(u, e_j) e_j = \sum_{j=1}^{\infty} l(e_j) e_j,$$

so that (4.66) converges in energy to u as $n \to \infty$. Since $a(v, w)$ is a coercive form, convergence in the ordinary norm follows from energy convergence, so that (4.68) also holds in the norm of H. From (4.67) we can conclude that

(4.69) $$J(u) = - \sum_{j=1}^{\infty} l^2(e_j).$$

Although (4.68) and (4.69) give explicit expressions for the solution of the minimum problem, they are not as useful as they appear, for it may be numerically awkward to construct (say, by the Gram-Schmidt procedure) an orthonormal energy basis.

In many applications, such as those of Example 1, (4.39)-(4.40), the minimum problem stems from the operator equation

$$(4.70) \qquad\qquad Au = f, \qquad u \in D_A,$$

where A is a coercive operator on D_A. In that case, as we have seen in (4.33) and (4.34), we let

$$(4.71) \qquad\qquad l(v) \doteq \langle v, f \rangle, \qquad v \in D_A,$$

$$(4.72) \qquad\qquad a(v, w) \doteq \langle Av, w \rangle \qquad \text{for } v, w \in D_A.$$

If f is such that (4.70) has a solution u in D_A, we can characterize u as the minimum of $J(v) \doteq a(v, v) - 2l(v)$ either on D_A or on M. Even in such a favorable case it is important to study the problem on M because the Ritz-Rayleigh procedure can then use approximating elements $h_1, \ldots, h_n$ which lie in M rather than D_A. Often calculations become simpler if the elements $\{h_i\}$ are only piecewise smooth (say, piecewise linear, such as the roof functions used in the finite element method) rather than being C^2 (as will usually be the case for elements of D_A). Another advantage of looking at the approximation problem on M is that the approximating elements $\{h_i\}$ need only satisfy the essential boundary conditions. The Ritz-Rayleigh approximations (4.66) and (4.67) then become

$$(4.73) \qquad u^{(n)} = \sum_{j=1}^{n} \langle f, e_j \rangle e_j, \qquad J(u^{(n)}) = - \sum_{j=1}^{n} \langle f, e_j \rangle^2,$$

and if $\{e_i\}$ is a basis for M, then

$$(4.74) \qquad u = \sum_{j=1}^{\infty} \langle f, e_j \rangle e_j, \qquad J(u) = - \sum_{j=1}^{\infty} \langle f, e_j \rangle^2.$$

We now illustrate the method through an example.

Example 5. Consider the BVP

$$(4.75) \qquad -u'' + ku = 1, \quad 0 < x < 1; \qquad u(0) = u'(1) = 0,$$

which is a particular case of (4.40) corresponding to the constant forcing function $f(x) = 1$. Although this problem can be solved in closed form, we shall try a Ritz-Rayleigh procedure with the approximating elements $h_1 = x, h_2 = x^2$. Both h_1 and h_2 belong to M since they satisfy the essential boundary condition at $x = 0$ (and of course they are more than smooth enough). From (4.71) and (4.72) we have

$$l(v) = \langle v, 1 \rangle = \int_0^1 v \, dx.$$

A simple calculation gives

$$l(h_1) = \tfrac{1}{2}, \qquad l(h_2) = \tfrac{1}{3},$$

$$a_{11} = 1 + \tfrac{k}{3}, \qquad a_{12} = a_{21} = 1 + \tfrac{k}{4}, \qquad a_{22} = \tfrac{4}{3} + \tfrac{k}{5},$$

where $a_{ij} = a(h_i, h_j)$.

If we use only a *one-term* approximation in terms of h_1, we find, from (4.61) and (4.64), that

(4.76)

$$\alpha_1^{(1)} = \frac{3}{2(k+3)}, \qquad u^{(1)} = \frac{3x}{2(k+3)}, \qquad J(u^{(1)}) = -\alpha_1^{(1)} l(h_1) = -\frac{3}{4(k+3)}.$$

For a *two-term* approximation using h_1 and h_2, we see from (4.61) that $\alpha_1^{(2)}$ and $\alpha_2^{(2)}$ satisfy the linear equations

(4.77)

$$\left(1 + \tfrac{k}{3}\right)\alpha_1^{(2)} + \left(1 + \tfrac{k}{4}\right)\alpha_2^{(2)} = \tfrac{1}{2},$$

$$\left(1 + \tfrac{k}{4}\right)\alpha_1^{(2)} + \left(\tfrac{4}{3} + \tfrac{k}{5}\right)\alpha_2^{(2)} = \tfrac{1}{3}.$$

For $k = 0$ we obtain

(4.78) $\quad \alpha_1^{(2)} = 1, \qquad \alpha_2^{(2)} = -\tfrac{1}{2}, \qquad u^{(2)} = x - \dfrac{x^2}{2}, \qquad J(u^{(2)}) = -\tfrac{1}{3},$

whereas for $k = 1$ we find that

$$\alpha_1^{(2)} = 0.726, \qquad \alpha_2^{(2)} = -0.375,$$

(4.79)

$$u^{(2)} = (0.726)x - (-0.375)x^2, \qquad J(u^{(2)}) = -(0.238).$$

For purposes of comparison we conveniently recall that problem (4.75) can be solved in closed form:

$$u(x) = \frac{1}{k} - \frac{1}{k \cosh \sqrt{k}} \cosh \sqrt{k} \, (x-1), \qquad k>0,$$

$$u(x) = x - \frac{x^2}{2}, \qquad\qquad\qquad k=0,$$

where the solution for $k=0$ can be derived either separately or by taking the limit as $k \to 0$ of the solution for $k>0$. The corresponding values of J are

$$J(u) = -\frac{1}{k} + (k^{-3/2}) \tanh \sqrt{k}, \qquad k>0,$$

$$J(u) = -\tfrac{1}{3}, \qquad\qquad\qquad k=0.$$

The two-term approximation is perfect for $k=0$ because, fortuitously, the exact solution of the BVP is a linear combination of h_1 and h_2. For $k=1$ we note that $[(d/dx)u^{(2)}]_{x=1} \neq 0$. The exact value of $J(u)$ when $k=1$ is

$$J = -1 + \tanh(1) = -0.238,$$

which to three decimal places agrees with (4.79), whereas the approximation for $u^{(2)}$ is not nearly as good.

Of course we can also use the Ritz-Rayleigh approximation $u^{(n)}$ for v in (4.37) to obtain a lower bound to $J(u)$. Additional pairs of bounds will be given in the next subsection. We observe here that we can also estimate the error between $u^{(n)}$ and u. From (3.51) we find that

$$J(u^{(n)}) = J(u) + \|u^{(n)} - u\|_a^2,$$

so that

$$\|u^{(n)} - u\|_a^2 = J(u^{(n)}) - J(u) \leqslant J(u^{(n)}) - m,$$

where m is a lower bound to $J(u)$ such as the one given in (4.37). With $v=u^{(n)}$ we then obtain

$$\|u^{(n)} - u\|_a^2 \leqslant \frac{1}{L_A} \|Au^{(n)} - f\|^2.$$

Complementary Variational Principles

Consider a linear BVP, possibly with inhomogeneous boundary conditions, for a function $u(x)$ on a domain Ω. Existence is not in question here, as we shall assume that the BVP has a solution. It is sometimes tedious to obtain even an approximate solution that is valid throughout the domain Ω. We may, however, be willing to settle for an approximate value of a single numerical quantity having physical significance as some sort of "average" of the solution. Such examples as capacity, torsional rigidity, and scattering cross section immediately come to mind. In each case a single number gives a great deal of information about a phenomenon whose details are governed by a BVP of some complexity. The number in question can usually be expressed as a functional of the solution $u(x)$ of the BVP. We shall try to obtain upper and lower bounds for this functional in terms of comparison functions $v(x)$ that may be required to satisfy certain constraints.

The proper setting for our work is a Hilbert space H with inner product $\langle\,,\,\rangle$ and norm $\|\ \|$. Let $a(v,w)$ be a *symmetric, nonnegative* bilinear form on the linear manifold D in H. Then $a(v,w)$ does not quite serve as a new inner product because the vanishing of $a(v,v)$ does not guarantee that v is the zero element. Nevertheless we shall see that the Schwarz inequality and the Pythagorean theorem remain true, and we shall therefore still refer to $a(v,w)$ as an energy inner product. The following simple inequalities suffice for our purposes:

$$(4.80) \qquad a(u,u) \geqslant 2a(u,v) - a(v,v),$$

$$(4.81) \qquad a(u,u) \geqslant \frac{a^2(u,v)}{a(v,v)} \qquad \text{if } a(v,v) \neq 0,$$

$$(4.82) \qquad a(u,u) \leqslant a(v,v) \qquad \text{if } a(v-u,u) = 0.$$

The first of these inequalities follows from $a(u-v,u-v) \geqslant 0$, the second from replacing v by $va(u,v)/a(v,v)$ in the first, and the third from

$$a(v,v) = a(u+v-u, u+v-u) = a(u,u) + 2a(u,v-u) + a(v-u,v-u)$$
$$= a(u,u) + a(v-u,v-u) \geqslant a(u,u).$$

Inequality (4.82) has a simple geometrical meaning. Since $v-u$ and u are orthogonal in the energy inner product, we may view v as the hypotenuse of a right triangle whose other two sides are u and $v-u$, so that the length of u cannot exceed that of v.

Our first application of these principles is to a BVP for the Poisson equation:

$$(4.83) \qquad -\Delta u = f(x), \quad x \in \Omega; \qquad u|_\Gamma = 0,$$

where Ω is a bounded domain in R_n with boundary Γ, and f is a given function. For simplicity we shall assume that f is smooth enough so that (4.83) has a classical solution. When $f(x) = 2$ and $n = 2$, (4.83) governs the torsion of a cylinder of cross section Ω. The function $u(x) = u(x_1, x_2)$ is then the stress function. The constant of proportionality relating the twisting moment and the angle of twist per unit length of the cylinder is known as the torsional rigidity and will be denoted by T. If, for convenience, we set the shear modulus equal to unity, it can be shown that

$$T = \int_\Omega 2u(x)\,dx = \int_\Omega |\text{grad}\,u|^2\,dx,$$

where u is the solution of (4.83) for $f(x) \equiv 2$.

Since it is just as easy to deal with the case of a general function in (4.83), we shall consider the problem of estimating

$$(4.84) \qquad a(u,u) \doteq \int_\Omega fu\,dx = \int_\Omega |\text{grad}\,u|^2\,dx,$$

where u is the solution of (4.83). The corresponding bilinear form is

$$(4.85) \qquad a(v,w) \doteq \int_\Omega \text{grad}\,v \cdot \text{grad}\,w\,dx,$$

defined on the set D of piecewise smooth functions on $\bar{\Omega}$.

We want to estimate (4.84) without solving (4.83) for u. To use (4.80) we will need to find comparison functions v for which $a(u,v)$ can be calculated without a knowledge of u. We note that

$$a(u,v) = \int_\Omega (\text{grad}\,u \cdot \text{grad}\,v)\,dx = -\int_\Omega v\Delta u\,dx + \int_\Gamma v\frac{\partial u}{\partial n}\,dS,$$

so that, since $-\Delta u = f$,

$$a(u,v) = \int_\Omega fv\,dx \qquad \text{if } v|_\Gamma = 0, \quad v \in D.$$

Thus we obtain

$$(4.86) \qquad a(u,u) \geqslant 2\int_\Omega fv\,dx - \int_\Omega |\text{grad}\,v|^2\,dx \qquad \text{if } v|_\Gamma = 0, \quad v \in D,$$

which is just the minimum principle (4.53) since $a(u,u) = -J(u)$. Applying (4.81), we find that

$$(4.87) \qquad a(u,u) \geqslant \frac{\left(\int_\Omega fv\,dx\right)^2}{\int_\Omega |\text{grad}\,v|^2\,dx} \qquad \text{if } v|_\Gamma = 0, \quad v \in D, v \not\equiv 0,$$

which is just the Schwinger-Levine principle (4.56) with a change of sign. Of course in (4.87) the maximum is achieved not just for the solution of (4.83) but also for any nonzero multiple of it.

A bound from the opposite direction can be obtained from (4.82) by finding comparison functions v which satisfy $a(v-u,u)=0$, where u is the unknown solution of (4.83). A simple calculation gives

$$a(v-u,u) = -\int_\Omega u\Delta(v-u)\,dx + \int_\Omega u\frac{\partial}{\partial n}(v-u)\,dS.$$

Since u is 0 on Γ, it suffices that $-\Delta v = f$ for $a(v-u,u)$ to vanish. We therefore find from (4.82) that

$$(4.88) \qquad a(u,u) \leqslant \int_\Omega |\text{grad}\,v|^2\,dx \qquad \text{for every } v \text{ satisfying } -\Delta v = f.$$

Because the comparison function v is required to satisfy Poisson's equation (without the boundary condition, however), it is clear that v must be restricted to a subset of D containing sufficiently smooth functions.

In the particular case of torsional rigidity, we have the bounds

$$(4.89) \qquad \frac{4\left(\int_\Omega w\,dx\right)^2}{\int_\Omega |\text{grad}\,w|^2\,dx} \leqslant T \leqslant \int_\Omega |\text{grad}\,v|^2\,dx,$$

where v satisfies $-\Delta v = 2$, and w vanishes on Γ.

We can easily use the inequalities in (4.89) to obtain explicit bounds for T in terms of simpler domain functionals. For an arbitrary choice of the origin, the function $v = -(x_1^2 + x_2^2)/2$ satisfies $-\Delta v = 2$, and therefore

$$T \leqslant \int_\Omega (x_1^2 + x_2^2)\, dx_1\, dx_2.$$

The right side is smallest if the origin is chosen at the center of gravity of the cross section Ω. The integral then becomes the polar moment of inertia I_p, so that

(4.90) $T \leqslant I_p.$

An improvement of this upper bound can be found in Exercise 4.4.

A different type of upper bound can be obtained by using level line coordinates. The solution u of the torsion problem

(4.91) $-\Delta u = 2, \quad x \in \Omega; \qquad u|_\Gamma = 0$

is positive in Ω by the maximum principle. Let us introduce a coordinate t such that $0 < t \leqslant u_m$, where u_m is the unknown maximum value of the solution. Let $A(t)$ be the area enclosed within the level curve $u = t$:

$$A(t) \doteq \int_{u > t} dx, \qquad \text{so that } A(u_m) = 0, \quad A(0) = \text{area of } \Omega.$$

By considering the annular domain between $u = t$ and $u = t + dt$, we find that

$$-A'(t) = \int_{u = t} \frac{dl}{|\operatorname{grad} u|},$$

and from the differential equation (4.91)

$$\int_{u = t} |\operatorname{grad} u|\, dl = 2A(t).$$

Multiplying the expressions for $2A$ and for $-A'$, we obtain, after using Schwarz's inequality,

$$-2AA' = \int_{u = t} \frac{dl}{|\operatorname{grad} u|} \int_{u = t} |\operatorname{grad} u|\, dl \geqslant L^2(t),$$

where $L(t)$ is the length of the curve $u=t$. Of course a curve of length L cannot enclose more area than a circle of the same length. Therefore

$$L^2(t) \geqslant 4\pi A(t) \quad \text{and} \quad -\frac{d}{dt}A^2 \geqslant 4\pi A.$$

We integrate the latter inequality from $t=0$ to $t=u_m$ to obtain

$$A^2(0) \geqslant 4\pi \int_0^{u_m} A\, dt = -4\pi \int_0^{u_m} tA'\, dt = -4\pi \int_0^{u_m} \psi'\, dt,$$

where

$$\psi(t) \doteq \int_{u>t} u\, dx, \qquad \psi(0) = \frac{T}{2}.$$

We therefore recover Pólya's isoperimetric inequality:

(4.92)
$$T \leqslant \frac{A^2}{2\pi},$$

the term "isoperimetric" referring to the fact that equality is achieved for a particular domain—in this case, a circular disk.

To find a lower bound we shall use a special class of comparison functions w whose level lines are similar to the curve Γ. We assume Γ is star-shaped with respect to an origin O within Ω; this means that every ray emanating from O intersects Γ in only one point whose polar coordinates are (r,θ) with $r=f(\theta)$, $0 \leqslant \theta < 2\pi$. Let $\phi(t)$ be a nonnegative function on $0 \leqslant t \leqslant 1$ with $\phi(0)=1$, $\phi(1)=0$. Choosing $w=\phi(r/f(\theta))$, we see that w is constant on the curves $r=cf(\theta)$, $0 \leqslant c \leqslant 1$, which are similar to the curve Γ, whose equation is $r=f(\theta)$. A simple calculation gives

$$\int_\Omega w\, dx = \int_0^{2\pi} d\theta \int_0^f \phi\left(\frac{r}{f}\right) r\, dr = \int_0^{2\pi} f^2(\theta)\, d\theta \int_0^1 c\phi(c)\, dc$$

$$= 2A \int_0^1 c\phi(c)\, dc,$$

where A is the area of Ω.

We also obtain

$$\int_\Omega |\operatorname{grad} w|^2\, dx = \int_0^{2\pi} d\theta \int_0^f \left\{ \left[\frac{\partial}{\partial r}\phi\left(\frac{r}{f}\right) \right]^2 + \frac{1}{r^2}\left[\frac{\partial}{\partial\theta}\phi\left(\frac{r}{f}\right) \right]^2 \right\} r\, dr$$

$$= \int_0^{2\pi} \left[1 + \frac{(f')^2}{f^2} \right] d\theta \int_0^1 c[\phi'(c)]^2\, dc.$$

The quantity $[1+(f')^2/f^2]d\theta$ can be written as $dl/h(\theta)$, where dl is an element of length along the boundary and $h(\theta)$ is the perpendicular distance from the origin to the line tangent to the curve at the point whose angular coordinate is θ. We therefore find from (4.89) that

$$(4.93) \quad T \geqslant \frac{16A^2 \left[\int_0^1 c\phi(c)\,dc \right]^2}{\int_0^{2\pi} [dl/h(\theta)] \int_0^1 c[\phi'(c)]^2 dc} \, ; \qquad \phi(0)=1, \quad \phi(1)=0.$$

Let us now choose $\phi(c)$ to be proportional to the actual stress function for a circular cylinder of radius R. The exact solution of (4.91) for a disk of radius R is

$$\tfrac{1}{2}(R^2 - r^2),$$

so that $\phi(c) = (1 - c^2)$ and (4.93) gives

$$(4.94) \qquad\qquad T \geqslant \frac{A^2}{\displaystyle\int_0^{2\pi} dl/h(\theta)}.$$

Next we consider a problem from electrostatics. A thin metallic conductor coincides with the closed surface Γ in R_3. We let Ω_e stand for the unbounded domain exterior to Γ, and Ω_i for the bounded interior domain. If a charge is placed on Γ so that the potential on Γ is unity while the potential at large distances tends to 0, the electrostatic potential satisfies the exterior Dirichlet problem [see (3.39)]

$$(4.95) \qquad \begin{aligned} &\Delta u = 0, \quad x \in \Omega_e; \qquad u|_\Gamma = 1, \\ &\lim_{|x| \to \infty} |x| u = 0, \qquad \lim_{|x| \to \infty} |x|^2 |\operatorname{grad} u| = 0. \end{aligned}$$

The capacity C of the conductor is the total charge on Γ, that is,

$$C = -\int_\Gamma \frac{\partial u}{\partial \nu}\, dS,$$

where ν is the normal to Γ outward from Ω_i. By integrating $u\Delta u$ over Ω_e and using the limiting behavior of u and $|\operatorname{grad} u|$ at infinity, we find that

$$C = -\int_\Gamma \frac{\partial u}{\partial \nu}\, dS = \int_{\Omega_e} |\operatorname{grad} u|^2 dx,$$

where the latter equality shows that the capacity is also equal to the electrostatic energy stored in the field.

To apply our variational principles we introduce the inner product

$$a(v,w) = \int_{\Omega_e} \operatorname{grad} v \cdot \operatorname{grad} w\, dx,$$

which is defined on the set D of functions v with continuous second derivatives in Ω_e and such that $|x|v$ and $|x|^2|\operatorname{grad} v|$ are bounded at infinity. Then the capacity C can be expressed as

$$C = a(u,u).$$

Since

$$a(v-u,u) = \int_{\Omega_e} \operatorname{grad} u \cdot \operatorname{grad}(v-u)\, dx = -\int_{\Gamma}(v-u)\frac{\partial u}{\partial \nu}\, dS,$$

we have $a(v-u,u)=0$ if $v=1$ on Γ. Therefore

(4.96) $$C \leqslant \int_{\Omega_e} |\operatorname{grad} v|^2\, dx \qquad \text{for } v \in D, \quad v|_{\Gamma}=1.$$

It is clear that the minimum of the right side is achieved for $v=u$. A simple calculation gives

$$a(u,v) = -\int_{\Omega_e} u\,\Delta v\, dx - \int_{\Gamma} u \frac{\partial v}{\partial \nu}\, dS,$$

so that

$$a(u,v) = -\int_{\Gamma}\frac{\partial v}{\partial \nu}\, dS \qquad \text{if } v \in D, \quad \Delta v = 0 \text{ in } \Omega_e.$$

From (4.81) we then obtain the lower bound

(4.97) $$C \geqslant \frac{\left[\int_{\Gamma}(\partial v/\partial \nu)\, dS\right]^2}{\int_{\Omega_e} |\operatorname{grad} v|^2\, dx}, \qquad v \in D, \quad v \not\equiv 0, \quad \Delta v = 0 \text{ in } \Omega_e.$$

The capacity problem can also be formulated as an integral equation on Γ. In fact, we can use (3.47) with $u_i(x) \equiv 1, u_e(x) = u(x)$, $\partial u_i / \partial v = 0$, so that

$$(4.98) \qquad\qquad 1 = \int_\Gamma \frac{1}{4\pi|s - \xi|} I(\xi) \, dS_\xi, \qquad s \in \Gamma,$$

where $I(\xi) = -\partial u / \partial v$ is just the charge density on Γ.

Once (4.98) has been solved for I, the potential $u(x)$ is given, as in (3.45), by

$$u(x) = \int_\Gamma \frac{1}{4\pi|x - \xi|} I(\xi) \, dS_\xi, \qquad x \in \Omega_e.$$

Rather than solving (4.98), we try to estimate the capacity. Since

$$C = \int_\Gamma I(x) \, dS_x,$$

we also obtain from (4.98)

$$C = \int_\Gamma \int_\Gamma \frac{1}{4\pi|x - \xi|} I(x) I(\xi) \, dS_x \, dS_\xi.$$

Since the integral operator appearing in (4.98) was shown to be positive at the end of Section 3, we can introduce the inner product

$$a(p, q) = \int_\Gamma \int_\Gamma \frac{1}{4\pi|x - \xi|} p(x) q(\xi) \, dS_x \, dS_\xi,$$

which is certainly well defined for continuous functions on Γ. Then, applying (4.81), we find that

$$C \geqslant \frac{\left[\int_\Gamma p(x) \, dS_x \right]^2}{\int_\Gamma \int_\Gamma dS_x \, dS_\xi \left[p(x) p(\xi) / 4\pi|x - \xi| \right]}, \qquad p(x) \in C(\Gamma), \ p \neq 0.$$

Unilateral Constraints

Boundary value problems with inequality constraints arise frequently in applications, particularly in solid mechanics, where they often go under the name of contact problems. As the barest introduction to the subject, we shall study only two simple examples in beam theory. Although we use the linearized differential equation for transverse deflections, the problems are nevertheless intrinsically nonlinear because of the nature of the constraints. As evidence of this nonlinearity, the general form of the superposition principle does not apply (see Dundurs for a beautiful treatment of this question).

In our first problem the unilateral constraint is on the boundary. Consider a beam $0 < x < 1$ subject to a transverse distributed load $f(x)$. The left end of the beam is built in, but its right end rests on weightless rollers near the edge of a smooth table, as in Figure 4.3. Depending on the nature of $f(x)$, the beam may remain in contact with the table or may be lifted from it. For instance, if $f = -1$ contact is maintained, but if $f = 1$ the right end is raised from its support. Of course if $f = 0$ the deflection vanishes and is therefore not equal to the sum of the deflections corresponding to the respective loads $f = 1$ and $f = -1$. For an arbitrary load $f(x)$ which changes sign along the beam, it is not clear at the outset whether or not contact is maintained, so that both possibilities must be taken into account in the formulation of the boundary conditions at the right end. In either case the moment vanishes at the right end, and therefore $u''(1) = 0$. It would seem at first that the only other information about the right endpoint is that $u(1) \geqslant 0$. However, if $u(1) = 0$ the reaction of the table on the beam must be upward, so that $u'''(1) \leqslant 0$; also, if $u(1) > 0$ the end is free and $u'''(1) = 0$. A suitable formulation of the BVP is therefore

(4.99)
$$\frac{d^4 u}{dx^4} = f(x), \quad 0 < x < 1;$$

$$u(0) = u'(0) = u''(1) = 0, \quad u(1)u'''(1) = 0, \quad u(1) \geqslant 0, \quad u'''(1) \leqslant 0.$$

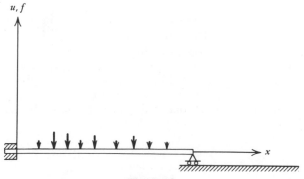

Figure 4.3

We would like to characterize the solution of (4.99) as the minimizing function of a variational principle. Our goal here is not to prove existence theorems but merely to show the equivalence between the BVP and the variational principle. The potential energy of the deflected beam is given by

$$\int_0^1 (u'')^2 \, dx - 2\int_0^1 fu \, dx.$$

This suggests that we consider the minimum, over a suitable set K of admissible functions, of the functional

(4.100) $J(v) \doteq a(v,v) - 2l(v),$

where

$$a(v,w) \doteq \int_0^1 v'' w'' \, dx,$$

$$l(v) \doteq \int_0^1 fv \, dx.$$

We see easily that $a(v,w)$ is a positive bilinear form on the subset D of functions in $C^2(0,1)$ satisfying the conditions $v(0) = v'(0) = 0$. Thus we can use $a(v,w)$ as an energy inner product on D. Leaving aside questions of smoothness, let us analyze the boundary conditions to be satisfied by the elements v of K. Clearly v must satisfy the two essential boundary conditions $v(0) = v'(0) = 0$ at the left end. Since $u''(1) = 0$ is a natural boundary condition, a function v in K does not have to satisfy that condition. As for the other conditions at the right end, it suffices to require that $v(1) \geqslant 0$, the other conditions being automatically satisfied by the solution of the variational problem.

Theorem 5.   Let K be the set of functions v in C^2 on $0 \leqslant x \leqslant 1$, satisfying the conditions $v(0) = v'(0) = 0$ and $v(1) \geqslant 0$. Then the necessary and sufficient condition for the problem

(4.101) $J(v) \to \min, \qquad v \in K,$

to have a solution $u \in K$ is that this element satisfies the *variational inequality*

(4.102) $a(u, v-u) - l(v-u) \geqslant 0 \qquad$ for all $v \in K.$

Remark. This variational inequality should be contrasted with the variational equation (4.20). The reason for the difference is that the set K is not a linear manifold but a *convex* set. A set K is said to be convex if it contains the straight line segments joining any two points in the set. Thus, if $v, w \in K$, then

$$v + t(w - v) \in K, \qquad 0 \leqslant t \leqslant 1.$$

Of course any linear manifold is a convex set but not vice versa.

Proof. Let u be an element of K that provides the minimum in (4.101). Then, if $v \in K$, so does $u + t(v - u)$ for $0 \leqslant t \leqslant 1$. Therefore

$$J[u + t(v - u)] \geqslant J(u), \qquad 0 \leqslant t \leqslant 1,$$

and hence

$$\left\{ \frac{d}{dt} J[u + t(v - u)] \right\}_{t=0} \geqslant 0,$$

from which (4.102) follows. Now suppose $u \in K$ satisfies (4.102). Then, if $v \in K$, we have

$$J(v) = J(u + v - u) = a(u + v - u, u + v - u) - 2l(u + v - u)$$
$$= J(u) + 2[a(u, v - u) - l(v - u)] + a(v - u, v - u) \geqslant J(u),$$

so that u furnishes the minimum in (4.101).

Furthermore, we note that $a(v, v) > 0$ for all $v \not\equiv 0$ satisfying the boundary conditions $v(0) = v'(0) = 0$. This guarantees that the solution of the equivalent problems is unique (assuming existence, of course). If instead of K we use its completion in the energy norm, we can also prove existence (see Exercise 4.5).

Next we must show that the solution u of the variational inequality (4.102) also satisfies the BVP (4.99). The solution u of (4.102) is an element of K for which

$$\int_0^1 u''(v'' - u'')\,dx - \int_0^1 f(v - u)\,dx \geqslant 0 \qquad \text{for every } v \in K.$$

Assuming we have shown that u actually belongs to C^4, we find, using integration by parts, that

(4.103)
$$\int_0^1 (u^{(4)} - f)(v - u)\, dx - u'''(1)[v(1) - u(1)] + u''(1)[v'(1) - u'(1)] \geq 0.$$

If ϕ is any sufficiently smooth function for which $\phi(0) = \phi'(0) = \phi(1) = \phi'(1) = 0$, then $v \doteq u + \phi$ belongs to K and (4.103) shows that $\int_0^1 (u^{(4)} - f)\phi\, dx = 0$ for every such ϕ and therefore certainly for every test function ϕ with compact support in $(0, 1)$. Hence u is a weak solution of the differential equation $u^{(4)} = f$. Such a weak solution is a classical solution if f is continuous or piecewise continuous. What boundary conditions does u satisfy? Since $u^{(4)} - f = 0$, we find from (4.103) that

$$-u'''(1)[v(1) - u(1)] + u''(1)[v'(1) - u'(1)] \geq 0 \qquad \text{for every } v \in K.$$

Now let ϕ satisfy $\phi(0) = 0, \phi'(0) = 0, \ \phi(1) = 0, \phi'(1) < 0$. Then $v = u + \phi \in K$, $v(1) - u(1) = 0, v'(1) - u'(1) < 0$, so that $u''(1) = 0$, which is the first of the boundary conditions at the right end. We finally have

$$u'''(1)[(v(1) - u(1)] \leq 0 \qquad \text{for every } v \in K.$$

If $u(1) = 0$, we choose $v(1) > 0$ to find that $u'''(1) \leq 0$. If $u(1) > 0$, we choose $v(1) > u(1)$ to conclude that $u'''(1) = 0$. Thus we have shown that the solution of (4.102) satisfies (4.99).

In our second example the unilateral constraint is active throughout the domain rather than on the boundary alone. A beam whose ends are clamped is subject to a transverse loading $f(x)$, and rests on an impenetrable obstacle whose upper boundary is the curve $y = h(x)$, where $h(x) \leq 0$, $0 \leq x \leq 1$ (see Figure 4.4). Under the load the beam may be in contact with the obstacle over some unknown portion of its length.

The potential energy of the deflected beam is then

$$\int_0^1 (u'')^2\, dx - 2\int_0^1 fu\, dx,$$

and we again introduce the functional $J(v)$ as in (4.100). Let K be the set of sufficiently smooth functions $v(x)$ satisfying $v(0) = v'(0) = v(1) = v'(1) = 0$ and $v(x) \geq h(x)$. Then K is a convex set, and by the same reasoning as was used in Theorem 5 we have that the necessary and sufficient condition for the problem

$$J(v) \rightarrow \min, \qquad v \in K,$$

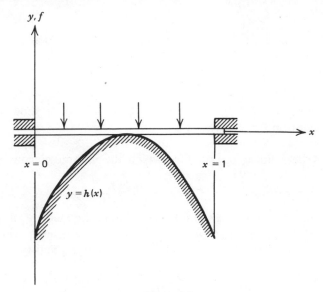

Figure 4-4

to have a solution $u \in K$ is that this element satisfies the variational inequality

$$a(u, v - u) - l(v - u) \geqslant 0 \qquad \text{for all } v \in K.$$

Thus u is characterized as the element in K for which

$$(4.104) \quad \int_0^1 u''(v'' - u'')\,dx - \int_0^1 f(v - u)\,dx \geqslant 0 \qquad \text{for every } v \in K.$$

What consequences can we derive from (4.104)? First, let $\phi(x)$ be a non-negative function in $C_0^\infty(0, 1)$; then $v = u + \phi$ belongs to K and $\int_0^1 (u''\phi'' - f\phi)\,dx \geqslant 0$, which is a weak form of the differential inequality $u^{(4)} \geqslant f$ on $0 < x < 1$.

At any point x where $u(x) > h(x)$, we can take variations of either sign about u so that the differential equation $u^{(4)} = f$ must be satisfied. In any interval where $u(x) = h(x)$, the inequality $u^{(4)} \geqslant f$ must hold, so that the additional reaction force due to the obstacle is in the positive u direction in keeping with physical requirements.

Exercises

4.1. Suppose w is a continuous function on $0 \leqslant x \leqslant 1$ such that

$$\int_0^1 w(x)\phi(x)\,dx = 0$$

for every $\phi \in C_0^\infty(0,1)$ satisfying $\int_0^1 \phi\,dx = 0$; then show that w is constant on $0 \leqslant x \leqslant 1$.

4.2. Let A be a linear operator for which the inhomogeneous equation

$$Au = f, \qquad u \in D_A,$$

has one and only one solution. Determine the element of the form $u^* = \sum_{i=1}^n c_i v_i$, where the v_i's are independent elements, which minimizes $\|Au^* - f\|^2$. Show that the coefficients $\{c_i\}$ satisfy

$$(4.105) \qquad \langle f, Av_j \rangle = \sum_{k=1}^n c_k \langle Av_k, Av_j \rangle, \qquad j = 1, \ldots, n.$$

The method just described is known as the *method of least squares*.

4.3. Let J be the functional (4.14), where $a(v,w)$ is coercive and $l(v)$ is bounded in energy [see (4.24)]. Then J is bounded below on D and has an infimum d. On the space M of elements of finite energy, J has a minimum. Show that this minimum coincides with d.

4.4. By considering trial elements of the form

$$v = -\frac{x_1^2 + x_2^2}{2} + A\left(x_1^2 - x_2^2\right),$$

show that (4.90) can be improved to

$$T \leqslant \frac{4I_1 I_2}{I_p},$$

where I_1 and I_2 are the moments of inertia about the x_1 and x_2 axes, respectively.

4.5. *Existence for unilateral problem.* Let $l(v)$ be bounded in energy, and let $a(v,w)$ be coercive on the manifold M, where M is complete in energy. Let K be a convex subset of M that is closed in energy (thus

K is itself a metric space complete in energy). Show that the minimum problem

$$J(v) \to \min, \qquad v \in K,$$

has one and only one solution. *Hint:* Since J is bounded below on K, construct a minimizing sequence $\{v_n\}$ in K and show that v_n converges to an element of K.

4.6. Show that the natural boundary conditions associated with the functional

$$J(u) \doteq \int_0^1 (u'')^2 \, dx - 2 \int_0^1 fu \, dx$$

are $u''(0) = u''(1) = u'''(0) = u'''(1) = 0$. Give the physical interpretation in beam theory.

4.7. Let Ω be a bounded domain in R_n with smooth boundary Γ. Consider the BVP

$$(4.106) \qquad -\Delta u = f, \quad x \in \Omega; \qquad -\frac{\partial u}{\partial n} = k(x)u, \quad x \in \Gamma,$$

where f is a given continuous function in $\bar{\Omega}$ and $k(x)$ is a positive function on Γ. Assuming that (4.106) has a classical solution $u(x)$, show that $u(x)$ can also be characterized as the element which minimizes

$$(4.107) \qquad J(v) \doteq \int_\Omega |\operatorname{grad} v|^2 \, dx - 2\int_\Omega fv \, dx + \int_\Gamma kv^2 \, dS$$

among functions which are sufficiently smooth but are not required to satisfy any boundary condition on Γ. Thus the boundary condition in (4.106) is natural for (4.107).

4.8. Consider the biharmonic operator $\Delta\Delta$ and the associated BVP

$$\Delta\Delta u = f, \quad x \in \Omega; \qquad u = \frac{\partial u}{\partial n} = 0, \quad x \in \Gamma,$$

where Ω is a bounded domain in R_3 with boundary Γ. Introduce the bilinear form

$$a(v, w) = \int_\Omega (\Delta v)(\Delta w) \, dx$$

to obtain two-sided bounds for the quantity

$$\alpha = \int_{\Omega} fu \, dx.$$

4.9. (a) Consider the integrodifferential equation (5.38), Chapter 6, repeated here:

$$-u'' + \int_{0}^{1} xy u(y) \, dy = f(x), \quad 0 < x < 1; \quad u(0) = u'(1) = 0,$$

which has one and only one solution $u(x)$. Show that

$$(4.108) \quad \int_{0}^{1} fu \, dx = \max_{v \in D_A} 2 \int_{0}^{1} fv \, dx - \int_{0}^{1} (v')^2 \, dx - \left(\int_{0}^{1} xv \, dx \right)^2,$$

where D_A is the set of functions in $C^2(0,1)$ satisfying $v(0) = v'(1) = 0$.

(b) Show further that the boundary condition $v'(1) = 0$ need not be imposed in the variational principle (4.108).

4.10. Consider the beam problem (4.99) when the applied load is $A \sin 2\pi x$. For what values of A is contact maintained? Find the deflection of the beam.

4.11. A beam with clamped ends is subject to a uniform load $f(x) = -1$. The beam lies a distance δ above a flat, impenetrable foundation. Find the deflection of the beam. *Hint:* If δ is sufficiently small, the beam will be in contact with the foundation along part of its length; there may be reaction forces at the end of the contact portion.

4.12. Consider the beam of Figure 4.4 when the obstacle has the equation $h(x) = -\alpha(x - \frac{1}{2})^4, \alpha > 0$. Show that, if $\alpha > \frac{1}{24}$, contact can occur only at isolated points along the beam and no contact can occur at $x = \frac{1}{2}$. What happens? Find the deflection.

4.13. Let Ω be a bounded plane domain with boundary Γ, and let $u(x)$ be positive on Ω and vanish on Γ. We introduce the level curves $u = t$, and, as in the discussion following (4.91), we define

$$A(t) \doteq \int_{u > t} dx, \quad J(t) \doteq \int_{u > t} |\text{grad } u|^2 \, dx.$$

Show that

$$-A' = \int_{u=t} \frac{dl}{|\text{grad}\, u|}, \qquad -J' = \int_{u=t} |\text{grad}\, u|\, dl, \qquad 4\pi A \leq A'J',$$

and therefore

(4.109)
$$\int_\Omega |\text{grad}\, u|^2\, dx \geq -4\pi \int_0^{u_m} \frac{A(t)}{A'(t)}\, dt,$$

where equality holds if and only if Ω is a disk and $u(x)$ depends only on the distance from the center.

For $u(x)$ given on Ω, we construct a circularly symmetrized function $u^*(x)$ on a disk Ω^* of area equal to Ω as follows:

$$u^* = t \qquad \text{on the circle } |x| = r \quad \text{where } \pi r^2 = A(t).$$

Clearly u^* depends only on the radial coordinate, and the function $A^*(t) = \int_{u^*>t} dx$ is identically equal to $A(t)$.

From (4.109) it follows that

(4.110)
$$\int_\Omega |\text{grad}\, u|^2\, dx \geq \int_{\Omega^*} |\text{grad}\, u^*|^2\, dx,$$

so that circular symmetrization decreases the Dirichlet integral. If h is an arbitrary function, we also find that

(4.111)
$$\int_\Omega h(u(x))\, dx = \int_{\Omega^*} h(u^*(x))\, dx.$$

4.14. (a) Let $u(x)$ be the solution of (4.91), and let u^* be the circularly symmetrized function (see Exercise 4.13). Show that, if T and T^* are the respective torsional rigidities for Ω and Ω^*, then

$$T = \frac{4\left(\int_\Omega u\, dx\right)^2}{\int_\Omega |\text{grad}\, u|^2\, dx} \leq \frac{4\left(\int_{\Omega^*} u^*\, dx\right)^2}{\int_{\Omega^*} |\text{grad}\, u^*|^2\, dx} \leq T^*,$$

where the first inequality follows from (4.110) and (4.111), and the second from the variational characterization of the torsional rigidity.

(b) Let $u(x)$ be the fundamental eigenfunction (chosen positive) of

(4.112) $-\Delta u = \lambda u, \quad x \in \Omega; \qquad u = 0, \quad x \in \Gamma,$

and let $u^*(x)$ be the circularly symmetrized function. Let λ^* be the fundamental eigenvalue for the disk Ω^*, and show that $\lambda^* \leq \lambda$.

Thus, of all domains of equal area, the disk has the highest torsional rigidity and the lowest fundamental frequency.

4.15. Let $u(x)$ be the fundamental eigenfunction of (4.112). Define $I(t) = \int_{u > t} u \, dx$. Show that

$$-\lambda I I' \geq 4\pi t A,$$

and therefore obtain the Payne-Rayner inequality

$$\left(\int_{\Omega} u \, dx \right)^2 \geq \frac{4\pi}{\lambda} \int_{\Omega} u^2 \, dx.$$

REFERENCES AND ADDITIONAL READING

Bergman, S. and Schiffer, M., *Kernel functions and elliptic differential equations in mathematical physics*, Academic Press, New York, 1953.

Coulson, C. A. and Jeffrey, A., *Waves*, Longman, London, 1977.

Courant, R. and Hilbert, D., *Methods of mathematical physics*, Vols. I and II, Interscience, New York, 1953, 1962.

Dundurs, J., Properties of elastic bodies in contact, in *The mechanics of the contact between deformable bodies*, Edited by A. D. de Pater and J. J. Kalker, Delft University Press, Delft, 1975.

Duvaut, G. and Lions, J. L., *Inequalities in mechanics and physics*, Springer, New York, 1976.

Friedman, A., *Partial differential equations*, Holt, Rinehart, and Winston, New York, 1969.

Garabedian, P. R., *Partial differential equations*, Wiley, New York, 1964.

Hsiao, G. C. and Wendland, W. L., A finite element method for some integral equations of the first kind, *J. Math. Anal. Appl.*, **58**, No. 3 (1977).

Hellwig, G., *Partial differential equations*, Blaisdell, New York, 1964.

Noble, B. and Sewell, M. J., On dual extremum principles in applied mathematics, *J. Inst. Math. Appl.*, **9**, No. 2 (1972).

Robinson, P. D., Complementary variational principles, in *Nonlinear functional analysis and applications*, Academic Press, New York, 1971.

Stakgold, I., *Boundary value problems of mathematical physics*, Vols. I and II, Macmillan, New York, 1967.

Velte, W., *Direkte Methoden der Variationsrechnung*, Teubner, Stuttgart, 1976.

Weinberger, H. F., *Variational methods for eigenvalue approximation*, SIAM, Philadelphia, 1974.

9
Nonlinear Problems

1. INTRODUCTORY CONCEPTS

One-Dimensional Equations

Although our principal interest is in nonlinear differential equations and integral equations whose appropriate setting is an infinite-dimensional Hilbert or Banach space, we begin modestly with a single nonlinear equation in one independent variable. Let A be a real valued function of a real variable $(A : R_1 \rightarrow R_1)$, and consider either of the nonlinear equations $A(u) = 0$ and $A(u) = u$, where it is clear that one form can be obtained from the other by a suitable change in the definition of the function A.

A popular method for solving $A(u) = 0$ is *Newton's method* (or the method of tangents), illustrated in Figure 1.1. To find a root α known to be located in $a \leqslant u \leqslant b$, we start with the initial approximation $u_0 = b$ and define the next approximation u_1 as the intersection of the u axis and the tangent line drawn to $A(u)$ at u_0. Proceeding in this way, we obtain a sequence

$$(1.1) \qquad u_n = u_{n-1} + \frac{A(u_{n-1})}{A'(u_{n-1})},$$

which, under favorable circumstances, will converge to α. The important feature of Newton's method is that at each step in the approximation procedure the curve is replaced by its tangent line (linearization). The calculations are greatly reduced if we use $A'(u_0)$ instead of $A'(u_{n-1})$ in (1.1). Geometrically we are then drawing a line *parallel* to the original tangent instead of drawing the new tangent. This simplified Newton's method is equivalent to the method of successive substitutions (or successive approximations) described below.

We now consider our nonlinear equation in the fixed-point form

$$(1.2) \qquad u = A(u).$$

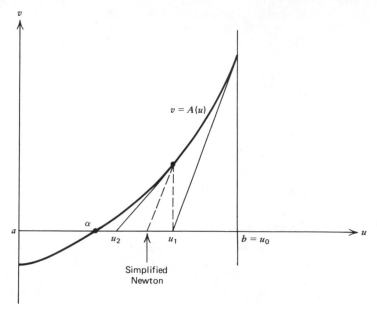

Figure 1.1

To solve this equation means to find the intersections of the straight line $y = u$ with the curve $y = A(u)$. If A is a contraction on the interval $a \leqslant u \leqslant b$, we know from Section 4, Chapter 4, that (1.2) has one and only one solution which is the limit of the iterative sequence $u_n = A(u_{n-1})$ independently of how the initial element u_0 is chosen in $[a, b]$. We now want to discard the contraction assumption and concentrate instead on the possible monotonicity of the iterative sequence. In monotone iteration we construct a pair of sequences, one converging to the solution from above and the other from below. The method we shall outline works for the solutions marked by a circle in Figure 1.2 but not for the one marked by a square.

Definition. The number u_0 is said to be a *lower solution* to (1.2) if

$$(1.3) \qquad\qquad\qquad u_0 \leqslant A(u_0),$$

and u^0 is an *upper solution* to (1.2) if

$$(1.4) \qquad\qquad\qquad u^0 \geqslant A(u^0).$$

Remark. A value of the abscissa for which the straight line lies below the curve is a lower solution, and one for which the straight line lies above the

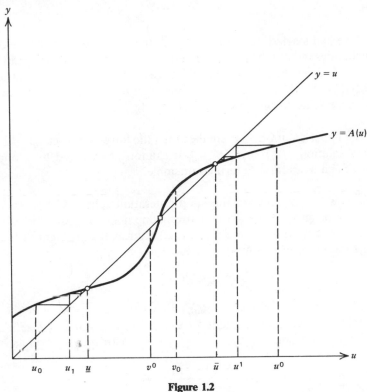

Figure 1.2

curve is an upper solution. Thus in Figure 1.2 u_0 and v_0 are lower solutions, whereas u^0 and v^0 are upper solutions.

The following result seems obvious from a glance at Figure 1.2, and will be proved in a more general setting later. Let the lower solution u_0 and the upper solution u^0 satisfy $u_0 \leqslant u^0$, and let A be *increasing* on the closed interval $[u_0, u^0]$. The sequence

$$(1.5) \qquad u_n = A(u_{n-1}), \qquad n = 1, 2, \ldots,$$

is increasing and converges to the minimal solution $\underline{u}$ of (1.2) on $[u_0, u^0]$. The sequence

$$(1.6) \qquad u^n = A(u^{n-1}), \qquad n = 1, 2, \ldots,$$

is decreasing and converges to the maximal solution $\bar{u}$ of (1.2) on $[u_0, u^0]$.

Remarks

1. If we had started iteration (1.6) with the upper solution v^0 instead of u^0, the iterates would have converged downward to the one and only solution of (1.2) in the interval $[u_0, v^0]$. Observe that A is not a contraction on either $[u_0, u^0]$ or $[u_0, v^0]$. In any event uniqueness can be guaranteed only by making additional assumptions on A.

2. The upper solution v^0 is smaller than the lower solution v_0. Although there is a solution of (1.2) in $[v^0, v_0]$, it cannot be obtained by monotone iteration. Such a solution is termed unstable.

3. If A is not increasing, the simple iteration schemes (1.5) and (1.6) may diverge or only converge in an alternating manner. However, addition of a large linear term to both sides of (1.2) reduces the problem to the case of increasing A. Indeed, suppose we have

$$u = S(u).$$

where all we know is that $u_0 \leqslant S(u_0), u^0 \geqslant S(u^0)$, and $u_0 \leqslant u^0$. Let the constant M be chosen large enough so that $S'(u) + M \geqslant 0$ in $[u_0, u^0]$. Then $S(u) + Mu$ is increasing on $[u_0, u^0]$, and $u = S(u)$ can be rewritten as

$$u = \frac{S(u) + Mu}{M+1} \doteq A(u),$$

where A is now increasing on $[u_0, u^0]$ and $u_0 \leqslant A(u_0), u^0 \geqslant A(u^0)$.

Another type of method for solving a nonlinear equation introduces an additional real parameter λ to provide a continuous transition between the desired problem and a simpler, *base problem* whose solution is *explicitly known*. Suppose that the equation to be solved is $A(u) = 0$ and that a solution u_0 is known for the base problem $B(u) = 0$. We construct a function $F(\lambda, u)$ depending smoothly on λ and u, such that $F(0, u) = B(u)$ and $F(1, u) = A(u)$. We then focus our attention on the equation

(1.7) $F(\lambda, u) = 0$

with the hope that (1.7) will determine implicitly a function $u(\lambda)$ on $0 \leqslant \lambda \leqslant 1$ with the property $u(0) = u_0$. If this is the case, $u(\lambda)$ is said to be the branch of (1.7) passing through $(0, u_0)$, and $u(1)$ will be a solution of the original problem, $A(u) = 0$. Two remarks should be made at this time. First, (1.7) often occurs in its own right with λ having a natural significance as a perturbation parameter. Second, the choice of the interval $0 \leqslant \lambda \leqslant 1$ is arbitrary and sometimes inconvenient; we could as well have λ vary between λ_0 and λ_1.

In view of these remarks we look at the following slightly more general problem: Given a solution (λ_0, u_0) of (1.7), that is, a point in the solution set of (1.7), can we construct a branch of solutions passing through this point? The implicit function theorem tells us that, if grad F is continuous and if $(\partial F / \partial u)(\lambda_0, u_0) \neq 0$, there will exist a branch through (λ_0, u_0) having the functional form $u = u(\lambda)$, but this branch's existence is guaranteed only for a small λ interval around λ_0. In the *method of continuity* (also known as an *imbedding method*) we extend $u(\lambda)$ by solving a succession of initial value problems obtained by differentiating (1.7) with respect to λ. Denoting partial derivatives by subscripts, we find that

$$(1.8) \qquad F_u(\lambda, u) u_\lambda + F_\lambda(\lambda, u) = 0,$$

and, if $F_u \neq 0$,

$$(1.9) \qquad u_\lambda = -\frac{F_\lambda(\lambda, u)}{F_u(\lambda, u)}.$$

At $\lambda - \lambda_0$ we have $u = u_0$, and the right side of (1.9) is therefore known; hence we have $u_\lambda(\lambda_0)$. For $\Delta\lambda$ sufficiently small we can write

$$(1.10) \qquad u(\lambda_0 + \Delta\lambda) = u(\lambda_0) + u_\lambda(\lambda_0)\, \Delta\lambda,$$

and we can then use (1.9) again to find $u_\lambda(\lambda_0 + \Delta\lambda)$. We can then repeat the procedure to find $u(\lambda_0 + 2\Delta\lambda)$ and so on. In this way we hope to construct the desired branch away from the initial point. Note that in (1.10) $\Delta\lambda$ can be either positive or negative, so that we can move forward or backward in λ. Assuming that grad F is well behaved throughout the $\lambda - u$ plane, the construction can run into trouble only if $F_u = 0$ somewhere along the branch. What goes wrong geometrically if $F_u = 0$?

The following two simple examples are typical of the principal kinds of difficulties encountered:

(a) $u^2 - \lambda^2 = 0$, (b) $u^2 - 1 + \lambda^2 = 0$.

For (a) the solution set consists of the two intersecting straight lines $u = \lambda$ and $u = -\lambda$. Pretending to be unaware of this but armed with the information that the point $P = (\lambda_0, u_0) \doteq (-1, 1)$ is a solution, we try to construct the branch through P (see Figure 1.3a). Equation (1.9) gives

$$(1.11) \qquad u_\lambda = \frac{\lambda}{u}$$

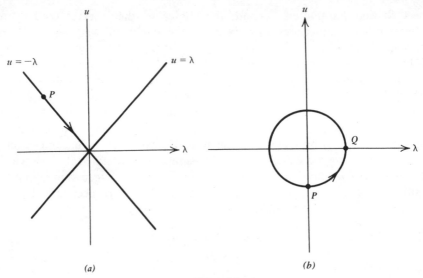

(a)   (b)

Figure 1.3

and $u_\lambda(\lambda_0) = -1$. No difficulty is encountered in integrating (1.11) numerically for $\lambda < 0$. At $\lambda = 0$ a new branch of solutions $(u = \lambda)$ intersects the branch on which P lies $(u = -\lambda)$. This phenomenon is signaled by the fact that the denominator on the right side of (1.11) vanishes. Admittedly the numerator also vanishes, so that careful numerical integration might enable one to cross 0 and continue along the original branch. It is certainly not clear how one would go about picking up the new branch at the origin.

For (b) the solution set is the unit circle in the λ-u plane shown in Figure 1.3b. Suppose we start from the solution $P = (0, -1) \doteq (\lambda_0, u_0)$ and try to construct the branch through P. Equation (1.9) becomes

$$(1.12) \qquad\qquad\qquad u_\lambda = -\frac{\lambda}{u}$$

and $u_\lambda(\lambda_0) = 0$. We have no difficulty in integrating (1.12) numerically until we reach a neighborhood of Q where the denominator of the right side of (1.12) vanishes. This time the numerator does *not* vanish, and u_λ becomes positively infinite. Obviously the branch becomes vertical and is no longer suitably parametrized by λ. This difficulty could be circumvented by making a change of variables (say, using u as the independent variable), but all we are interested in at this time is to characterize the nature of the difficulty.

To recapitulate, we have found that the vanishing of F_u along a branch may signal (a) intersection with a new branch, or (b) the fact that the branch becomes vertical and may then turn back so that further extension in λ is impossible. A third possibility is that $u(\lambda)$ tends to infinity as we approach the point where $F_u = 0$ (say, for $F = \lambda u - 1$ at $\lambda = 0$).

A closely related method for constructing the branch through a solution (λ_0, u_0) is the *perturbation method*. Here we attempt to calculate higher derivatives of u at the initial point by further differentiation of (1.7) or (1.8) with respect to λ. For instance, $u_{\lambda\lambda}$ is determined from

$$(1.13) \qquad u_{\lambda\lambda}F_u + u_\lambda(2F_{u\lambda} + u_\lambda F_{uu}) + F_{\lambda\lambda} = 0.$$

We then write

$$(1.14) \qquad u(\lambda) = u(\lambda_0) + (\lambda - \lambda_0)u_\lambda(\lambda_0) + \frac{(\lambda - \lambda_0)^2}{2}u_{\lambda\lambda}(\lambda_0) + \ldots,$$

where the three dots can be interpreted in two ways. Either one hopes that the infinite series converges, or one substitutes a remainder for which an estimate has to be found. The perturbation method is more difficult to justify mathematically and numerically since it is not clear that information pertaining only to the immediate neighborhood of the initial point on the branch can predict the behavior at values of λ appreciably distant. The compensating advantage is that the differentiations, however complicated, are all performed at the same point (λ_0, u_0). We shall illustrate these various methods when treating nonlinear problems in function space.

Linear Versus Nonlinear: Buckling of a Rod

The phenomenon of buckling is a familiar one. When a flexible ruler is compressed, it retains its straight shape until the compressive force reaches a critical value which causes the ruler to bend (buckle) rather suddenly with appreciable transverse deflection. The phenomenon is essentially nonlinear since the transverse deflection is not proportional to the applied compressive load P, whereas in a linear theory such as elementary beam theory the deflection is in fact proportional to the applied transverse load.

Figure 1.4 shows a homogeneous thin rod whose ends are pinned, the left end being fixed and the right end free to move along the x axis. In its unloaded state (reference configuration) the axis of the rod coincides with the portion of the x axis between 0 and l. Under a compressive load P a possible state for the rod is that of pure compression, but experience shows that, for sufficiently large P, transverse deflections occur. Assuming that

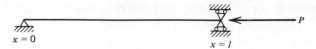

Figure 1.4

this buckling takes place in the x-y plane, we investigate the equilibrium of forces on a portion of the rod including its left end. The free body diagram shown in Figure 1.5 uses the same sign convention as in Figure 4.1, Chapter 0, so that T and M must turn out to be negative. Note that, by taking a free body diagram of the entire rod, we conclude that the only reaction at the left end is the force P shown.

A particle occupying the position $(S,0)$ before loading has moved to the position $(u(S),v(S))$. We let ϕ be the angle between the tangent to the buckled rod and the x axis, and let s be the arclength measured from the left end. Although we could proceed from (4.6) and (4.7), Chapter 0, it is simpler to start afresh for our particular problem. The equilibrium equations are

(1.15) $$M = -Pv,$$

(1.16) $$T = -P\cos\phi,$$

(1.17) $$Q = P\sin\phi,$$

whose form suggests the use of v and ϕ as dependent variables rather than v and u. The strain measures δ and μ of (4.17) and (4.18), Chapter 0, can be expressed in terms of ϕ and v:

(1.18) $$\mu = \frac{d\phi}{dS}, \qquad \delta\sin\phi = \frac{ds}{dS}\frac{dv}{ds} = \frac{dv}{dS}.$$

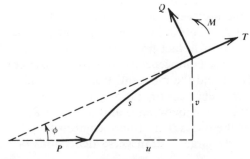

Figure 1.5

Since the rod is homogeneous, the constitutive laws of (4.19), Chapter 0, take on the simpler form

$$(1.19) \qquad \delta = \hat{\delta}(T), \qquad \mu = \hat{\mu}(M),$$

where $\hat{\delta}$ and $\hat{\mu}$ are prescribed functions having the general properties shown in Figure 4.2, Chapter 0. The function $\hat{\delta}$ is positive and monotonically decreasing, with $\hat{\delta}(0) = 1$. The function $\hat{\mu}$ is odd and increasing, with $\hat{\mu}(0) = 0$. Using (1.15), (1.16), (1.18), and (1.19), we obtain the system of two nonlinear differential equations in the unknown functions $\phi(S)$ and $v(S)$:

$$(1.20) \qquad \frac{d\phi}{dS} = \hat{\mu}(-Pv), \qquad \frac{dv}{dS} = \hat{\delta}(-P\cos\phi)\sin\phi.$$

The two boundary conditions associated with (1.20) can be taken either as $v(0) = v(l) = 0$ or as $\phi'(0) = \phi'(l) = 0$. The fact that the coordinate S refers to the undeformed rod is the reason why these boundary conditions are so easy to impose.

A particularly simple (yet important) case occurs when the rod is *inextensible* and obeys the *Euler-Bernoulli* bending law. In our notation this means that

$$(1.21) \qquad \delta(T) \equiv 1, \qquad \hat{\mu}(M) = \frac{M}{EI},$$

where the constants E and I are defined in (4.10), Chapter 0. Then $s = S$, and we could use s as an independent variable but shall not do so. The system (1.20) becomes

$$(1.22) \qquad -\phi' = \alpha v, \quad v' = \sin\phi, \quad 0 < S < l; \qquad v(0) = v(l) = 0,$$

where $\alpha \doteq P/EI$. The boundary conditions could also have been written as $\phi'(0) = \phi'(l) = 0$.

Note that (1.22) is still *nonlinear* even though linear constitutive relations have been used. The nonlinearity stems from the fact that we have used the exact expression for the curvature. In the analysis of (1.22) an important role is played by the *linearized system* derived on the assumption that $|\phi|$ is small. Either by using the linearization procedure of Chapter 0 or, more simply, by replacing $\sin\phi$ by ϕ, we obtain

$$-\phi' = \alpha v, \qquad v' = \phi,$$

or the single equation

$$(1.23) \qquad v'' + \alpha v = 0, \quad 0 < S < l; \qquad v(0) = v(l) = 0,$$

which could have been derived directly by using the approximate form v'' for the curvature $d\phi/dS$.

According to (1.23), a nontrivial transverse deflection is possible only for certain discrete values of α (that is, for discrete values of the compressive load). These eigenvalues are

$$(1.24) \qquad\qquad \alpha_n = \frac{n^2\pi^2}{l^2}, \qquad n = 1, 2, \ldots,$$

with corresponding deflection $A\sin(n\pi S/l)$. Thus only the shape, not the size, of the deflection is determined. This is depicted in Figure 1.6, where we have chosen $v'(0)$ as a measure of the deflection. The indeterminacy in $v'(0)$ at $\alpha = \alpha_n$ reflects the indeterminacy in the deflection $v(S)$. These results are physically unrealistic. The deflection should be determinate, and the rod should not return to its undeflected form when the load is slightly increased beyond the critical level. We shall see how the *nonlinear system* (1.22) resolves these inconsistencies.

A solution of (1.22) is a pair of functions (v, ϕ) satisfying the coupled differential equations and the boundary conditions. If (v, ϕ) is a solution, so is $(v, \phi + 2n\pi)$ for any integer n, but these new solutions give rise to the same physical deflection v. If (v, ϕ) is a solution, so is $(-v, -\phi)$, which corresponds to a deflection that is the mirror image of the deflection v about the undeformed axis of the rod. System (1.22) always has the solutions $(0, 2n\pi)$, which all yield the trivial physical solution $v \equiv 0$, easily seen to be the only solution for $\alpha = 0$. To obtain a nontrivial deflection v we must have $\phi(0) \neq 2n\pi$ [otherwise we would have $v(0) = 0$ and $\phi(0) = 2n\pi$,

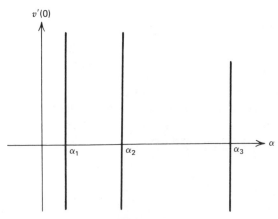

Figure 1.6

so that, by the uniqueness theorem for the initial value problem, the solution would be $v(S)\equiv 0$, $\phi(S)\equiv 2n\pi$, which corresponds to a trivial deflection]. Thus any solution of (1.22) that yields a nontrivial deflection must satisfy the IVP

(1.25) $-\phi'=\alpha v, \quad v'=\sin\phi; \qquad v(0)=0, \quad \phi(0)=\phi_0\neq 2n\pi.$

With α and ϕ_0 given, the IVP (1.25) has one and only one solution $(v(S,\alpha,\phi_0),\phi(S,\alpha,\phi_0))$ because the nonlinear term satisfies the Lipschitz condition of Exercise 4.6, Chapter 4. This unique solution will usually *not* satisfy the condition $v(l)=0$ unless α and ϕ_0 are suitably related. In any event our previous discussion shows that it suffices to *consider only initial values satisfying* $0<\phi_0<\pi$.

For some purposes it is convenient to reduce system (1.25) to a single differential equation of the second order for v or ϕ. The simpler of these is

(1.26) $\phi''+\alpha\sin\phi=0, \qquad \phi(0)=\phi_0, \qquad \phi'(0)=0,$

which, for $\alpha>0$, also describes the motion of a simple pendulum (here S is the time, and ϕ the angle between the pendulum and the downward vertical; initially the bob is displaced by an angle ϕ_0 and released from rest). For $\alpha>0$ we shall see that the solution of (1.25) (or (1.26)) is periodic in S, tracing out the orbit in the v-ϕ plane shown in Figure 1.7. The equation of this orbit is obtained from (1.25) by adding the results of

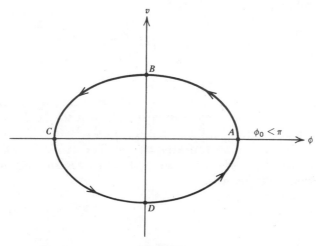

Figure 1.7

multiplying the first equation by v' and the second by ϕ':

$$\frac{\alpha}{2}\frac{d}{dS}v^2 - \frac{d}{dS}\cos\phi = 0,$$

or, using the initial conditions,

(1.27) $$v^2 = \frac{2}{\alpha}(\cos\phi - \cos\phi_0).$$

From (1.27) we have $|\phi| < \phi_0$. At the beginning of the motion we are at A with $v = 0$, $\sin\phi > 0$, and hence $v'(0) > 0$. For small positive s we can therefore infer from (1.25) that v is positive and ϕ' negative; these properties continue to hold until we reach B. At that point $\phi = 0$ and $v > 0$, which we may regard as new initial values for the next "time" interval. Since the transformation $s \to -s$, $\phi \to -\phi$, $v \to v$ keeps the equations invariant, the path starting at B will be the mirror image about the v axis of the path from A to B. Similar arguments show that the path CDA is the reflection about the ϕ axis of ABC. We have returned to A with the same initial values of ϕ and v, so that we will just keep going around the same orbit. If $\alpha < 0$, it can easily be seen that the motion is *not* periodic.

The period of the orbit (the time it takes to go around the orbit once) depends on α and ϕ_0 and will be denoted by $T(\alpha, \phi_0)$. For the linearized problem

(1.28) $$-\phi' = \alpha v, \quad v' = \phi; \quad v(0) = 0, \quad \phi(0) = \phi_0,$$

the explicit solution is

$$\phi = \phi_0 \cos\sqrt{\alpha}\,S, \qquad v = \frac{\phi_0}{\sqrt{\alpha}}\sin\sqrt{\alpha}\,S,$$

with period $2\pi/\sqrt{\alpha}$ independent of ϕ_0. For the nonlinear problem with small ϕ_0 the period will tend to $2\pi/\sqrt{\alpha}$, but, as we increase ϕ_0, the period increases. This can be seen by comparing (1.26) with the corresponding linear problem and noting that $\sin\phi < \phi$ for $0 < \phi < \pi$.

With this information we can analyze the BVP (1.22) qualitatively. A solution of (1.25) will satisfy $v(l) = 0$ if and only if l is an integral multiple of a half-period $T/2$, that is,

(1.29) $$\frac{2l}{n} = T(\alpha, \phi_0).$$

Since $T(\alpha, \phi_0)$ is an increasing function of ϕ_0 with $T(\alpha, 0+) = 2\pi/\sqrt{\alpha}$ and

$T(\alpha, \pi-) = \infty$, nontrivial solutions will be possible for a *given* l only if $k > \pi^2 / l^2$. If

(1.30) $$\frac{n^2 \pi^2}{l^2} < \alpha < \frac{(n+1)^2 \pi^2}{l^2},$$

there will be exactly n solutions of (1.22), each having a different value of ϕ_0 between 0 and π. If we also take into account the possibility of deflections which are mirror images about the x axis, we have established the existence of n pairs of solutions of (1.22) in the interval (1.30). All of this is confirmed by writing T as a complete elliptic integral. From (1.27) we find, for $\alpha > 0$,

$$(\phi')^2 = 4\alpha \left(\sin^2 \frac{\phi_0}{2} - \sin^2 \frac{\phi}{2} \right).$$

Along the path AB of Figure 1.7, ϕ decreases from ϕ_0 to 0, while S increases from 0 to $T/4$, so that

$$2\sqrt{\alpha} \, \frac{dS}{d\phi} = - \left(\sin^2 \frac{\phi_0}{2} - \sin^2 \frac{\phi}{2} \right)^{-1/2}$$

and, on integration from $S = 0$ to $S = T/4$,

$$T \frac{\sqrt{\alpha}}{2} = \int_0^{\phi_0} \left(\sin^2 \frac{\phi_0}{2} - \sin^2 \frac{\phi}{2} \right)^{-1/2} d\phi.$$

The permissible change of variables

$$\sin z = \frac{\sin(\phi/2)}{\sin(\phi_0/2)}$$

then gives

(1.31) $$T = \frac{4}{\sqrt{\alpha}} \int_0^{\pi/2} \frac{dz}{\sqrt{1 - p^2 \sin^2 z}}, \qquad p = \sin \frac{\phi_0}{2},$$

from which it is apparent that $T(\alpha, \phi_0)$ increases with ϕ_0 and that $T(\alpha, 0+) = 2\pi / \sqrt{\alpha}$, $T(\alpha, \pi-) = \infty$.

The *branching diagram* (solid lines) in Figure 1.8 summarizes the results of the nonlinear analysis and contrasts them with those for the linearized problem of Figure 1.6. There are still some important unanswered questions, the principal being as follows: Which of the mathematical buckled

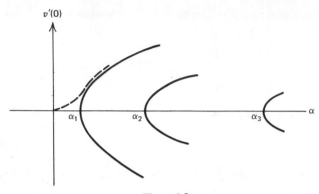

Figure 1.8

states does the rod actually choose? This is a problem of *stability* and can be answered only by studying the dynamical equations. Sometimes arguments involving potential energy can be used to discuss the stability of the various states, but such apparently "static" arguments work only because a suitable connection has been made between potential energy and stability. What is found in our particular problem is that the zero solution is stable for $\alpha < \alpha_1$ but not for $\alpha > \alpha_1$. For $\alpha > \alpha_1$ only the branches emanating from $\alpha = \alpha_1$ are stable. This still leaves the minor ambiguity of sign in the stable buckled state. No model based on a perfectly straight homogeneous rod, centrally compressed, can resolve this point. If, however, we assume that the load P is applied with a slight eccentricity or that the column is slightly imperfect, the ambiguity is removed. A typical load-deflection curve for such a case is shown as the dotted line in Figure 1.8.

Nonlinear Operators

We shall deal with nonlinear operators on a *real* Hilbert space H (with an occasional excursion into a Banach space of continuous functions). The class of nonlinear operators is understood to include linear operators as a subset. Some of the ideas below have already been encountered in Section 4, Chapter 4.

Examples

(a) $H = R_1$. An arbitrary real-valued function of a real variable maps R_1 into itself and is also said to be a nonlinear operator A on H. The value of A at u is written as Au or $A(u)$. A linear operator is one whose value at every u is obtained by multiplying u by a fixed, real constant.

(b) $H = R_n$. A nonlinear operator A maps R_n into itself. Thus A operates on a vector $u = (u_1, \ldots, u_n)$ to give an image vector $A_u \doteq v = (v_1, \ldots, v_n)$. Each component v_i of Au is the value at u of a real-valued function A_i of the n variables $u_1, \ldots, u_n$: $v_i = A_i u = A_i(u_1, \ldots, u_n)$. For a linear operator L each A_i is linear, so that $A_i(u) = \sum_j a_{ij} u_j$; the real n by n matrix (a_{ij}) then completely specifies the linear operator L.

(c) $H = L_2^{(r)}(0,1)$. The elements u of H are funcions of the real variable x, $0 < x < 1$. The operator A maps functions into functions. Even simple nonlinear operators such as $Au = u^2$ are not defined on the whole of H [if $u(x)$ is square integrable, $u^2(x)$ may not be]. This is a minor inconvenience which we dismiss by restricting the domain of A either explicitly or implicitly to elements of H for which Au is also in H. Perhaps a more satisfactory approach would be to consider operators on the Banach space $C(0,1)$ of continuous functions with the uniform metric, but we usually prefer to enjoy the comparative familiarity of Hilbert spaces. A more interesting example of a nonlinear operator is the *Hammerstein integral operator*, defined as follows:

$$(1.32) \qquad v(x) = \int_0^1 k(x,\xi) f(u(\xi)) \, d\xi \qquad \text{or} \qquad v = Tu,$$

where f is a real-valued function of a real variable and k is a real-valued function on R_2. Note that

$$(1.33) \qquad\qquad\qquad\qquad T = KF,$$

where F is the nonlinear operator mapping the element u into the element $f(u)$, and K is the real linear integral operator with kernel $k(x,\xi)$. T can be considered as an operator on $C(0,1)$ or on $L_2(0,1)$.

(d) $H = L_2^{(r)}(\Omega)$, where Ω is a domain in R_n. Consider the set of sufficiently smooth functions $u(x)$ which vanish on the boundary Γ of Ω, and set

$$(1.34) \qquad\qquad v(x) = -\Delta u(x) + f(u(x)) \doteq Au,$$

where Δ is the Laplacian in R_n. This operator A occurs in problems of combined diffusion and reaction.

(e) Consider the space of pairs of real-valued functions $(v(x), \phi(x))$ square integrable on $0 \leqslant x \leqslant l$. This is a real Hilbert space H under the inner product

$$\langle(v_1,\phi_1),(v_2,\phi_2)\rangle = \int_0^l (v_1 v_2 + \phi_1\phi_2)\, dx.$$

If v and ϕ are continuously differentiable, we can define a nonlinear operator A on H by

$$A(v,\phi) = (-\phi' - \alpha v, v' - \sin\phi).$$

Except for the boundary conditions, this is the operator that appears in (1.22).

Linearization

Again we start with a real-valued function of a real variable. Suppose we want to calculate $A(u)$ for all u close to a fixed u_0. We try to write $A(u) - A(u_0)$ as the sum of a principal part *linear in the increment* $h = u - u_0$ and of a remainder that is *small in comparison with the linear term as the increment tends to* 0. We therefore want a formula of the type

$$(1.35) \qquad A(u) - A(u_0) = Lh + r, \qquad \lim_{h\to 0} \frac{r}{h} = 0,$$

where the notation suppresses the dependence of r on h. Of course (1.35) is just the necessary and sufficient condition for A to be differentiable at u_0 with derivative L. The linear operator L acts on the increment $h = u - u_0$. Although L is defined for all h, (1.35) is normally useful only if h is small; we can then evaluate $A(u)$ for all u near u_0 by the linearized approximation

$$A(u) - A(u_0) = Lh.$$

These ideas are easily extended to the case of an operator A on an arbitrary Hilbert or Banach space.

Definition.   The operator A is *linearizable* at u_0 if there exists a *bounded linear operator* L such that

$$(1.36) \qquad Au - Au_0 = Lh + r, \qquad \lim_{h\to 0} \frac{\|r\|}{\|h\|} = 0,$$

where $h = u - u_0$.

Remarks

1. If A is linearizable at u_0, the operator L is uniquely determined and is known as the *Fréchet derivative* of A at u_0. We sometimes write $L = A'(u_0)$. If A is linearizable at many (or all) points u_0, the dependence of L on u_0 will usually *not* be linear. The remainder r in (1.36) depends on h and u_0. When we wish to emphasize the dependence of r on h we write $r = Rh$, where R is a nonlinear operator.

2. If A is a bounded linear operator, $A'(u_0) = A$ for all u_0.

3. The notion of linearization can be extended to operators from one Banach space to another (hence to functionals on either a Banach or a Hilbert space; these ideas are pivotal in the calculus of variations).

4. If A is linearizable at u_0, the following method is often useful in calculating $L = A'(u_0)$. We write $h = \varepsilon\eta$ where ε is a real number and η is a fixed element in H. Then (1.36) states that

(1.37) $$\lim_{\varepsilon \to 0} \frac{A(u_0 + \varepsilon\eta) - Au_0}{\varepsilon} = L\eta \qquad \text{for each } \eta \in H.$$

Examples

(a) Let A be an operator on R_n. Let $\tilde{u} = (\tilde{u}_1, \dots, \tilde{u}_n)$ be a fixed point in R_n (the terminology u_0 would be confusing), and let $u = (u_1, \dots, u_n)$ be an arbitrary point, $h = (h_1, \dots, h_n) = u - \tilde{u}$. The value of A at u can be expressed as

$$Au \doteq v = (v_1(u), \dots, v_n(u))$$

where all $v_i(u)$ are assumed to be continuously differentiable. Then, by Taylor's theorem in n variables, we find that

$$v_i(u) - v_i(\tilde{u}) = \sum_{j=1}^{n} \frac{\partial v_i}{\partial u_j}(\tilde{u})h_j + r_i, \qquad \lim_{h \to 0} \frac{r_i}{\|h\|} = 0.$$

Hence

$$Au - A\tilde{u} = Lh + r, \qquad \lim_{h \to 0} \frac{\|r\|}{\|h\|} = 0,$$

where L is the *Jacobian* matrix with entries $(\partial v_i/\partial u_j)(\tilde{u})$ and r is the vector $(r_1, \dots, r_n)$.

(b)   To linearize the integral operator (1.32) at u_0, we use (1.37):

$$(1.38) \qquad L\eta = \lim_{\varepsilon \to 0} \int_0^1 k(x,\xi) \left[\frac{f(u_0(\xi) + \varepsilon\eta(\xi)) - f(u_0(\xi))}{\varepsilon} \right] d\xi$$

$$= \int_0^1 k(x,\xi) f_u(u_0(\xi)) \eta(\xi) \, d\xi.$$

Thus L is a linear integral operator with kernel $k(x,\xi) f_u(u_0(\xi))$.

(c)   Let f be a real-valued, nonlinear functional on H. We say that f is linearizable at u_0 if there exists a bounded linear functional l such that

$$(1.39) \qquad f(u) - f(u_0) = l(h) + r, \qquad \lim_{\|h\| \to 0} \frac{|r|}{\|h\|} = 0.$$

We also write $l = f'(u_0)$. A *critical point* for f is a point u_0 at which $f' = 0$, that is, $l = 0$, and then (1.39) shows that $f(u_0 + h) - f(u_0)$ is small in comparison with h for h near 0. A necessary condition for u_0 to be a critical point of f is condition (4.13), Chapter 8, with J replaced by f and u by u_0.

If f is linearizable at u_0, we can use the Riesz representation theorem to write $l(h) = \langle h, z \rangle$, where z is uniquely determined from l. If f is linearizable at every u_0, the element z will depend on u_0 and we will have

$$z = A u_0,$$

where A is a *nonlinear operator* known as the *gradient* of f. It is useful to proceed in the opposite direction. Suppose A is an operator for which there happens to exist a functional f such that

$$(1.40) \qquad f(u_0 + h) - f(u_0) = \langle h, A u_0 \rangle + r \qquad \lim_{h \to 0} \frac{|r|}{\|h\|} = 0.$$

We then say that A is a *gradient operator* (and of course A is the gradient of f) and call f a (scalar) *potential* of A. The potential is determined up to an arbitrary additive constant element. In particular, the potential which vanishes at $u = 0$ is

$$(1.41) \qquad f(u) = \int_0^1 \langle u, A(tu) \rangle \, dt.$$

If A is linear and symmetric, it is a gradient and its potential is $\frac{1}{2}\langle Au, u \rangle$. If A is a nonlinear, continuously differentiable operator with $A'(u_0)$ symmetric, then A is a gradient. Thus, if

$$\langle A'(u_0)h_1, h_2 \rangle = \langle h_1, A'(u_0)h_2 \rangle \qquad \text{for all } u_0, h_1, h_2,$$

then A is a gradient.

Monotone Iteration on $C(R)$

The idea of monotone iteration introduced at the beginning of this section can easily be extended to function spaces. To be specific consider the space $C(R)$ of real-valued, continuous functions $u(x)$ on the closed region R in R_n. We order the elements of $C(R)$ in the natural way: $u \leqslant v$ if $u(x) \leqslant v(x)$ for every x in R. A similar definition holds for $u \geqslant v$. If $u \leqslant v$, we denote by $[u, v]$ the set of all elements w such that $u \leqslant w \leqslant v$, and we refer to $[u, v]$ as an *order interval*. Of course not all elements of $C(R)$ are comparable; there exist u, v for which neither of the order relations $u \leqslant v$ and $u \geqslant v$ holds.

A typical nonlinear boundary value problem can often be translated into an equation of the form

$$(1.42) \qquad\qquad u = Tu,$$

where T is a nonlinear integral operator (usually of the Hammerstein type) on $C(R)$. Thus solving the BVP is equivalent to finding the fixed point(s) of T. Often we shall look for fixed points in the order interval $[u_0, u^0]$. The operator T is said to be increasing on $[u_0, u^0]$ if, whenever $u \leqslant v$ and $u, v \in [u_0, u^0]$, then $Tu \leqslant Tv$. In all our applications T will be a *compact* operator, that is, continuous and mapping bounded sets into relatively compact sets (see Section 7, Chapter 5). Since our norm is the uniform norm, convergence in $C(R)$ means uniform convergence on R. A fixed point u of T is said to be *maximal* in $[u_0, u^0]$ if for any fixed point v of T in $[u_0, u^0]$ we have $u \geqslant v$. A similar definition is used for *minimal* fixed point.

Theorem 1. Let $u_0 \leqslant Tu_0$, $u^0 \geqslant Tu^0$, and $u_0 \leqslant u^0$. If T is compact and increasing on $[u_0, u^0]$, the sequence $u_0, u_1, \ldots,$ where

$$(1.43) \qquad\qquad u_n = Tu_{n-1}, \qquad n = 1, 2, 3, \ldots,$$

is increasing and converges to the minimal solution $\underline{u}$ of (1.42) on $[u_0, u^0]$. The sequence $u^0, u^1, \ldots,$ where

$$(1.44) \qquad\qquad u^n = Tu^{n-1}, \qquad n = 1, 2, 3, \ldots,$$

is decreasing and converges to the maximal solution $\bar{u}$ of (1.42) on $[u_0, u^0]$.

Proof. If $u_0 \leqslant u \leqslant u^0$, then $Tu_0 \leqslant Tu \leqslant Tu^0$, so that $u_0 \leqslant Tu \leqslant u^0$ and Tu lies in the order interval $[u_0, u^0]$. Thus T maps $[u_0, u^0]$ into itself, and the sequences $\{u_n\}$ and $\{u^n\}$ are well defined. Since $u_0 \leqslant u_1$, we have $Tu_0 \leqslant Tu_1$, that is, $u_1 \leqslant u_2$. Hence $\{u_n\}$ is increasing, and, similarly, $\{u^n\}$ is decreasing. Moreover, $u_0 \leqslant u^0$ implies that $u_1 \leqslant u^1$ and, recursively, $u_n \leqslant u^n$. Thus the two sequences can be ordered as

$$u_0 \leqslant u_1 \leqslant \ldots \leqslant u_n \leqslant \ldots \leqslant u^n \leqslant \ldots \leqslant u^1 \leqslant u^0.$$

The sequence $\{u_n\}$ is bounded, so that $\{Tu_n\}$ contains a converging sub-sequence. Since $Tu_n = u_{n+1}$, the sequence $\{u_n\}$ contains a converging sub-sequence. Because $\{u_n\}$ is increasing, the whole sequence actually converges, to $\underline{u}$, say. Since T is continuous and $Tu_n = u_{n+1}$, we pass to the limit to obtain $T\underline{u} = \underline{u}$. Similarly it can be shown that u^n converges to $\bar{u}$ where $T\bar{u} = \bar{u}$. Clearly $\underline{u} \leqslant \bar{u}$.

Let us now show that $\underline{u}$ is the minimal solution of (1.42) in $[u_0, u^0]$. Let v be any fixed point of T in $[u_0, u^0]$. Then $u_0 \leqslant v$ and $Tu_0 \leqslant Tv$ or, since $Tv = v, u_1 \leqslant v$. Continuing in this way, we find that $u_n \leqslant v$ and hence $\underline{u} \leqslant v$, so that $\underline{u}$ is the minimal solution in $[u_0, u^0]$. Similarly it can be shown that $\bar{u}$ is the maximal solution in the order interval.

Extensive treatments of monotone methods can be found in the works of Amann and of Sattinger.

2. BRANCHING THEORY

Guided by our limited experience of Section 1, we shall now investigate the equation

$$(2.1) \qquad\qquad\qquad F(\lambda, u) = 0,$$

where λ is a real number, but u is now an *element of a real Hilbert space H* (in most cases of interest H will be a function space and hence infinite-dimensional). The operator F is an arbitrary nonlinear operator mapping ordered pairs (λ, u) in $R_1 \times H$ into elements of H. Thus, despite its simple appearance, (2.1) may disguise a difficult nonlinear partial differential equation or integral equation. Strictly speaking, solutions of (2.1) are ordered pairs (λ, u), but we shall often refer to an element u as a solution corresponding to a particular λ. Throughout we work in λ-u space, that is, in $R_1 \times H$; we unabashedly use geometrical language and visualization that would be appropriate if H were one-dimensional instead of infinite-dimensional. Some care will have to be exercised in translating the conceptual image into numerical calculations.

Often we shall deal with a slightly simpler version of (2.1):

(2.2) $Au - \lambda u = 0,$

where A is a nonlinear operator mapping H into itself. Note that (2.2) includes as special cases both linear homogeneous and linear inhomogeneous problems (in the first instance set $A = L$; in the second let $Au = Lu + f$). We can imitate this distinction when A is nonlinear by defining the problem to be *unforced* if $A0 = 0$, and *forced* if $A0 \neq 0$. If the problem is unforced, we always have the *basic solution branch* $(\lambda, 0)$ extending from $\lambda = -\infty$ to $\lambda = \infty$.

The problem of finding all solutions of (2.1) or (2.2) is much too difficult to entertain in any kind of generality (although there will be particular examples where the complete solution set can be exhibited). Branching theory deals with some particular aspects of (2.1) and (2.2). Given a solution (λ_0, u_0), can we use the method of continuity (or some other method) to construct a branch of solutions through (λ_0, u_0)? If so, when does the extension run into difficulties? Does the branch sprout twigs? Does the branch turn around at some limiting value of λ? As a special case consider the *unforced* problem (2.2). We then have at our disposal the basic branch $(\lambda, 0)$, which can be viewed as the infinitely-long trunk of a tree. There is no problem of extending this solution since it already goes from $-\infty$ to ∞, nor does the solution turn around at some value of λ. Therefore the principal remaining task is to locate the branch-points at which other branches join the trunk and to obtain qualitative information about the shape of these branches near their intersections with the trunk.

We shall also make some observations about a problem which is not strictly within the province of branching theory. There are good physical grounds for trying to find *positive* solutions of (2.1) or (2.2) (obviously we are thinking of H as a space where this notion makes sense). Positive solutions can often be constructed by monotone methods (see Section 4).

Without completely abandoning our arboreal analogy, we turn now to more mathematical forms of expression.

Definition. Consider (2.2) with $A0 = 0$. We say that $\lambda = \lambda^0$ is a *branch-point* (of the basic solution) if in every neighborhood of $(\lambda^0, 0)$ in $R_1 \times H$ there exists a solution (λ, u) of (2.2) with $\|u\| \neq 0$.

Remarks

1. The definition places the burden on small neighborhoods of $(\lambda^0, 0)$. Observe that each of the numbers $n^2 \pi^2 / l^2$ is a branch-point in both the

linear and the nonlinear problems in Figures 1.6 and 1.8. In this case the branch-points of the nonlinear problem coincide with those values of λ (eigenvalues) for which the linearized problem has nontrivial solutions.

2.   Our definition is clearly equivalent to the existence of a sequence (λ^n, u^n) of solutions of (2.2) with $\|u^n\| \neq 0$ and $(\lambda^n, u^n) \to (\lambda^0, 0)$.

3.   The definition does not guarantee the existence of a *continuous* branch of solutions. Such a continuous branch will be formed if the boundary of each sufficiently small, open ball in $R_1 \times H$ with center at $(\lambda^0, 0)$ contains a solution (λ, u) with $\|u\| \neq 0$.

We expect the branch-points of the nonlinear problem to be related to the spectrum of the linearized problem. Suppose we consider (2.2) in the *unforced* case, and let us look for the branch-points of the basic solution. The Fréchet derivative of $A - \lambda I$ at $u = 0$ is the operator $L - \lambda I$, where $L = A'(0)$. In one dimension the vanishing of the number $L - \lambda I$ alerted us to the possibility of a branch-point; in a general Hilbert space the equivalent condition is that the operator $L - \lambda I$ fails to have a bounded inverse (that is, 0 is in the spectrum of $L - \lambda I$, or, in other words, λ is in the spectrum of L).

Theorem 1.   The number λ^0 can be a branch-point of the basic solution of (2.2) only if it is in the spectrum of $L = A'(0)$.

Proof.   Let λ^0 be a branch-point. There must therefore exist a sequence $(\lambda^n, u^n) \to (\lambda^0, 0)$ with $\|u^n\| \neq 0$ and $Au^n - \lambda^n u^n = 0$. Using (1.38), we can write the last equation as

$$Lu^n - \lambda^0 u^n = (\lambda^n - \lambda^0)u^n - Ru^n, \qquad \lim_{n \to \infty} \frac{\|Ru^n\|}{\|u^n\|} = 0.$$

Setting $z^n = u^n / \|u^n\|$, we find that $Lz^n - \lambda^0 z^n = f^n$, where $\lim_{n \to \infty} \|f^n\| = 0$ and $\|z^n\| = 1$. Thus $L - \lambda^0 I$ is not bounded away from 0 and hence cannot have a bounded inverse. Therefore λ^0 must in the spectrum of L.

Remark.   Theorem 1 tells us that branch-points of the nonlinear problem are to be found in the spectrum of the linearized problem. In the problems we shall consider, L will be either a compact self-adjoint operator or its inverse. Thus the spectrum of L will consist solely of eigenvalues (except perhaps for $\lambda = 0$). Even so, we cannot guarantee that an eigenvalue of L will be a branch-point of the nonlinear problem. In Example 4(a) below there is no branching from an eigenvalue of the linearized problem. Theorems 2 and 3, however, give some sufficient conditions for an eigenvalue of the linearized problem to be a branch-point of the nonlinear problem.

Theorem 2   (Leray-Schauder-Krasnoselskii).   If A is compact [see Section 1, above (1.43)] and $\lambda^0 \neq 0$ is an eigenvalue of odd multiplicity for L, then λ^0 is a branch-point of the basic solution of (2.2).

Remark.   The proof is based on topological degree theory and is omitted. The multiplicity referred to in the theorem is the algebraic multiplicity. For a self-adjoint operator this is the same as the geometric multiplicity, which is the dimension of the eigenspace corresponding to that eigenvalue. In applying Theorem 2 one is limited by the need for predicting the multiplicity of the eigenvalues of a linear problem. The only available theorems are those which state that certain eigenvalues are simple. This is the case, for instance, for Sturm-Liouville problems of ordinary differential equations, for certain special kinds of matrices and integral operators, and for the lowest eigenvalue of elliptic BVPs.

Theorem 3   (Krasnoselskii).   Let A be a compact operator which is the gradient of a uniformly differentiable functional, and let A have a second derivative at the origin. Then any nonzero eigenvalue of L is a branch-point.

Remark.   The proof, which is based on the category theory of Liusternik and Schnirelmann, is omitted. For generalizations, see Browder. We shall not clarify the hypotheses of uniform differentiability and existence of a second derivative; suffice it to say that the hypotheses are met in all but pathological cases. Note that, if A is a gradient, problem (2.2) is equivalent to finding the critical points of $f(u) - \lambda \|u\|^2 / 2$, where f is the potential of A.

We now take up some simple examples. The reader may be disappointed that these examples are chosen from finite-dimensional spaces. Some of the qualitative features of branching, however, are already present in these examples. We turn to physically more important applications in Section 4.

Example 1.   With $H = R_1$ consider the nonlinear operator

$$(2.3) \qquad\qquad A(u) = Lu + cu^2,$$

where L and c are given real numbers, $c \neq 0$. Of course Lu is just the value at u of a linear operator on R_1. Equation (2.2) becomes

(2.4) $$Lu + cu^2 = \lambda u,$$

which always admits the solution $u = 0$. For $\lambda \neq L$ we also have the solution

$$u = \frac{\lambda - L}{c}.$$

The branching diagram consists of the intersecting straight lines in Figure 2.1 (dotted lines represent unstable solutions; see Section 5), where c and L have been chosen positive. The only branch-point of the basic solution is at $\lambda = L$. The linearized problem about $u = 0$ is $Lu = \lambda u$, which has $\lambda = L$ as its only eigenvalue.

Let us now alter (2.3) slightly to generate a forced problem:

$$A(u, \varepsilon) = Lu + cu^2 - \varepsilon, \qquad \varepsilon \neq 0.$$

Equation (2.2) now has the form

(2.5) $$Lu + cu^2 - \varepsilon = \lambda u \qquad \text{or} \qquad cu^2 - \varepsilon = (\lambda - L)u,$$

whose solutions are easily found graphically by plotting the parabola $cu^2 - \varepsilon$ and the straight line $(\lambda - L)u$. The branching diagrams for $\varepsilon > 0$ and $\varepsilon < 0$ are both shown in Figure 2.1. In the case $\varepsilon > 0$ there is no branch-point along either solution branch and no limiting value of λ; it is easy to

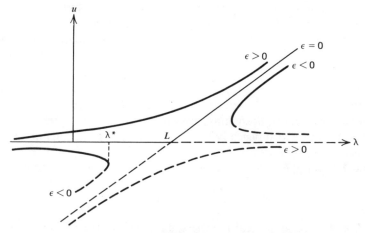

Figure 2.1

check that along both branches

$$\frac{d}{du}\left[A(u,\varepsilon)-\lambda u\right]=A_u-\lambda\neq0.$$

The case $\varepsilon<0$ exhibits two solutions branches, each of which has a limiting value of λ where $A_u(u,\varepsilon)-\lambda$ vanishes.

If $c<0$, there is no need to redraw the diagram. It is essentially enough to reverse the λ direction (that is, to regard λ as increasing from right to left) to obtain the appropriate diagrams.

It turns out that branching diagrams similar to those of Figures 2.1 and 2.2 occur in many stability problems in elasticity. Here λ is a loading parameter, ε an imperfection parameter, and u a measure of the deflection. Thus the case $\varepsilon=0$ corresponds to a perfect system (no eccentricity in the loading or imperfection in the structure), whereas $\varepsilon\neq0$ corresponds to an imperfect system. A dangerous feature of the slightly imperfect system with $\varepsilon<0$ in Figure 2.1 is that a critical load (limit load) λ^* is reached much earlier than expected for the perfect system. If λ is increased beyond this critical value, the system usually snaps into another static solution [not shown in our diagram but presumably present if a more realistic model than (2.5) is used] or else experiences large oscillations.

Such problems were studied by Koiter and more recently by Thompson and Hunt, who have also related their theory to a modern topological approach (Thom's catastrophe theory, see Golubitsky).

Example 2. With $H=R_1$ we now consider the second canonical nonlinear operator

(2.6) $$A(u)=Lu+cu^3,\qquad c\neq0,$$

which differs from (2.3) in that the nonlinear term is cubic rather than quadratic. Equation (2.2) becomes

$$Lu+cu^3=\lambda u,$$

which has the solution $u=0$ for all λ. In addition, if $c>0$, we find two nontrivial solutions for $\lambda>L$, given by

(2.7) $$u=\pm\sqrt{(\lambda-L)/c}\ .$$

The branching diagram is shown in Figure 2.2. Again $\lambda=L$ is the only branch-point, and it is also the only eigenvalue of the linearized problem $Lu=\lambda u$. If c were negative, the branching would be to the left (again it suffices to reverse the λ direction).

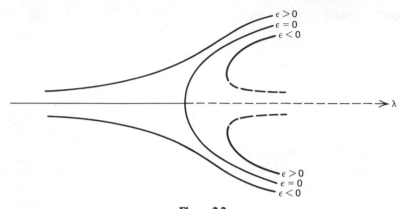

Figure 2.2

For the forced problem

$$(2.8) \qquad\qquad Lu + cu^3 - \varepsilon = \lambda u, \qquad \varepsilon \neq 0,$$

the branching diagram is also shown in Figure 2.2. The situation is now quite symmetric in ε. The solution for $\varepsilon > 0$ consists of two branches, one of which increases smoothly from $\lambda = -\infty$ to $+\infty$, whereas the other exhibits a turning point.

Example 3. With $H = R_2$ the problem is two-dimensional, and hence the linearized problem can have at most two eigenvalues. *Suppose these eigenvalues are distinct*, say $\lambda_1 < \lambda_2$. Each eigenvalue is simple, so Theorem 2 guarantees that λ_1 and λ_2 are actually branch-points of the basic solution (we are considering an *unforced problem*). The structure of the solution set can now be more complicated, even for relatively simple-looking problems. All the cases we shall study are of the form

$$(2.9) \qquad\qquad Au = Lu + Ru,$$

where

$$(2.10) \qquad\qquad Lu = (\lambda_1 u_1, \lambda_2 u_2), \qquad \lambda_1 < \lambda_2,$$

and R, to be specified later, is always *homogeneous of the third degree*:

$$(2.11) \qquad\qquad R(\alpha u) = \alpha^3 Ru.$$

We see that L is the linearization of A at $u = 0$ and branching will occur at the eigenvalues λ_1, λ_2 of L.

   (a)   If

$$(2.12) \qquad\qquad Ru = \left(c_1 u_1^3, c_2 u_2^3 \right),$$

then (2.2) becomes the pair of uncoupled equations

$$(2.13) \qquad\qquad \begin{aligned} \lambda_1 u_1 + c_1 u_1^3 &= \lambda u_1, \\ \lambda_2 u_2 + c_2 u_2^3 &= \lambda u_2. \end{aligned}$$

Thus, as in Example 2, a parabolic branch lying in the λ-u_1 plane emanates from λ_1, and a parabolic branch lying in the λ-u_2 plane emanates from $\lambda = \lambda_2$. If $c_1 < 0, c_2 > 0$, the first parabola points the left and the other to the right; there are then no other solutions of (2.13). If, however, c_1 and c_2 are of the same sign (positive, say), (2.13) yields additional solutions with neither u_1 nor $u_2 = 0$: the intersection of the two parabolic cylinders $\lambda - \lambda_1 = c_1 u_1^2, \lambda - \lambda_2 = c_2 u_2^2$. As is easily seen, the intersection consists of two curves which represent secondary branching from the parabolic branch that lies in the λ-u_1 plane. Exercise 2.1 gives an even more picturesque example of secondary branching.

   (b)   Example 4(a) below gives an instance where there is no branching from a double eigenvalue ($\lambda = 1$) of the linearized problem. A slight perturbation splits the double eigenvalue into two near, simple eigenvalues $\lambda_1 = 1 - \varepsilon$ and $\lambda_2 = 1 + \varepsilon$. By Theorem 2 we must now have branching from both these eigenvalues. How do the branches disappear as $\varepsilon \to 0$? In (2.9) we take

$$(2.14) \qquad\qquad Ru = \left(u_2^3, -u_1^3 \right),$$

so that (2.2) becomes

$$u_2^3 = [\lambda - (1 - \varepsilon)] u_1, \qquad -u_1^3 = [\lambda - (1 + \varepsilon)] u_2.$$

Nontrivial solutions are possible only for $1 - \varepsilon < \lambda < 1 + \varepsilon$. It can be seen that this solution is a closed loop connecting the branch-points $\lambda_1 = 1 - \varepsilon$ and $\lambda_2 = 1 + \varepsilon$. The loop does not wander far from the λ axis since the value of $u_1^2 + u_2^2$ does not exceed 2ε. As ε tends to 0, all dimensions of the loop tend to 0. This gives a geometrical picture of what can happen when two simple eigenvalues coalesce into a multiple eigenvalue for which no branching occurs.

Example 4. With $H = R_2$ we consider branching from a double eigenvalue of the linearized problem. We still use operators satisfying (2.9), (2.10), and (2.11), but now we take $\lambda_1 = \lambda_2 = 1$. We confine ourselves to two extreme examples showing the variety of phenomena that can now occur.

(a) R is given by (2.14), so that (2.2) becomes

$$u_2^3 = (\lambda - 1)u_1, \qquad -u_1^3 = (\lambda - 1)u_2.$$

Multiply these equations by u_2 and u_1, respectively, and subtract to obtain

$$u_1^4 + u_2^4 = 0,$$

whose only solution is $u_1 = u_2 = 0$. Thus there is only the basic solution branch with no branching at $\lambda = 1$ (even though this is an eigenvalue of the linearized system). This example does not violate Theorem 2, because $\lambda = 1$ has even multiplicity, Theorem 3, because A is not a gradient operator.

(b) Taking

$$Ru = (u_1 \|u\|^2, u_2 \|u\|^2),$$

we find that (2.2) has the form

$$(\lambda - 1)u_1 = u_1(u_1^2 + u_2^2),$$
$$(\lambda - 1)u_2 = u_2(u_1^2 + u_2^2),$$

which is equivalent to the single equation $u_1^2 + u_2^2 = \lambda - 1$ describing a paraboloid of revolution about the λ axis. Thus a whole surface of solutions branches off from the basic solution at $\lambda = 1$.

Before attempting a study of branching problems in function spaces, we develop in the next section some perturbation theory for linear problems.

Exercises

2.1. Consider the problem $Au - \lambda u = 0$, with A given by (2.9), (2.10), (2.11), and

$$Ru = (c_1 u_1 \|u\|^2, c_2 u_2 \|u\|^2), \qquad c_1 > c_2 > 0.$$

Find the entire solution set, and show that there are two parabolic branches joined by a circular hoop.

2.2. In R_2 consider the equation $Au - \lambda u = 0$, with

$$Au = (u_1 + 2u_1 u_2, u_2 + u_1^2 + 2u_2^2).$$

Show that $\lambda = 1$ is a double eigenvalue of the linearized problem but that there is only a single branch emanating from the basic solution at $\lambda = 1$.

2.3. In R_2 let $Au = (u_1^3 + u_2, u_1^2 u_2 - u_1^3)$, and consider the equation $Au = \lambda u$. Show that the linearized problem about $u = 0$ has only one eigenvalue and that this eigenvalue has geometric multiplicity 1 and algebraic multiplicity 2. Show that there is no branching from the basic solution.

2.4. Construct integral equations with separable kernels that lead to the branching diagrams of Examples 1 and 2.

3. PERTURBATION THEORY FOR LINEAR PROBLEMS

The methods used in this section can be loosely grouped under the heading of perturbation theory. With an eye to the nonlinear applications in Sections 4 and 5, we first employ these approximate methods on a variety of linear problems. Our purpose is not to carry out detailed calculations but rather to show how different types of problems can be treated in a fairly systematic manner. We deal successively with inhomogeneous problems, eigenvalue perturbations, change in boundary conditions, and domain perturbations.

Inhomogeneous Problems

For a given continuous $f(x)$ we shall consider the class of problems

$$(3.1) \quad P(\varepsilon): -u'' + (1 + \varepsilon x^2)u = f, \quad 0 < x < 1; \quad u(0) = u(1) = 1.$$

We may be interested in solving the problem for a particular value of $\varepsilon \neq 0$ or in studying the dependence of the solution on ε. In any event we shall show that for a suitable range of ε the problem has one and only one solution, denoted by $u(x, \varepsilon)$. The *base problem* $P(0)$ has the solution

$$(3.2) \quad u(x, 0) = w_0(x) + F(x),$$

where

$$w_0(x) = \frac{\sinh x + \sinh(1 - x)}{\sinh 1}, \qquad F(x) = \int_0^1 g(x, \xi) f(\xi) \, d\xi,$$

and $g(x, \xi)$ is Green's function for $-D^2 + 1$ with vanishing boundary conditions:

(3.3)

$$g(x,\xi) = \frac{\sinh x_< \sinh(1-x_>)}{\sinh 1}; \qquad x_< \doteq \min(x,\xi), \quad x_> \doteq \max(x,\xi).$$

One way of determining $u(x,\varepsilon)$ approximately is by expanding u in a Taylor series about $\varepsilon = 0$. The coefficients in this expansion are derivatives of u with respect to ε evaluated at $\varepsilon = 0$. These coefficients satisfy BVPs which can be obtained by differentiating (3.1) with respect to ε. Denoting ε derivatives by subscripts and x derivatives by primes, we find that

(3.4) $\quad -u_\varepsilon'' + (1 + \varepsilon x^2) u_\varepsilon = -x^2 u, \quad 0 < x < 1; \qquad u_\varepsilon|_{x=0} = u_\varepsilon|_{x=1} = 0.$

On setting $\varepsilon = 0$ and using (3.3), we obtain

$$u_\varepsilon(x,0) = -\int_0^1 g(x,\xi)\xi^2 u(\xi,0)\,d\xi,$$

which, together with (3.2), gives the approximation

(3.5)

$$u(x,\xi) \sim u(x,0) + \varepsilon u_\varepsilon(x,0) = w_0(x) + F(x) - \varepsilon \int_0^1 g(x,\xi)\xi^2 [w_0(\xi) + F(\xi)]\,d\xi.$$

Alternatively, by moving the term $\varepsilon x^2 u$ to the right side, (3.1) can be written as the integral equation

(3.6) $\qquad u(x,\xi) = w_0(x) + F(x) - \varepsilon \int_0^1 g(x,\xi)\xi^2 u(\xi,\varepsilon)\,d\xi.$

With ε fixed, (3.6) can be regarded as a fixed-point problem $u = Tu$ in the Banach space $C(0,1)$. With $\|\ \|_\infty$ denoting the uniform norm, we find that

$$\|Tu_1 - Tu_2\|_\infty \leqslant |\varepsilon|\,\|u_1 - u_2\|_\infty \max_{0 < x < 1} \int_0^1 \xi^2 g(x,\xi)\,d\xi,$$

where no absolute value sign is needed since the integrand is positive. Using (3.3), one can show that the mapping T is a contraction for $|\varepsilon| \leqslant 6$. Thus (3.1) has one and only one solution $u(x,\varepsilon)$ for $|\varepsilon| \leqslant 6$. We can find $u(x,\varepsilon)$ by successive substitutions:

$$u(x,\varepsilon) = \lim_{n\to\infty} u_n(x,\varepsilon), \qquad u_n(x,\varepsilon) \doteq Tu_{n-1}(x,\varepsilon),$$

where $u_0(x,\varepsilon)$ is arbitrary. If we use for $u_0(x,\varepsilon)$ the solution (3.2) of the base problem, then $u_1(x,\varepsilon) \doteq Tu_0(x,\varepsilon)$ is just (3.5).

The problem $P(\varepsilon)$ can also be treated by monotone methods, even though their main efficacy is for nonlinear problems. Since the technique does not rely on small ε, we shall consider the particular value $\varepsilon = 1$ and drop further reference to ε. The BVP under consideration is

$$(3.7) \qquad -u'' + (1+x^2)u = f(x), \quad 0 < x < 1; \qquad u(0) = u(1) = 1.$$

To apply Theorem 1 of Section 1, we rewrite (3.7) as

$$(3.8) \qquad -u'' + 2u = f + (1-x^2)u, \quad 0 < x < 1; \qquad u(0) = u(1) = 1,$$

where now the right side of the differential equation is an increasing function of u for each x in $0 \leqslant x \leqslant 1$. The basis for the iteration scheme is the equation

$$(3.9) \qquad -w'' + 2w = f + (1-x^2)v, \quad 0 < x < 1; \qquad w(0) = w(1) = 1,$$

whose one and only solution can be written as

$$(3.10) \qquad w = Tv \doteq 1 + \int_0^1 g(x,\xi)\big[\, f(\xi) - 2 + (1-\xi^2)v(\xi)\,\big]\, d\xi,$$

where g is Green's function for $-D^2 + 2$ with vanishing boundary conditions. The BVP (3.7) is then equivalent to the fixed-point problem $u = Tu$.

To verify the hypotheses of Theorem 1 we shall need the following maximum principle (see also Exercise 3.1, Chapter 1): If $z(x)$ satisfies

$$(3.11) \qquad -z'' + 2z \geqslant 0, \quad 0 < x < 1; \qquad z(0) \geqslant 0, \quad z(1) \geqslant 0,$$

then $z(x) \geqslant 0$ for $0 \leqslant x \leqslant 1$. Indeed, if $z < 0$ at an interior point, there is an interior negative minimum where $z < 0$ and $z'' \geqslant 0$; but these conditions violate the differential inequality. We can now show easily that T is an increasing operator, that is, $v_1 \leqslant v_2$ implies that $Tv_1 \leqslant Tv_2$ [we write $v_1 \leqslant v_2$ if $v_1(x) \leqslant v_2(x)$ on $0 \leqslant x \leqslant 1$]. It suffices to set $z = Tv_2 - Tv_1$ and use (3.9) to obtain

$$-z'' + 2z = (1-x^2)(v_2 - v_1) \geqslant 0, \quad 0 < x < 1; \qquad z(0) \geqslant 0, \quad z(1) \geqslant 0,$$

so that the maximum principle gives $z(x) \geqslant 0$ in $0 \leqslant x \leqslant 1$, that is, $Tv_1 \leqslant Tv_2$.

Starting with elements $u_0(x)$, $u^0(x)$ satisfying $u_0 \leqslant u^0$ and

$$(3.12) \quad -u_0'' + (1+x^2)u_0 \leqslant f(x), \quad 0 < x < 1; \quad u_0(0) \leqslant 1, \quad u_0(1) \leqslant 1,$$

$$(3.13) \quad -(u^0)'' + (1+x^2)u^0 \geqslant f(x), \quad 0 < x < 1; \quad u^0(0) \geqslant 1, \quad u^0(1) \geqslant 1,$$

we use (3.10) or, equivalently, (3.9) to obtain the sequences $\{u_n\}$ and $\{u^n\}$ needed in Theorem 1. We must show that $u_0 \leqslant Tu_0$ and $u^0 \geqslant Tu^0$. Let us prove the first of these, the second yielding to a similar argument. Since $Tu_0 = u_1$, we have

$$-u_1'' + 2u_1 = f + (1 - x^2)u_0, \quad 0 < x < 1; \quad u_1(0) = u_1(1) = 1,$$

and, combining with (3.12), we find that

$$-(u_1 - u_0)'' + 2(u_1 - u_0) \geqslant 0, \quad 0 < x < 1;$$

$$u_1(0) - u_0(0) \geqslant 0, \quad u_1(1) - u_0(1) \geqslant 0,$$

so that $u_1(x) - u_0(x)$ satisfies our maximum principle and hence $u_1 \geqslant u_0$ or $Tu_0 \geqslant u_0$.

Theorem 1 of Section 1 then tells us that there is at least one solution of (3.7) in the order interval $[u_0, u^0]$. That there exist elements that satisfy (3.12) and (3.13) is easily verified. We can choose u_0 and u^0 to be the *constants* given by

$$u_0 = \min\left(1, \frac{m}{2}, m\right), \qquad u^0 = \max\left(1, \frac{M}{2}, M\right),$$

where $m \leqslant f(x) \leqslant M$ on $0 \leqslant x \leqslant 1$. Better choices can often be made for u_0 and u^0, but this need not concern us now. Elementary arguments (unrelated to monotone methods) show that the solution of (3.7) is unique. This unique solution is sandwiched between the iterative sequences $\{u_n\}$ and $\{u^n\}$.

Eigenvalue Perturbation

For each ε satisfying $|\varepsilon| < a$, suppose that $L(\varepsilon)$ is a self-adjoint compact operator on the real Hilbert space H. We shall assume that $L(\varepsilon)$ depends continuously on ε, that is, $\lim_{\Delta\varepsilon \to 0} \|L(\varepsilon + \Delta\varepsilon) - L(\varepsilon)\| = 0$. The eigenvalue problem for $L(\varepsilon)$ is

$$(3.14) \qquad\qquad\qquad L(\varepsilon)u = \lambda u,$$

and we would like to relate the spectrum of the *perturbed operator* $L(\varepsilon)$ to the presumably known spectrum of the *base operator* $L \doteq L(0)$.

Theorem. Let $\lambda_n \neq 0$ be a simple eigenvalue of L with normalized eigenvector e_n. Then, in some neighborhood of $\varepsilon = 0$, there exists an eigenpair $(\lambda(\varepsilon), e(\varepsilon))$ of (3.14) with the properties

$$\lim_{\varepsilon \to 0} \lambda(\varepsilon) = \lambda_n, \qquad \lim_{\varepsilon \to 0} e(\varepsilon) = e_n.$$

Proof. We now sketch the proof of this theorem by a method which can be adapted to nonlinear problems. We rewrite (3.14) as

$$(3.15) \quad Lu - \lambda_n u = \delta u - R(\varepsilon)u, \qquad \delta \doteq \lambda - \lambda_n, \qquad R(\varepsilon) \doteq L(\varepsilon) - L.$$

Setting $f = \delta u - R(\varepsilon)u$, we see that (3.15) has the form $Lu - \lambda_n u = f$. From Chapter 6 we know that this equation has solutions if and only if $\langle f, e_n \rangle = 0$. With this solvability condition satisfied, the equation has many solutions, differing from one another by a multiple of e_n, but there is only one solution orthogonal to e_n. This particular solution is denoted by Tf, where T is known as the *pseudo inverse* of $L - \lambda_n I$. From (4.12), Chapter 6, we obtain easily the explicit expansion

$$(3.16) \qquad Tf = -\frac{Pf}{\lambda_n} + \sum_{k \neq n} \frac{1}{\lambda_k - \lambda_n} \langle f, e_k \rangle e_k,$$

where P is the projection on the null space of L. Note that T is defined as a bounded operator on all of H, but that Tf solves the equation $Lu - \lambda_n u = f$ only if $\langle f, e_n \rangle = 0$.

Thus, if u is to satisfy (3.15), we must have simultaneously

$$(3.17) \qquad \langle \delta u - R(\varepsilon)u, e_n \rangle = 0$$

and

$$(3.18) \qquad u = ce_n + T[\delta u - R(\varepsilon)u].$$

It is also clear that any solution of (3.17) and (3.18) will satisfy (3.15). We can proceed further if $|\delta|$ and $|\varepsilon|$ are small ($|\varepsilon|$ small means that the perturbation is small, and $|\delta|$ small gives the eigenvalue branch passing through λ_n). Our goal is to obtain a relation between δ and ε. With δ, ϵ, c fixed and $|\delta|, |\varepsilon|$ small we shall show that (3.18) has a unique solution which depends linearly on c and can therefore be written as $u = c\bar{u}(\delta, \varepsilon)$. We then substitute in (3.17), canceling c since we are looking for nontrivial solutions, to find

(3.19) $$\delta = \langle R(\varepsilon)\bar{u}(\delta,\varepsilon), e_n \rangle = \langle \bar{u}(\delta,\varepsilon), R(\varepsilon)e_n \rangle,$$

which is the desired relation between δ and ε.

Let us carry out these steps. The right side of (3.18) can be regarded as mapping an element $u \in H$ into the element $v \doteq Bu$ given by

$$v = ce_n + T[\delta u - R(\varepsilon)u].$$

Fixed points of B coincide with the solutions of (3.17)-(3.18). If u_1, u_2 are arbitrary elements of H,

$$\|Bu_2 - Bu_1\| \leqslant \|T\| \{|\delta| + \|R(\varepsilon)\|\} \|u_2 - u_1\|,$$

so that B is a contraction whenever

(3.20) $$|\delta| + \|R(\varepsilon)\| < \frac{1}{\|T\|}.$$

Assuming $|\delta|, |\varepsilon|$ small enough so that (3.20) is satisfied, we find that B has a unique fixed point given by

(3.21) $$u = c \sum_{k=0}^{\infty} \{T[\delta I - R(\varepsilon)]\}^k e_n \doteq cS(\delta,\varepsilon)e_n,$$

where S is a linear operator depending nonlinearly on the parameters δ and ε, and having the property $S(\delta,0) = I$. Substitution in (3.17) gives (3.19) or

(3.22) $$\delta = \langle S(\delta,\varepsilon)e_n, R(\varepsilon)e_n \rangle,$$

which is a *single nonlinear equation in two real variables*.

Writing (3.22) as $F(\delta,\varepsilon) = 0$, we observe that $F(\delta,0) = \delta$, so that $(0,0)$ is a solution of (3.22) and $\partial F/\partial \delta = 1$ at the origin. Under very mild restrictions on $R(\varepsilon)$ we can show that there is a neighborhood of $(0,0)$ where grad F is continuous and $\partial F/\partial \delta \neq 0$. The implicit function theorem then guarantees the existence of a unique solution of (3.22) in the form $\delta = \delta(\varepsilon)$, at least for $|\delta|$ and $|\varepsilon|$ sufficiently small. The first-order change in λ_n is obtained by using $S(\delta,0) = I$ in (3.22):

(3.23) $$\lambda(\varepsilon) - \lambda_n \sim \langle e_n, R(\varepsilon)e_n \rangle,$$

a well-known formula that could have been derived in other ways. The eigenvector of $L(\varepsilon)$ corresponding to $\lambda(\varepsilon) = \lambda_n + \delta(\varepsilon)$ is $\bar{u}(\delta(\varepsilon), \varepsilon)$, where the normalization $\langle \bar{u}, e_n \rangle = 1$ (that is, $c = 1$) has been used. To first order, we have

(3.24) $\qquad \bar{u} \sim e_n + T(\delta e_n) - TR(\varepsilon)e_n = e_n - TR(\varepsilon)e_n.$

Substitution of (3.16) in (3.24) gives

(3.25) $\qquad \bar{u} \sim e_n + \dfrac{PR(\varepsilon)e_n}{\lambda_n} + \sum_{k \neq n} \dfrac{1}{\lambda_k - \lambda_n} \langle R(\varepsilon)e_n, e_k \rangle e_k.$

Once the theory has been established by the rigorous method outlined above, simpler ways can be found to perform the calculations. Let us see how the method of continuity (parametric differentiation) can be used in the BVP

(3.26) $\qquad -u'' + (1 + \varepsilon x^2)u = \mu u, \quad 0 < x < 1; \qquad u(0) = u(1) = 0.$

For $\varepsilon = 0$ the eigenvalues are all simple: $\mu_n = n^2\pi^2 + 1$, $n = 1, 2, \ldots$, with corresponding normalized eigenfunctions $e_n(x) = \sqrt{2}\,\sin n\pi x$, where we have reverted to functional notation. To apply the earlier results we would have to translate (3.26) into an integral equation by using an appropriate Green's function. For $\varepsilon = 0$ the integral equation has the eigenvalues $\lambda_n = 1/\mu_n$ and the eigenfunctions $e_n(x)$. From our treatment of compact operators we know that there is a branch of eigenpairs $(\lambda(\varepsilon), u(x, \varepsilon))$ of the perturbed problem with the property $\lambda(\varepsilon) \to \lambda_n$, $u(x, \varepsilon) \to e_n(x)$ as $\varepsilon \to 0$. Therefore, for (3.26), there is a branch $(\mu(\varepsilon), u(x, \varepsilon))$ tending to $(\mu_n, e_n(x))$ as $\varepsilon \to 0$. To calculate the perturbed eigenpair, we assume that we can differentiate (3.26) with respect to ε to obtain

(3.27) $\qquad -u_\varepsilon'' + (1 + \varepsilon x^2)u_\varepsilon - \mu u_\varepsilon = \mu_\varepsilon u - x^2 u; \qquad u_\varepsilon = 0 \quad \text{at } x = 0, 1.$

Equation (3.27) should be viewed as an inhomogeneous equation for $u_\varepsilon(x, \varepsilon)$ with a right side that involves $\mu_\varepsilon(\varepsilon)$ and $u(x, \varepsilon)$. Setting $\varepsilon = 0$ gives an inhomogeneous equation for $u_\varepsilon(x, 0)$:

(3.28) $\qquad -u_\varepsilon'' + u_\varepsilon - \mu_n u_\varepsilon = \mu_\varepsilon(0)e_n(x) - x^2 e_n(x); \qquad u_\varepsilon = 0 \quad \text{at } x = 0, 1.$

At first we are troubled by the presence of the unknown $\mu_\varepsilon(0)$, which makes it appear that we cannot solve (3.28) for $u_\varepsilon(x, 0)$. Note, however, that the corresponding homogeneous problem has the nontrivial solution $e_n(x)$, so that (3.28) must obey the solvability condition

(3.29) $\qquad \langle \mu_\varepsilon(0)e_n(x) - x^2 e_n(x), e_n(x) \rangle = 0,$

which serves to determine

(3.30) $\qquad \mu_\varepsilon(0) = \int_0^1 x^2 e_n^2(x)\,dx, \qquad \mu(\varepsilon) \sim \mu_n + \varepsilon \int_0^1 x^2 e_n^2(x)\,dx.$

Just as in (3.23), the first-order change in the eigenvalue can be calculated independently of the change in the eigenfunction. Exercise 3.2 compares (3.23) and (3.30).

Now that (3.28) is consistent, we can solve for $u_\varepsilon(x,0)$. Of course the solution will contain an additive term proportional to $e_n(x)$. There is, however, a unique solution $w(x)$ that is orthogonal to $e_n(x)$:

$$w(x) = Q\big[\,\mu_\varepsilon(0)e_n(x) - x^2 e_n(x)\,\big],$$

where Q is the pseudo inverse of the differential operator in (3.28) with the corresponding boundary conditions. This pseudo inverse is just the integral operator whose kernel is the modified Green's function of Section 5, Chapter 3. For any f in H we have

$$Qf = \sum_{m \neq n} \frac{\langle f, e_m \rangle}{\mu_m - \mu_n} e_m(x),$$

so that

$$(3.31) \quad w(x) = -\sum_{m \neq n} \frac{\langle x^2 e_n(x), e_m(x) \rangle}{(m^2 - n^2)\pi^2} e_m(x), \qquad e_m(x) = \sqrt{2}\,\sin m\pi x.$$

Therefore we have the first-order change in the eigenfunction

$$(3.32) \qquad\qquad u_n(x,\varepsilon) - e_n(x) = \varepsilon w(x),$$

where the normalization $\langle u_n, e_n \rangle = 1$ has been used.

Change in Boundary Conditions

Consider the class of eigenvalue problems

$$(3.33) \quad -u'' - \lambda u = 0, \quad 0 < x < 1; \qquad u(0) = 0, \quad \varepsilon u'(1) + u(1) = 0,$$

where the parameter ε occurs in the boundary condition. [The problem has already been discussed fully; see (1.27), Chapter 7.] For $\varepsilon = 0$ we have a base problem whose eigenvalues are $\lambda_n = n^2\pi^2$ with normalized eigenfunctions $e_n(x) = \sqrt{2}\,\sin n\pi x$. We want to calculate the perturbed eigenpair $(\lambda(\varepsilon), u(x,\varepsilon))$, which tends to $(\lambda_n, e_n(x))$ as ε approaches 0. As usual, we differentiate (3.33) with respect to ε to obtain

$$(3.34) \quad -u_\varepsilon'' - \lambda u_\varepsilon = \lambda_\varepsilon u; \qquad u_\varepsilon(0,\varepsilon) = 0, \quad \varepsilon u_\varepsilon'(1,\varepsilon) + u_\varepsilon(1,\varepsilon) = -u'(1,\varepsilon),$$

where note should be made of the additional contribution in the boundary term [stemming from the presence of ε in the boundary condition at $x=1$ in (3.33)]. Setting $\varepsilon=0$ gives the following BVP for $u_\varepsilon(x,0)$:

$$(3.35) \quad -u_\varepsilon'' - \lambda_n u_\varepsilon = \lambda_\varepsilon(0) e_n(x); \qquad u_\varepsilon(0,0)=0, \quad u_\varepsilon(1,0)=-e_n'(1,0).$$

Since the homogeneous problem has a nontrivial solution, there will be a solvability condition to be satisfied. In view of the inhomogeneous boundary condition in (3.35), the solvability condition is found as follows. We multiply the equation for e_n by $u_\varepsilon(x,0)$ and (3.35) by $e_n(x)$, subtract, and integrate from $x=0$ to 1. This gives

$$\lambda_\varepsilon(0) = \int_0^1 e_n^2(x)\,dx = -\left[e_n'(1)\right]^2 = -2n^2\pi^2,$$

and, to first order,

$$(3.36) \qquad \lambda_n(\varepsilon) \sim n^2\pi^2 - 2\varepsilon n^2\pi^2,$$

in general agreement with Figure 1.2, Chapter 7, where $\varepsilon=0$ corresponds to $\beta=0$, $\varepsilon=0+$ to $\beta=0+$, $\varepsilon=0-$ to $\beta=\pi-$. Note that approximation (3.36) becomes progressively worse as n increases; in fact Figure 1.2 shows that $|\sqrt{\lambda_n(\varepsilon)} - \sqrt{\lambda_n}|$ cannot exceed $\pi/2$, whereas (3.36) gives $|(1-\sqrt{1-2\varepsilon})n\pi|$, which, for fixed $\varepsilon\neq0$, tends to ∞ with n.

Another interesting feature of the problem is that the perturbed eigenfunctions obtained in this way do not form a basis for $\varepsilon<0$. We saw in Chapter 7 that for $\varepsilon=0-$ (which corresponds to $\beta=\pi-$) there is a large negative eigenvalue (not arising from one of our perturbation branches) whose eigenfunction is therefore not obtainable by perturbing the $\{e_n(x)\}$ of the base problem. This is not surprising when we observe that the physical problem is drastically changed as ε changes from 0 to $0-$ (this corresponds to β going from the value 0 to a value just below π, which is a totally different problem, as was explained in Chapter 7). If, however, we had considered instead the boundary condition $u'(1)+\tau u(1)=0$, the problem for $|\tau|$ small is close to the one for $\tau=0$.

Domain Perturbations

Consider first the problem of finding the natural frequencies of a vibrating membrane of elliptical shape. The base problem is the eigenvalue problem for a disk. Let $\Omega(\varepsilon)$ represent a continuously varying domain depending on ε, with $\Omega(0)$ being the unit disk and $\Omega(\varepsilon_0)$ the desired domain [we could, for instance, let $\Omega(\varepsilon)$ be an ellipse with semiaxes $1-\varepsilon$ and 1]. The eigenvalue

problem is

(3.37) $\Delta u + \lambda u = 0, \quad x \in \Omega(\varepsilon); \qquad u = 0 \text{ on } \Gamma(\varepsilon),$

where $\Gamma(\varepsilon)$ is the boundary of $\Omega(\varepsilon)$.

We shall construct an eigenvalue branch $\lambda(\varepsilon)$ which coincides for $\varepsilon = 0$ with some particular simple eigenvalue, say λ_n, of the base problem. The corresponding eigenfunction (suitably normalized) $u(x, \varepsilon)$ is defined in $\Omega(\varepsilon)$. Differentiating the differential equation in (3.37) is permissible at every interior point since such a point remains in the interior for sufficiently small changes in ε. This gives the differential equation $\Delta u_\varepsilon + \lambda u_\varepsilon = -\lambda_\varepsilon u$ in $\Omega(\varepsilon)$.

The question of the boundary condition for u_ε on $\Gamma(\varepsilon)$ is more subtle; it is true that u remains 0 on the changing boundary, but u_ε refers only to the change of u at a fixed point and does not take into account the motion of the boundary. Thus it is the substantial derivative of u with respect to ε that vanishes on $\Gamma(\varepsilon)$, and not u_ε. Following the approach suggested by Joseph and Fosdick, we shall perform an invertible coordinate transformation which maps each perturbed domain onto the base domain $\Omega(0)$. Let x be the point in $\Omega(\varepsilon)$ that is mapped into the point ξ in $\Omega(0)$; we can then write $\xi = \xi(x, \varepsilon)$ for the mapping that takes $\Omega(\varepsilon)$ into $\Omega(0)$, and $x = x(\xi, \varepsilon)$ for the inverse. An explicit form for the function $\xi(x, \varepsilon)$ will not be needed, as it must play no role in the final answer [the exact eigenvalue $\lambda(\varepsilon)$ is presumably determinate and cannot depend on an artificial domain transformation introduced for mathematical convenience]. With $u(x, \varepsilon)$ the eigenfunction on $\Omega(\varepsilon)$, we see that $u(x(\xi, \varepsilon), \varepsilon) = \bar{u}(\xi, \varepsilon)$ is defined on $\Omega(0)$ for each ε and vanishes when ξ is on $\Gamma(0)$. Thus we can write that

$$\frac{d\bar{u}}{d\varepsilon} = 0, \qquad \xi \text{ on } \Gamma(0),$$

and hence

(3.38) $\dfrac{\partial u}{\partial \varepsilon} + \operatorname{grad} u \cdot \dfrac{dx}{d\varepsilon} = 0, \qquad x \text{ on } \Gamma(\varepsilon).$

The boundary condition appears to depend on the nature of the transformation $x(\xi, \varepsilon)$, but this is deceptive. We know that $x(\xi, \varepsilon)$ takes $\Omega(0)$ into $\Omega(\varepsilon)$, while $x(\xi, \varepsilon + d\varepsilon)$ takes $\Omega(0)$ into the domain $\Omega(\varepsilon + d\varepsilon)$. The vector dx therefore takes a point in $\Omega(\varepsilon)$ into a point of $\Omega(\varepsilon + d\varepsilon)$. On the boundary, dx takes a point of $\Gamma(\varepsilon)$ into a point of $\Gamma(\varepsilon + d\varepsilon)$; we can conveniently regard this motion as taking place in a direction normal to $\Gamma(\varepsilon)$. Let x be a point in $\Gamma(\varepsilon)$, and let $\hat{\nu}$ be the outward normal to $\Gamma(\varepsilon)$ at that point. This

normal intersects $\Gamma(\varepsilon+d\varepsilon)$ at a point $x+\hat{\nu}\,d\nu$, where $d\nu=d\nu(x,\varepsilon)$ is a number which may be positive or negative. Thus (3.38) takes the form

$$u_\varepsilon+\frac{\partial u}{\partial \nu}\frac{d\nu}{d\varepsilon}=0,$$

and $u_\varepsilon(x,\varepsilon)$ therefore satisfies the BVP

$$\Delta u_\varepsilon+\lambda u_\varepsilon=-\lambda_\varepsilon u,\quad x\in\Omega(\varepsilon);\qquad u_\varepsilon=-\frac{\partial u}{\partial \nu}\frac{d\nu}{d\varepsilon}\text{ on }\Gamma(\varepsilon).$$

We now set $\varepsilon=0$ to obtain the equation relating $u_\varepsilon(x,0)$ and $\lambda_\varepsilon(0)$:

(3.39)

$$\Delta u_\varepsilon+\lambda_n u_\varepsilon=-\lambda_\varepsilon(0)e_n(x),\quad x\in\Omega(0);\qquad u_\varepsilon=-\frac{\partial e_n}{\partial \nu}\left(\frac{d\nu}{d\varepsilon}\right)_{\varepsilon=0}\text{ on }\Gamma(0).$$

Since the homogeneous problem has a nontrivial solution, a solvability condition must be satisfied. Combining the equation $\Delta e_n+\lambda_n e_n=0$ with (3.39) in the usual way, we find that

$$\int_{\Gamma(0)}\left(u_\varepsilon\frac{\partial e_n}{\partial \nu}-e_n\frac{\partial u_\varepsilon}{\partial \nu}\right)dS=\lambda_\varepsilon(0)\int_{\Omega(0)}e_n^2(x)\,dx=\lambda_\varepsilon(0),$$

or, since e_n vanishes on $\Gamma(0)$ and $u_\varepsilon=-(\partial e_n/\partial \nu)(d\nu/d\varepsilon)_{\varepsilon=0}$ on $\Gamma(0)$,

$$\lambda_\varepsilon(0)=-\int_{\Gamma(0)}\left(\frac{\partial e_n}{\partial \nu}\right)^2\left(\frac{d\nu}{d\varepsilon}\right)_{\varepsilon=0}dS.$$

The first-order change in the eigenvalue λ_n is then given by

(3.40)
$$\lambda(\varepsilon)-\lambda_n\sim-\varepsilon\int_{\Gamma(0)}\left(\frac{\partial e_n}{\partial \nu}\right)^2\left(\frac{d\nu}{d\varepsilon}\right)_{\varepsilon=0}dS.$$

As a concrete example let us calculate the fundamental eigenvalue of an elliptical membrane with semiaxes $1-\varepsilon$ and 1, where ε is small. For $\varepsilon=0$ we have $\lambda_1=\beta^2$, where β is the first zero of $J_0(x)$, and $e_1(x)=cJ_0(\beta r), r=|x|,c$ being a normalization constant chosen so that $2\pi c^2\int_0^1 rJ_0^2(\beta r)\,dr=1$. We then find from (3.40) that

$$\lambda(\varepsilon)\sim\beta^2-\varepsilon\int_0^{2\pi}d\theta\,c^2\beta^2[J_0'(\beta)]^2(-\cos^2\theta)$$

$$=\beta^2+\varepsilon c^2\beta^2[J_0'(\beta)]^2\pi.$$

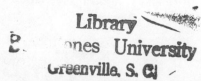

As another illustration of domain perturbation, we derive Hadamard's formula (see Bergman and Schiffer, reference in Chapter 8) for the variation of Green's function. Let $\Omega(0)$ be a domain for which Green's function $g(x,\xi;0)$ is known, and let $\Omega(\varepsilon)$ be a domain for which $g(x,\xi;\varepsilon)$ is sought. For simplicity we deal with the negative Laplacian, but the idea applies to other operators as well. Then $g(x,\xi;\varepsilon)$ satisfies

$$-\Delta g = \delta(x-\xi), \quad x,\xi \in \Omega(\varepsilon); \qquad g=0 \text{ for } x \text{ on } \Gamma(\varepsilon).$$

Proceeding as before, we find that

$$-\Delta g_\varepsilon = 0, \quad x,\varepsilon \in \Omega(\varepsilon); \qquad g_\varepsilon = -\frac{\partial g}{\partial \nu}\frac{d\nu}{d\varepsilon} \text{ on } \Gamma(\varepsilon).$$

On setting $\varepsilon=0$, we see that $g_\varepsilon(x,\xi;0)$ satisfies

$$-\Delta g_\varepsilon = 0, \quad x,\varepsilon \in \Omega(0); \qquad g_\varepsilon = -\frac{\partial g}{\partial \nu}\left(\frac{d\nu}{d\varepsilon}\right)_0 \text{ on } \Gamma(0),$$

from which it follows that

$$(3.41) \qquad g_\varepsilon(\eta,\xi;0) = \int_{\Gamma_0}\left[\frac{\partial g}{\partial \nu}(x,\xi;0)\frac{\partial g}{\partial \nu}(x,\eta;0)\right]\frac{d\nu}{d\varepsilon}(x,0)\,dS_x.$$

Exercises

3.1. Prove that the mapping T defined by (3.6) is a contraction on $C(0,1)$ for $|\varepsilon| \leqslant 6$. Hence show that the smallest eigenvalue of

$$-u'' + u = \lambda x^2 u, \quad 0 < x < 1; \qquad u(0) = u(1) = 0$$

must exceed 6.

3.2. Compare (3.23) and (3.30).

3.3. Consider steady heat conduction in the strip $= -\infty < x < \infty, |y| < 1$. The boundary temperatures are $u(x,\pm 1) = x$. The strip is filled with a mixture of water and ice. Where the temperature is positive, there is water; where it is negative, there is ice. If the conductivity of water and ice were equal, the interface between the phases would be $x=0$. By domain perturbation find an approximate expression for the interface position if the thermal conductivities in water and ice are $1+\varepsilon$ and 1, respectively.

3.4. The geometry is the same as in Exercise 3.3, but the boundary temperatures are $u(x, 1) = 1$, $u(x, -1) = -1$. If the conductivities of water and ice were equal, the interface would be at $y = 0$. For unequal conductivities find the interface by domain perturbation and also by an exact method.

4. TECHNIQUES FOR NONLINEAR PROBLEMS

A Branching Problem

Many of the methods introduced in Section 3 for linear problems have nonlinear analogs. We shall illustrate their application to concrete examples, beginning with a branching problem:

$$(4.1) \qquad -u'' = \alpha \sin u, \quad 0 < x < 1; \qquad u'(0) = u(1) = 0,$$

which differs from the buckling problem (1.22) for $\phi(x)$ only in the boundary condition at $x = 1$. Every solution ϕ of (1.22) with $\alpha > 0$ and $|\phi(0)| < \pi$ generates a solution of (4.1) by eliminating the last quarter-wave and rescaling; conversely, every solution of (4.1) can be extended into a solution ϕ of (1.22). The translation of (4.1) into an integral equation is a little simpler than for $\phi(x)$ in (1.22) because the latter boundary value problem has a nontrivial solution when $\alpha = 0$.

Problem (4.1) has the basic solution $u \equiv 0$, and linearization about $u = 0$ gives the linear BVP

$$(4.2) \qquad -u'' = \alpha u, \quad 0 < x < 1; \qquad u'(0) = u(1) = 0,$$

whose eigenvalues and normalized eigenfunctions are

$$(4.3) \quad \alpha_n = \frac{(2n-1)^2 \pi^2}{4}, \qquad e_n(x) = \frac{\sqrt{2} \cos(2n-1)\pi x}{2}, \quad n = 1, 2, \dots.$$

Since 0 is not an eigenvalue, we can construct Green's function $g(x, \xi)$ satisfying

$$(4.4) \qquad -g'' = \delta(x - \xi), \quad 0 < x, \xi < 1; \qquad g'(0, \xi) = g(1, \xi) = 0,$$

and a straightforward calculation gives

$$(4.5) \qquad g(x, \xi) = 1 - \max(x, \xi).$$

We can therefore write (4.1) as the equivalent integral equation

$$(4.6) \qquad \lambda u(x) = \int_0^1 g(x,\xi) \sin u(\xi)\, d\xi \doteq Au, \qquad \lambda = \frac{1}{\alpha}.$$

Here A is a nonlinear Hammerstein integral operator (see (1.32)) whose linearization at $u=0$ is the linear operator G with kernel $g(x,\xi)$. The equation $Gu = \lambda u$ has eigenvalues $\lambda_n = 1/\alpha_n$ and the same eigenfunctions as (4.3). Since A is compact, and the eigenvalues of $G = A'(0)$ are simple, Theorem 2 of Section 2 guarantees branching from the basic solution at each λ_n. Before calculating the initial shape of these branches by various methods (and at the same time proving the existence independently of Theorem 2, Section 2), let us draw some preliminary conclusions from (4.1) and (4.6).

Multiplication of the differential equation in (4.1) by u' and integration from $x=0$ to $x=1$ yields

$$\left[u'(1) \right]^2 = 2\alpha \left[1 - \cos u(0) \right],$$

from which it follows that $\alpha > 0$ unless $u'(1) = 0$ (which in turn implies that (4.1) has only the trivial solution). Therefore nontrivial solutions of (4.1) and (4.6) are possible only for $\alpha > 0$ and $\lambda > 0$, respectively. We can say more from (4.6) by taking the norm of both sides:

$$|\lambda| \, \|u\| \leqslant \|G\| \, \|\sin u\| \leqslant \|G\| \, \|u\|.$$

Now $\|G\|$ is the largest eigenvalue $\lambda_1 = 4/\pi^2$ of the linear integral operator G. We therefore conclude that (4.6) can have nontrivial solutions only if $0 < \lambda \leqslant 4/\pi^2$ [and (4.1) only if $\alpha \geqslant \pi^2/4$]. This is a global result and tells us, among other things, that any branch emanating from the basic solution at $\lambda = \lambda_n$ cannot wander beyond the segment $0 < \lambda \leqslant 4/\pi^2$.

We now turn to the branching analysis of (4.1). We shall use three methods;

1. The Liapunov-Schmidt method, which deals rigorously with the integral equation (4.6) but has computational disadvantages.

2. The Poincaré-Keller method, which exploits the connection between the boundary value problem (4.1) and the initial value problem for the same equation.

3. The monotone iteration method, which, unlike the other two, is not based on a local analysis near the branch-points, but has the disadvantage that it can be used easily only for the branch of positive (or negative) solutions, that is, the branch emanating from λ_1.

The Liapunov-Schmidt Method

The operator A of (4.6) has the property $A0=0$ and $A'(0)=G$. In light of (1.38), where we set $u_0=0$ and $r=Ru$, (4.6) becomes

$$(4.7) \qquad \lambda u = Au = Gu + Ru, \qquad \lim_{u \to 0} \frac{\|Ru\|}{\|u\|} = 0,$$

where

$$(4.8) \quad Gu = \int_0^1 g(x,\xi)u(\xi)\,d\xi, \qquad Ru = \int_0^1 g(x,\xi)\big[\sin u(\xi) - u(\xi)\big]\,d\xi.$$

It is useful to further split the remainder Ru into a leading homogeneous term Cu of third degree and a higher order term Du:

$$(4.9)$$

$$Ru = Cu + Du = -\int_0^1 g(x,\xi)\frac{u^3(\xi)}{6}\,d\xi + \int_0^1 g(x,\xi)\left(\sin u - u + \frac{u^3}{6}\right)d\xi.$$

With a view toward studying branching from the null solution at $\lambda = \lambda_n$, we set

$$(4.10) \qquad \delta = \lambda - \lambda_n, \qquad u = ce_n + w, \qquad \langle w, e_n \rangle = 0,$$

so that (4.7) becomes

$$(4.11) \qquad\qquad Gu - \lambda_n u = \delta u - Ru.$$

On the left side of (4.11) appears the linear self-adjoint integral operator G; λ_n is a simple eigenvalue of G. Therefore (4.11) will be consistent if and only if

$$\langle \delta u - Ru, e_n \rangle = 0.$$

If this solvability condition is satisfied, (4.11) has a unique solution orthogonal to e_n. This solution has been called w in (4.10) and can be expressed as $T(\delta u - Ru)$, where T is the familiar pseudo inverse given by (3.16). Therefore (4.11) is equivalent to the pair of equations $\langle \delta u - Ru, e_n \rangle = 0$, $w = T(\delta u - Ru)$, which, since $Te_n = 0$, can be simplified to

$$(4.12) \qquad\qquad \delta c = \langle R(ce_n + w), e_n \rangle,$$

$$(4.13) \qquad\qquad w = T[\delta w - R(ce_n + w)],$$

which should be compared to (3.18) for linear perturbation theory. Our operator R is now nonlinear, and there appears to be no natural perturbation parameter ε; it turns out that c in (4.12)-(4.13) will play that role. We know that nontrivial solutions branching from the null solution begin with small norm, so that c and w will both be small for small δ; this is quite different from (3.18), where c was merely an arbitrary normalization constant (since for a linear problem the eigenfunction is only determined up to a multiplicative constant).

We now turn our attention to the pair of equations (4.12)-(4.13), where the numbers δ and c and the element w are unknown. A solution with δ, c, w small would represent branching from the null solution near λ_n. To find all such solutions, the procedure (whose details will not be carried out) is as follows. With δ and c fixed and small, it can be shown by contraction methods that (4.13) has one and only one small solution $w = \overline{w}(c, \delta)$. This solution is then substituted in (4.12) to yield a *single nonlinear equation* in the two real variables c, δ. Of course none of this is ever carried out exactly. Instead, one obtains from (4.13) asymptotic information on $\overline{w}$, which is then used to simplify (4.12). In our particular problem it can be shown that, for small $|c|$, $\overline{w}$ is of the order $|c|^3$ uniformly in a δ interval, $|\delta| \leqslant \delta_0$. This tells us, as expected, that near the branch-point the solution u of the nonlinear problem is *essentially proportional to* e_n, the eigenfunction of the linearized problem. The order information on w and the specific forms of C and D enable us to write (4.12) to the first order as

$$\delta c = \langle C(ce_n), e_n \rangle = c^3 \langle Ce_n, e_n \rangle.$$

From the definition of C in (4.9) we have

$$\langle Ce_n, e_n \rangle = -\int_0^1 e_n(x)\,dx \int_0^1 g(x, \xi) \frac{e_n^3(\xi)}{6}\,d\xi$$

$$= -\frac{1}{6}\int_0^1 e_n^3(\xi)\,d\xi \int_0^1 g(x, \xi) e_n(x)\,dx$$

$$= -\tfrac{1}{6}\lambda_n \int_0^1 e_n^4(\xi)\,d\xi = -\frac{1}{\pi^2(2n-1)^2},$$

which yields

(4.14) $$\delta = \lambda - \lambda_n \sim -\frac{c^2}{\pi^2(2n-1)^2}$$

and

(4.15) $$\alpha - \alpha_n = \frac{1}{\lambda} - \frac{1}{\lambda_n} \sim -\frac{\delta}{\lambda_n^2} = \frac{c^2\pi^2(2n-1)^2}{16}.$$

Thus the branching diagram (c vs. α) consists of initially parabolic branches opening to the right as in Figure 1.8.

The Poincaré-Keller Method

Here we work directly with formulation (4.1) and a related initial value problem. First let us observe that any nontrivial solution of (4.1) must have $u(0) \neq 0$. Along the branch emanating from the trivial solution at $\alpha = \alpha_n$, we expect that the amplitude $u(0)$ will serve as a suitable perturbation parameter. We therefore set

$$(4.16) \qquad u(x) = az(x), \qquad z(0) = 1,$$

and consider the IVP obtained by substituting (4.16) into (4.1) and temporarily omitting the boundary condition at the right end:

$$(4.17) \qquad z'' + \alpha \frac{\sin az}{a} = 0; \qquad z(0) = 1, \qquad z'(0) = 0.$$

The one and only solution of (4.17) is denoted by $z(x, \alpha, a)$ and depends analytically on α and a. As $a \to 0$, we have a linear problem whose solution is

$$(4.18) \qquad z(x, \alpha, 0) = \cos \sqrt{\alpha}\, x.$$

Any solution of (4.17) for $a \neq 0$ that happens to vanish at $x = 1$ will, through (4.16), provide a nontrivial solution of (4.1). If α and a are chosen randomly in (4.17), the corresponding z will not usually vanish at $x = 1$; we are looking for the relationship between α and a that will make

$$(4.19) \qquad 0 = z(1, \alpha, a) \doteq b(\alpha, a).$$

Of course we already know that $b(\alpha_n, 0) = 0$, but we wish to find solutions with $a \neq 0$, a near 0, α near α_n. The function $b(\alpha, a)$ is not usually explicitly calculable (although in our case it can be expressed in terms of elliptic functions); nevertheless we want to establish the existence and nature of the solutions (α, a) of (4.19) near $(\alpha_n, 0)$. The only weapon at our disposal is the implicit function theorem. If $b_\alpha(\alpha_n, 0) \neq 0$, then locally the solution of (4.19) can be expressed as $\alpha = \alpha(a)$; if $b_a(\alpha_n, 0) \neq 0$, the solution can be expressed as $a = a(\alpha)$. It is clear from (4.17) that $z(x, \alpha, a)$ is an even function of a, so that $z_a(x, \alpha, 0) = 0$ and hence $b_a(\alpha_n, 0) = 0$ [this result could also be obtained by differentiating (4.17) with respect to a]. We can calculate $b_\alpha(\alpha_n, 0)$ from (4.18):

$$b_\alpha(\alpha_n, 0) = z_\alpha(1, \alpha_n, 0) = -\frac{1}{2\sqrt{\alpha_n}} \sin \sqrt{\alpha_n} = \frac{(-1)^n}{\pi(2n-1)} \neq 0.$$

If the explicit expression (4.18) were not available, one could differentiate (4.17) with respect to α to obtain the desired result.

The implicit function theorem guarantees the existence of a unique function $\alpha = \alpha_n(a)$, defined near $a = 0$, such that $\alpha_n(0) = \alpha_n$ and $b(\alpha_n(a), a) = 0$. This function $\alpha_n(a)$ generates a solution $az(x, \alpha_n(a), a)$ of (4.1) which completely describes the branching at $\alpha = \alpha_n$.

The function $z(x, \alpha_n(a), a)$ satisfies (4.17) and also the boundary condition at $x = 1$. To find $\alpha_n(a)$ and z for a near zero, we differentiate (4.17) totally with respect to a. To simplify the notation we let $z(x, \alpha_n(a), a) = Z(x, a)$ and drop the index on $\alpha_n(a)$. Then $Z(x, a)$ satisfies

$$ Z'' + \alpha(a) \frac{\sin aZ}{a} = 0; \qquad Z(0, a) = 1, \quad Z(1, a) = 0, \quad Z'(0, a) = 0. $$

Taking the a derivative, we find that $Z_a(x, 0) = 0$, $\alpha_a(0) = 0$. Further differentiation gives, after setting $a = 0$, an equation for $Z_{aa}(x, 0)$:

$$ Z_{aa}'' + \alpha_n Z_{aa} = \tfrac{1}{3}\alpha_n Z^3(x, 0) - \alpha_{aa}(0)Z(x, 0), \qquad 0 < x < 1; $$

$$ Z_{aa}(0, a) = Z_{aa}(1, a) = 0, \qquad Z_{aa}'(0, a) = 0, $$

where $Z(x, 0) = \cos \sqrt{\alpha_n}\, x$. The solvability condition then gives

$$ \alpha_{aa}(0) = \frac{\alpha_n}{3} \frac{\displaystyle\int_0^1 \cos^4 \sqrt{\alpha_n}\, x\, dx}{\displaystyle\int_0^1 \cos^2 \sqrt{\alpha_n}\, x\, dx} = \frac{\alpha_n}{4}. $$

Thus, to the first nonvanishing order, we have

$$ \alpha_n(a) - \alpha_n \sim \alpha_{aa}(0) \frac{a^2}{2} = \frac{(2n-1)^2 \pi^2}{32} a^2, $$

which is in agreement with (4.15) if we take into account the relationship between c and a. In fact, c is the coefficient of the normalized eigenfunction $\sqrt{2}\cos \sqrt{\alpha_n}\, x$, whereas a is the coefficient of $\cos \sqrt{\alpha_n}\, x$; thus $c = a/\sqrt{2}$, as is needed to reconcile the two formulas.

It is worth noting that once we have determined that a is a suitable expansion parameter, and that $\alpha_n(a)$ and $Z(x, a)$ are even functions of a, we can merely substitute series expansions in a^2 in (4.17) and calculate the coefficients by equating like powers of a^2.

Monotone Iteration Method

The two preceding methods used for (4.1) are essentially local in nature. We have been able to predict that near $\alpha = \alpha_n$ there is branching to the right from the basic solution; there are two nontrivial solutions of small norm for α slightly larger than α_n, and these solutions (*to the first order*) are $\pm a \cos \sqrt{\alpha_n}\, x$, where

$$a = \frac{\sqrt{32}}{(2n-1)\pi} \sqrt{\alpha - \alpha_n}\,.$$

The functions $\{\cos \sqrt{\alpha_n}\, x\}$, being eigenfunctions of a Sturm-Liouville system, have well-defined nodal properties. The fundamental eigenfunction (corresponding to $n=1$) is of one sign (and can therefore, of course, be chosen positive); the eigenfunction corresponding to α_n has $n-1$ zeros in $0<x<1$. It is interesting to speculate whether these properties carry over to the nonlinear problem. Near the branch-point α_n this is certainly true from the fact that the linearized problem dominates; but as we move along the branch, it is not obvious that the nodal properties are maintained. (For proofs, see Pimbley and also Crandall and Rabinowitz.) We shall content ourselves with showing that the branching from α_1 provides a *positive solution for all $\alpha > \alpha_1$* (of course the "lower" half of the parabolic branch yields a negative solution). The monotone method used is not based on a branching argument. We therefore take α in (4.1) as *fixed* and larger than α_1. Our principal result is contained in the following theorem.

Theorem.   For each $\alpha > \alpha_1$, (4.1) has one and only one positive solution, that is, there is one and only one solution $u(x,\alpha)$ with the property $u(x,\alpha)>0$, $0<x<1$.

Proof.   We now proceed in the rest of the subsection to prove this theorem by applying Theorem 1 of Section 1.
   By adding αu to both sides of (4.1), we obtain the equivalent BVP

$$-u'' + \alpha u = \alpha(\sin u + u), \quad 0<x<1; \quad u'(0)=u(1)=0,$$

where $\alpha(\sin u + u)$ is an increasing function of u on the real line. For a given $v(x)$ the solution of the linear BVP

$$(4.20) \qquad -u'' + \alpha u = \alpha(\sin v + v), \quad 0<x<1; \quad u'(0)=u(1)=0$$

can be written as

$$(4.21) \qquad u = Tv \doteq \int_0^1 g(x,\xi)\left[\alpha \sin v(\xi) + \alpha v(\xi)\right] d\xi,$$

where $g(x,\xi)$ is Green's function satisfying

$$(4.22) \quad -g'' + \alpha g = \delta(x-\xi), \quad 0<x, \xi<1; \quad g'|_{x=0} = g|_{x=1} = 0.$$

Clearly T can be regarded as a nonlinear operator on the space $C(0,1)$ of real-valued continuous functions on $0 \leqslant x \leqslant 1$. Problem (4.1) is therefore equivalent to the equation $u = Tu$ or

$$u(x) = \int_0^1 g(x,\xi)\left[\alpha \sin u(\xi) + \alpha u(\xi)\right] d\xi, \qquad 0 < x < 1,$$

so that solutions of (4.1) are fixed points of T and vice versa. Since T is an integral operator, it is hardly surprising that it is compact, but we shall not give a proof.

For $u, v \in C(0,1)$ we write $u \leqslant v$ if $u(x) \leqslant v(x)$ on $0 \leqslant x \leqslant 1$. We shall prove that T is increasing; that is, if $u \leqslant v$, then $Tu \leqslant Tv$. One proof follows from the fact that $g(x,\xi)$ is positive and that $\alpha \sin u + \alpha u$ is an increasing function of u on the real line. Another proof is based on the following extension of the maximum principle.

Lemma. If v satisfies the differential inequality

$$-v'' + \alpha v \geqslant 0, \quad 0 < x < 1; \qquad -v'(0) \geqslant 0, \quad v(1) \geqslant 0,$$

then $v(x) \geqslant 0$ on $0 \leqslant x \leqslant 1$, that is, $v \geqslant 0$.

Indeed, if $v(x) < 0$ for some x, then either v has an interior negative minimum or v has a negative minimum at $x = 0$. The differential inequality rules out the first of these possibilities. The second implies that $v(0) < 0$, so that, by the differential inequality, $v''(0) \leqslant 0$; these conditions, together with the boundary condition $v'(0) \leqslant 0$, imply that $v(x)$ decreases with x at $x = 0$, hence $x = 0$ cannot be a minimum.

Now let $v \leqslant w$, and set $z = Tw - Tv$. Then, from (4.20) and (4.21), we find that

$$-z'' + \alpha z = \alpha\left[w - v + \sin(w - v)\right], \quad 0 < x < 1; \qquad z'(0) = z(1) = 0.$$

Since $w \geqslant v$, the right side of the differential equation is nonnegative, and the lemma gives $z \geqslant 0$, that is, $Tw \geqslant Tv$.

The iterative sequences (1.43) and (1.44) of Theorem 1, Section 1, are defined in terms of the operator T in (4.21) or, equivalently, in terms of the BVP (4.20). For instance, $\{u_n\}$ satisfies the equivalent iteration schemes

$$-u_n'' + \alpha u_n = \alpha(\sin u_{n-1} + u_{n-1}), \quad 0 < x < 1; \qquad u_n'(0) = u_n(1) = 0,$$

or

$$u_n = Tu_{n-1} = \int_0^1 g(x,\xi)\left[\alpha \sin u_{n-1}(\xi) + \alpha u_{n-1}(\xi)\right] d\xi,$$

and similarly for the sequence $\{u^n\}$.

As starting elements for the iterations we need a lower solution u_0 for T ($u_0 \leqslant Tu_0$) and an upper solution u^0 for $T(u^0 \geqslant Tu^0)$ with $u_0 \leqslant u^0$. We shall show that these properties follow for elements $u_0 \leqslant u^0$ defined by the inequalities below, which are in terms of the original BVP:

$$(4.23) \qquad -u_0'' \leqslant \alpha \sin u_0, \quad 0 < x < 1; \qquad -u_0'(0) \leqslant 0, \quad u_0(1) \leqslant 0,$$

$$(4.24) \quad -(u^0)'' \geqslant \alpha \sin u^0, \quad 0 < x < 1; \qquad -(u^0)'(0) \geqslant 0, \quad u^0(1) \geqslant 0.$$

To show that $u_0 \leqslant Tu_0$, we note that $u_1 = Tu_0$, and

$$-u_1'' + \alpha u_1 = \alpha(\sin u_0 + u_0), \quad 0 < x < 1; \qquad u_1'(0) = u_1(0) = 0,$$

which, when combined with (4.23), gives

$$-(u_1 - u_0)'' + \alpha(u_1 - u_0) \geqslant 0, \quad 0 < x < 1;$$
$$-[u_1'(0) - u_0'(0)] \geqslant 0, \quad u_1(1) - u_0(1) \geqslant 0.$$

Our lemma then shows that $u_1 \geqslant u_0$. A similar proof yields $u^0 \geqslant Tu^0$.

Next we note that $u_0(x) \equiv 0$ and $u^0(x) \equiv \pi$ satisfy (4.23) and (4.24), respectively. Since T is compact and increasing on $[0, \pi]$, we can apply Theorem 1 to guarantee the existence of a minimal solution u and a maximal solution $\bar{u}$ in the order interval $[0, \pi]$. Thus $0 \leqslant u(x) \leqslant \bar{u}(x) \leqslant \pi$, but u and $\bar{u}$ might both vanish identically. We must try to find another lower solution $u_0(x)$ which is positive in the interval $0 < x < 1$, so that $u(x)$ will be positive in $0 < x < 1$. Physical insight helps; for α slightly above α_1 we expect a positive solution which is nearly proportional to the eigenfunction $e_1(x)$ of the linearized problem. Therefore we try the lower solution

$$u_0 = \varepsilon e_1, \qquad \varepsilon > 0,$$

which clearly satisfies the boundary inequalities in (4.23). Moreover, we also have

$$-u_0'' - \alpha \sin u_0 = \varepsilon \alpha_1 e_1 - \alpha \sin \varepsilon e_1 = -\alpha \varepsilon e_1 \left(\frac{\sin \varepsilon e_1}{\varepsilon e_1} - \frac{\alpha_1}{\alpha} \right).$$

With α fixed and larger than α_1, we can choose $\varepsilon > 0$ sufficiently small so that the term in parentheses is positive and hence $-u_0'' \leqslant \alpha \sin u_0$ as required. Therefore Theorem 1 now guarantees solutions $u(x)$ and $\bar{u}(x)$ satisfying

$$\varepsilon e_1(x) \leqslant u(x) \leqslant \bar{u}(x) \leqslant \pi, \qquad \alpha > \alpha_1.$$

Strictly speaking, the solution $\underline{u}$ should be denoted as $\underline{u}_\varepsilon$ since it might depend on ε. It is easily shown, however, that in fact this solution is independent of ε for all small positive ε satisfying $(\sin \varepsilon e_1)/\varepsilon e_1 \geqslant \alpha_1/\alpha$. We can therefore legitimately drop the ε subscript. Since e_1 is positive for $0 \leqslant x \leqslant 1$, so is $\underline{u}$, and $\underline{u}$ is the minimal solution in the order interval $[\varepsilon e_1, \pi]$ for every small positive ε. Remarkably this does not guarantee that $\underline{u}$ is the minimal nontrivial solution in $[0, \pi]$. It is conceivable that there exists a solution ϕ of (4.1) which vanishes at some point $0 \leqslant x < 1$ or has a vanishing derivative at $x = 1$. If this were the case, we could not squeeze εe_1 under ϕ no matter how small a positive value of ε were chosen. Fortunately this situation cannot occur. One can show explicitly that Green's function as defined by (4.22) is positive in $0 \leqslant x < 1$ and that $g'(1, \xi) < 0$. It then follows from the equation $\phi = T\phi$ that $\phi > 0$ in $0 \leqslant x < 1$ and that $\phi'(1) < 0$, so that a positive ε can be chosen small enough to satisfy $\varepsilon e_1 \leqslant \phi$. Thus $\underline{u}$ is the minimal nontrivial solution in $[0, \pi]$.

We show that $\underline{u} = \bar{u}$, so that the positive solution is unique. From (4.1), satisfied by $\underline{u}$ and $\bar{u}$, we find that

$$- \underline{u}'' \bar{u} + \bar{u}'' \underline{u} = \alpha(\bar{u} \sin \underline{u} - \underline{u} \sin \bar{u}),$$

and, by Green's theorem and the boundary conditions,

(4.25)
$$0 = \int_0^1 \left[\frac{\sin \underline{u}}{\underline{u}} - \frac{\sin \bar{u}}{\bar{u}} \right] \underline{u} \bar{u} \, dx.$$

We already know that $\underline{u}(x) \leqslant \bar{u}(x)$ and that $\underline{u}(x) > 0$ on $0 \leqslant x < 1$. Let D be the set on which $\underline{u} < \bar{u}$. Then the integral in (4.25) reduces to an integral over D. Since $(\sin u)/u$ is strictly decreasing on $0 \leqslant u \leqslant \pi$, we have

$$\frac{\sin \underline{u}}{\underline{u}} > \frac{\sin \bar{u}}{\bar{u}} \quad \text{on } D \qquad \text{and} \qquad \bar{u} \underline{u} > 0 \quad \text{on } D.$$

This, however, contradicts the fact that the integral over D vanishes.

Combined Diffusion and Reaction

The boundary value problem

(4.26)
$$- \Delta u = f(u(x)), \quad x \in \Omega; \qquad u = 0, \quad x \in \Gamma,$$

where Ω is a bounded domain in R_n and Γ is its boundary, occurs in chemical reactor analysis, nuclear reactor analysis, Joule heating, biochemical processes, and so on. In many of these applications the function $u(x)$ must be nonnegative, and we shall be interested in finding *positive solutions*, that is, *nontrivial* solutions which are $\geqslant 0$. In view of the boundary condition a positive solution $u(x)$ must take on values between 0 and some positive number u_M. If the diffusion coefficient is itself nonlinear, the differential equation has the form $\operatorname{div}(k(u)\operatorname{grad} u) = f(u)$, which can be reduced to type (4.26) by the change of variables $v = \int_0^u k(z)\,dz$.

Positive solutions of (4.26) will exist only if f satisfies certain conditions whose nature can be surmised by examining the *much simpler* problem of finding positive numbers u such that

$$(4.27) \qquad lu = f(u),$$

where $l > 0$. Note that l is a positive linear operator on R_1, just as $-\Delta$ with vanishing boundary conditions is a positive linear operator on $L_2(\Omega)$. All that is required for (4.27) to have positive solution(s) is for the curve $f(u)$ to intersect the straight line lu for $u > 0$. No such solution can exist if f lies either entirely below or entirely above the straight line. Remarkably, the same nonexistence result holds for (4.26) if we identify l with the lowest (fundamental) eigenvalue α_1 of

$$(4.28) \qquad -\Delta u = \alpha u, \quad x \in \Omega; \qquad u = 0, \quad x \in \Gamma.$$

We recall that α_1 is simple and that the corresponding eigenfunction $e_1(x)$ can (and will) be chosen to be strictly positive in Ω. Moreover, e_1 is normalized so that its maximum is 1. To avoid the multivaluedness associated with linear problems we shall make the following *nonlinearity assumption*:

The functions $f(z)$ and $\alpha_1 z$ are not identically equal on any interval $(0, z_1)$ with $z_1 > 0$.

We are now in a position to state the following theorem, which is a nonexistence theorem for positive solutions.

Theorem. Positive solutions of (4.26) are possible only if the real-valued function $f(z) - \alpha_1 z$ changes sign for $z > 0$. Moreover, if $f(z) - \alpha_1 z < 0$ for $z > 0$, a solution of (4.26) cannot be positive at any point in Ω.

Remark. If $f(z) - \alpha_1 z > 0$ for $z > 0$, there may exist solutions that are positive in part of Ω and negative elsewhere in Ω.

Proof. Write (4.26) as

(4.29) $\qquad -\Delta u - \alpha_1 u = f(u) - \alpha_1 u, \quad x \in \Omega; \qquad u = 0, \quad x \in \Gamma.$

The solvability condition for (4.29) is

$$\int_\Omega [f(u(x)) - \alpha_1 u(x)] e_1(x)\, dx = 0.$$

Since $e_1(x) > 0$ in Ω, either $f(z) - \alpha_1 z \equiv 0$, $0 \leqslant z \leqslant u_M$, or $f(z) - \alpha_1 z$ must change sign in $0 < z < u_M$. The first alternative is impossible by the nonlinearity assumption, so that $f(z) - \alpha_1 z$ must change sign for $z > 0$. Next we want to prove that, if $f(z) - \alpha_1 z < 0$ for $z > 0$, then $u(x)$ cannot be positive anywhere in Ω. If $u(x_0) > 0$, there is a domain Ω^* surrounding x_0 for which $u(x) > 0$ with $u = 0$ on the boundary Γ^* of Ω^*. Let (α_1^*, e_1^*) be a fundamental eigenpair for (4.28) over Ω^*; since Ω^* is contained in Ω, it follows from the variational characterization of eigenvalues that $\alpha_1^* \geqslant \alpha_1$ and therefore $f(z) - \alpha_1^* z < 0$ for $z > 0$, which, by the first part of the theorem, means that $u(x)$ cannot be a positive solution on Ω^*. We have therefore arrived at a contradiction, and the theorem is proved.

Now that we have excluded some possibilities, when do there actually exist positive solutions? Let us confine ourselves to the unforced case: $f(0) = 0$ (see Exercise 4.1 for the forced case). We shall establish the existence of a positive solution by a method similar to that used for (4.1) and again based on Theorem 1, Section 1. To begin we suppose that there exist functions $u_0(x) \leqslant u^0(x)$, continuous on the closure $\bar{\Omega}$ of Ω and satisfying

(4.30) $\qquad -\Delta u_0 \leqslant f(u_0), \quad x \in \Omega; \qquad u_0 \leqslant 0, \quad x \in \Gamma,$

(4.31) $\qquad -\Delta u^0 \geqslant f(u^0), \quad x \in \Omega; \qquad u^0 \geqslant 0, \quad x \in \Gamma.$

Once such elements have been found, we pick M large enough so that

(4.32) $\qquad\qquad f'(u) + M \geqslant 0 \qquad \text{for} \quad m_0 \leqslant u \leqslant m^0,$

where

$$m_0 = \min_{x \in \bar{\Omega}} u_0(x), \qquad m^0 = \max_{x \in \bar{\Omega}} u^0(x).$$

Such a choice of M is possible if f' exists and is bounded on bounded intervals. By adding Mu to both sides of the differential equation, the BVP

(4.26) becomes

$$(4.33) \qquad -\Delta u + Mu = f(u) + Mu, \quad x \in \Omega; \qquad u = 0, \quad x \in \Gamma,$$

or, equivalently,

$$(4.34) \qquad u = Tu \doteq \int_\Omega g(x,\xi)\big[\, f(u(\xi)) + Mu(\xi)\,\big]\, d\xi,$$

where $g(x,\xi)$ is Green's function satisfying

$$(4.35) \qquad -\Delta g + Mg = \delta(x - \xi), \quad x, \xi \in \Omega; \qquad g = 0, \quad x \in \Gamma.$$

The operator T is a compact operator on the space $C(\overline{\Omega})$ of real-valued continuous functions on $\overline{\Omega}$. Our notation is the same as in the discussion preceding Theorem 1, with Ω playing the role of R. In view of the definition (4.34) of T, the iteration schemes $u^n = Tu^{n-1}$ and $u_n = Tu_{n-1}$ can equally well be expressed in terms of BVPs. For instance, the scheme $u^n = Tu^{n-1}$ is equivalent to

$$(4.36) \quad -\Delta u^n + Mu^n = f(u^{n-1}) + Mu^{n-1}, \quad x \in \Omega; \qquad u^n = 0, \quad x \in \Gamma.$$

To use Theorem 1, it remains to show that $u_0 \leqslant Tu_0, u^0 \geqslant Tu^0$, and T is increasing on $[u_0, u^0]$. Let $u_0 \leqslant v \leqslant w \leqslant u^0$, and set $z = Tw - Tv$. Then we find that

$$-\Delta z + Mz = f(w) - f(v) + M(w - v), \quad x \in \Omega; \qquad z = 0, \quad x \in \Gamma.$$

The right side of the differential equation is nonnegative by virtue of (4.32). By the maximum principle we conclude that $z \geqslant 0$, that is, $Tw \geqslant Tv$, so that T is increasing on $[u_0, u^0]$. Next we show that $u^0 \geqslant Tu^0$ (a similar argument holding for lower solutions). By (4.31) and (4.36) with $n = 1$, we obtain

$$-\Delta(u^0 - u^1) + M(u^0 - u^1) \geqslant 0, \quad x \in \Omega; \qquad u^0 - u^1 \geqslant 0, \quad x \in \Gamma.$$

The maximum principle again shows that $u^0(x) \geqslant u^1(x)$ in Ω. Since $u^1 = Tu^0$, we have the desired result.

To prove that the minimal and maximal solutions $\underline{u}$ and $\bar{u}$ guaranteed by Theorem 1 are actually positive in Ω, we must find u_0 satisfying (4.30) and $u_0(x) > 0$ in Ω. Of course, we must also find u^0 satisfying (4.31) with $u^0 \geqslant u_0$.

For the lower solution we try $u_0 = \varepsilon e_1$, where ε is a positive number and e_1 is the normalized eigenfunction introduced in connection with (4.28).

We obtain

(4.37)
$$-\Delta u_0 - f(u_0) = \varepsilon e_1 \left[\alpha_1 - \frac{f(\varepsilon e_1)}{\varepsilon e_1} \right].$$

For ε small and positive, $f(\varepsilon e_1)/\varepsilon e_1$ is nearly $f'(0)$; therefore if $f'(0) > \alpha_1$ it will be possible to choose $\varepsilon > 0$ so small that the right side of (4.37) will be negative. Since e_1 vanishes on Γ, the boundary inequality in (4.30) is clearly satisfied, so that u_0 is a lower solution if $f'(0) > \alpha_1$. Moreover, u_0 is positive in Ω. Let us try to find an upper solution of the same form with ε so *large* that $\alpha_1 - f(\varepsilon e_1)/\varepsilon e_1 \geqslant 0$ in Ω; the controlling factor should be the behavior of $f(z)/z$ for large z, but unfortunately the fact that e_1 vanishes on Γ means that, for x near Γ, $f(\varepsilon e_1)/\varepsilon e_1$ will be close to $f'(0)$ even if ε is large. There is a simple remedy if $f(z)$ satisfies the condition

(4.38)
$$\limsup_{z \to \infty} \frac{f(z)}{z} = \beta < \alpha_1.$$

We then consider a domain Ω^* similar to but slightly larger than Ω, with $\overline{\Omega}$ contained in Ω^*. The fundamental eigenvalue α_1^* of (4.28) for the domain Ω^* is slightly smaller than α_1. We can choose Ω^* so that $\beta < \alpha_1^* < \alpha_1$. The corresponding eigenfunction e_1^* is positive on $\overline{\Omega}$ (say, $e_1^* \geqslant \delta$). Trying an upper solution $u^0 = A e_1^*$, we find that

$$-\Delta u^0 - f(u^0) = A e_1^* \left[\alpha_1^* - \frac{f(A e_1^*)}{A e_1^*} \right],$$

whose right side, in view of (4.38), is nonnegative in $\overline{\Omega}$ for A sufficiently large. Simpler upper solutions can be found in some other cases (see Exercise 4.2). For instance, if $f(\gamma) = 0$ for some $\gamma > 0$, then $u^0 = \gamma$ is an upper solution.

Assuming that $f(z)$ satisfies (4.38) and $f'(0) > \alpha_1$, we have constructed a positive lower solution u_0 and an upper solution u^0. Thus Theorem 1, Section 1, guarantees the existence of a positive minimal solution $\underline{u}(x)$ and of a maximal solution $\bar{u}(x)$ in $[\varepsilon e_1, u^0]$. Again one can show that $\underline{u}$ is independent of ε for small ε and that for every positive solution ϕ of (4.34) there exists an $\varepsilon > 0$ such that $\varepsilon e_1 \leqslant \phi$. Therefore $\underline{u}(x)$ is the minimal nontrivial solution in $[0, u^0]$. Clearly $\underline{u}(x) \leqslant \bar{u}(x)$ for x in Ω; uniqueness follows by the same arguments as were used for (4.1) if $f(z)/z$ is strictly decreasing (or if $f'' < 0$).

Our results take on a clearer meaning when we introduce a parameter α in the nonlinear term. Consider the problem

(4.39)
$$-\Delta u = \alpha h(u(x)), \quad x \in \Omega; \qquad u = 0, \quad x \in \Gamma,$$

where $h(z)$ satisfies

(4.40) $h(0)=0,$ $h'(0)=1,$ $\displaystyle\limsup_{z\to\infty}\frac{h(z)}{z}=\beta<1.$

Then (4.39) has at least one positive solution when

(4.41) $\alpha_1<\alpha<\dfrac{\alpha_1}{\beta}.$

If, furthermore, $h(z)/z$ is strictly decreasing for $z>0$, the positive solution is unique and there is no positive solution for $\alpha\leqslant\alpha_1$ and $\alpha\geqslant\alpha_1/\beta$ [since then the curves α_1z and $\alpha h(z)$ do not intersect for $z>0$].

Remark. If $\beta=0$ in (4.40), then α_1/β should be interpreted as $+\infty$. Thus, if $h(z)$ is bounded on the positive real axis, a positive solution exists for all $\alpha>\alpha_1$.

Exercises

4.1. Consider the *forced* problem

(4.42) $-\Lambda u=\alpha h(u),$ $x\in\Omega;$ $u=0,$ $x\in\Gamma,$

with $h(0)>0$ and $\limsup_{z\to\infty}[h(z)/z]=\beta$. Show that there exists a positive solution for $0<\alpha<\alpha_1/\beta$. *Hint:* We can now use $u_0=0$ as a lower solution that yields $u_1(x)>0$ in Ω and hence $\underline{u}(x)>0$ in Ω (for the unforced problem we cannot use $u_0=0$ to construct a positive solution since all iterates vanish identically). Use the same upper solution as for the unforced problem. Under what conditions for h can you obtain uniqueness?

4.2. Consider the unforced problem (4.26) with $f'(0)>\alpha_1, f(z)/z$ strictly decreasing, and $f(\gamma)/\gamma=\alpha_1$ for some $\gamma>0$. Then we have proved that (4.26) has one and only one positive solution $u(x)$. Show that $u(x)\geqslant\gamma e_1(x)$. If, furthermore, there is a value δ for which $f(\delta)=0$, show that $u(x)\leqslant\delta$. In this way obtain bounds on the branch of positive solutions of

$$-\Delta u=\alpha\sin u,\quad x\in\Omega;\qquad u=0,\quad x\in\Gamma,$$

$$-\Delta u=\alpha(u-u^3),\quad x\in\Omega;\qquad u=0,\quad x\in\Gamma,$$

$$-\Delta u=\alpha u-u^3,\quad x\in\Omega;\qquad u=0,\quad x\in\Gamma.$$

In which case does $\|u\|\to\infty$ when $\alpha\to\infty$?

4.3. Let $f(z) < \alpha_1 z$ for $z > 0$ and $f(z) > \alpha_1 z$ for $z < 0$. Show that (4.26) has only the trivial solution. What can you say if both inequalities are reversed?

4.4. Prove the theorem preceding (4.29) for the case when the boundary condition is $(\partial u / \partial \nu) + hu = 0$, where h is a positive constant.

4.5. If Ω is a unit ball in R_3 and $h(u) = e^u$, the BVP (4.42) can be shown to have an infinite number of positive solutions when $\alpha = 2$ (see Joseph and Lundgren). In the one-dimensional case the problem can be solved explicitly, say on the interval $-\frac{1}{2} < x < \frac{1}{2}$. Show that there exists α_c (approximately equal to 3.52) such that there are two positive solutions for $\alpha < \alpha_c$, one for $\alpha = \alpha_c$, and none for $\alpha > \alpha_c$.

4.6. Problem (4.39) with

$$h(u) = \begin{cases} 0, & 0 \leqslant u \leqslant 1, \\ u - 1, & u \geqslant 1, \end{cases}$$

is a simplified version of one that occurs in plasma containment. Show that there exists no positive solution for $\alpha < \alpha_1$. Solve explicitly the one-dimensional problem on the interval $-1 < x < 1$, and show that there exists a positive solution for $\alpha > \alpha_1$, but that the norm of this solution tends to infinity as $\alpha \to \alpha_1 +$ (*branching from infinity*).

4.7. Consider (4.26) when $f(z) < 0$ for $z > \gamma$. Show that *every* positive solution satisfies $u(x) \leqslant \gamma$. Show that, if $f(z) = \alpha z(1 - z)$, there is one and only one positive solution for $\alpha > \alpha_1$ and this solution satisfies $u(x) \leqslant 1$. The problem occurs in population studies (see Murray).

4.8. Consider again the satellite web coupling problem (4.22), Chapter 4, repeated here for convenience:

$$-u'' = -\lambda u^4, \quad 0 < x < 1; \qquad u(0) = u(1) = 1; \qquad \lambda \geqslant 0.$$

(a) Prove that there is at most one positive solution. (Assume $u_2 > u_1$ on a subdomain, and then use the maximum principle for $u_2 - u_1$ on that subdomain to obtain a contradiction.)

(b) Show that $u_0 = 0$ is a lower solution and $u^0 = 1$ an upper solution.

(c) Show that the iteration scheme $w = Tv$, defined by

$$-w'' + 4\lambda w = -\lambda v^4 + 4\lambda v, \quad 0 < x < 1; \qquad w(0) = w(1) = 1,$$

meets the requirements of Theorem 1, Section 1, so that a nontrivial positive solution satisfying $0 \leqslant u(x) \leqslant 1$ is guaranteed. Uniqueness follows from (a).

4.9. Let C be the space of real-valued continuous functions on the closed interval $0 \leqslant x \leqslant 1$. As usual, C is equipped with the uniform norm. Let the symmetric kernel $k(x,y)$ be positive in the interior of the unit square, and assume that

$$\int_0^1 k(x,y)\,dy \leqslant M \qquad \text{for all } x, \quad 0 \leqslant x \leqslant 1.$$

Consider the nonlinear integral equation

(4.43) $$u = Tu \doteq \int_0^1 k(x,y)f(u(y))\,dy,$$

where

$$f(u) \geqslant 0, \quad 0 \leqslant u \leqslant A; \qquad f'(0) \geqslant 0;$$

$$f''(u) < 0, \quad 0 < u < A; \qquad \frac{f(A)}{A} \leqslant \frac{1}{M}.$$

(a) Show that $u^0 = A$ is an upper solution of (4.43) and that $u_0 = 0$ is a lower solution. Prove the existence of at least one solution in the order interval $[0, A]$, and show that this solution is nontrivial if $f(0) > 0$.

(b) We want to prove the existence of a nontrivial solution in $[0, A]$ if $f(0) = 0$. Let (λ_1, e_1) be the fundamental eigenpair of the linear problem

$$\int_0^1 k(x,y)e(y)\,dy = \lambda e(x), \qquad 0 \leqslant x \leqslant 1,$$

and assume that λ_1 is simple and $e_1(x) > 0$ in $0 < x < 1$. By considering candidates for lower solutions of the form εe_1, with ε small and positive, show that, if $f'(0) > 1/\lambda_1$, there exists a minimal solution of (4.43) in $[\varepsilon e_1, A]$.

(c) Prove that the minimal solution in (b) is independent of the choice of ε for sufficiently small ε. It can be shown (but not without some difficulty) that the solution is the minimal nontrivial solution in $[0, A]$. The reason why the proof is hard is that we do not have the lovely properties of Green's function!

(d) Prove that there is a unique nontrivial solution in $[0, A]$.

4.10. *Chain reaction in a rod.* We shall consider a one-dimensional version of a neutron transport problem investigated by Pazy and Rabinowitz. A thin rod, $0 < x < a$, is regarded as a simple nuclear reactor. Between collisions with nuclei, neutrons are assumed to move with unit velocity along the rod. The outcome of a collision is instantaneous replacement of the original neutron by $0, 1, \ldots, N$ neutrons, the respective probabilities being denoted by $c_0, c_1, \ldots, c_N$ with $c_k \geq 0, \Sigma_{k=0}^{N} c_k = 1$. The neutrons arising from a collision are equally likely to be moving to the left or the right with unit velocity. The expected number of neutrons arising from a collision is the *multiplying factor* $c = \Sigma_{k=1}^{N} k c_k$.

Taking the collision cross section to be σ (see Section 5, Chapter 0), we find that the probability of no collision in a path of length b is $e^{-\sigma b}$. When a neutron reaches the end of the rod, it is lost.

A single neutron moving to the right is introduced at the point x at $t = 0$. We want to calculate the probability $u(x)$ that at least one neutron is alive at $t = \infty$. The quantity $v(x)$ is similarly defined for a neutron moving to the left. There are two roads to extinction: (i) the neutron reaches an end of the rod without collision, and (ii) the neutron has its first collision in some interval $(y, y + dy)$ with all products ultimately dying.

Show that

$$1 - u(x) = e^{-(a-x)\sigma} + \int_x^a e^{-(y-x)\sigma} \sigma p(y)\, dy,$$

$$1 - v(x) = e^{-x\sigma} + \int_0^x e^{-(x-y)\sigma} \sigma p(y)\, dy,$$

where $p(y)$ is the probability of all products of a collision in $(y, y + dy)$ becoming extinct. Show that

$$p(y) = \sum_{k=0}^{N} c_k \left(\frac{1}{2}\right)^k \sum_{j=0}^{k} \binom{k}{j} [1 - u(y)]^j [1 - v(y)]^{k-j}$$

and that $z(x) \doteq [u(x) + v(x)]/2$ satisfies the nonlinear integral equation

$$(4.44) \qquad z(x) = Tz \doteq \int_0^a E(x,y) G[z(y)]\, dy, \qquad 0 \leq x \leq a,$$

where

$$E(x,y) \doteq \frac{\sigma}{2} e^{-\sigma|x-y|},$$

$$G(z) \doteq cz - \sum_{k=2}^{N} c_k \big[(1-z)^k - 1 + kz\big] = 1 - \sum_{k=0}^{N} c_k(1-z)^k.$$

By using the results of Exercise 4.9, show that there exists a unique nontrivial solution of (4.44) in the order interval $[0,1]$ if $c > 1/\lambda_1$, where λ_1 is the fundamental eigenvalue of

$$\int_0^a E(x,y)e(y)\,dy = \lambda e(x), \qquad 0 \leqslant x \leqslant a.$$

Prove that, for $c < 1/\lambda_1$, (4.44) has only the trivial solution. Thus a chain reaction is possible only if $c > 1/\lambda_1$. Show that $1/\lambda_1 > 1$, so that, even if the multiplying factor is somewhat larger than 1, no chain reaction is possible. Explain.

5.   THE STABILITY OF THE STEADY STATE

The problems of Section 4 can be viewed as steady-state versions of time-dependent problems. In the steady state we study a boundary value problem in the space coordinates (denoted collectively by x); the time-dependent problem is of the initial value type in time and of the boundary value type in space. Although we normally think of the solution as a function $u(x,t)$ in which the variables x and t have more or less equal status, another point of view will be more useful to us. At each time t the solution u is an element or point in a Hilbert space H (admittedly such an element is a function of the space coordinates, but we prefer to regard this function in toto as a point in Hilbert space); as t changes, the element $u = u(t)$ moves in Hilbert space according to the *evolution equation*

$$(5.1) \qquad\qquad \frac{du}{dt} = F(u), \quad t > 0; \quad u|_{t=0} = u_0,$$

where u_0 is the initial value of u and F is a transformation (generally, nonlinear) of H into itself. The derivatives, if any, with respect to space coordinates are contained in the operator F. Let us illustrate our point of view for the case of source-free heat conduction in a rod of unit length whose initial temperature is $u_0(x)$ and whose ends are kept at 0 temperature. At each time t the temperature (or "state" of the rod) is an element of

$H = L_2(0, 1)$; the state u changes in time according to law (5.1), where the operator F is defined as

$$Fu \doteq \frac{d^2 u}{dx^2}$$

on the subset of H consisting of twice-differentiable functions that vanish at $x = 0$, $x = 1$. In our case F is not defined on all of H, but this is a technicality which does not obscure the general idea. If H is a finite-dimensional space, (5.1) reduces to a system of nonlinear ordinary differential equations. Two remarks should be made at this time. First, with no additional difficulty we could take the operator F to depend explicitly on t, but we shall have no need for this generalization. (When, as in our case, F does not depend explicitly on t, the equation is said to be *autonomous*). The second remark has to do with equations which, in their classical form, are of higher order in time. Suppose we are dealing with the wave equation

$$\frac{\partial^2 u}{\partial t^2} = \frac{\partial^2 u}{\partial x^2} \qquad \text{on} \qquad 0 < x < 1,$$

for instance. We set $\partial u / \partial t = v$ and consider the Hilbert space of pairs $(u, v) = z$, where u and v are elements of $L_2(0, 1)$. Then the "state" z satisfies an evolution equation of type (5.1). We are familiar with this idea for finite-dimensional problems; it is the way in which Newton's laws are reduced to Hamiltonian form.

One of the principal questions related to (5.1) is that of the stability of steady states. Let $\tilde{u}$ be an element of H satisfying

$$F(\tilde{u}) = 0.$$

Thus $\tilde{u}$ does not depend on t and is therefore also a solution of (5.1) with initial value $\tilde{u}$. We call $\tilde{u}$ a *steady-state solution* or an *equilibrium state*. One is often interested in the effect of small disturbances on an equilibrium state. Does such a disturbance die out, grow, or remain pretty much the same size? The analysis consists of studying (5.1) with initial value u_0 near $\tilde{u}$. We shall say that $\tilde{u}$ is *stable* if there exists $\varepsilon > 0$ such that the solution $u(t)$ of (5.1) satisfies

(5.2) $\lim_{t \to \infty} \| u - \tilde{u} \| = 0$ whenever $\| u_0 - \tilde{u} \| \leqslant \varepsilon.$

For problems of undamped vibration, (5.2) is obviously too stringent a definition of stability; one would be quite satisfied with avoiding resonance (that is, $u - \tilde{u}$ should remain of the same order as $u_0 - \tilde{u}$). However, for the problems we shall consider, (5.2) is adequate. Note that we cannot expect stability for disturbances of arbitrary size. It is only in linear problems that the question of stability is independent of the size of the disturbance.

Let $u(t)$ be a solution of (5.1) corresponding to the initial value u_0 close to $\tilde{u}$; setting

$$(5.3) \qquad\qquad\qquad u = \tilde{u} + h$$

and substituting in (5.1), we find that

$$(5.4) \qquad\qquad \frac{dh}{dt} = F(\tilde{u} + h) - F(\tilde{u}) = F_u(\tilde{u})h + r,$$

where $F_u(\tilde{u})$ is the Fréchet derivative of F at $\tilde{u}$ (that is, the linearization of F at $\tilde{u}$) and r is of higher order in h: $\lim_{h \to 0}(\|r\| / \|h\|) = 0$. Since the initial value for h is small, it would appear that the behavior of the solution of (5.4) is determined by the *linearized equation*

$$(5.5) \qquad\qquad\qquad \frac{dh}{dt} = F_u(\tilde{u})h.$$

We shall assume the following *principle of linearized stability*: if all solutions of (5.5) decay exponentially in t, so will solutions of (5.4) for $h(0)$ sufficiently small; if (5.5) has some solution that is exponentially increasing, there will exist a small initial state for which the solution of (5.4) will increase rapidly at first (until nonlinearity takes over).

Remark. This principle is well established for ordinary differential equations and for some classes of partial differential equations. In any event a study of (5.4) begins with (5.5).

The question is, then, when does (5.5) have all solutions exponentially decaying? If we try a solution of the form $h(t) = ke^{\mu t}$, where k is an element of H that is independent of time, we find that

$$(5.6) \qquad\qquad\qquad F_u(\tilde{u})k = \mu k,$$

so that k is an eigenvector of the linear operator $F_u(\tilde{u})$ with eigenvalue μ. *If* *the spectrum of $F_u(\tilde{u})$ is in the left half of the complex μ plane*, all solutions of (5.5) decay [and $\tilde{u}$ is a stable solution of (5.1)]. Obviously, if $F_u(\tilde{u})$ is compact and self-adjoint, the solution of (5.5) can be expanded in terms of the eigenvectors of $F_u(\tilde{u})$ and there is no difficulty in showing that all solutions of (5.5) decay.

In many applications the operator F in (5.1) also depends on a real parameter λ which can often be controlled. The differential equation in (5.1) then becomes

$$(5.7) \qquad\qquad \frac{du}{dt} = F(\lambda, u).$$

Suppose we are given a branch $\tilde{u}(\lambda)$ of steady solutions of (5.7) [often $\tilde{u}(\lambda) \equiv 0$ will be the branch in question], that is, $F(\lambda, \tilde{u}(\lambda)) = 0$. This equilibrium solution will usually be stable for a certain range of λ (which, by redefining λ if necessary, can be taken to be an interval of the form $\lambda < \lambda_0$) and unstable outside this range ($\lambda \geqslant \lambda_0$, say). Let the physical system be in the equilibrium state for some $\lambda < \lambda_0$, and suppose we slowly increase the value of the parameter λ (which, depending on the application, may be a flow rate, a compressive load, a Reynolds number, etc.). On passing through the *critical value* λ_0, the equilibrium solution becomes unstable and the physical system can no longer follow it. Although there are many possible behaviors when an equilibrium state becomes unstable, we shall consider only some of these.

A new steady-state solution may branch off from $\tilde{u}(\lambda)$ at $\lambda = \lambda_0$, and the physical system may then follow this new equilibrium solution for $\lambda > \lambda_0$. If the original equilibrium solution turns around at $\lambda = \lambda_0$, or if the new branch starts off to the left ($\lambda < \lambda_0$), the physical system may experience a *snap-through* to another steady state not close to the original branch. Another possibility is that a new *time-periodic* solution (initially of small amplitude) will appear at $\lambda = \lambda_0$. Further on we shall give an example of each of these three cases (which do not exhaust all the possibilities).

The mathematical criterion for a critical value λ_0 is straightforward. For $\lambda < \lambda_0$ the spectrum of $F_u(\lambda, \tilde{u})$ is in the left half of the complex plane, and for $\lambda > \lambda_0$ there is at least one point of the spectrum in the right half of the μ plane. Thus, as λ passes through λ_0, the front of the spectrum of F_u crosses into the right half of the μ plane. Two important cases occur (a) when a simple eigenvalue of F_u crosses through the origin, and (b) when a pair of simple, complex conjugate, eigenvalues cross the imaginary axis. The first of these possibilities is related to the appearance of a new steady state; the second, to a time-periodic solution.

Example 1. Consider the operation of a continuously stirred tank reactor. Replacing u by $u+1$ in (3.10), we find that $u(t)$ satisfies the single nonlinear differential equation

(5.8)
$$\frac{du}{dt} = -u + \frac{1}{\delta}(\beta - u)e^{-\gamma/u+1},$$

where δ, β, γ are positive constants, and u is restricted to $0 \leqslant u \leqslant \beta$. The constant δ is the *flow rate*.

The steady-state equation is

(5.9)
$$(\beta - u)e^{-\gamma/u+1} = \delta u.$$

The graph of the left side has the general shape of the curve in Figure 5.1; for other values of β and γ the curve may be somewhat flatter. The steady-state solutions are the intersections of the straight line of slope δ and the curve. For δ small there is a single intersection at A. As δ increases we go through the continuous sequence A, B, C, D. If δ is increased above the value δ^* corresponding to D, the only intersection is at a point near H.

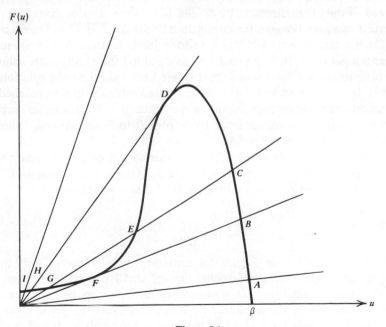

Figure 5.1

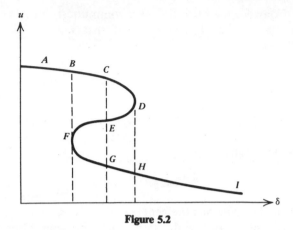

Figure 5.2

This corresponds to a violent quenching of the reaction (the ratio of the concentration of the reactant to the feed concentration is given by $1-(u/\beta)$, so that u near 0 means that very little reaction is taking place, whereas u near β implies nearly complete conversion). If, however, we proceed in the other direction by decreasing δ from a value corresponding to state I, we pass through the continuous sequence I, H, G, F. On reaching F a further decrease in δ brings a sudden jump to state B with its much higher temperature. In Figure 5.2 we have plotted the steady-state solution as a function of the flow rate δ, that is, we have exhibited the solution set of (5.9). It does not appear that a steady state such as E can be reached by continuous increase or decrease in the flow rate; it will come as no surprise that the steady states on the inner loop from D to F are *unstable*, whereas the other steady states are *stable*.

The stability analysis is very simple in our case. Let $\tilde{u}$ be a steady state on the upper part of the curve in Figure 5.2 (say a point such as C, for definiteness). Setting $u(t) = \tilde{u} + h(t)$, we find that $h(t)$ satisfies

$$(5.10) \qquad \frac{dh}{dt} = -h + \frac{1}{\delta} \left[f(\tilde{u}+h) - f(\tilde{u}) \right],$$

where f is the curve on Figure 5.1 and the initial value of h is taken small in conformity with the assumption of small disturbances from the steady state. Linearization is unnecessary because we can deduce our results directly from the nonlinear equation (5.10). At the point C we see from Figure 5.1 that $f(\tilde{u}+h) - f(\tilde{u})$ is negative for small $h > 0$ and positive for small $h < 0$. Thus (5.10) shows a tendency to return to equilibrium, and we have stability. Obviously all that is required for stability is that $-1+(1/\delta)f'(\tilde{u})<0$, and therefore the steady states on the lower part of the curve in Figure 5.2 are also stable (states such as G,H,I). A similar analysis shows that the states on the inner loop (such as E) are unstable.

Example 2. The problem of stability of an equilibrium solution can sometimes be related to the existence or nonexistence of monotone iteration schemes for constructing the equilibrium solution itself. It seems puzzling at first that purely steady considerations could establish dynamic stability, but we can perhaps see the connection by examining the preceding example. Any steady state $\tilde{u}$ of (5.8) is a solution of

$$(5.11) \qquad u=f(u), \text{ where } f(u)=\frac{1}{\delta}(\beta-u)e^{-\gamma/u+1}.$$

Such a steady state is stable if, for $|u-\tilde{u}|$ small,

$$f(u)-f(\tilde{u})-(u-\tilde{u})<0, \qquad u>\tilde{u},$$
$$f(u)-f(\tilde{u})-(u-\tilde{u})>0, \qquad u<\tilde{u}.$$

This merely states that the straight line lies above the curve to the right of $\tilde{u}$ and below the curve to the left of $\tilde{u}$. But that is exactly the condition needed to establish a monotone scheme for the solution $\tilde{u}$ of (5.11).

Without proof we quote two theorems (see Sattinger) of this type for the evolution equation

$$(5.12) \qquad \frac{du}{dt}-\Delta u-f(u)=0, \quad t>0, x\in\Omega; \qquad u=c(x), \quad x\in\Gamma, t>0;$$
$$u \text{ given in } \Omega \text{ at } t=0.$$

Theorem. Let $\hat{u}$ be a solution of the steady problem

$$(5.13) \qquad -\Delta\hat{u}-f(\hat{u})=0, \quad x\in\Omega; \qquad \hat{u}=c(x), \quad x\in\Gamma,$$

and let $u_0(x)$ and $u^0(x)$ be lower and upper solutions, respectively, for this steady problem; assume that $u_0 \leqslant u^0$ and that the iteration schemes starting from u_0 and u^0 both converge to $\hat{u}$. Then $\hat{u}$ is a stable solution of (5.12), and any solution of (5.12) with initial data $v(x)$ satisfying $u_0(x)\leqslant v(x)\leqslant u^0(x)$ tends to $\hat{u}(x)$ as $t\to\infty$.

Remark. The theorem not only tells us about the stability on the basis of purely steady considerations but also gives an indication of the extent of the stability domain. Recall that a lower solution u_0 satisfies

$$-\Delta u_0 \leqslant f(u_0), \quad x \in \Omega; \qquad u_0 \leqslant c(x), \quad x \in \Gamma,$$

and for an upper solution u^0 the inequalities are reversed.

Theorem. Let $\hat{u}$ be a solution of (5.13), and suppose that for each $\varepsilon > 0$ there exists an upper solution $u^0(x, \varepsilon)$ lying *below* $\hat{u}$ with $\hat{u} - u^0 < \varepsilon$ and a lower solution u_0 lying *above* $\hat{u}$ with $u_0 - \hat{u} < \varepsilon$. Then u is an *unstable* solution of (5.12).

As illustration of these theorems, let $c(x) = 0$ and $f = \alpha g$, where g satisfies the conditions imposed below (4.39) for the existence of a unique positive solution. Then this solution is stable.

If one abandons the fairly severe restrictions imposed on g, one can still prove that supercritical branches are stable and subcritical ones are unstable in the neighborhood of the branch-point. Consider, for instance, the equation

$$(5.14) \qquad -\Delta u = \alpha g(u), \quad x \in \Omega; \qquad u = 0, \quad x \in \Gamma,$$

when $g(0) = 0, g'(0) = 1, g''(u) < 0$ on some interval $-a \leqslant u \leqslant a$. (A typical example would be $g = u - u^2$ or any g whose Maclaurin series begins with $u - cu^2, c > 0$.) First we show that the null solution is stable for $\alpha < \alpha_1$ and unstable for $\alpha > \alpha_1$.

We have

$$(5.15) \qquad -\Delta(\varepsilon e_1) - \alpha g(\varepsilon e_1) = \alpha \varepsilon e_1 \left[\frac{\alpha_1}{\alpha} - \frac{g(\varepsilon e_1)}{\varepsilon e_1} \right],$$

so that, for $\alpha < \alpha_1$ and $\varepsilon > 0$, εe_1 is an upper solution, whereas $-\varepsilon e_1$ is a lower solution. Thus the null solution (which is the only solution lying between $-\varepsilon e_1$ and εe_1 for all $\varepsilon > 0$) must be *stable*. For $\alpha > \alpha_1$ the situation is reversed: for every small $\varepsilon > 0$, εe_1 is a lower solution and $-\varepsilon e_1$ is an upper solution, so that the null solution is unstable. What about the stability of the branch emanating from the simple eigenvalue α_1? In this case the branching looks as in Figure 2.1, at least near the value $\alpha = \alpha_1$ (which corresponds to L). This branch represents a positive solution for values of α slightly larger than α_1 and a negative solution for α just below α_1. The part of the branch for $\alpha > \alpha_1$ is stable for α near α_1; the part of the branch for $\alpha < \alpha_1$ is unstable for α near α_1. Let us prove the first of these statements; if $\alpha > \alpha_1$, then εe_1 with ε small and positive is a lower solution.

We shall look for an upper solution of the form Ae_1^*, where e_1^* is defined below (4.38). The same calculation gives

$$-\Delta(Ae_1^*) - \alpha g(Ae_1^*) = Ae_1^* \alpha \left[\frac{\alpha_1^*}{\alpha} - \frac{g(Ae_1^*)}{Ae_1^*} \right].$$

By construction α_1^* is only slightly smaller than α_1. The ratio α_1^*/α is therefore only slightly smaller than 1 if α is near α_1. Since $g''(u) < 0$ near $u = 0$, A can be chosen larger than ε, yet sufficiently small so that $g(Ae_1^*)/Ae_1^*$ is smaller than α_1^*/α. One can show that there is a unique positive solution in this alpha range with $u \leqslant Ae_1^*$, so that the iteration schemes starting from εe_1 and Ae_1^* both converge to this positive solution, which must therefore be stable.

Example 3. We give a simple example in which the incipient instability of a steady state is accompanied by the appearance of a branch of time-periodic solutions. Consider the pair of nonlinear ordinary differential equations

(5.16a) $$\dot{u}_1 = \lambda u_1 - u_2 - u_1(u_1^2 + u_2^2),$$

(5.16b) $$\dot{u}_2 = \lambda u_2 + u_1 - u_2(u_1^2 + u_2^2),$$

where the dot indicates differentiation with respect to t. Obviously $u_1 = u_2 = 0$ is an equilibrium solution for all λ; let us examine its stability. The linearized problem is

(5.17) $$\left. \begin{array}{l} \dot{h}_1 = \lambda h_1 - h_2 \\ \dot{h}_2 = \lambda h_2 + h_1 \end{array} \right\} \quad \text{or} \quad \dot{h} = Ah.$$

The spectrum of A consists of the eigenvalues μ_1, μ_2 of the matrix

$$\begin{pmatrix} \lambda & -1 \\ 1 & \lambda \end{pmatrix},$$

from which we conclude that

$$(\lambda - \mu)^2 = -1 \quad \text{or} \quad \mu = \lambda \pm i.$$

Thus, if $\lambda < 0$, the null solution is stable, whereas for $\lambda > 0$ it is unstable. The front of the spectrum (the whole spectrum in this case) crosses the imaginary axis at the conjugate points $\pm i$ at the critical value $\lambda_0 = 0$.

The full nonlinear system can be solved explicitly by introducing polar coordinates $R^2 = u_1^2 + u_2^2$, $\tan\theta = u_2/u_1$. Multiply (5.16a) and (5.16b) by u_1 and u_2, respectively, and add to obtain

$$\text{(5.18)} \qquad \frac{1}{2}\frac{d}{dt}R^2 = \lambda R^2 - R^4.$$

Multiply (5.16b) by u_1 and (5.16a) by u_2, and subtract to find

$$\frac{d}{dt}\left(\frac{u_2}{u_1}\right) = 1 + \left(\frac{u_2}{u_1}\right)^2 \qquad \text{or} \qquad \frac{d}{dt}\tan\theta = 1 + \tan^2\theta.$$

Since straightforward differentiation also gives $(d/dt)\tan\theta = \dot\theta(1 + \tan^2\theta)$, we conclude that

$$\text{(5.19)} \qquad\qquad \dot\theta = 1.$$

For $\lambda < 0$ it is clear from (5.18) that, for every initial condition, R decreases with t and tends to 0 (thus we have global stability of the null solution); and in view of (5.19) we proceed at uniform angular velocity along a spiral which converges to the origin. For $\lambda > 0$ the solution of (5.18) and (5.19) converges to the stable *periodic* solution $R = \sqrt{\lambda}$, $\dot\theta = 1$.

Exercise

5.1. *Blow-up in finite time.* The BVP

$$-\Delta u = \alpha e^u, \quad x \in \Omega; \qquad u = 0, \quad x \in \Gamma,$$

can have a positive solution only if the straight line $y = \alpha_1 u$ intersects the curve αe^u. Show therefore that a positive solution is possible only for $\alpha < \alpha_1/e$. The absence of a positive solution has implications for the nonlinear heat equation

$$\text{(5.20)} \qquad \frac{\partial u}{\partial t} - \Delta u = \alpha e^u, \quad x \in \Omega, t > 0; \qquad u = 0, \quad x \in \Gamma, t > 0;$$

$$u(x,0) = v(x) \geqslant 0.$$

Let (α_1, e_1) be the fundamental eigenpair of (4.28). Show that

$$\frac{dE}{dt} + \alpha_1 E = \int_\Omega \alpha e^u e_1(x)\,dx,$$

where $E(t) \doteq \int_\Omega u(x,t)e_1(x)\,dx$. The convexity of e^u enables one to use

Jensen's inequality to obtain

$$\int_{\Omega} e^{u} e_1(x)\, dx \geqslant e^{E}.$$

Show therefore that, if $\alpha > \alpha_1/E$,

$$t \leqslant \int_{E(0)}^{E(t)} \frac{ds}{\alpha e^{s} - \alpha_1 s}$$

and hence that the solution of (5.20) must blow up in finite time.

REFERENCES AND ADDITIONAL READING

Amann, H., Fixed point equations and nonlinear eigenvalue problems in ordered Banach spaces, *SIAM Rev.* **18** (1976), 620.

Aris, H., *The mathematical theory of diffusion and reaction in permeable catalysts*, Vol. 2, Oxford University Press, Oxford, 1975.

Berger, M., *Nonlinearity and functional analysis*, Academic Press, New York, 1976.

Browder, F. E., *Functional analysis and related fields*, Springer-Verlag, New York, 1970.

Cohen, D. S. and Keller, H. B., Some positone problems suggested by nonlinear heat generation, *J. Math. Mech.* **16** (1967), 1361.

Crandall, M. G. and Rabinowitz, P. H., Bifurcation, perturbation of simple eigenvalues, and linearized stability, *Arch. Ration. Mech. Anal.* **52** (1973), 161.

Golubitsky, M., An introduction to catastrophe theory and its applications, *SIAM Rev.* **20** (1978).

Hale, J. K., *Generic bifurcation with applications in nonlinear analysis and mechanics*, Vol. I (R. K. Knops, editor), Research notes in mathematics 17, Pitman, London, 1977.

Hastings, S. P., Some mathematical problems from neurobiology, *Am. Math. Mon*, **82**, No. 9 (1975).

Joseph, D. D. and Fosdick, R. L., The free surface on a liquid between cylinders rotating at different speeds, part I, *Arch. Ration. Mech. Anal.* **(49)**, 1973, 321.

Joseph, D. D. and Lundgren, T. S., Quasilinear Dirichlet problems driven by positive sources, *Arch. Ration. Mech. Anal.* **49** (1973), 241.

Keller, H. B. and Keener, J. P., Perturbed bifurcation theory, *Arch. Ration. Mech. Anal.* **50** (1973), 159.

Keller, J. B. and Antman, S., *Bifurcation theory and nonlinear eigenvalue problems*, Benjamin, New York, 1969.

Koiter, W. T., Elastic stability and post-buckling behavior, in *Nonlinear problems*, (R. E. Langer, editor), University of Wisconsin Press, Madison, 1963.

Kolodner, I. I., Equations of Hammerstein type in Hilbert spaces, *J. Math. Mech.* **13** (1964).

Martin, R. H., jr., *Nonlinear operators and differential equations in Banach spaces*, Wiley-Interscience, New York, 1976.

Matkowsky, B. J. and Reiss, E. L., Singular perturbation of bifurcations, *SIAM J. Appl. Math.* **33** (1977).

Murray, J. D., *Nonlinear-differential-equation models in biology*, Oxford University Press, Oxford, 1977.

Pazy, A. and Rabinowitz, P. H., A nonlinear integral equation with application to neutron transport theory, *Arch. Ration. Mech. Anal.* **32** (1969), 226.

Pimbley, G. H., Jr., *Eigenfunction branches of nonlinear operators, and their bifurcation*, Springer, New York, 1969.

Poston, T. and Stewart, I., *Catastrophe theory and its applications*, Pitman, London, 1978.

Sattinger, D. H., *Topics in stability and bifurcation theory*, Lecture notes in mathematics 309, Springer-Verlag, Heidelberg, 1973.

Stakgold, I., Branching of solutions of nonlinear equations, *SIAM Rev.* **13** (1971), 289.

Stakgold, I., Joseph, D. D., and Sattinger, D. H., *Nonlinear problems in the physical sciences and biology*, Lecture notes in mathematics 322, Springer Verlag, Heidelberg, 1973.

Thompson, J. M. T. and Hunt, G. W., *A general theory of elastic stability*, Wiley, New York, 1973.

Wake, G. C. and Fradkin, L., The critical explosion parameter in thermal ignition, *J. Inst. Math. Appl.* **20** (1977).

INDEX

IVAR STAKGOLD received his Ph.D. degree in applied mathematics from Harvard University. He has served as Professor and Chairman in the Department of Engineering Sciences at Northwestern University, and was a Visiting Professor at both University College, London, and Oxford University, Oxford. He is presently Professor and Chairman of the Department of Mathematical Sciences at the University of Delaware. Dr. Stakgold is Chairman of the Board of Trustees of the Society for Industrial and Applied Mathematics; a Committeeman of the National Research Council; and Associate Editor of the *American Mathematical Monthly, Journal of Integral Equations, Applicable Analysis,* and *International Journal of Engineering Science.*